Pharmaceutical Industry Practices on Genotoxic Impurities

CHROMATOGRAPHIC SCIENCE SERIES

A Series of Textbooks and Reference Books

Editor:
Nelu Grinberg

Founding Editor:
Jack Cazes

1. Dynamics of Chromatography: Principles and Theory, J. Calvin Giddings
2. Gas Chromatographic Analysis of Drugs and Pesticides, Benjamin J. Gudzinowicz
3. Principles of Adsorption Chromatography: The Separation of Nonionic Organic Compounds, Lloyd R. Snyder
4. Multicomponent Chromatography: Theory of Interference, Friedrich Helfferich and Gerhard Klein
5. Quantitative Analysis by Gas Chromatography, Josef Novák
6. High-Speed Liquid Chromatography, Peter M. Rajcsanyi and Elisabeth Rajcsanyi
7. Fundamentals of Integrated GC-MS (in three parts), Benjamin J. Gudzinowicz, Michael J. Gudzinowicz, and Horace F. Martin
8. Liquid Chromatography of Polymers and Related Materials, Jack Cazes
9. GLC and HPLC Determination of Therapeutic Agents (in three parts), Part 1 edited by Kiyoshi Tsuji and Walter Morozowich, Parts 2 and 3 edited by Kiyoshi Tsuji
10. Biological/Biomedical Applications of Liquid Chromatography, edited by Gerald L. Hawk
11. Chromatography in Petroleum Analysis, edited by Klaus H. Altgelt and T. H. Gouw
12. Biological/Biomedical Applications of Liquid Chromatography II, edited by Gerald L. Hawk
13. Liquid Chromatography of Polymers and Related Materials II, edited by Jack Cazes and Xavier Delamare
14. Introduction to Analytical Gas Chromatography: History, Principles, and Practice, John A. Perry
15. Applications of Glass Capillary Gas Chromatography, edited by Walter G. Jennings
16. Steroid Analysis by HPLC: Recent Applications, edited by Marie P. Kautsky
17. Thin-Layer Chromatography: Techniques and Applications, Bernard Fried and Joseph Sherma
18. Biological/Biomedical Applications of Liquid Chromatography III, edited by Gerald L. Hawk
19. Liquid Chromatography of Polymers and Related Materials III, edited by Jack Cazes

20. Biological/Biomedical Applications of Liquid Chromatography, edited by Gerald L. Hawk
21. Chromatographic Separation and Extraction with Foamed Plastics and Rubbers, G. J. Moody and J. D. R. Thomas
22. Analytical Pyrolysis: A Comprehensive Guide, William J. Irwin
23. Liquid Chromatography Detectors, edited by Thomas M. Vickrey
24. High-Performance Liquid Chromatography in Forensic Chemistry, edited by Ira S. Lurie and John D. Wittwer, Jr.
25. Steric Exclusion Liquid Chromatography of Polymers, edited by Josef Janca
26. HPLC Analysis of Biological Compounds: A Laboratory Guide, William S. Hancock and James T. Sparrow
27. Affinity Chromatography: Template Chromatography of Nucleic Acids and Proteins, Herbert Schott
28. HPLC in Nucleic Acid Research: Methods and Applications, edited by Phyllis R. Brown
29. Pyrolysis and GC in Polymer Analysis, edited by S. A. Liebman and E. J. Levy
30. Modern Chromatographic Analysis of the Vitamins, edited by André P. De Leenheer, Willy E. Lambert, and Marcel G. M. De Ruyter
31. Ion-Pair Chromatography, edited by Milton T. W. Hearn
32. Therapeutic Drug Monitoring and Toxicology by Liquid Chromatography, edited by Steven H. Y. Wong
33. Affinity Chromatography: Practical and Theoretical Aspects, Peter Mohr and Klaus Pommerening
34. Reaction Detection in Liquid Chromatography, edited by Ira S. Krull
35. Thin-Layer Chromatography: Techniques and Applications, Second Edition, Revised and Expanded, Bernard Fried and Joseph Sherma
36. Quantitative Thin-Layer Chromatography and Its Industrial Applications, edited by Laszlo R. Treiber
37. Ion Chromatography, edited by James G. Tarter
38. Chromatographic Theory and Basic Principles, edited by Jan Åke Jönsson
39. Field-Flow Fractionation: Analysis of Macromolecules and Particles, Josef Janca
40. Chromatographic Chiral Separations, edited by Morris Zief and Laura J. Crane
41. Quantitative Analysis by Gas Chromatography, Second Edition, Revised and Expanded, Josef Novák
42. Flow Perturbation Gas Chromatography, N. A. Katsanos
43. Ion-Exchange Chromatography of Proteins, Shuichi Yamamoto, Kazuhiro Naka-nishi, and Ryuichi Matsuno
44. Countercurrent Chromatography: Theory and Practice, edited by N. Bhushan Man-dava and Yoichiro Ito
45. Microbore Column Chromatography: A Unified Approach to Chromatography, edited by Frank J. Yang
46. Preparative-Scale Chromatography, edited by Eli Grushka
47. Packings and Stationary Phases in Chromatographic Techniques, edited by Klaus K. Unger

48. Detection-Oriented Derivatization Techniques in Liquid Chromatography, edited by Henk Lingeman and Willy J. M. Underberg
49. Chromatographic Analysis of Pharmaceuticals, edited by John A. Adamovics
50. Multidimensional Chromatography: Techniques and Applications, edited by Hernan Cortes
51. HPLC of Biological Macromolecules: Methods and Applications, edited by Karen M. Gooding and Fred E. Regnier
52. Modern Thin-Layer Chromatography, edited by Nelu Grinberg
53. Chromatographic Analysis of Alkaloids, Milan Popl, Jan Fähnrich, and Vlastimil Tatar
54. HPLC in Clinical Chemistry, I. N. Papadoyannis
55. Handbook of Thin-Layer Chromatography, edited by Joseph Sherma and Bernard Fried
56. Gas–Liquid–Solid Chromatography, V. G. Berezkin
57. Complexation Chromatography, edited by D. Cagniant
58. Liquid Chromatography–Mass Spectrometry, W. M. A. Niessen and Jan van der Greef
59. Trace Analysis with Microcolumn Liquid Chromatography, Milos Krejcl
60. Modern Chromatographic Analysis of Vitamins: Second Edition, edited by André P. De Leenheer, Willy E. Lambert, and Hans J. Nelis
61. Preparative and Production Scale Chromatography, edited by G. Ganetsos and P. E. Barker
62. Diode Array Detection in HPLC, edited by Ludwig Huber and Stephan A. George
63. Handbook of Affinity Chromatography, edited by Toni Kline
64. Capillary Electrophoresis Technology, edited by Norberto A. Guzman
65. Lipid Chromatographic Analysis, edited by Takayuki Shibamoto
66. Thin-Layer Chromatography: Techniques and Applications: Third Edition, Revised and Expanded, Bernard Fried and Joseph Sherma
67. Liquid Chromatography for the Analyst, Raymond P. W. Scott
68. Centrifugal Partition Chromatography, edited by Alain P. Foucault
69. Handbook of Size Exclusion Chromatography, edited by Chi-San Wu
70. Techniques and Practice of Chromatography, Raymond P. W. Scott
71. Handbook of Thin-Layer Chromatography: Second Edition, Revised and Expanded, edited by Joseph Sherma and Bernard Fried
72. Liquid Chromatography of Oligomers, Constantin V. Uglea
73. Chromatographic Detectors: Design, Function, and Operation, Raymond P. W. Scott
74. Chromatographic Analysis of Pharmaceuticals: Second Edition, Revised and Expanded, edited by John A. Adamovics
75. Supercritical Fluid Chromatography with Packed Columns: Techniques and Applications, edited by Klaus Anton and Claire Berger
76. Introduction to Analytical Gas Chromatography: Second Edition, Revised and Expanded, Raymond P. W. Scott
77. Chromatographic Analysis of Environmental and Food Toxicants, edited by Takayuki Shibamoto

78. Handbook of HPLC, edited by Elena Katz, Roy Eksteen, Peter Schoenmakers, and Neil Miller
79. Liquid Chromatography–Mass Spectrometry: Second Edition, Revised and Expanded, Wilfried Niessen
80. Capillary Electrophoresis of Proteins, Tim Wehr, Roberto Rodríguez-Díaz, and Mingde Zhu
81. Thin-Layer Chromatography: Fourth Edition, Revised and Expanded, Bernard Fried and Joseph Sherma
82. Countercurrent Chromatography, edited by Jean-Michel Menet and Didier Thiébaut
83. Micellar Liquid Chromatography, Alain Berthod and Celia García-Alvarez-Coque
84. Modern Chromatographic Analysis of Vitamins: Third Edition, Revised and Expanded, edited by André P. De Leenheer, Willy E. Lambert, and Jan F. Van Bocxlaer
85. Quantitative Chromatographic Analysis, Thomas E. Beesley, Benjamin Buglio, and Raymond P. W. Scott
86. Current Practice of Gas Chromatography–Mass Spectrometry, edited by W. M. A. Niessen
87. HPLC of Biological Macromolecules: Second Edition, Revised and Expanded, edited by Karen M. Gooding and Fred E. Regnier
88. Scale-Up and Optimization in Preparative Chromatography: Principles and Biopharmaceutical Applications, edited by Anurag S. Rathore and Ajoy Velayudhan
89. Handbook of Thin-Layer Chromatography: Third Edition, Revised and Expanded, edited by Joseph Sherma and Bernard Fried
90. Chiral Separations by Liquid Chromatography and Related Technologies, Hassan Y. Aboul-Enein and Imran Ali
91. Handbook of Size Exclusion Chromatography and Related Techniques: Second Edition, edited by Chi-San Wu
92. Handbook of Affinity Chromatography: Second Edition, edited by David S. Hage
93. Chromatographic Analysis of the Environment: Third Edition, edited by Leo M. L. Nollet
94. Microfluidic Lab-on-a-Chip for Chemical and Biological Analysis and Discovery, Paul C.H. Li
95. Preparative Layer Chromatography, edited by Teresa Kowalska and Joseph Sherma
96. Instrumental Methods in Metal Ion Speciation, Imran Ali and Hassan Y. Aboul-Enein
97. Liquid Chromatography–Mass Spectrometry: Third Edition, Wilfried M. A. Niessen
98. Thin Layer Chromatography in Chiral Separations and Analysis, edited by Teresa Kowalska and Joseph Sherma
99. Thin Layer Chromatography in Phytochemistry, edited by Monika Waksmundzka-Hajnos, Joseph Sherma, and Teresa Kowalska
100. Chiral Separations by Capillary Electrophoresis, edited by Ann Van Eeckhaut and Yvette Michotte
101. Handbook of HPLC: Second Edition, edited by Danilo Corradini and consulting editor Terry M. Phillips
102. High Performance Liquid Chromatography in Phytochemical Analysis, edited by Monika Waksmundzka-Hajnos and Joseph Sherma

103. Hydrophilic Interaction Liquid Chromatography (HILIC) and Advanced Applications, edited by Perry G. Wang and Weixuan He
104. Hyphenated and Alternative Methods of Detection in Chromatography, edited by R. Andrew Shalliker
105. LC-NMR: Expanding the Limits of Structure Elucidation, Nina C. Gonnella
106. Thin Layer Chromatography in Drug Analysis, edited by Łukasz Komsta, Monika Waksmundzka-Hajnos, and Joseph Sherma
107. Pharmaceutical Industry Practices on Genotoxic Impurities, edited by Heewon Lee

Pharmaceutical Industry Practices on Genotoxic Impurities

EDITED BY

Heewon Lee

Boehringer Ingelheim Pharmaceuticals, Inc.,
Ridgefield, Connecticut, USA

CRC Press
Taylor & Francis Group
Boca Raton London New York

CRC Press is an imprint of the
Taylor & Francis Group, an **informa** business

CRC Press
Taylor & Francis Group
6000 Broken Sound Parkway NW, Suite 300
Boca Raton, FL 33487-2742

ISBN-13: 978-1-4398-7420-2 (hbk)

Library of Congress Cataloging-in-Publication Data

Pharmaceutical industry practices on genotoxic impurities / editor, Heewon Lee.
pages cm. -- (Chromatographic science series)
Includes bibliographical references and index.
ISBN 978-1-4398-7420-2 (hardback)
1. Genetic toxicology. 2. Drugs--Toxicology. I. Lee, Heewon, editor.

RA1224.3.P43 2015
615.9'02--dc23 2014034796

Visit the Taylor & Francis Web site at
http://www.taylorandfrancis.com

and the CRC Press Web site at
http://www.crcpress.com

To my parents,

To my husband,

To my children,

I thank you for the love and opportunities you gave me.

"Start, and you are halfway there."

A Korean wise old saying

Contents

Preface

According to the World Health Organization's International Agency for Research on Cancer (IARC) (Technical Publication No. 24, 1994), genotoxicity is defined as structural changes at the level of a gene. It is also defined as a broad term that refers to any deleterious change in the genetic material regardless of the mechanism by which the change is induced, as stated by ICH S2(R1), "Guidance on Genotoxicity Testing and Data Interpretation for Pharmaceuticals Intended for Human Use." Genotoxic compounds have the theoretical potential to damage DNA at any level of exposure, leading to tumor development. This makes the toxicological assessment and determination of acceptable limits of genotoxic impurities (GTIs) difficult.

GTIs have become an intensely debated topic in the pharmaceutical industry in recent years. My experiences in the internal and external discussions in meetings and conferences indicate that there is a great deal of confusion and uncertainty over identification, assessment, and control of GTIs. GTIs require different rationale and strategies from "regular" or "ordinary" impurities. Proper management of GTIs entails appropriate guidance and strategies, effective communication between various disciplines, sharing information and results, and understanding of the overall implications. The goal of this book is to present the rationale, strategies, methods, interpretation, practices, and case studies in the pharmaceutical industry in order to facilitate scientific and systematic consensus establishment on GTIs.

The book chapters are organized in the sequence of regulatory guidelines, safety aspects, quality aspects, and a pharmaceutical industry survey.

Chapter 1 provides an overview of regulatory guidelines that are relevant to impurities, including GTIs. Chapter 2 reports well-known structural alerts in the literature for genotoxicity and carcinogenicity to orient readers in the chemical functionalities. Chapter 3 illustrates calculations of the acceptable exposure values of impurities based on the regulatory guidelines under different circumstances. Also included in Chapter 3 are selected compounds from the Carcinogenic Potency Database that are relevant to pharmaceutical development.

Safety aspects of GTIs are discussed in Chapters 4 through 6. Chapter 4 describes cancer risk assessment and in silico, in vitro, and in vivo tests of GTIs. Chapter 5 provides an in-depth report on computational assessment of the GTI evaluation. Chapter 6 explains preclinical assessment of GTIs with an overview of regulatory guidance and description of available genotoxicity assays for risk characterization.

Quality aspects of GTIs are presented in Chapters 7 through 13. Chapter 7 describes purification efforts to provide quality lead compounds for toxicological profiling. Chapter 8 reviews GC/MS analyses of GTIs in pharmaceuticals. Chapter 9 reports a case study of controlling chloroethane from raw materials to drug substances. Diverse and numerous case studies of quantitative analyses of GTIs in active pharmaceutical ingredients (API) are presented in Chapter 10. Case studies of analytical testing and control of GTIs in drug substances are reported in Chapter 11. Chapter 12 describes various approaches to assess, analyze, and control GTIs in drug substances with case studies. Chapter 13 reports control and analysis of GTIs including the use of Quality by Design.

Quality aspects of GTIs from different perspectives continue in Chapters 14 through 16. Chapter 14 reports diverse considerations involved in salt formation and selection and associated GTI risks. Chapter 15 presents stability and stress testing strategies regarding genotoxic degradation products. Chapter 16 discusses identification and qualification of GTIs as leachables and extractables in drug products.

The book concludes with a pharmaceutical industry survey on GTIs and related topics in Chapter 17. This is a summary of the API Development GMP Forum, an informal and interactive conference, over the period of 2008 through 2012.

Colleagues in various disciplines from nine major pharmaceutical companies (AbbVie, Amgen, Boehringer Ingelheim, Eli Lilly, Genentech, Merck, Novartis, Pfizer, and Sanofi) contributed to this book, presenting pharmaceutical industry practices. I asked the chapter contributors to report their rationale and practices without being concerned about overlapping with other chapters in the book. An overlapped discussion indicates that the specific practice is common in the industry, which is important and valuable to recognize.

My sincere appreciation goes to contributors from the nine pharmaceutical companies. This book does not exist without the contributors' dedication and commitment to provide quality manuscripts within the defined timeline, despite their busy schedules and challenging situations. Lastly, I would like to thank the editorial staff of Taylor & Francis for their professional support and thoughtful patience from inception to reality.

Heewon Lee
Boehringer Ingelheim Pharmaceuticals Inc.

Editor

Dr. Heewon Lee is a senior associate director in the Chemical Development department at Boehringer Ingelheim Pharmaceuticals Inc., in Ridgefield, Connecticut. She earned a BS degree in chemistry and an MS degree in physical chemistry at the Seoul National University in Seoul, Korea. She earned her PhD in analytical chemistry from the University of Michigan in Ann Arbor, Michigan. After postdoctoral positions, she worked at ArQule, a biotechnology company in Massachusetts. Subsequently, Dr. Lee joined the Medicinal Chemistry department at Boehringer Ingelheim and then moved to the Chemical Development department.

In her current position, Dr. Lee leads the Analytical Research Group and is responsible for analytical method development, in-process control for process research and development, and quality control of outsourced materials. Dr. Lee is an active participant in the Genotoxic Impurity (GTI) Council for identification, assessment, and control of GTIs in drug substances and drug products. In addition, she is involved in Process Analytical Technology (PAT), Continuous Process Improvement (CPI), and regulatory compliance. She is a member of the Pharmaceutical IQ Consortium Working Group.

List of Contributors

Mohammad Al-Sayah
Merck & Co.
Rahway, New Jersey

Alexander Amberg
Sanofi
Frankfurt, Germany

Douglas J. Ball
Pfizer, Inc.
Groton, Connecticut

Wesley W. Barnhart
Amgen Inc.
Thousand Oaks, California

William P. Beierschmitt
Pfizer, Inc.
Groton, Connecticut

Xiaodong Bu
Merck & Co.
Rahway, New Jersey

Zhi Chen
Novartis Pharmaceuticals
East Hanover, New Jersey

Andreas Czich
Sanofi
Frankfurt, Germany

Krista L. Dobo
Pfizer, Inc.
Groton, Connecticut

Kyung H. Gahm
Amgen Inc.
Thousand Oaks, California

Simon Hamilton
Merck & Co.
Hoddesdon, United Kingdom

Roy Helmy
Merck & Co.
Rahway, New Jersey

Debie Hoivik
Boehringer Ingelheim Pharmaceuticals, Inc.
Ridgefield, Connecticut

Zheng Hua
Amgen Inc.
Thousand Oaks, California

Soojin Kim
Boehringer Ingelheim Pharmaceuticals, Inc.
Ridgefield, Connecticut

Claire Lee
Merck & Co.
Rahway, New Jersey

Heewon Lee
Boehringer Ingelheim Pharmaceuticals, Inc.
Ridgefield, Connecticut

Jane Li
Genentech, Inc.
South San Francisco, California

Frances Liu
Novartis Pharmaceuticals
East Hanover, New Jersey

Cynthia J. Magee
Pfizer, Inc.
Groton, Connecticut

Todd J. Page
Eli Lilly and Company
Indianapolis, Indiana

J. Barry Phelps
Eli Lilly and Company
Indianapolis, Indiana

Qin Ji
Novartis Pharmaceuticals
East Hanover, New Jersey

Fenghe Qiu
Boehringer Ingelheim
Pharmaceuticals, Inc.
Ridgefield, Connecticut

Vincent L. Reynolds
Eli Lilly and Company
Indianapolis, Indiana

Adam P. Schellinger
AbbVie Inc.
North Chicago, Illinois

Sophie Strickfuss
Merck & Co.
Hoddesdon, United Kingdom

Véronique Thybaud
Sanofi
Vitry-sur-seine, France

Johanna Ubben
Boehringer Ingelheim
Pharmaceuticals, Inc.
Ridgefield, Connecticut

Esther Vock
Boehringer Ingelheim
Pharma GmbH & Co., KG
Biberach, Germany

Tiebang Wang
Merck & Co.
Rahway, New Jersey

Christopher Welch
Merck & Co.
Rahway, New Jersey

Larry Wigman
Genentech, Inc.
South San Francisco, California

Shan Xiao
Novartis Pharmaceuticals
East Hanover, New Jersey

William A. Yakush
Boehringer Ingelheim
Pharmaceuticals, Inc.
Ridgefield, Connecticut

Bing-Shiou Yang
Boehringer Ingelheim
Pharmaceuticals, Inc.
Ridgefield, Connecticut

Jianbing Zhang
Novartis Pharmaceuticals
East Hanover, New Jersey

Jie Zheng
AbbVie Inc.
North Chicago, Illinois

1 Overview of Regulatory Guidelines on Impurities and Genotoxic Impurities

Heewon Lee and Debie Hoivik

CONTENTS

1.1 INTRODUCTION

Impurities in drug substances are defined as "any component of the new drug substance that is not the chemical entity defined as the new drug substance" according to the International Conference on Harmonisation of Technical Requirements for Registration of Pharmaceuticals for Human Use (ICH) Q3A(R2) guideline [1]. In the case of drug products, impurities are defined as "any component of the new drug product that is not the drug substance or an excipient in the drug product" according to the ICH Q3B(R2) guideline [2]. A subset of the impurities is genotoxic, presenting a safety concern to clinical trial subjects and patients. This chapter provides an overview of the regulatory guidelines that are relevant to impurities and genotoxic impurities.

The ICH [3] is a joint initiative involving both regulators and research-based industry focusing on the technical requirements for medicinal products containing new drugs. The ICH brings together the regulatory authorities and pharmaceutical industry of Europe, Japan, and the United States to discuss scientific and technical aspects of drug registration. Since its inception in 1990, the ICH has evolved, through its ICH Global Cooperation Group, to respond to the increasingly global face of drug development, so that the benefits of international harmonization for better global health can be realized worldwide. ICH's mission is to achieve greater harmonization to ensure that safe, effective, and high-quality medicines are developed and registered in the most resource-efficient manner. The main focus of the ICH process is the preparation of harmonized guidelines that are adopted in the three ICH regions: the European Union (EU), the United States, and Japan. Countries outside the ICH may also use the ICH guidelines within their own countries. Generally, the ICH guidelines are accepted as the industry standard. ICH guideline topics are divided into four categories (Q, S, E, and M), and ICH topic codes are assigned according to these categories. The current description of the four categories is quoted as follows:

Q (quality guidelines): harmonization achievements in the quality area include pivotal milestones such as the conduct of stability studies, defining relevant thresholds for impurities testing, and a more flexible approach to pharmaceutical quality based on good manufacturing practice (GMP) risk management.

S (safety guidelines): the ICH has produced a comprehensive set of safety guidelines to uncover potential risks like carcinogenicity, genotoxicity, and reprotoxicity. A recent breakthrough has been a nonclinical testing strategy for assessing the QT interval prolongation liability: the single most important cause of drug withdrawals in recent years.

E (efficacy guidelines): the work carried out by the ICH under the efficacy heading is concerned with the design, conduct, safety, and reporting of clinical trials. It also covers novel types of medicines derived from biotechnological processes and the use of pharmacogenetics/genomics techniques to produce better targeted medicines.

M (multidisciplinary guidelines): these are the cross-cutting topics that do not fit uniquely into one of the quality, safety, and efficacy categories. This category includes ICH medical terminology (MedDRA), the common technical document, and development of Electronic Standards for the Transfer of Regulatory Information.

The following sections provide a summary of pertinent guidelines with the focus on aspects of impurities and genotoxic impurities. In order not to change the meaning of the guidelines, the original content has been used as closely as possible. References cited in the guidelines are not included in this chapter for the sake of conciseness.

1.2 IMPURITIES IN NEW DRUG SUBSTANCES, ICH Q3A(R2)

ICH Q3A(R2) [1] is intended to provide guidance for registration applications on the content and quantification of impurities in new drug substances produced by chemical syntheses and not previously registered in a region or member state. It is not intended to apply to new drug substances used during the clinical research stage of development. Although this guideline is not intended to be applied during the clinical research stage of development, in later stages of development the thresholds in this guideline can be useful in evaluating new impurities observed in the drug substance batches prepared by the proposed commercial process.

Impurities in new drug substances are addressed from two perspectives:

- *Chemistry aspects* include classification and identification of impurities, report generation, listing of impurities in specifications, and a brief discussion on analytical procedures.
- *Safety aspects* include specific guidance for qualifying those impurities that were not present, or were present at substantially lower levels, in batches of new drug substances used in safety and clinical studies.

Impurities can be classified into the following categories:

- Organic impurities (process and drug related)
- Inorganic impurities
- Residual solvents

Organic impurities can arise during the manufacturing process and/or storage of the new drug substances. They can be identified or unidentified, volatile or nonvolatile, and include the following:

- Starting materials
- By-products
- Intermediates
- Degradation products
- Reagents, ligands, and catalysts

Inorganic impurities can result from the manufacturing process. They are normally known and identified and include the following:

- Reagents, ligands, and catalysts
- Heavy metals or other residual metals
- Inorganic salts
- Other materials (e.g., filter aids and charcoal)

Solvents are inorganic or organic liquids used as vehicles for the preparation of solutions or suspensions in the synthesis of a new drug substance. Since these are generally of known toxicity, the selection of appropriate controls is easily accomplished (see ICH guideline Q3C on residual solvents [3]).

Details of the actual and potential organic impurities most likely to arise during the synthesis, purification, and storage of a new drug substance should be summarized and included in the registration application. The summary should be based on a sound scientific appraisal of the chemical reactions involved in the synthesis, impurities associated with raw materials that could contribute to the impurity profile of the new drug substance, and possible degradation products. This discussion can be limited to those impurities that might reasonably be expected based on the knowledge of the chemical reactions and conditions involved.

Table 1.1 shows the reporting, identification, and qualification thresholds. Any impurity at a level greater than (>) the identification threshold given in Table 1.1 (e.g., calculated using the response factor of the drug substance) in any batch manufactured by the proposed commercial process should be identified. The studies conducted to characterize the structure of actual impurities present in the new drug substance at a level greater than (>) the identification threshold should be described. In addition, any degradation product observed in stability studies at recommended storage conditions at a level greater than (>) the identification threshold should be identified. The note in Table 1.1 stating that "lower thresholds can be appropriate if the impurity is unusually toxic" is applicable to genotoxic impurities.

TABLE 1.1
Reporting, Identification, and Qualification Thresholds for Impurities in New Drug Substances, ICH Q3A

Maximum Daily Dose[a] (g/day)	Reporting Threshold[b,c] (%)	Identification Threshold[c]	Qualification Threshold[c]
≤2	0.05	0.10% or 1.0 mg/day Intake (whichever is lower)	0.15% or 1.0 mg/day Intake (whichever is lower)
>2	0.03	0.05%	0.05%

Source: *Impurities in New Drug Substances*, ICH Q3A(R2), ICH Harmonised Tripartite Guideline, International Conference on Harmonisation of Technical Requirements for Registration of Pharmaceuticals for Human Use, 2006.

[a] The amount of drug substance administered per day.

[b] Higher reporting thresholds should be scientifically justified.

[c] Lower thresholds can be appropriate if the impurity is unusually toxic.

Identification of impurities present at an apparent level not more than (≤) the identification threshold is generally not considered necessary. However, analytical procedures should be developed for those potential impurities that are expected to be unusually potent, producing toxic or pharmacological effects at a level not more than (≤) the identification threshold. Genotoxic impurities are considered to be unusually potent impurities in this respect.

Documented evidence that the analytical procedures are validated and suitable for the detection and quantification of impurities should be included in the registration application. Technical factors (e.g., manufacturing capability and control methodology) can be considered part of the justification for selection of alternative thresholds based on manufacturing experience with the proposed commercial process. The use of two decimal places for thresholds (Table 1.1) does not necessarily reflect the precision of the analytical procedure used for routine quality control purposes. The quantitation limit for the analytical procedure should be not more than (≤) the reporting threshold. Any impurity at a level greater than (>) the reporting threshold and total impurities observed in these batches of the new drug substance should be reported with the analytical procedures indicated. Below 1.0%, the results should be reported to two decimal places (e.g., 0.06% and 0.13%); at and above 1.0%, the results should be reported to one decimal place (e.g., 1.3%). Results should be rounded using conventional rules. All impurities at a level greater than (>) the reporting threshold should be summed and reported as total impurities.

For impurities known to be unusually potent or to produce toxic or unexpected pharmacological effects, the quantitation/detection limit of the analytical procedures should be commensurate with the level at which the impurities should be controlled. Genotoxic impurities belong to this category.

Acceptance criteria should be set no higher than the level that can be justified by safety data and should be consistent with the level achievable by the manufacturing process and the analytical capability. Where there is no safety concern, impurity acceptance criteria should be based on data generated on batches of the new drug substance manufactured by the proposed commercial process, allowing sufficient latitude to deal with normal manufacturing and analytical variations and the stability characteristics of the new drug substance.

Qualification is the process of acquiring and evaluating data that establishes the biological safety of an individual impurity or a given impurity profile at the levels specified. A rationale for establishing impurity acceptance criteria that includes safety considerations should be provided in the registration application. The level of any impurity present in a new drug substance that has been adequately tested in safety and/or clinical studies would be considered qualified. Impurities that are also significant metabolites present in animal and/or human studies are generally considered qualified.

Figure 1.1 describes considerations for the qualification of impurities when thresholds are exceeded. In some cases, decreasing the level of impurity to not more than the threshold can be simpler than providing safety data. Alternatively, adequate data could be available in the scientific literature to qualify an impurity. If neither is the case, additional safety testing should be considered. The studies considered appropriate to qualify an impurity will depend on a number of factors, including the patient population, daily dose, and route and duration of drug administration. Such studies can be conducted on the new drug substance containing the impurities to be controlled, although studies using isolated impurities can sometimes be appropriate.

1.3 IMPURITIES IN NEW DRUG PRODUCTS, ICH Q3B(R2)

ICH Q3B(R2) [2] provides guidance for registrations applications on the content and qualification of impurities in new drug products produced from chemically synthesized new drug substances. This guideline is complementary to the ICH Q3A(R) guideline. The ICH Q3C guideline on residual solvents should also be consulted, if appropriate.

This guideline addresses only those impurities in new drug products classified as degradation products of the drug substance or reaction products of the drug substance with an excipient and/or immediate container closure system (collectively referred to as "degradation products" in this guideline). Generally, impurities present in the new drug substance need not be monitored or specified in the new drug product unless they are also degradation products. Impurities arising from excipients present in the new drug product or extracted or leached from the container closure system are not covered by this guideline. This guideline also does not apply to new drug products used during the clinical research stages of development.

The applicant should summarize the degradation products observed during manufacture and/or stability studies of the new drug product. This summary should be based on a sound scientific appraisal of potential degradation pathways in the new drug product and impurities arising from the interaction with excipients and/or the immediate container closure system. In addition, the applicant should summarize any laboratory studies conducted to detect degradation products in the new drug product. This summary should also include test results of batches manufactured during the development process and batches representative of the proposed commercial process.

Any degradation product observed in stability studies conducted at the recommended storage condition should be identified when present at a level

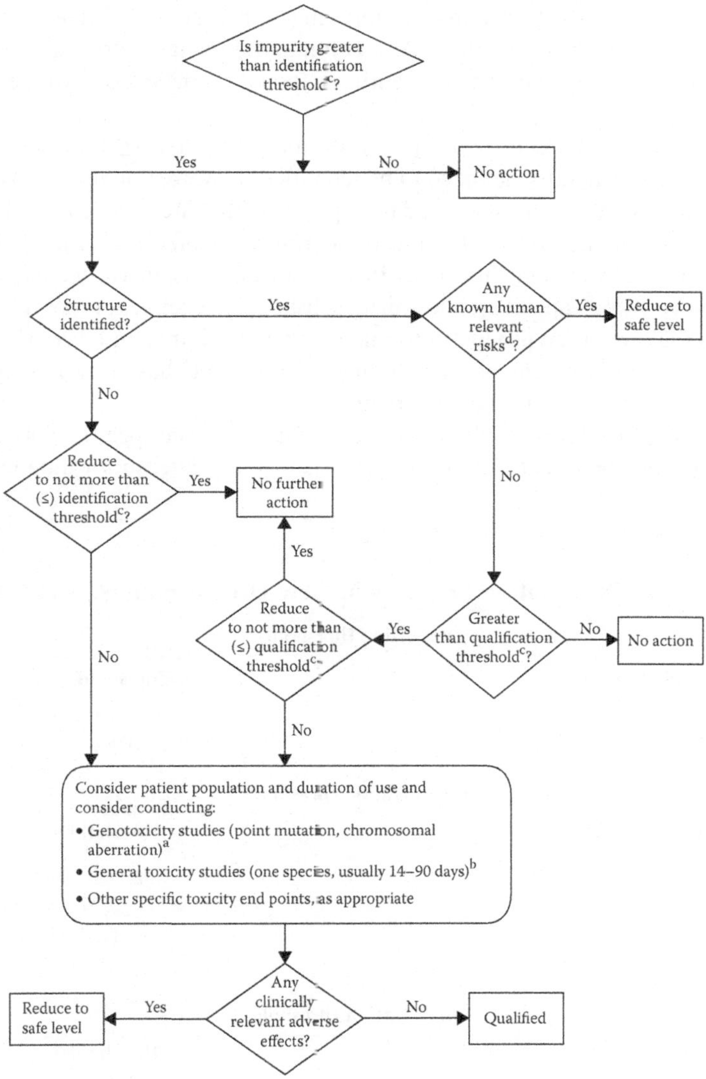

FIGURE 1.1 Decision tree for identification and qualification, ICH Q3A.

[a]If considered desirable, a minimum screen (e.g., genotoxic potential) should be conducted. A study to detect point mutation and another to detect chromosomal aberrations, both in vitro, are considered an appropriate minimum screen.

[b]If general toxicity studies are desirable, one or more studies should be designed to allow the comparison of unqualified to qualified material. The study duration should be based on available relevant information and the study performed in the species most likely to maximize the potential to detect the toxicity of an impurity. On a case-by-case basis, single-dose studies can be appropriate, especially for single-dose drugs. In general, a minimum of 14 days and a maximum duration of 90 days would be considered appropriate.

[c]Lower thresholds can be appropriate if the impurity is unusually toxic.

[d]For example, do known safety data for this impurity or its structural class preclude human exposure at the concentration present?

greater than (>) the identification thresholds given in Table 1.2. When identification of a degradation product is not feasible, a summary of the laboratory studies demonstrating the unsuccessful efforts to identify it should be included in the registration application.

Degradation products present at a level not more than (≤) the identification threshold would generally not need to be identified. However, analytical procedures should be developed for those degradation products that are suspected to be unusually potent, producing toxic or significant pharmacological effect at levels not more than (≤) the identification threshold. In unusual circumstances, technical factors (e.g., manufacturing capability, a low drug substance to excipient ratio, or the use of excipients that are crude products of animal or plant origin) can be considered part of the justification for selection of alternative thresholds based on manufacturing experience with the proposed commercial process.

Analytical procedures should be validated to demonstrate specificity for the specified and unspecified degradation products. As appropriate, this validation should

TABLE 1.2
Thresholds for Degradation Products in New Drug Products, ICH Q3B

Reporting Thresholds	
Maximum Daily Doses[a]	**Threshold**[b,c]
≤1 g	0.1%
>1 g	0.05%
Identification Thresholds	
Maximum Daily Doses[a]	**Threshold**[b,c]
≤1 mg	1.0% or 5 μg TDI, whichever is lower
1–10 mg	0.5% or 20 μg TDI, whichever is lower
>10 mg to 2 g	0.2% or 2 mg TDI, whichever is lower
>2 g	0.10%
Qualification Thresholds	
Maximum Daily Doses[a]	**Threshold**[b,c]
≤10 mg	1.0% or 50 μg TDI, whichever is lower
10–100 mg	0.5% or 200 μg TDI, whichever is lower
>100 mg to 2 g	0.2% or 3 mg TDI, whichever is lower
>2 g	0.15%

Source: *Impurities in New Drug Products*, ICH Q3B(R2), ICH Harmonised Tripartite Guideline, International Conference on Harmonisation of Technical Requirements for Registration of Pharmaceuticals for Human Use, 2006.

[a] The amount of drug substance administered per day.

[b] Thresholds for degradation products are expressed either as a percentage of the drug substance or as total daily intake (TDI) of the degradation product. Lower thresholds can be appropriate if the degradation product is unusually toxic.

[c] Higher thresholds should be scientifically justified.

include samples stored under relevant stress conditions: light, heat, humidity, acid/base hydrolysis, and oxidation. The quantitation limit for the analytical procedure should be not more than (≤) the reporting threshold.

The specification for a new drug product should include a list of degradation products expected to occur during manufacture of the commercial product and under recommended storage conditions. Stability studies, knowledge of degradation pathways, product development studies, and laboratory studies should be used to characterize the degradation profile. The selection of degradation products in the new drug product specification should be based on the degradation products found in batches manufactured by the proposed commercial process. The individual degradation products with specific acceptance criteria included in the specification for the new drug product are referred to as "specified degradation products" in this guideline. Specified degradation products can be identified or unidentified. A rationale for the inclusion or exclusion of degradation products in the specification should be presented. This rationale should include a discussion on the degradation profiles observed in the safety and clinical development batches and in stability studies, together with a consideration of the degradation profile of batches manufactured by the proposed commercial process.

For degradation products known to be unusually potent or to produce toxic or unexpected pharmacological effects, the quantitation/detection limit of the analytical procedures should be commensurate with the level at which the degradation products should be controlled. For a given degradation product, its acceptance criterion should be established by taking into account its acceptance criterion in the drug substance, its qualified level, its increase during stability studies, and the proposed shelf life and recommended storage conditions for the new drug product.

Where there is no safety concern, degradation product acceptance criteria should be based on data generated from batches of the new drug product manufactured by the proposed commercial process, allowing sufficient latitude to deal with normal manufacturing and analytical variations and the stability characteristics of the new drug product.

Qualification is the process of acquiring and evaluating data that establishes the biological safety of an individual degradation product or a given degradation profile at the levels specified. The applicant should provide a rationale for establishing degradation product acceptance criteria that includes safety considerations. The level of any degradation product present in the new drug product that has been adequately tested in safety and/or clinical studies would be considered qualified. Therefore, it is useful to include any available information on the actual content of degradation products in the relevant batches at the time of use in safety and/or clinical studies. Degradation products that are also significant metabolites present in animal and/or human studies are generally considered qualified. Degradation products could be considered qualified at levels higher than those administered in safety studies based on a comparison between the actual doses given in the safety studies and the intended dose of the new drug product.

Higher or lower thresholds for qualification of degradation products can be appropriate for some individual new drug products based on scientific rationale and level of concern, including drug class effects and clinical experience. For example, qualification can be especially important when there is evidence that such degradation products in certain new drug products or therapeutic classes have previously been associated with adverse reactions in patients. In these instances, a lower qualification threshold can be appropriate. Conversely, a higher qualification threshold can be appropriate for individual new drug products when the level of concern for safety is less than usual based on similar considerations (e.g., patient population, drug class effects, and clinical considerations). Proposals for alternative thresholds would be considered on a case-by-case basis.

Figure 1.2 describes considerations for the qualification of degradation products when thresholds are exceeded. In some cases, reducing the level of degradation products (e.g., use of a more protective container closure or modified storage conditions) to not more than (≤) the threshold can be simpler than providing safety data. Alternatively, adequate data could be available in the scientific literature to qualify a degradation product. If neither is the case, additional safety testing should be considered. The studies considered appropriate to qualify a degradation product will depend on a number of factors, including the patient population, daily dose, and route and duration of new drug product administration. Such studies can be conducted on the new drug product or substance containing the degradation products to be controlled, although studies using isolated degradation products can sometimes be appropriate.

Although this guideline is not intended to be applied during the clinical research stage of development, in later stages of development the thresholds in this guideline can be useful in evaluating new degradation products observed in the new drug product batches prepared by the proposed commercial process. Any new degradation product observed in the later stages of development should be identified if its level is greater than (>) the identification threshold given in Table 1.2. Similarly, qualification of the degradation product should be considered if its level is greater than (>) the qualification threshold given in Table 1.2. Safety studies should provide a comparison of the results of safety testing of the new drug product or drug substance containing a representative level of the degradation products with previously qualified material, although studies using the isolated degradation products can also be considered.

1.4 IMPURITIES: GUIDELINE FOR RESIDUAL SOLVENTS, Q3C(R5)

The objective of Q3C (R5) [4] is to recommend acceptable amounts for residual solvents in pharmaceuticals for the safety of the patient. The guideline recommends the use of less toxic solvents and describes levels considered to be toxicologically acceptable for some residual solvents.

Residual solvents in pharmaceuticals are defined here as organic volatile chemicals that are used or produced in the manufacture of drug substances or excipients, or in the preparation of drug products. The solvents are not completely removed by practical manufacturing techniques. Appropriate selection of the solvent for the synthesis of drug substances may enhance the yield or determine characteristics such as crystal form, purity, and solubility. Therefore, the solvent may sometimes be a critical parameter in the synthetic process. This guideline does not address solvents deliberately used as excipients nor does it address solvates. However, the content of solvents in such products should be evaluated and justified.

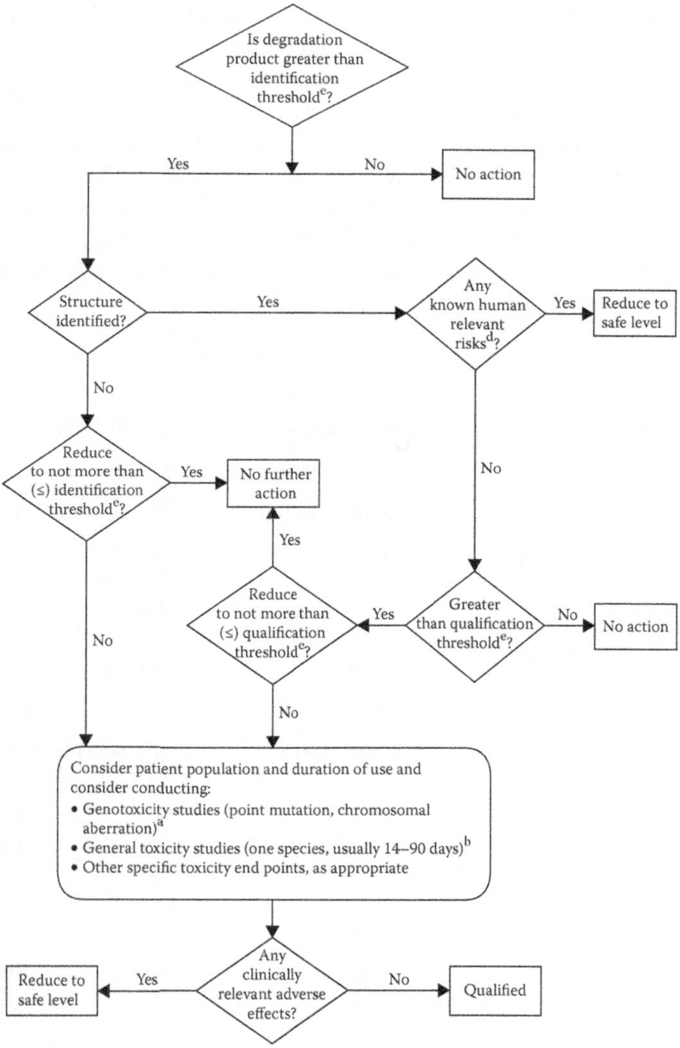

FIGURE 1.2 Decision tree for identification and qualification of a degradation product, ICH Q3B.

[a]If considered desirable, a minimum screen (e.g., genotoxic potential) should be conducted. A study to detect point mutation and another to detect chromosomal aberrations, both in vitro, are considered an appropriate minimum screen.

[b]If general toxicity studies are desirable, one or more studies should be designed to allow the comparison of unqualified to qualified material. The study duration should be based on available relevant information and performed in the species most likely to maximize the potential to detect the toxicity of an impurity. On a case-by-case basis, single-dose studies can be appropriate, especially for single-dose drugs. In general, a minimum of 14 days and a maximum duration of 90 days would be considered appropriate.

[c]Lower thresholds can be appropriate if the degradation product is unusually toxic.

[d]For example, do known safety data for this degradation product or its structural class preclude human exposure at the concentration present?

Since there is no therapeutic benefit from residual solvents, all residual solvents should be removed to the extent possible to meet product specifications, GMPs, or other quality-based requirements. Drug products should contain no higher levels of residual solvents than can be supported by safety data. Some solvents that are known to cause unacceptable toxicities (class 1, Table 1.3) should be avoided in the production of drug substances, excipients, or drug products unless their use can be strongly justified in a risk–benefit assessment. Some solvents associated with less severe toxicity (class 2, Table 1.3) should be limited to protect patients from potential adverse effects. Ideally, less toxic solvents (class 3, Table 1.3) should be used where practical.

TABLE 1.3
Classification of Solvents in Pharmaceutical Products, ICH Q3C

	Class 1 Solvents	
Solvent	**Concentration Limit (ppm)**	**Concern**
Benzene	2	Carcinogen
Carbon tetrachloride	4	Toxic and environmental hazard
1,2-Dichloroethane	5	Toxic
1,1-Dichloroethene	8	Toxic
1,1,1-Trichloroethane	1500	Environmental hazard
	Class 2 Solvents	
Solvent	**PDE (mg/day)**	**Concentration Limit (ppm)**
Acetonitrile	4.1	410
Chlorobenzene	3.6	360
Chloroform	0.6	60
Cumene	0.7	70
Cyclohexane	38.8	3880
1,2-Dichloroethene	18.7	1870
Dichloromethane	6.0	600
1,2-Dimethoxyethane	1.0	100
N,*N*-Dimethylacetamide	10.9	1090
N,*N*-Dimethylformamide	8.8	880
1,4-Dioxane	3.8	380
2-Ethoxyethanol	1.6	160
Ethyleneglycol	6.2	620
Formamide	2.2	220
Hexane	2.9	290
Methanol	30.0	3000
2-Methoxyethanol	0.5	50
Methylbutyl ketone	0.5	50
Methylcyclohexane	11.8	1180
N-Methylpyrrolidone	5.3	530
Nitromethane	0.5	50
Pyridine	2.0	200
Sulfolane	1.6	160

TABLE 1.3
Classification of Solvents in Pharmaceutical Products, ICH Q3C

Class 2 Solvents		
Solvent	**PDE (mg/day)**	**Concentration Limit (ppm)**
Tetrahydrofuran	7.2	720
Tetralin	1.0	100
Toluene	8.9	890
1,1,2-Trichloroethene	0.8	80
Xylene	21.7	2170
Class 3 Solvents (PDE 50 mg/day)		
Acetic acid	Ethyl acetate	Methylethyl ketone
Acetone	Ethyl ether	Methylisobutyl ketone
Anisole	Ethyl formate	2-Methyl-1-propanol
1-Butanol	Formic acid	Pentane
2-Butanol	Heptane	1-Pentanol
Butyl acetate	Isobutyl acetate	1-Propanol
tert-Butylmethyl ether	Isopropyl acetate	2-Propanol
Dimethyl sulfoxide	Methyl acetate	Propyl acetate
Ethanol	3-Methyl-1-butanol	

Source: Impurities: Guideline for Residual Solvents, ICH Q3C (R5), ICH Harmonised Tripartite Guideline, International Conference on Harmonisation of Technical Requirements for Registration of Pharmaceuticals for Human Use, 2011.

Recommended limits of class 1 and 2 solvents or classification of solvents may change as new safety data become available. Supporting safety data in a marketing application for a new drug product containing a new solvent may be based on the concepts in this guideline or the concept of qualification of impurities as expressed in the guideline for drug substances (Q3A) or drug products (Q3B), or all three guidelines.

Testing should be performed for residual solvents when production or purification processes are known to result in the presence of such solvents. It is only necessary to test for the solvents that are used or produced in the manufacture or purification of drug substances, excipients, or drug products. Although manufacturers may choose to test the drug product, a cumulative method may be used to calculate the residual solvent levels in the drug product from the levels in the ingredients used to product the product. If the calculation results in a level equal to or below that recommended in this guideline, no testing of the drug product for residual solvents needs to be considered. If, however, the calculated level is above the recommended level, the drug product should be tested to ascertain whether the formulation process has reduced the relevant solvent level to within the acceptable amount. The drug product should also be tested if a solvent is used during its manufacture.

This guideline applies to all dosage forms and routes of administration. Higher levels of residual solvents may be acceptable in certain cases such as short-term (30 days or less) or topical application. Justification for these levels should be made on a case-by-case basis.

The term "tolerable daily intake" is used by the International Program on Chemical Safety (IPCS) to describe the exposure limits of toxic chemicals, and "acceptable daily intake" (ADI) is used by the World Health Organization (WHO) and other national and international health authorities and institutes. The new term "permitted daily exposure" (PDE) is defined in the present guideline as a pharmaceutically acceptable intake of residual solvents to avoid confusion of differing values for ADIs of the same substance.

Residual solvents were evaluated for their possible risk to human health and placed into one of three classes as follows (see Table 1.3):

Class 1 solvents: solvents to be avoided (known human carcinogens, strongly suspected human carcinogens, and environmental hazards). The solvents in class 1 should not be employed in the manufacture of drug substances, excipients, and drug products because of their unacceptable toxicity or their deleterious environmental effects. However, if their use is unavoidable to produce a drug product with a significant therapeutic advance, then their levels should be restricted as shown in Table 1.3, unless otherwise justified.

Class 2 solvents: solvents to be limited (nongenotoxic animal carcinogens or possible causative agents of other irreversible toxicity such as neurotoxicity or teratogenicity). Class 2 solvents should be limited in pharmaceutical products because of their inherent toxicity. PDEs are given to the nearest 0.1 mg/day, and concentrations are given to the nearest 10 ppm. The stated values do not reflect the necessary analytical precision of determination. Precision should be determined as part of the validation of the method.

Class 3 solvents: solvents with low toxic potential (solvents with low toxic potential to humans; no health-based exposure limit is needed. Class 3 solvents have PDEs of 50 mg or more per day). The solvents in class 3 may be regarded as less toxic and having lower risk to human health. Class 3 includes no solvent known as a human health hazard at levels normally accepted in pharmaceuticals. However, there are no long-term toxicity or carcinogenicity studies for many of the solvents in class 3. Available data indicate that they are less toxic in acute or short-term studies and negative in genotoxicity studies. It is considered that amounts of these residual solvents of 50 mg/day or less would be acceptable without justification. Higher amounts may also be acceptable provided they are realistic in relation to manufacturing capability and GMPs.

Two options are available when setting limits for class 2 solvents:

Option 1: the concentration limits in parts per million stated in Table 1.3 can be used. They were calculated using the following equation by assuming a product mass of 10 g administered daily:

$$\text{Concentration (ppm)} = \frac{1000 \times \text{PDE (mg/day)}}{\text{Dose (g/day)}}$$

These limits are considered acceptable for all substances, excipients, or products. Therefore, this option may be applied if the daily dose is not known or fixed. If all excipients and drug substances in a formulation meet the limits given in option 1, then these components may be used in any proportion. No further calculation is necessary provided the daily dose does not exceed 10 g. Products that are administered in doses greater than 10 g/day should be considered under option 2.

Option 2: it is not considered necessary for each component of the drug product to comply with the limits given in option 1. The PDE in terms of milligrams per day as stated in Table 1.3 can be used with the known maximum daily dose and the aforementioned equation to determine the concentration of the residual solvent allowed in the drug product. Such limits are considered acceptable provided that it has been demonstrated that the residual solvent has been reduced to the practical minimum. The limits should be realistic in relation to analytical precision, manufacturing capability, and reasonable variation in the manufacturing process, and the limits should reflect contemporary manufacturing standards. Option 2 may be applied by adding the amounts of a residual solvent present in each of the components of the drug product. The sum of the amounts of solvent per day should be less than that given by the PDE.

The method used to establish PDEs for residual solvents is presented in Section 1.4.1. Summaries of the toxicity data that were used to establish limits are published in *Pharmeuropa*, Vol. 9, No. 1, Supplement, April 1997.

1.4.1 Methods for Establishing Exposure Limits

The Gaylor–Kodell method of risk assessment (Gaylor, D. W. and Kodell, R. L.: Linear interpolation algorithm for low dose assessment of toxic substance. *J Environ. Pathology*, 4, 305, 1980) is appropriate for class 1 carcinogenic solvents. Only in cases where reliable carcinogenicity data are available should extrapolation by the use of mathematical models be applied to setting exposure limits. Exposure limits for class 1 solvents could be determined with the use of a large safety factor (i.e., 10,000 to 100,000) with respect to the no-observed-effect level (NOEL). The detection and quantitation of these solvents should be by state-of-the-art analytical techniques.

Acceptable exposure levels in this guideline for class 2 solvents were established by calculation of PDE values according to the procedures for setting exposure limits in pharmaceuticals (*Pharmacopeial Forum*, November–December 1989), and the method adopted by the IPCS for assessing the human health risk of chemicals (Environmental Health Criteria 170, WHO, Geneva, 1994). These methods are similar to those used by the U.S. Environmental Protection Agency (EPA) (Integrated Risk Information System) and the U.S. Food and Drug Administration (FDA) (*Redbook*) and others. The method is outlined here to give a better understanding of the origin of PDE values.

PDE is derived from NOEL, or the lowest-observed-effect level (LOEL) in the most relevant animal study, as follows:

$$\text{PDE} = \frac{\text{NOEL} \times \text{Weight Adjustment}}{\text{F1} \times \text{F2} \times \text{F3} \times \text{F4} \times \text{F5}}$$

The PDE is derived preferably from a NOEL. If no NOEL is obtained, a LOEL may be used. The modifying factors proposed here, for relating the data to humans, are the same kind of "uncertainty factors" used in Environmental Health Criteria

(Environmental Health Criteria 170, WHO, Geneva, 1994), and "modifying factors" or "safety factors" used in the *Pharmacopeial Forum* (November–December 1989). The assumption of 100% systemic exposure is used in all calculations regardless of the route of administration.

The modifying factors are as follows:

F1 = a factor to account for extrapolation between species
F1 = 5 for extrapolation from rats to humans
F1 = 12 for extrapolation from mice to humans
F1 = 2 for extrapolation from dogs to humans
F1 = 2.5 for extrapolation from rabbits to humans
F1 = 3 for extrapolation from monkeys to humans
F1 = 10 for extrapolation from other animals to humans

F1 takes into account the comparative surface area: body weight ratios for the species concerned and for humans. The surface area (S) is calculated as follows:

$$S = kM^{0.67}$$

where M = body mass, and the constant k has been taken to be 10. The body weights used in the equation are the ones shown in Table 1.4.

TABLE 1.4
Values Used for PDE Calculations, ICH Q3C

Rat body weight	425 g
Pregnant rat body weight	330 g
Mouse body weight	28 g
Pregnant mouse body weight	30 g
Guinea pig body weight	500 g
Rhesus monkey body weight	2.5 kg
Rabbit body weight (pregnant or not)	4 kg
Beagle dog body weight	11.5 kg
Rat respiratory volume	290 L/day
Mouse respiratory volume	43 L/day
Rabbit respiratory volume	1,440 L/day
Guinea pig respiratory volume	430 L/day
Human respiratory volume	28,800 L/day
Dog respiratory volume	9,000 L/day
Monkey respiratory volume	1,150 L/day
Mouse water consumption	5 mL/day
Rat water consumption	30 mL/day
Rat food consumption	30 g/day

Source: *Impurities: Guideline for Residual Solvents*, ICH Q3C (R5), ICH Harmonised Tripartite Guideline, International Conference on Harmonisation of Technical Requirements for Registration of Pharmaceuticals for Human Use, 2011.

F2 = a factor of 10 to account for variability between individuals. A factor of 10 is generally given for all organic solvents, and the factor 10 is used consistently in this guideline.

F3 = a variable factor to account for toxicity studies of short-term exposure
- F3 = 1 for studies that last at least one half-lifetime (1 year for rodents or rabbits; 7 years for cats, dogs, and monkeys)
- F3 = 1 for reproductive studies in which the whole period of organogenesis is covered
- F3 = 2 for a 6-month study in rodents, or a 3.5-year study in nonrodents
- F3 = 5 for a 3-month study in rodents, or a 2-year study in nonrodents
- F3 = 10 for studies of a shorter duration

In all cases, the higher factor has been used for study durations between the time points, for example, a factor of 2 for a 9-month rodent study.

F4 = a factor that may be applied in cases of severe toxicity, for example, nongenotoxic carcinogenicity, neurotoxicity, or teratogenicity. In studies of reproductive toxicity, the following factors are used:
- F4 = 1 for fetal toxicity associated with maternal toxicity
- F4 = 5 for fetal toxicity without maternal toxicity
- F4 = 5 for a teratogenic effect with maternal toxicity
- F4 = 10 for a teratogenic effect without maternal toxicity

F5 = a variable factor that may be applied if the no-effect level was not established. When only an LOEL is available, a factor of up to 10 could be used depending on the severity of the toxicity.

The weight adjustment assumes an arbitrary adult human body weight for either sex of 50 kg. This relatively low weight provides an additional safety factor against the standard weights of 60 or 70 kg that are often used in this type of calculation. It is recognized that some adult patients weigh less than 50 kg; these patients are considered to be accommodated by the built-in safety factors used to determine a PDE. If the solvent was present in a formulation specifically intended for pediatric use, an adjustment for a lower body weight would be appropriate.

As an example of the application of this equation, consider a toxicity study of acetonitrile in mice that is summarized in *Pharmeuropa*, Vol. 9, No. 1, Supplement, April 1997, page S24. The NOEL is calculated to be 50.7 mg/kg/day. The PDE for acetonitrile in this study is calculated as follows:

$$\text{PDE} = \frac{(50.7 \text{ mg/kg/day}) \times 50 \text{ kg}}{12 \times 10 \times 5 \times 1 \times 1} = 4.22 \text{ mg/day}$$

In this example,

F1 = 12 to account for the extrapolation from mice to humans
F2 = 10 to account for differences between individual humans
F3 = 5 because the duration of the study was only 13 weeks

F4 = 1 because no severe toxicity was encountered
F5 = 1 because the no-effect level was determined

In the Glossary section of the ICH Q3C guideline, the following terms were defined that are relevant to genotoxic impurities:

Genotoxic carcinogens: carcinogens that produce cancer by affecting genes or chromosomes
LOEL: the lowest dose of substance in a study or group of studies that produces biologically significant increases in frequency or severity of any effects in the exposed humans or animals
NOEL: the highest dose of substance at which there are no biologically significant increases in frequency or severity of any effects in the exposed humans or animals
PDE: the maximum acceptable intake per day of residual solvent in pharmaceutical products
Strongly suspected human carcinogen: a substance for which there is no epidemiological evidence of carcinogenesis but there are positive genotoxicity data and clear evidence of carcinogenesis in rodents

1.5 GUIDELINE ON THE LIMITS OF GENOTOXIC IMPURITIES, THE EUROPEAN MEDICINES AGENCY

The toxicological assessment of genotoxic impurities and the determination of acceptable limits for such impurities in active substances is a difficult issue and not addressed in sufficient detail in the existing ICH Q3X guidances. The data set usually available for genotoxic impurities is quite variable and is the main factor that dictates the process used for the assessment of acceptable limits. In the absence of the data usually needed for the application of one of the established risk assessment methods, that is, data from long-term carcinogenicity studies or data providing evidence for a threshold mechanism of genotoxicity, implementation of a generally applicable approach as defined by the threshold of toxicological concern (TTC) is proposed. A TTC value of 1.5 μg/day intake of a genotoxic impurity is considered to be associated with an acceptable risk (excess cancer risk of <1 in 100,000 over a lifetime) for most pharmaceuticals. From this threshold value, a permitted level in the active substance can be calculated based on the expected daily dose. Higher limits may be justified under certain conditions such as short-term exposure periods.

A general concept of qualification of impurities is described in the guidelines for active substances (Q3A) or medicinal products (Q3B), whereby qualification is defined as the process of acquiring and evaluating data that establish the biological safety of an individual impurity or a given impurity profile at the levels specified. In the case of impurities with a genotoxic potential, determination of acceptable dose levels is generally considered a particularly critical issue, which is not specifically covered by the existing guidelines.

Guideline on the Limits of Genotoxic Impurities [5] describes a general framework and practical approaches on how to deal with genotoxic impurities in new active substances. It also relates to new applications for existing active substances, where assessment of the route of synthesis, process control, and impurity profile does not provide reasonable assurance that no new or higher levels of genotoxic impurities are introduced compared to products currently authorized in the EU containing the same active substances. The same also applies to variations to existing marketing authorizations pertaining to the synthesis. The guideline does, however, not need to be applied retrospectively to authorized products unless there is a specific cause for concern.

In the current context, the classification of a compound (impurity) as genotoxic in general means that there are positive findings in established in vitro or in vivo genotoxicity tests with the main focus on DNA-reactive substances that have a potential for direct DNA damage. Isolated in vitro findings may be assessed for in vivo relevance in adequate follow-up testing. In the absence of such information, in vitro genotoxicants are usually considered as presumptive in vivo mutagens and carcinogens.

According to current regulatory practice, it is assumed that (in vivo) genotoxic compounds have the potential to damage DNA at any level of exposure and that such damage may lead/contribute to tumor development. Thus, for genotoxic carcinogens it is prudent to assume that there is no discernible threshold and that any level of exposure carries a risk. However, the existence of mechanisms leading to biologically meaningful threshold effects is increasingly acknowledged also for genotoxic events. This holds true in particular for compounds interacting with non-DNA targets and also for potential mutagens, which are rapidly detoxified before coming into contact with critical targets. The regulatory approach to such chemicals can be based on the identification of a critical NOEL and the use of uncertainty factors.

Even for compounds that are able to react with the DNA molecules, extrapolation in a linear manner from effects in high-dose studies to very-low-level (human) exposure may not be justified due to several protective mechanisms operating effectively at low doses. However, at present it is extremely difficult to experimentally prove the existence of a threshold for the genotoxicity of a given mutagen. Thus, in the absence of appropriate evidence supporting the existence of a threshold for a genotoxic compound making it difficult to define a safe dose it is necessary to adopt a concept of a level of exposure that carries an acceptable risk.

Guided by existing genotoxicity data or the presence of structural alerts, potential genotoxic impurities should be identified. When a potential impurity contains structural alerts, additional genotoxicity testing of the impurity, typically in a bacterial reverse mutation assay, should be considered. Although according to the Q3A guideline such studies can usually be conducted on the drug substance containing the impurity to be controlled, studies using isolated impurities are much more appropriate for this purpose and highly recommended.

1.5.1 Genotoxic Compounds with Sufficient (Experimental) Evidence for a Threshold-Related Mechanism

Examples of mechanisms of genotoxicity that may be demonstrated to lead to nonlinear or thresholded dose–response relationships include interaction with the spindle apparatus of cell division leading to aneuploidy, topoisomerase inhibition, inhibition of DNA synthesis, overloading of defense mechanisms, metabolic overload, and physiological perturbations (e.g., induction of erythropoiesis and hyper- or hypothermia).

For classes of compounds with clear evidence for a thresholded genotoxicity, exposure levels that are without appreciable risk of genotoxicity can be established according to the procedure outlined for class 2 solvents in Q3C (see Section 1.4), which is derived from the NOEL, or the LOEL in the most relevant animal study using uncertainty factors.

1.5.2 Genotoxic Compounds without Sufficient Evidence for a Threshold-Related Mechanism

The assessment of acceptability of genotoxic impurities for which no threshold mechanisms are identified should include both pharmaceutical and toxicological evaluations.

Pharmaceutical assessment: in general, pharmaceutical measurements should be guided by a policy of controlling levels to "as low as reasonably practicable" (the ALARP principle), where avoiding is not possible. A justification needs to be provided that no viable alternative exists, including alternative routes of synthesis or formulations or different starting materials. This might, for instance, include cases where the structure, which is responsible for the genotoxic and/or carcinogenic potential, is equivalent to that needed in chemical synthesis (e.g., alkylation reactions). If a genotoxic impurity is considered to be unavoidable in a drug substance, technical efforts (e.g., purification steps) should be undertaken to reduce the content of the genotoxic residues in the final product in compliance with safety needs or to a level as low as reasonably practicable. Data on the chemical stability of reactive intermediates, reactants, and other components should be included in this assessment. Detection and/or quantification of these residues should be done by state-of-the-art analytical techniques.

Toxicological assessment: the impossibility of defining a safe exposure level (the zero risk concept) for genotoxic carcinogens without a threshold and the realization that complete elimination of genotoxic impurities from drug substances is often unachievable require the implementation of a concept of an acceptable risk level, that is, an estimate of daily human exposure at and below which there is negligible risk to human health. However, these approaches require the availability of adequate data from long-term carcinogenicity studies. In most cases of toxicological assessment of genotoxic impurities only limited data from in vitro studies with the impurity (e.g., the Ames test and the chromosomal aberration test) are available, and thus established approaches to determine acceptable intake levels cannot be applied. The calculation of "safety multiples" from in vitro data (e.g., the Ames test) is considered inappropriate for justifying acceptable limits. Moreover, negative carcinogenicity and genotoxicity data with the drug substance containing the impurity at low parts-per-million levels do not provide sufficient assurance for setting acceptable limits for the impurity due to the lack of sensitivity of this testing approach. Even potent mutagens and carcinogens are most likely to remain undetected when tested as part of the drug substance, that is, at very low exposure levels. A pragmatic approach is therefore needed that recognizes that the presence of very low levels of genotoxic impurities is not associated with an unacceptable risk.

1.5.3 Threshold of Toxicological Concern

A TTC has been developed to define a common exposure level for any unstudied chemical that will not pose a risk of significant carcinogenicity or other toxic effects. This TTC value was estimated to be 1.5 μg/person/day. The TTC, originally developed as a "threshold of regulation" at the FDA for food-contact materials, was established based on the analysis of 343 carcinogens from a carcinogenic potency database (Gold et al., *Environ. Health Perspect.* 1984, 58, 9–319) and was repeatedly confirmed by evaluations expanding the database to more than 700 carcinogens. The probability distribution of carcinogenic potencies has been used to derive an estimate of a daily exposure level (micrograms per person) of most carcinogens that would give rise to less than a 1 in 10^6 (1×10^{-6}) upper-bound lifetime risk of cancer ("virtually safe dose"). Further analysis of subsets of high-potency carcinogens led to the suggestion of a ten-fold lower TTC (0.15 μg/day) for chemicals with structural alerts that raise concern for potential genotoxicity. However, for the application of a TTC in the assessment of acceptable limits of genotoxic impurities in drug substances, a value of 1.5 μg/day, corresponding to a 10^{-5} lifetime risk of cancer, can be justified as a benefit exists for pharmaceuticals. It should be recognized in this context that the methods on which the TTC value is based are generally considered very conservative since they involve a simple linear extrapolation from the dose giving a 50% tumor incidence (TD50) to a 1 in 10^6 incidence, using TD50 data from the most sensitive species and most sensitive site (several worst-case assumptions).

Some structural groups were identified to be of such high potency that intakes even below the TTC would be associated with a high probability of a significant carcinogenic risk. This group of high-potency genotoxic carcinogens comprises aflatoxin-like, *N*-nitroso-, and azoxy- compounds that have to be excluded from the TTC approach. Risk assessment of members of such groups requires compound-specific toxicity data.

There may be reasons to deviate from the TTC value based on the profile of genotoxicity results. Positive results from in vitro studies may only allow exemption of an impurity from limitation at TTC level if a lack of in vivo relevance of the findings is convincingly demonstrated based on a weight-of-evidence approach (see ICH S2 guidelines). This approach will usually need negative results with the impurity from some additional in vitro and/or appropriate in vivo testing.

A TTC value higher than 1.5 μg/day may be acceptable under certain conditions, for example, short-term exposure, for the treatment of a life-threatening condition, when life expectancy is less than 5 years, or where the impurity is a known substance and human exposure will be much greater from other sources (e.g., food). Genotoxic impurities that are also significant metabolites may be assessed based on the acceptability of the metabolites.

The concentration limits in parts per million of a genotoxic impurity in a drug substance derived from the TTC can be calculated based on the expected daily dose to the patient using the following equation:

$$\text{Concentration (ppm)} = \frac{\text{TTC } (\mu\text{g/day})}{\text{Dose (g/day)}}$$

The TTC concept should not be applied to carcinogens where adequate toxicity data (long-term studies) are available and allow for a compound-specific risk assessment. It has to be emphasized that the TTC is a pragmatic risk management tool using a probabilistic methodology, that is, there is a high probability that a 10^{-5} lifetime cancer risk will not be exceeded if the daily intake of a genotoxic impurity with unknown carcinogenic potential/potency is below the TTC value. The TTC concept should not be interpreted as providing absolute certainty of no risk.

The decision tree for assessing the acceptability of genotoxic impurities is shown in Figure 1.3.

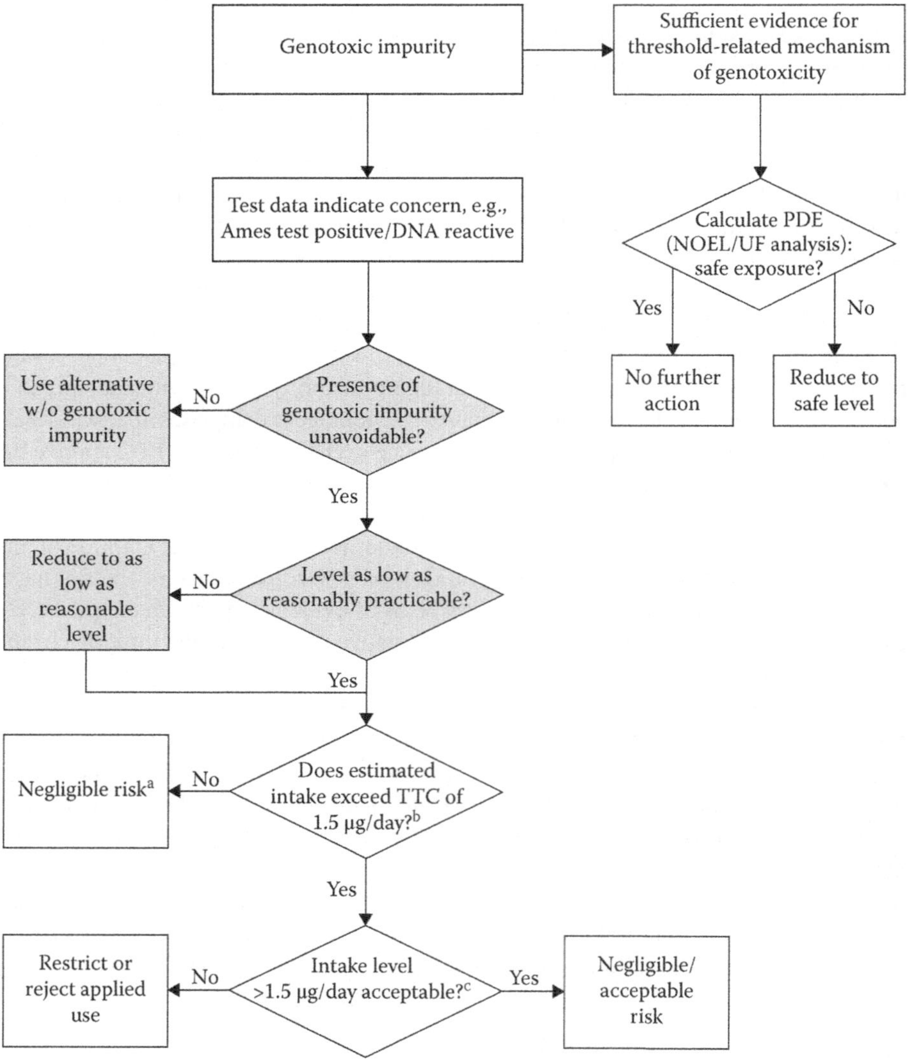

FIGURE 1.3 Decision tree for assessment of acceptability of genotoxic impurities (shaded boxes show pharmaceutical assessment, and white boxes show toxicological assessment). NOEL/UF; no-observed-effect level/uncertainty factor, PDE; permitted daily exposure, TTC; threshold of toxicological concern.

[a]Impurities with structural relationship to high-potency carcinogens (see Section 1.5.3) are to be excluded from the TTC approach.

[b]If carcinogenicity data are available, does intake exceed calculated 10–5 cancer lifetime risk?

[c]Case-by-case assessment should include duration of treatment, indication, patient population, and so on.

1.6 QUESTIONS AND ANSWERS ON THE "GUIDELINE ON THE LIMITS OF GENOTOXIC IMPURITIES," EUROPEAN MEDICINES AGENCY

The aim of this questions and answers document [6] is to provide clarification and harmonization of the *Guideline on the Limits of Genotoxic Impurities* published in 2006.

Question 1: The guideline does not need to be applied retrospectively to authorized products unless there is a specific cause for concern. What might constitute a cause for concern in terms of application to currently marketed products?

Answer: If a manufacturing procedure for an active pharmaceutical ingredient (API) remains essentially unchanged, a reevaluation with respect to the presence of potentially genotoxic impurities is generally not needed. However, new knowledge may indicate a previously unknown cause for concern. One example is the mesylate salt drug substances for which a concern regarding the potential for the formation of genotoxic alkyl mesylates was raised a few years ago. This concern resulted in the production statement requesting a specific evaluation of the potential for the formation of these highly toxic products, which is now included as part of the *European Pharmacopoeia* monographs for all mesylate salts.

Question 2: The guideline indicates that it is necessary to reduce a known or suspected mutagenic impurity to ALARP even if the level is below the TTC, that is, in the decision tree ALARP considerations precede the TTC considerations. If the level is already known to be below the TTC (which is a very conservative value), why is it necessary to reduce it even further? On a practical basis, this would often involve developing analytical methods that have limits of quantitation around or below 1 ppm. The effort and resource this can take, and the possibility that it is not actually achievable, does not seem commensurate with the perceived risk.

Answer: If the level of a mutagenic impurity is below the TTC (equivalent to a clinical dose $\leq$ 1.5 μg/day), it is not necessary to apply ALARP considerations unless it is a structure of very high concern, for example, *N*-nitroso, aflatoxin-like, and azoxy compounds.

Question 3: The guideline states: "When a potential impurity contains structural alerts, additional genotoxicity testing of the impurity, typically in a bacterial reverse mutation assay, should be considered."

i: If an impurity triggers a mutagenic structural alert, will a negative result in an Ames test on the impurity (conducted to regulatory acceptable standards) be sufficient to conclude that the compound is of no concern with respect to genotoxicity and no further qualification studies will be required?

Answer: Yes, a negative Ames test (conducted to regulatory acceptable standards) will overrule the structural alerts and no further studies would be required providing the level remains below ICH Q3A/B limits.

ii. Is the absence of a structural alert sufficient to show that the impurity is of no concern?

Answer: Yes, the absence of a structural alert based on a well-performed assessment (e.g., through the application of a commonly used quantitative structure–activity relationship (QSAR) assessment software such as DEREK of MultiCASE) will be sufficient to conclude that the impurity is of no concern with respect to genotoxicity and no further qualification studies of justification will be required.

iii. Is it acceptable to control impurities with a structural alert by assuming they will be positive (without resorting to any testing) and ensuring the level remains below the appropriate TTC value?

Answer: Yes, genotoxicity testing is not obligatory when a potential genotoxic impurity is controlled at the TTC level, unless it belongs to a class of very potent genotoxic carcinogens (*N*-nitroso and azoxy compounds, or an aflatoxin-like compound).

Question 4: What would be an appropriate strategy to qualify a new impurity that arises during phase III or with a commercial product? For example, would it be acceptable that a new unidentified impurity discovered at the 0.05%–0.09% range requires no action to be taken? Would an impurity discovered in the 0.10%–0.15% range, even if it triggers a structural alert, be fully qualified by testing the active ingredient containing this impurity in an Ames test?

Answer: In line with the ICH guideline, no action is generally required for a new unidentified impurity found at levels below the ICH identification threshold. When an impurity is found above the ICH identification threshold but below the qualification threshold, and the structure gives rise to a structural alert, this can be negated by carrying out an Ames test on the active ingredient containing the impurity as long as the impurity is present at a minimum concentration of 250 μg/plate (estimated detection limit for most relevant mutagens in the Ames test). If the structure cannot be elucidated, then no action is generally required. Above the ICH qualification threshold, the ICH guidance should be followed.

Question 5: The guideline states that "a TTC value higher than 1.5 μg/day may be acceptable under certain conditions, e.g., short-term exposure ..." Is the Committee for Medicinal Products for Human Use (CHMP) in agreement with the principles of "staged TTC" dependent on the duration of clinical exposure, that is, higher levels allowed for marketed drugs used to treat acute conditions, for example, antibiotics? If "yes," what levels would be acceptable?

Answer: TTC values for genotoxic impurities above 1.5 μg/day will be treated on a case-by-case basis. For short-duration treatments, the acceptability of higher levels will be in line with the principles outlined in the CHMP's answer to Question 6.

Question 6: The wording of the guideline implies that it is intended to cover an issue not covered specifically by ICH Q3A/B guidelines. The scope of the ICH guidelines does not include clinical development prior to new drug submission or marketing authorization application (MAA) submission. However, experience across a range of pharmaceutical companies indicates that some regulatory authorities are applying the genotoxic impurity guideline during clinical development. In case the scope should include application to clinical development, can the CHMP advise on how genotoxic impurities should be controlled during the development stages?

Answer: According to the scope of the guideline, it mainly applies to genotoxic impurities in new active substances, which include new active substances in both MAA submissions and clinical trial applications. In fact, the CHMP *Guideline on the Requirements to the Chemical and Pharmaceutical Quality Documentation Concerning Investigational Medicinal Products in Clinical Trials* (CHMP/QWP/185401/2004) states that "for impurities in IMPs, a justification that the product is safe for its intended use, considering the anticipated exposure of volunteers and patients, respectively, will be required" and further that "a brief justification of the specifications and acceptance criteria for impurities ... should be provided based on safety and toxicity data."

The CHMP agrees with the use of a staged TTC concept during clinical development. The acceptable limits for daily intake of genotoxic impurities are 5, 10, 20, and 60 μg/day for durations of exposure of 6–12 months, 3–6 months, 1–3 months, and less than 1 month, respectively. For a single dose, an intake of up to 120 μg is acceptable. Compared to the proposal of a staged TTC in the paper by Mueller and others (*Reg Tox & Pharm*, 2006, 44, 198–211), these values incorporate a dose rate correction factor of 2 to account for deviations from the linear extrapolation model.

	Duration of Exposure				
	Single Dose	**≤1 month**	**≤3 months**	**≤6 months**	**≤12 months**
Allowable daily intake (μg)	120	60	20	10	5

Question 7: The guideline text implies that the indicated TTC value of 1.5 μg/day is a level that can be applied to each individual genotoxic impurity in a drug substance. Could this be confirmed?

Answer: When more than one genotoxic impurity is present in the drug substance, the TTC value of 1.5 μg/day can be applied to each individual impurity only if the impurities are structurally unrelated.

In case of structural similarity, it can be assumed that the impurities act by the same genotoxic mode of action and have the same molecular target and thus might exert effects in an additive manner. In such a situation, a limitation of the sum of the genotoxic impurities at 1.5 μg/day is recommended. This might be practically unachievable with reasonable efforts, in particular when the maximum daily dose is very high and thus may demand the application of lower group limits. Justifications should be made on a case-by-case basis taking into consideration issues such as the following:

- Maximum daily dose of the active substance
- Therapeutic indication
- Step of the synthesis at which the genotoxic impurities arise
- Capability of the manufacturing process (purification steps) to eliminate these impurities
- Capability of the analytical procedure to control these impurities

In cases where routine use of more powerful detection methods is difficult, one could consider using such methods during the development or testing of the first commercial batches, to demonstrate that the actual values are sufficiently below the TTC. In such a case, skip testing could be considered instead of routine testing, provided that the competent authorities, based on a risk assessment, consider the approach as acceptable.

Question 8: The European Pharmacopoeia Commission Policy for dealing with potentially genotoxic impurities was issued in March 2008. This policy is to be applied during elaboration and revision of pharmacopoeial monographs. It provides very pragmatic guidance on how the CHMP guideline can be applied to active substances described in existing monographs. Can the same principles, as outlined within this policy, apply to established drug substances that do not have pharmacopoeial monographs?

Answer: Yes. For active substances included in medical products authorized by the competent authorities before implementation of the CHMP guideline, the specifications as described in the dossier for marketing authorization should be followed. Action is needed only where there is study data demonstrating genotoxicity of the impurity. The existence of structural alerts alone is considered insufficient to trigger follow-up measures unless it is a structure of very high concern, for example, *N*-nitroso, aflatoxin-like, and azoxy compounds. If a new synthetic route is used that may give rise to different potentially genotoxic impurities or to higher levels of previously recognized potentially genotoxic impurities, then the situation should be discussed with the competent authority.

Question 9: What is a reasonable policy for setting specifications for potentially genotoxic impurities that are theoretical or actual impurities in a drug substance–manufacturing process?

Answer: Different possible scenarios can be identified, and the policies to be applied for each of them are described in the following examples:

Example 1: A potential genotoxic impurity

The definition of a potential genotoxic impurity is derived from the definition of "potential impurity," an impurity that can theoretically be formed during manufacture or storage. It may or may not actually appear in the (new) drug substance (ICH Q3A, Glossary).

If a potential genotoxic impurity is just a theoretical impurity, that is, based on theoretical considerations but not found in practice as demonstrated by studies during development or manufacture, the impurity does not need to be included in the drug substance specification.

Example 2: A (potentially) genotoxic impurity actually formed or introduced prior to the final step of the synthesis

If a (potentially) genotoxic impurity is formed or introduced in a step before the final synthesis step, it is considered possible to not include this impurity in the drug substance specification if it is controlled by a suitable limit in a synthesis intermediate and if it is unambiguously demonstrated by analysis results (use of spiking experiments are encourage) that the presence of this impurity does not exceed 30% of the limit, derived from either TTC or otherwise defined acceptable limit and so on, in the drug substance. If these conditions are not fulfilled, this impurity has to be included in the drug substance specification and the test has to be carried out on a routine basis. Should a genotoxic impurity not be controlled at the intermediate stage, then the scenario of Example 3 applies.

Example 3: A (potentially) genotoxic impurity is formed or introduced in the last synthesis step

If a (potentially) genotoxic impurity is formed or introduced in the final synthesis step, it should be included in the specifications. However, it is considered possible to apply skip testing if the level of the impurity does not exceed 30% of the limit, derived from either TTC or otherwise defined acceptable limit and so on, in the drug substance. Data should be presented for at least six consecutive pilot scale or three consecutive production scale batches. If this condition is not fulfilled, a routine test in the drug substance specification is needed.

The following definitions apply:

Genotoxic impurity: an impurity that has been demonstrated to be genotoxic in an appropriate genotoxicity test model, for example, the bacterial gene mutation (Ames) test.

Potentially genotoxic impurity: an impurity that shows structural alerts for genotoxicity but has not been tested in an experimental test model. Here, "potentially" relates to genotoxicity, not to the presence or absence of this impurity.

1.7 GENOTOXIC AND CARCINOGENIC IMPURITIES IN DRUG SUBSTANCES AND PRODUCTS: RECOMMENDED APPROACHES, DRAFT GUIDANCE, THE FOOD AND DRUG ADMINISTRATION

This guidance [7] is intended to inform pharmaceutical manufacturers of the FDA's current thinking regarding genotoxic and carcinogenic impurities in drug substances and drug products, including biologic products that are regulated by the Center for Drug Evaluation and Research (CDER). This guidance provides recommendations on how to evaluate the safety of these impurities during clinical development (investigational new drug [IND] applications) and for marketing applications (new drug applications [NDAs]), biologics license applications (BLAs), and abbreviated new drug applications (ANDAs). This guidance provides recommended exposure thresholds on the clinical exposure to genotoxic or carcinogenic impurities. Additional testing and exposure threshold recommendations for situations where there are known or theoretical safety concerns based on available data, structural alerts, and/or assessment of the synthetic pathway are also provided.

This guidance is intended as an adjunct to the ICH guidances for industry, *Q3A(R2) Impurities in New Drug Substances, Q3B(R2) Impurities in New Drug Products,* and *Q3C(R3) Impurities: Residual Solvents*, that deal with the topic of impurities in a more general fashion. This guidance provides specific recommendations regarding the safety qualification of impurities with known or suspected genotoxic or carcinogenic potential, whereas the ICH guidances provide only a general direction. It addresses synthetic impurities and degradants in drug substances but does not otherwise address the genotoxicity or carcinogenicity of actual drug substances or intended drug product ingredients. It also applies to known starting materials or anticipated reaction products.

This guidance describes a variety of ways to characterize and reduce the potential lifetime cancer risk associated with patient exposure to genotoxic and carcinogenic impurities both during clinical development and after approval. These approaches include the following:

- Changing the synthetic and/or purification routes to minimize the formation and/or maximize the removal of the relevant impurity.
- Allowing a maximum daily exposure target of 1.5 μg/day for the relevant impurity as a general target for marketed products, although higher levels may be acceptable during clinical development. Certain impurities with structural alerts suggesting particularly high genotoxic and carcinogenic potential would not be appropriate for this general threshold approach and would need to be evaluated on a case-by-case basis.
- Further characterizing the genotoxic and carcinogenic risk via mechanism of action or weight-of-evidence approaches, or through additional studies to better support appropriate impurity specifications.

This guidance also applies to drug products approved before the issuance of this guidance, but only in the presence of a specific safety signal that suggests the potential for an increased carcinogenic risk associated with the presence of an impurity or a degradant, or with respect to a supplemental application for a previously approved drug product that proposes a significant change in the drug product's approved labeling that suggests the potential for an increased carcinogenic risk associated with the presence of an impurity or a degradant (e.g., new indication, new dosage regimen, or longer duration of use). Applicants should also take these recommendations into consideration when preparing supplemental manufacturing submissions to NDAs, BLAs, and ANDAs, such as submissions proposing new formulations or new synthetic routes. Although this guidance applies to impurities present in biologic products regulated by the CDER, it is noted that, in most cases, the genotoxicity assays conducted for small-molecule pharmaceuticals are not applicable to biopharmaceuticals. Likewise, the standard assessment of the genotoxic potential of impurities in biopharmaceuticals may not be appropriate in many cases since they may include residual host cell proteins and nucleic material, fermentation components, and bacterial and viral components and do not include the organic chemicals typically found in small-molecule manufacturing.

FDA's guidance documents, including this guidance, do not legally establish enforceable responsibilities. Instead, the guidances describe the agency's current thinking on a topic and should be viewed only as recommendations, unless specific regulatory or statutory requirements are cited. The use of the word "should" in agency guidances means that something is suggested or recommended but not required.

Compounds that have been demonstrated to induce genetic mutations, chromosomal breaks, and/or chromosomal rearrangements are considered genotoxic and have the potential to cause cancer in humans. Exposures to even low levels of these impurities may be of significant concern. Therefore, the identification limits provided in ICH Q3A(R2) and ICH Q3B(R2) may not be acceptable for genotoxic or carcinogenic impurities. Although genotoxic and carcinogenic properties can be acceptable for some APIs depending on clinical circumstances (e.g., cancer chemotherapies), impurities in drug substances and drug products generally do not have beneficial effects and may impose a risk without any associated benefit. Therefore, manufacturers should strive to achieve the lowest levels of genotoxic or carcinogenic impurities that are technically feasible and/or levels that convey no significant cancer risk.

ICH Q3A(R2) and ICH Q3B(R2) recommend when, after consideration of factors such as patient population and duration of use, qualification studies of an impurity are appropriate. Part of the battery of tests used to qualify an impurity could include assays to determine whether the impurity is genotoxic. These guidances also recommend that, when considered appropriate, assays to assess genotoxic potential include the "minimum screen" of in vitro assays: a gene mutation assay and a chromosomal aberration assay. ICH Q3A(R2) indicates that "such studies can be conducted on the new drug substance containing the impurities to be controlled, although studies using isolated impurities can sometimes be appropriate." A similar recommendation is included in ICH Q3B(R2). It should be noted, however, that allowing genotoxicity assessment of the impurity as it is present with the drug substance, rather than in isolation, renders the genotoxicity assessment much less sensitive. For example, the potent mutagens that are typically used as positive controls in the bacterial mutation assay, such as 9-aminoanthracene and methyl methanesulfonate, when present with a noncytotoxic drug substance at the minimal level for qualification, would not be detected by these genotoxicity assays because the maximum concentration of the impurity at the limit concentration of the drug substance would not be sufficient to produce a genotoxic response in the assays. If the drug substance is cytotoxic, this approach of assessing the impurity as it is present with the drug substance would be even more insensitive, since the drug's toxicity would further limit the level at which the impurity could be tested.

Although the ICH guidances provide some recommendations on the types of tests that should be conducted, the guidances do not provide specific recommendations on how to proceed if one or both of the genetic toxicology tests are positive; they simply state that additional testing, removal of the impurity, or lowering the level of the impurity should be considered.

ICH Q3C(R3) recommends acceptable concentration limits or permissible daily exposures for various classes of solvents, which are one type of impurity. The guidance does not, however, include a recommendation on limiting exposure based on concerns for genotoxic potential. The guidance recommends only that mathematical models be used for setting exposure limits in cases where reliable carcinogenicity data are available. The ICH guidances on impurities and residual solvents do not apply to drug substances or drug products used during the clinical research stages of development.

In June 2006, the European Medicines Evaluation Agency's (EMEA's) CHMP published a guideline on the limits of genotoxic impurities in support of a marketing application. A subsequent CHMP safety working party published a questions and answers document to provide clarification on the 2006 guideline. This guideline recommends dichotomizing genotoxic impurities into those for which there is "sufficient (experimental) evidence for a threshold-related mechanism" and those "without sufficient (experimental) evidence for a threshold-related mechanism." The genotoxic impurities with sufficient evidence for a threshold-related mechanism would be addressed using the methods outlined in ICH Q3C(R3) for class 2 solvents. This approach calculates a PDE, which is derived using the NOEL or, alternatively, the LOEL, from the most relevant animal study and incorporating a variety of uncertainty factors. Examples of genotoxic compounds that might fall into this category include compounds that induce aneuploidy by interfering with the mitotic spindle, compounds that interfere with the activity of topoisomerase, and/or compounds that inhibit DNA synthesis.

For genotoxic impurities without sufficient evidence for a threshold-related mechanism, the guideline proposes a policy of controlling levels to as low as reasonably practicable (called the ALARP principle). The ALARP approach specifies that every effort should be made to prevent the formation of such impurities during drug substance synthesis and, if this is not possible, technical effort should be made post synthesis to reduce the impurities (e.g., purification steps). Compounds that fall into this category are those that interact with DNA either directly or indirectly, such as alkylating agents, intercalating agents, or agents that can generate free radicals. Since any exposure to these agents can convey some level of carcinogenic risk, and since complete elimination of genotoxic impurities from drug substances is often unachievable, the presence of a concerning impurity requires the implementation of a concept of an acceptable risk level. Methods for the derivation of acceptable risk levels are discussed in ICH Q3C(R3), in reference to class 1 carcinogenic solvents.

Although the aforementioned approach is acceptable, in most instances mechanistic data sufficient to allow an assessment of whether there is a threshold mechanism are lacking. Furthermore, it is relatively uncommon for there to be sufficient data to allow a quantitative risk assessment. The EMEA guideline recognizes these limitations and, therefore, proposes the use of a TTC for genotoxic impurities.

The approach taken in the EMEA guideline for setting an exposure limit for genotoxic or carcinogenic impurities in drug products in support of a marketing application is reasonable. However, issues regarding the presence of genotoxic or carcinogenic impurities often occur during the clinical development stages. Therefore, this guidance provides recommendations for acceptable exposure thresholds during clinical development as well as for marketing applications.

If adequate data characterizing genotoxic and carcinogenic potential are not already available, impurities identified in drug substances or drug products at levels exceeding the stated qualification thresholds in the relevant ICH guidances should be assessed for genotoxic potential in an initial minimal screen. Assays conducted with the impurity in isolation are recommended. However, studies with the drug substance containing, or spiked with, the impurity can be considered in cases where it can be demonstrated that synthesizing sufficient amounts of the impurity is infeasible.

If an impurity that is present at levels below the ICH qualification thresholds is identified, the impurity should be evaluated for genotoxicity and carcinogenicity based on structure–activity relationship (SAR) assessments (i.e., whether there is a structural alert). This evaluation can be conducted via a review of the available literature or through a computational toxicology assessment; commonly used software includes MDL-QSAR, MC4PC, and DEREK for Windows. An in vitro mutation assay (i.e., the bacterial reverse mutation assay) would generally be an acceptable initial screen for impurities with an identified alert, since positive signals in computational toxicology programs are often derived from the results of bacterial mutation assays and mutagenic carcinogens are considered to operate through nonthreshold-related mechanisms. An assessment in a mammalian cell assay may be needed for impurities with specific structural groups, such as carbamates, that are not well characterized in bacterial assays or for compounds that are toxic to *Escherichia coli* and *Salmonella*, such as antibiotics.

If the initial evaluation of the genotoxic potential of an impurity is negative, no further genotoxicity studies are recommended and the impurity should be considered to be adequately qualified regarding its genotoxic potential. It should be noted that in cases where it is necessary from a feasibility standpoint to conduct the assays with the drug substance containing, or spiked with, the impurity the proposed acceptance criterion should be commensurate with the level of the impurity observed in clinical, stability, and/or production batches, taking into consideration the manufacturing and analytical variability. This acceptance criterion should not exceed the level present in the drug batch used in the genotoxicity assay and should be supported by the relevant qualification thresholds discussed in the ICH guidances or supporting general toxicity information. In some cases, the structure of an impurity leading to the structural alert is shared with the API. The genotoxic potential of such an impurity can be evaluated through the standard testing of the API if the chemical environment for the alerting structure of the compounds is deemed comparable for the reactivity potential.

Positive results in one or more genotoxicity assays or other information indicating a carcinogenic potential, such as positive data from a carcinogenicity study with the impurity, should be addressed further. Recommended approaches for handling genotoxic or carcinogenic impurities are described in this section. Table 1.5 summarizes the recommended approaches for characterizing the presence and addressing the safety of genotoxic and carcinogenic impurities depending on the clinical development stage. A decision tree is included in Figure 1.4.

Since drug-related impurities presumably provide limited, if any, therapeutic benefits and because of their potential to cause cancer in humans, every feasible technical effort should be made to prevent the formation of genotoxic or carcinogenic compounds during drug substance synthesis or drug product manufacture. However, we recognize that completely preventing the formation of or removing an impurity of concern may not be possible in many cases.

In lieu of completely preventing the formation of a genotoxic or carcinogenic impurity, steps to reduce the level of the impurity present in the drug substance or drug product should be considered. Sections 1.7.1 and 1.7.2 discuss acceptable

TABLE 1.5
Recommended Approaches Based on Development Stage, FDA Draft Guidance

Clinical Development Stage	Recommended Approach
IND	• Evaluate identified impurities for genotoxic and carcinogenic risk via SAR assessment • Conduct assay for the presence of anticipated genotoxic and carcinogenic impurities • If an impurity with genotoxic and carcinogenic potential is identified: –Modify synthetic pathway to eliminate the impurity, if possible OR –Conduct genotoxicity assays to characterize the genotoxic potential if not already known AND/OR –Set specification to that associated with a potential daily impurity exposure supported by compound-specific risk assessment or relevant qualification threshold (see Table 1.6)
Marketing application (NDA, BLA, or ANDA)	• Evaluate identified impurities for genotoxic and carcinogenic risk via SAR assessment • If impurity with genotoxic and carcinogenic potential is identified: –Conduct genotoxicity assays to characterize the genotoxic potential if not already known AND/OR –Set specification to that associated with a potential daily impurity exposure supported by compound-specific risk assessment or 1.5 μg per day threshold

Source: Center for Drug Evaluation and Research, *Guidance for Industry, Genotoxic and Carcinogenic Impurities in Drug Substances and Products: Recommended Approaches*, Draft Guidance, Food and Drug Administration, 2008.

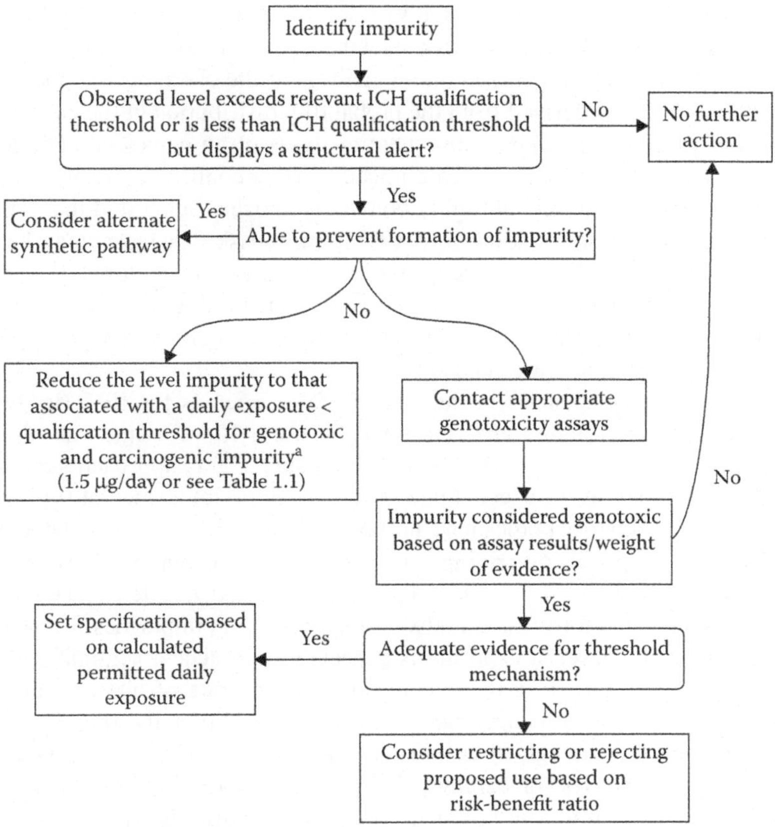

FIGURE 1.4 Decision tree flow diagram, Food and Drug Administration draft guidance. [a]Safety threshold approach for genotoxic and carcinogenic impurities is not applicable to compounds with adequate data to derive compound-specific risk assessment or for those with structure–activity relationships to high-potency carcinogens. In addition, the approach may not be appropriate for some routes of administration (e.g., dermal and ophthalmic) because of the lack of a relevant database from which a threshold limit can be derived.

thresholds to support safety for a marketing application and during clinical development, respectively. Analytical methodologies should be used that can adequately identify impurities of concern at levels associated with the relevant qualification thresholds. This threshold approach should be applied only in the absence of adequate qualification data (data that establish the biological safety of an impurity at the level specified) for the given impurity.

1.7.1 Acceptable Levels to Support Marketing Applications

In general, an exposure level of 1.5 μg/person/day for each impurity can be considered an acceptable qualification threshold for supporting a marketing application. Any impurity found at a level below this threshold generally should not need further safety qualification for genotoxicity and carcinogenicity concerns. The threshold is an estimate of daily exposure expected to result in an upper-bound lifetime risk of cancer of less than 10^{-6} (1 in 10^6), a risk level that is thought to pose negligible safety concerns. The threshold was based on an analysis of the carcinogenic potencies of 477 chemicals and was derived from the probability distribution of the carcinogenic potencies of those compounds. Subsequent analyses of an expanded carcinogenic potency database of more than 700 carcinogens further confirmed the threshold. An additional analysis of subsets of highly potent carcinogens suggested that a threshold of 0.15 μg/day, corresponding to a 10^{-6} lifetime risk of cancer, may be more appropriate for chemicals with structural alerts for potential genotoxicity. However, there are some compounds containing certain structural groups (aflatoxin-like, *N*-nitroso-, and azoxy- structures) that have extremely high carcinogenic potency and are excluded from the threshold approach.

Federal regulatory agencies in the United States, such as the EPA (in the context of ambient water quality criteria), typically use a 10^{-6} lifetime risk of cancer to determine negligible risk from chemical exposures. This approach supports an acceptable threshold level for genotoxic or carcinogenic impurities of 0.15 μg/day. However, other regulatory bodies have proposed a 10^{-5} level as an acceptable cancer risk. Given that there is an overriding expected benefit of an approved drug product, a daily exposure level of 1.5 μg/day, associated with a 10^{-5} lifetime risk of cancer, can be acceptable for most genotoxic or carcinogenic impurities for a marketing application. This level of exposure is expected to produce a negligible increase in carcinogenic risk based on the existing background rate of human cancer and the conservative nature of cancer risk assessments. Additionally, this threshold is considered to be low enough to ensure that the presence of a compound with an uncharacterized genotoxic or carcinogenic potential would not significantly alter the risk–benefit ratio of a drug product, even if the impurity is later shown to be a carcinogen.

The database from which the exposure threshold for genotoxic or carcinogenic impurities is derived includes studies that primarily use oral administration, although a smaller number use the inhalation route. Although the recommended threshold approach applies to all drug products regardless of the intended route of administration, the qualification threshold of 1.5 μg/day may not be appropriate for some routes (e.g., dermal or ophthalmic) because of the lack of a relevant database from which an exposure threshold can be derived. Applicants should contact specific drug review divisions regarding acceptable approaches in these cases.

As part of this threshold approach, applicants can conduct and provide to the FDA a SAR assessment to identify structural similarities to known carcinogens. In cases where significant structural similarities to a known carcinogen are identified, an estimate of the potential human cancer risk can be calculated based on the available information for the confirmed carcinogen. This assessment can result in an increase in the acceptable exposure threshold for impurities that are highly similar to carcinogens with relatively low potency, or a reduction in the limit for impurities that are highly similar to relatively potent carcinogens.

The EPA guidance *Supplemental Guidance for Assessing Susceptibility from Early-Life Exposure to Carcinogens* (EPA/630/R-03/003F) regarding cancer susceptibility in pediatric populations indicates that children exposed to mutagenic carcinogens between age 0 (birth) and 16 years have an increased cancer risk over a 70-year lifetime when compared to adults. The EPA concludes that cancer risks are generally higher from early-life exposure than from similar exposure durations later in life and recommends the application of adjustment factors to risk calculations to account for this observation. The EPA recommends an adjustment factor of 10 for exposures before 2 years of age (i.e., spanning a 2-year time interval from the first day after birth up until a child's second birthday), which represents an approximation of the weighted geometric mean tumor incidence ratio from juvenile or adult exposures in repeated dosing studies. In the absence of data to calculate a specific dose–response adjustment factor for exposures between 2 years and less than 16 years of age, the EPA recommends an adjustment factor of 3, which represents an intermediate level of adjustment and reflects a midpoint between the ten-fold adjustment for the first 2 years of life and no adjustment (i.e., one-fold) for adult exposures. However, the EPA guidance acknowledges that the resultant increases in cancer risk are relatively small for exposures that continue with fair uniformity over a lifetime. We recommend that this increase in susceptibility to carcinogens in pediatric populations be considered when determining the acceptable impurity level for a given drug product.

The threshold approach for genotoxic or carcinogenic impurities limits the likelihood that any individual impurity in a given drug product will present more than a 10^{-5} excess cancer risk, but the approach is not intended to ensure an aggregate excess cancer risk of less than 10^{-5}. This means the threshold approach to individual impurities is not intended to limit the overall excess cancer risk to 10^{-5} from all impurities in a single drug product or from multiple drug products concomitantly administered. As discussed earlier, this approach is consistent with the approaches taken by various regulatory bodies such as the EPA, WHO, and EMEA in implementing threshold levels for carcinogenic risk when no benefit from the expected exposure is perceived. However, in cases where a class or family of structurally similar impurities is identified and is expected to have similar mechanisms resulting in their genotoxic or carcinogenic potential, the total daily exposure to the related compounds should be evaluated relative to the recommended threshold exposure.

We recognize that drug products are often indicated for short-term use. However, for most drugs these threshold considerations still apply since a drug may be used multiple times by the same individual or may be used outside its approved indication. A detailed rationale should be provided to the FDA to support limits higher than that generally considered appropriate for a marketing application.

1.7.2 Acceptable Levels during Clinical Development

Section 1.7.1 describes the qualification threshold for genotoxic or carcinogenic impurities in support of a marketing application. Issues related to genotoxic impurities can also arise during a drug product's clinical development period and can affect the assessment of safety for conducting the program. Some flexibility in the previously described threshold level can be applied during the investigational stages, since clinical trials vary widely in duration from short term (single dose to 4 weeks) to years and the qualification threshold for a marketing application is based on lifetime risk estimates. On the other hand, it should be recognized that during early clinical development a benefit of the drug cannot be assumed. We recognize that the ability to identify and control drug-related impurities during early developmental stages is limited because of issues related to the scale and maturity of production processes. Taking all these considerations into account, higher daily levels of exposure to potentially genotoxic impurities may be acceptable during the clinical development of the drug product compared to what is appropriate for a marketed drug product.

Bos and others reviewed the derived cancer risk from short-term, high-dose exposure to a genotoxic carcinogen relative to the same cumulative dose distributed over a lifetime (virtually safe dose). Briefly, the authors state that only a limited number of animal studies have assessed the comparative tumor incidence from short-term versus long-term exposures with similar cumulative doses. From the studies that do exist, dose rate correction factors (factors by which a specific dose of a chemical carcinogen at long-term, low-dose rates should be multiplied to derive the expected tumor incidence from short-term, high-dose rates) ranged from 1 to 8.3. The authors conclude that the most pragmatic approach to calculate acceptable short-term exposures to known genotoxic carcinogens is to linearly extrapolate the short-term exposure from the acceptable lifetime exposure or virtually safe dose.

ADIs of genotoxic impurities during clinical development are presented in Table 1.6, based on the linear extrapolation approach described by Bos and others. The impurity threshold exposures for exposure durations of up to 12 months are based on a 10^{-6} cancer risk level (0.15 μg/day for lifetime exposure), since these trials often include healthy subjects for whom there is no expected health benefit and the efficacy of the drug may still be uncertain. The values are derived from a linear extrapolation from the qualification threshold using the maximum duration of dosing for each time period specified in Table 1.6. In addition, these values incorporate an uncertainty factor of 2 to allow for deviations from the linear extrapolation model. For trials greater than 1 year in duration, the threshold value is identical to the threshold for a marketing application and is based on a 10^{-5} cancer risk level (1.5 μg/day derived from lifetime exposures); subjects in these trials generally have the condition or disease being studied and are more certain to derive benefit from the treatment than subjects

TABLE 1.6
Acceptable Qualification Thresholds for Genotoxic and Carcinogenic Impurities, FDA Draft Guidance

	Duration of Clinical Trial Exposure					
	<14 days	14 days to 1 month	1–3 months	3–6 months	6–12 months	>12 months
Genotoxic and carcinogenic impurity threshold (μg/day)	120	60	20	10	5	1.5

Source: Center for Drug Evaluation and Research, *Guidance for Industry, Genotoxic and Carcinogenic Impurities in Drug Substances and Products: Recommended Approaches*, Draft Guidance, Food and Drug Administration, 2008.

in early trials. When determining the acceptable impurity threshold exposure, the specifics of the patient population in the clinical trial should be evaluated.

In cases where attempts to prevent the formation of an impurity of concern and/or to reduce the amount of the impurity to an acceptable level as per Table 1.6 are not possible, further characterization of the genotoxic and carcinogenic potential should be conducted. The *Guidance for Industry and Review Staff Recommended Approaches to Integration of Genetic Toxicology Study Results* describes the FDA's current thinking regarding appropriate additional evaluations that can be conducted. Briefly, these concepts include consideration of the mechanism of action, weight of evidence, or conduct of additional supportive studies. These concepts can also be considered relevant for genotoxic impurities. In addition to the aforementioned considerations, the conduct of a SAR evaluation of an impurity may provide useful information. When a significant structural similarity to a known carcinogen is identified, the drug substance and drug product acceptance criteria (typically in units of parts per million or percentage) can be set at a level that is commensurate with the risk assessment specific to that of the known compound. As noted previously, the proposed factors should be considered in light of manufacturing batch data.

The previous discussions are intended to be general recommendations to consider when developing a drug product in which a potentially genotoxic or carcinogenic impurity is identified. We recognize that these approaches may not necessarily apply to every development program, and flexibility in the application of these recommendations may be appropriate. When applying the recommendations, consideration should be given to the drug product's clinical development stage, maximum duration of drug administration at that stage, proposed indication (e.g., treatment of a life-threatening condition versus a less serious condition), patient population (e.g., adults versus children), and structural similarity of an impurity to a compound of known carcinogenic potency. In some of these cases, acceptance criteria higher than the recommended thresholds can be supported in the presence of a potential pharmacological benefit to patients. In rare cases, such as in the presence of highly potent carcinogens, decreases in the threshold also may be warranted. The appropriateness of a flexible approach should be informed by the feasibility of controlling impurity levels and the capabilities of the current process.

1.8 ASSESSMENT AND CONTROL OF DNA-REACTIVE (MUTAGENIC) IMPURITIES IN PHARMACEUTICALS TO LIMIT POTENTIAL CARCINOGENIC RISK, ICH M7 DRAFT CONSENSUS GUIDELINE

The synthesis of drug substances involves the use of reactive chemicals, reagents, solvents, catalysts, and other processing aids. As a result of chemical synthesis or subsequent degradation, impurities reside in all drug substances and associated drug products. While ICH Q3A(R2), *Impurities in New Drug Substances*, and Q3B(R2), *Impurities in New Drug Products*, provide guidance for qualification and control for the majority of impurities, only limited guidance is provided for the impurities that are DNA reactive. The purpose of the ICH M7 draft guideline [8] is to provide a practical framework that can be applied for the identification, categorization, qualification, and control of these mutagenic impurities to limit potential carcinogenic risk. This guideline is intended to complement ICH Q3A(R2); Q3B(R2) (Note 1); and ICH M3(R2), *Nonclinical Safety Studies for the Conduct of Human Clinical Trials and Marketing Authorizations for Pharmaceuticals*. This guideline emphasizes considerations of both safety and quality risk management in establishing levels of mutagenic impurities that are expected to pose negligible carcinogenic risk. It outlines recommendations for assessment and control of mutagenic impurities that reside or are reasonably expected to reside in the final drug substance or product, taking into consideration the intended conditions of human use.

This document is intended to provide guidance for new drug substances and new drug products during their clinical development and subsequent applications for marketing. It also applies to new marketing applications and postapproval submissions for marketed products, in both cases only where

- Changes to the drug substance synthesis result in new impurities or increased acceptance criteria for existing impurities
- Changes in the formulation, composition, or manufacturing process result in new degradants or increased acceptance criteria for existing degradants
- Changes in indication or dosing regimen are made that significantly affect the acceptable cancer risk level

This guideline does not apply to drug substances and drug products intended for advanced cancer indications as defined in the scope of ICH S9. Additionally, there may be some cases where a drug substance intended for other indications is itself genotoxic at therapeutic concentrations and may be expected to be associated with an increased cancer risk. Exposure to a mutagenic impurity in these cases would not significantly add to the cancer risk of the drug substance, and impurities could be controlled at acceptable levels for nonmutagenic impurities.

Excipients used in existing marketed products and flavoring agents are excluded from this guideline. The application of this guideline to leachables associated with drug product packaging is not intended, but the safety risk assessment principles outlined in this guideline for limiting potential carcinogenic risk can be used if warranted. The safety risk assessment principles of this guideline can be used if warranted for impurities in excipients that are used for the first time in a drug product and are chemically synthesized.

The focus of this guideline is on DNA-reactive substances that have a potential to directly cause DNA damage when present at low levels, leading to mutations and, therefore, potentially causing cancer. This type of mutagenic carcinogen is usually detected in a bacterial reverse mutation (mutagenicity) assay. Other types of genotoxicants that are nonmutagenic typically have thresholded mechanisms and usually do not pose carcinogenic risk in humans at the level ordinarily present as impurities. Therefore, to limit a possible human cancer risk associated with the exposure to potentially mutagenic impurities, the bacterial mutagenicity assay is used to assess the mutagenic potential/effect and the need for controls. Structure-based assessments are useful for predicting bacterial mutagenicity outcomes based on the established knowledge base. There are a variety of approaches to conduct this evaluation including a review of the available literature, and/or computational toxicology assessment.

A TTC concept was developed to define an acceptable intake for any unstudied chemical that will not pose a risk of carcinogenicity or other toxic effects. For the application of a TTC in the assessment of acceptable limits of mutagenic impurities in drug substances and drug products, a value of 1.5 μg/day corresponding to a theoretical 10^{-5} excess lifetime risk of cancer can be justified. The methods on which the TTC is based are generally considered very conservative since they involve a simple linear extrapolation from the dose giving a 50% tumor incidence (TD50) to a 1 in 10^6 incidence, using TD50 data for the most sensitive species and most sensitive sites of tumor induction (several worst-case assumptions). Some structural groups were identified to be of such high potency that intakes even below the TTC would theoretically be associated with a potential for significant carcinogenic risk. This group of high-potency mutagenic carcinogens ("cohort of concern") comprises aflatoxin-like-, *N*-nitroso-, and azoxy- compounds.

During clinical development, it is expected that control strategies and approaches will be less developed in earlier phases where overall development experience is limited. This guideline bases acceptable intakes for mutagenic impurities on established risk assessment strategies. Acceptable risk during the early development phase is set at a theoretically calculated level of approximately 1 additional cancer per 10^6. For later stages in development and marketed products when efficacy has been shown, acceptable increased cancer risk is set at a theoretically calculated level of approximately 1 in 100,000. These risk levels represent a small theoretical increase in risk compared to human overall lifetime incidence of developing any type of cancer, which is greater than one in three. It is noted that established cancer risk assessments are based on lifetime exposures. Less-than-lifetime (LTL) exposures during both development and marketing can have higher acceptable intakes of impurities and still maintain comparable risk levels. The use of a numerical cancer risk value (1 in 100,000) and its translation into risk-based doses (TTC) is a highly hypothetical concept that should not be regarded as a realistic indication of the actual risk. The TTC concept provides an estimate of safe exposures for any mutagenic compound. However, exceeding the TTC is not necessarily associated with an increased cancer risk given the conservative assumptions employed in the derivation of the TTC value. The most likely increase in cancer incidence is actually much less than 1 in 100,000. In addition, in cases where a mutagenic compound is a noncarcinogen in a rodent bioassay there would be no predicted increase in cancer risk. Based on these considerations, any exposure to an impurity that is later identified as a mutagen is not necessarily associated with an increased cancer risk for patients already exposed to the impurity. A risk assessment would determine whether any further actions are taken.

Where a potential risk has been identified for an impurity, an appropriate control strategy leveraging process understanding and/or analytical controls should be developed to ensure that the mutagenic impurity is at or below the acceptable cancer risk level. There may be cases when an impurity is also a metabolite of the drug substance. In such cases, the impurity is considered qualified provided that exposure to the metabolite in appropriate nonclinical studies of the drug substance is higher than what would be achieved from the impurity in the administered drug substance (ICH Q3A/Q3B).

Although this guideline is not intended to be applied retrospectively (i.e., to products marketed prior to the adoption of this guideline), some types of postapproval changes warrant a reassessment of safety relative to mutagenic impurities. This section is intended to be applied to products marketed prior to, or after, the adoption of this guideline.

Postapproval submissions involving the drug substance chemistry, manufacturing, and controls (changes to the route of synthesis, reagents, solvents, process conditions, etc.) should include an evaluation of the potential risk impact associated with mutagenic impurities. Specifically, changes should be evaluated to determine if the changes result in any new mutagenic impurities or higher acceptance criteria for existing mutagenic impurities. Reevaluation of impurities not impacted by the change is not required. Regulatory submissions associated with such changes should include a summary of the assessment and, if appropriate, an updated control strategy. Changes to the site of manufacture would typically not require a reassessment of mutagenic impurity risk. When a new drug substance supplier is proposed, evidence that the drug substance produced by this supplier (using same route of synthesis) has been approved for an existing drug product marketed in the assessor's region is considered to be sufficient evidence of acceptable risk/benefit regarding mutagenic impurities and an assessment per this guideline is not required. If this is not the case, then an assessment per this guideline is expected.

Postapproval submissions involving the drug product (e.g., change in composition, manufacturing process, and dosage form) should include an evaluation of the potential risk associated with any new mutagenic degradants or higher acceptance criteria for existing mutagenic degradants. If appropriate, the regulatory submission would include an updated control strategy. Reevaluation of the drug substance associated with drug products is not required or expected provided there are no changes to the drug substance. Changes to the site of manufacture would typically not require a reassessment of mutagenic impurity risk.

Changes to the clinical use of marketed products that typically may require a reevaluation of the mutagenic impurity limits include a significant increase in clinical dose, an increase in duration of use (in particular when a mutagenic impurity was controlled above the lifetime acceptable intake for a previous indication that may no longer be appropriate for the longer treatment duration associated with the new indication), or for a change in indication from a serious or life-threatening condition where higher acceptable intakes were justified to an indication for a less serious condition where the existing impurity acceptable intakes may no longer be appropriate. Changes to the clinical use of marketed products associated with new routes of administration or expansion into patient populations that include pregnant women and/or pediatrics typically would not require a reevaluation, assuming no changes in daily dose or duration of treatment.

Application of this guideline may be warranted to marketed products if there is a specific cause for concern. The existence of impurity structural alerts alone is considered insufficient to trigger follow-up measures, unless it is a structure in the cohort of concern. However, a specific cause for concern would be new relevant impurity hazard data (classified as class 1 or 2) generated after the overall control strategy and specifications for market authorization are established. These new relevant impurity hazard data should be derived from high-quality scientific studies consistent with relevant regulatory testing guidelines, with data records or reports readily available to marketing application holders. When the applicant becomes aware of these new relevant impurity hazard data, an evaluation should be conducted and if it is concluded by the applicant to affect the acceptable cancer risk/benefit notification to regulatory authorities with a proposed contemporary control strategy is warranted.

Actual and potential impurities that are likely to arise during the synthesis, work-up, and storage of a new drug substance and during the manufacturing and storage of a new drug product should be assessed.

Impurity assessment is a two-stage process. First, actual impurities that have been identified should be considered for their mutagenic potential. In parallel, an assessment of potential impurities likely to be present in the final drug substance is carried out to determine if further evaluation of their mutagenic potential is required.

Actual impurities include those observed in the drug substance above the ICH Q3A reporting thresholds. Identification of actual impurities is expected when the levels exceed the identification thresholds outlined by ICH Q3A. It is acknowledged that some impurities below the identification threshold may also have been identified. Potential impurities arising from the synthesis of the drug substance could include starting materials, reagents and intermediates, identified impurities in starting materials and intermediates, and reasonably expected reaction by-products based on knowledge of the chemical reactions and conditions involved. Knowledge of the starting material synthesis, in particular the use of mutagenic reagents, is an important factor in understanding potential impurities in the starting materials, especially when there is a reasonable expectation that such impurities may be carried through the synthesis to the drug substance. All impurities (actual and potential), where the structures are known, should be evaluated for mutagenic potential.

Actual drug substance degradation products include those observed above the ICH Q3B reporting threshold during storage of the drug product in the proposed long-term storage conditions and primary and secondary packaging, and also include those impurities that arise during the manufacture of the drug product. Identification of actual degradation products is expected when the levels exceed the identification thresholds outlined by ICH Q3A/Q3B. It is acknowledged that some degradation products below the identification threshold may also have been identified. Potential degradants in the drug substance and drug product are those that may reasonably be expected to form during long-term-storage conditions. Potential degradants include those that form above the ICH Q3A/B identification threshold during accelerated stability studies (e.g., 40°C/75% relative humidity for 6 months) and confirmatory photostability studies as described in ICH Q1B but are yet to be confirmed in the drug substance or drug product in the primary packaging. Knowledge of relevant degradation pathways can be used to help guide decisions on the selection of potential degradation products to be evaluated for mutagenicity, for example, from degradation chemistry principles, relevant stress testing studies, and development stability studies. Actual and potential degradants likely to be present in the final drug substance or drug product and where the structure is known should be evaluated for mutagenic potential.

For products in clinical development, the thresholds outlined in ICH Q3A/B do not apply and it is acknowledged that the thresholds for actual impurities and degradants will typically be higher than those outlined in ICH Q3A/B.

1.8.1 Hazard Assessment Elements

Hazard assessment involves an initial analysis of actual and potential impurities by conducting database and literature searches for carcinogenicity and bacterial mutagenicity data to classify them as class 1, 2, or 5 according to Table 1.7. If data for such a classification are not available, an assessment of SARs that focuses on bacterial mutagenicity predictions should be performed. This could lead to a classification into class 3, 4, or 5.

A computational toxicology assessment should be performed using QSAR methodologies that predict the outcome of a bacterial mutagenicity assay. Two QSAR prediction methodologies that complement each other should be applied. One methodology should be expert rule based, and the second methodology should be statistical. QSAR models utilizing these prediction methodologies should follow the

TABLE 1.7
Impurities Classification with Respect to Mutagenic and Carcinogenic Potential and Resulting Control Actions, ICH M7 Draft Guideline

Class	Definition	Proposed Action for Control
1	Known mutagenic carcinogens	Control at or below compound-specific acceptable limit
2	Known mutagens with unknown carcinogenic potential (bacterial mutagenicity positive[a], no rodent carcinogenicity data)	Control at or below acceptable limits (generic or adjusted TTC)
3	Alerting structure, unrelated to the structure of the drug substance; no mutagenicity data	Control at or below acceptable limits (generic or adjusted TTC) or do bacterial mutagenicity assay: If nonmutagenic = class 5 If mutagenic = class 2
4	Alerting structure, same alert in drug substance that has been tested and is nonmutagenic	Treat as nonmutagenic impurity
5	No structural alerts, or alerting structure with sufficient data to demonstrate lack of mutagenicity	Treat as nonmutagenic impurity

Source: *Assessment and Control of DNA Reactive (Mutagenic) Impurities in Pharmaceuticals to Limit Potential Carcinogenic Risk*, ICH M7, ICH Harmonised Tripartite Guideline, International Conference on Harmonisation of Technical Requirements for Registration of Pharmaceuticals for Human Use, 2013.

[a] Or other relevant positive mutagenicity data indicative of DNA reactivity–related induction of gene mutations (e.g., positive findings in in vivo gene mutation studies).

validation principles set forth by the Organisation for Economic Co-operation and Development (OECD). The outcome of any computer system-based analysis should be reviewed with the use of expert knowledge to provide additional supportive evidence on the relevance of any positive or negative prediction and to elucidate underlying reasons in case of conflicting results.

The absence of structural alerts from two complementary QSAR methodologies (expert rule based and statistical) is sufficient to conclude that the impurity is of no concern and no further testing is required (class 5 in Table 1.7).

To follow up on a structural alert (class 3 in Table 1.7), a bacterial mutagenicity assay can be applied. An appropriately conducted negative bacterial mutagenicity assay (Note 2) would overrule any structure-based concern, and no further genotoxicity assessments would be required (Note 1). These impurities (class 5 in Table 1.7) should be considered as nonmutagenic impurities. A positive bacterial mutagenicity result would warrant further hazard assessment and/or control measures (class 2 in Table 1.7). Alternatively, adequate control measures in the case of a positive structural alert alone could be applied in place of bacterial mutagenicity testing. An impurity with a structural alert that is shared with the drug substance (e.g., same structural alert in the same position and environment in the impurity and the drug substance) can be considered nonmutagenic (class 4 in Table 1.7) if the testing of the drug substance in the bacterial mutagenicity assay was negative.

Further hazard assessment of an impurity with a positive bacterial mutagenicity result (class 2 in Table 1.7) may be appropriate, for instance, when levels of the impurity cannot be controlled at an appropriate acceptable limit. To understand the relevance of the bacterial mutagenicity assay result under in vivo conditions, it is recommended that the impurity be tested in an in vivo gene mutation assay. The selection of other in vivo genotoxicity assays should be scientifically justified based on knowledge of the mechanism of action of the impurity and its organ site of contact (Note 3). In vivo studies should be designed taking into consideration existing guidance as per ICH S2(R1). Negative results in the appropriate in vivo assay may support setting of impurity limits in excess of the acceptable limits.

1.8.2 Risk Characterization

As a result of hazard assessment described in Section 1.8.1, each impurity will be assigned to one of the five classes in Table 1.7. For impurities belonging to classes 1, 2, and 3 (class 3 only if the presence of a structural alert is not followed up in a bacterial mutagenicity assay), the principles of risk characterization used to derive acceptable intakes are described in this section.

Compound-specific risk assessments to derive acceptable intakes should be applied instead of the TTC-based acceptable intakes where sufficient carcinogenicity data exist (class 1 in Table 1.7). For a known mutagenic carcinogen, a compound-specific acceptable intake can be calculated based on carcinogenic potency and linear extrapolation as a default approach. Alternatively, other established risk assessment practices such as those used by international regulatory bodies may be applied either to calculate acceptable intakes or to use already existing values published by the regulatory bodies (Note 4). Compound-specific calculations for acceptable intakes can be applied case by case for impurities that are chemically similar to a known carcinogen compound class (class-specific acceptable intakes) provided that a rationale for chemical similarity and supporting data can be demonstrated (Note 5).

The existence of mechanisms leading to a dose response that is nonlinear or has a practical threshold is increasingly recognized, not only for compounds that interact with non-DNA targets but also for DNA-reactive compounds, whose effects may be modulated by, for example, rapid detoxification before coming into contact with DNA or by effective repair of induced damage. The regulatory approach to such compounds can be based on the identification of a critical NOEL and use of uncertainty factors [ICH Q3C(R5)] when data are available (Note 6). The acceptable intakes derived from compound-specific risk assessments can be adjusted for shorter term use in the same proportions as defined below.

The TTC-based acceptable intake of 1.5 μg/day is considered to be protective for a lifetime of daily exposure. To address LTL exposures to mutagenic impurities in pharmaceuticals, an approach is applied in which the acceptable cumulative lifetime dose (1.5 μg/day × 25,550 days = 38.3 mg) is uniformly distributed over the total number of exposure days during LTL exposure. This would allow a higher daily intake of mutagenic impurities than would be the case for lifetime exposure and still maintain comparable risk levels for daily and nondaily treatment regimens. In the case of intermittent (nondaily) dosing, the acceptable intake will be capped by the total cumulative dose or the maximum acceptable intake (i.e., 120 μg/day), whichever is lower. Table 1.8 illustrates the acceptable intakes for LTL to lifetime exposures for clinical development and marketing.

Using this LTL concept, acceptable intakes of mutagenic impurities are recommended for limited treatment periods during clinical development of up to 1 month, 1–12 months, and more than 1 year up to the completion of phase III clinical trials (Table 1.8). These adjusted acceptable intake values maintain a 10^{-6} risk level in early clinical development when benefit has not yet been established and a 10^{-5} risk level for later stages in development (Note 7). An alternative approach to the strict use of an adjusted acceptable intake for any mutagenic impurity could be applied for phase I clinical trials of up to 14 days. Only impurities that are known mutagenic carcinogens (class 1) and known mutagens of unknown carcinogenic potential (class 2), as well as impurities in the cohort of concern chemical class, should be controlled to

TABLE 1.8
Acceptable Intakes for an Individual Impurity, ICH M7 Draft Guideline

Duration of treatment	≤1 month	>1–12 months	>1–10 years	>10 years to lifetime
Daily intake (μg/day)	120	20	10	1.5

Source: *Assessment and Control of DNA Reactive (Mutagenic) Impurities in Pharmaceuticals to Limit Potential Carcinogenic Risk*, ICH M7, ICH Harmonised Tripartite Guideline, International Conference on Harmonisation of Technical Requirements for Registration of Pharmaceuticals for Human Use, 2013.

acceptable limits. All other impurities would be treated as nonmutagenic impurities. These include impurities that contain structural alerts (class 3), which alone would not trigger action for an assessment for this limited phase I duration.

Standard risk assessments of known carcinogens operate under the assumption that cancer risk increases as a function of cumulative dose. Thus, the cancer risk of a continuous low dose over a lifetime would be equivalent to the cancer risk associated with an identical cumulative exposure averaged over a shorter duration or lifetime average daily dose. For marketed product treatments with cumulative intakes of less than 10 years (continuous or total intermittent treatments), the acceptable intake can be adjusted to less than 10 μg/day. For marketed products with much shorter treatment duration indications, the acceptable intake values of Table 1.8 can be applied. The proposed intakes would all comply with the principle of not exceeding a 10^{-5} cancer risk level (Note 7).

The TTC-based acceptable intakes should be applied to each individual impurity. When there are multiple mutagenic impurities specified on the drug substance specification, total mutagenic impurities should be limited as described in Table 1.9 for clinical development and marketed products. Only impurities that are specified on the drug substance specification contribute to the calculation for total. Degradants that form in the drug product would be controlled individually, and a total limit would not be applied. The aforementioned approach is supported by a detailed analysis of the effect of combining multiple impurities that are in similar or different chemical classes and by the conservative assumptions incorporated into the TTC, and the low likelihood of synergistic carcinogenic effects at very low mutagenic impurity levels.

The following are exceptions and flexibility in approaches:

- Higher acceptable intakes may be justified when human exposure to the impurity will be much greater from other sources, for example, food, or endogenous metabolism (e.g., formaldehyde).
- Case-by-case exceptions to the use of the appropriate acceptable intake can be justified in cases of severe disease, of reduced life expectancy, of late onset but chronic disease, or with limited therapeutic alternatives.

TABLE 1.9
Acceptable Intakes for Total Impurities, ICH M7 Draft Guideline

Duration of treatment	≤1 month	>1–12 months	>1–10 years	>10 years to lifetime
Daily intake (μg/day)	120	60	10 (30[a])	5

Source: *Assessment and Control of DNA Reactive (Mutagenic) Impurities in Pharmaceuticals to Limit Potential Carcinogenic Risk*, ICH M7, ICH Harmonised Tripartite Guideline, International Conference on Harmonisation of Technical Requirements for Registration of Pharmaceuticals for Human Use, 2013.

[a] For clinical development up to 3 years. Similar principles could be applied to marketed products with justification.

- A disproportionally high number of members of some structural classes of mutagens, that is, aflatoxin-like-, *N*-nitroso-, and azoxy structures, of which some may occur as impurities in pharmaceuticals, display extremely high carcinogenic potency. Acceptable intakes for these high-potency carcinogens would likely be significantly lower than the acceptable intakes defined in this guideline. Although the principles of this guideline can be used, a case-by-case approach using, for example, carcinogenicity data from closely related structures, if available, usually needs to be developed to justify acceptable intakes for pharmaceutical development and marketed products.

The aforementioned risk approaches are applicable to all routes of administration, and no corrections to acceptable intakes are generally warranted. Exceptions to consider may include situations where data justify route-specific concerns that need to be evaluated in a case-by-case manner. These approaches are also applicable to all patient populations based on the conservative nature of the risk approaches being applied.

1.8.3 Control

A control strategy is a planned set of controls derived from current product and process understanding that ensures process performance and product quality (ICH Q10). A control strategy can include, but is not limited to, the following:

- Controls on material attributes (including raw materials, starting materials, intermediates, reagents, solvents, and primary packaging materials)
- Facility and equipment operating conditions
- Controls implicit in the design of the manufacturing process
- In-process controls (including in-process tests and process parameters)
- Controls on drug substance and drug product (e.g., release testing)

When an impurity has been characterized as mutagenic, it is important to develop a control strategy that ensures that the level of this impurity in the drug substance and drug product is below the acceptable limit. A thorough knowledge of the chemistry associated with the drug substance manufacturing process and the drug product manufacturing process, along with an understanding of the overall stability of the drug substance and the drug product, is fundamental to developing the appropriate controls. Developing a strategy to mitigate mutagenic impurities in the drug product is consistent with the risk management processes identified in ICH Q9. A control strategy that is based on product and process understanding and utilization of risk management principles will lead to a combination of process design and control and appropriate analytical testing, which can also provide an opportunity to shift controls upstream and minimize the need for end-product testing.

There are four potential approaches to the development of a control strategy for process-related impurities in drug substances:

Option 1: include a test for the impurity in the drug substance specification with an acceptance criterion at or below the acceptable limit using an appropriate analytical procedure. It is considered possible to apply periodic (verification) testing per ICH Q6A.

Option 2: include a test for the impurity in the specification for a raw material, starting material or intermediate, or as an in-process control, with an acceptance criterion at or below the acceptable limit using an appropriate analytical procedure.

Option 3: include a test for the impurity in the specification for a raw material, starting material or intermediate, or as an in-process control, with an acceptance criterion above the acceptable limit using an appropriate analytical procedure coupled with demonstrated understanding of fate and purge and associated process controls that ensure that the level in the drug substance is below the acceptable limit without the need for any additional testing.

Option 4: understanding of process parameters and impact on residual impurity levels (including fate and purge knowledge) with sufficient confidence that the level of the impurity in the drug substance is below the acceptable limit such that no analytical testing is needed for this impurity.

A control strategy that relies on process controls in lieu of analytical testing (option 4) can be appropriate if the process chemistry and process parameters that impact levels of mutagenic impurities are understood and the risk of an impurity residing in the final drug substance or drug product above the acceptable limit is determined to be negligible. Elements of a scientific risk assessment/chemistry rationale should include an assessment of various factors that influence the fate and purge of an impurity, including chemical reactivity, solubility, volatility, ionizability, and any physical process steps designed to remove impurities. This option is especially useful for those impurities that are inherently unstable (e.g., thionyl chloride that reacts rapidly and completely with water) or for those impurities that are introduced early in the synthesis and are effectively purged.

For option 4 approaches where justification based on scientific principles alone is not considered sufficient, as well as for option 3 approaches, analytical data to support the control approach are expected. This could include appropriate information on the structural changes to the impurity caused by downstream chemistry ("fate"); analytical data on pilot-scale batches; and, in some cases, laboratory-scale studies with intentional addition of the impurity ("spiking studies"). In these cases, it is important to demonstrate that the fate–purge argument for the impurity is robust and will consistently ensure a negligible probability of an impurity residing in the final drug substance above the acceptable limit. Where the purge factor is based on developmental data, it is important to address the expected scale dependence or independence. In the case that the small-scale model used in the development stage is considered to not represent the commercial scale, confirmation of suitable control in pilot-scale and/or initial commercial batches is necessary. The need for data from pilot/commercial batches is influenced by the magnitude of the purge factor calculated from laboratory- or pilot-scale data, point of entry of the impurity, and knowledge of downstream process purge points. If options 3 and 4 cannot be justified, then a test for the impurity on the specification for a raw material, starting material or intermediate, or as an in-process control (option 2) for drug substance (option 1) at the acceptable limit should be included. For impurities introduced in the last synthetic step, an option 1 control approach would be expected unless otherwise justified.

The application of ALARP is not necessary if the level of the mutagenic impurity is below the acceptable limits. Similarly, it is not necessary to demonstrate that alternate routes of synthesis have been explored. In cases where control efforts cannot reduce the level of the mutagenic impurity to below the acceptable limit and levels are as low as reasonably practical, a higher limit may be justified based on a risk–benefit analysis.

The aforementioned options include situations in which a test is recommended to be included in the specification but routine measurement for the release of every batch may not be necessary. This approach referred to as periodic or skip testing in ICH Q6A could also be called "periodic verification testing." This approach may be appropriate when it can be demonstrated that processing subsequent to impurity formation/introduction clears the impurity. It should be noted that allowing periodic verification testing is contingent on the use of a process that is under a state of control (i.e., produces a quality product that consistently meets specifications and conforms to an appropriately established facility, equipment, processing, and operational control regimen). If, on testing, the drug substance or drug product fails an established specification, the drug producer should immediately revert to full testing (i.e., testing of every batch for the attribute specified) until the cause of the failure has been conclusively determined, corrective action has been implemented, and the process is again documented to be in a state of control. As noted in ICH Q6A, regulatory authorities should be notified of a periodic verification test failure to evaluate the risk/benefit of previously released batches that were not tested.

It is recognized that product and process knowledge increases over the course of development and, therefore, it is expected that data to support control strategies in the clinical development trial phases will be less than those at the marketing registration phase. A risk-based approach based on process chemistry fundamentals is encouraged to prioritize analytical efforts on those impurities with the highest likelihood of being present in the drug substance or drug product. Analytical data may not be needed to support early clinical development when the likelihood of an impurity being present is low, but in a similar situation analytical data may be needed to support the control approach for a marketing application. It is also recognized that commercial formulation design occurs later in clinical development and, therefore, efforts associated with drug product degradants will be limited in the earlier phases.

1.8.4 Documentation for Clinical Development Trial Application

- It is expected that the number of structures assessed for mutagenicity and the collection of analytical data will both increase throughout the clinical development period.
- For phase I clinical trials of 14 days or less, a summary of efforts to mitigate risks of mutagenic impurities focused on class 1 and 2 impurities and those in the cohort of concern should be included.

- For other clinical development trials including phase I studies longer than 14 days, a list of the structures assessed by QSAR should be included and any class 1, 2, or 3 actual and potential impurities should be described along with plans for control. The in silico QSAR systems used to perform the assessments should be stated.
- Chemistry arguments may be appropriate instead of analytical data for potential impurities that have a low likelihood of being present.

1.8.5 Documentation for Common Technical Document (Marketing Application)

For all actual and potential process-related impurities and degradants where assessments according to this guideline are conducted, the mutagenic impurity classification and rationale for this classification should be provided:

- This would include the results and description of in silico QSAR systems used and, as appropriate, supporting information to arrive at the overall conclusion for class 4 and 5 impurities.
- When bacterial mutagenicity assays are performed on impurities, all results and the study reports should be provided for any bacterial mutagenicity–negative impurities.
- Justification for the proposed specification and the approach to control should be provided (e.g., ICH Q11 example 5b). For example, this information could include the acceptable intake and the location and sensitivity of relevant routine monitoring. For option 3 and 4 control approaches, a summary of knowledge of the purge factor and the identification of factors providing control (e.g., process steps, solubility in wash solutions, etc.) are important.

Note 1: The ICH M7 guideline recommendations provide a state-of-the-art approach for assessing the potential of impurities to induce point mutations and ensure that such impurities are controlled to safe levels so that below or above the qualification threshold no further qualification for mutagenic potential is required. This includes the initial use of QSAR tools to predict bacterial mutagenicity. In cases where the amount of the impurity exceeds 1 mg daily dose for chronic administration, evaluation of genotoxic potential as recommended in ICH Q3A/B could be considered.

Note 2: To assess the mutagenic potential of impurities, a single bacterial mutagenicity assay can be carried out with a fully adequate protocol according to ICH S2(R1) and OECD 471 guidelines. The assays are expected to be performed in compliance with good laboratory practice (GLP) regulations; however, it is noted that the test article may not be prepared or analyzed in compliance with GLP regulations. Lack of full GLP compliance does not necessarily mean that the data cannot be used to support clinical trials and marketing authorizations. Such deviations should be described in the study report. In some cases, the selection of bacterial tester strains may be limited to those proven to be sensitive to an alert. For degradants that are not feasible to isolate or synthesize or when compound quantity is limited, it may not be possible to achieve the highest test concentrations recommended for an ICH-compliant bacterial mutagenicity assay according to the current testing guidelines. In this case, bacterial mutagenicity testing could be carried out using a miniaturized assay format with proven high concordance to the ICH-compliant assay to enable testing at higher concentrations with justification. Confidence in the detection of mutagens requires testing concentrations at levels greater than or equal to 250 μg/plate.

Note 3: Tests to investigate the in vivo relevance of in vitro mutagens (positive bacterial mutagenicity)

In Vivo Test	Mechanistic Data to Justify Choice of Test as Fit for Purpose
Transgenic mutation assays	• For any bacterial mutagenicity positive. Justify selection of assay tissue/organ
Pig-a assay (blood)	• For directly acting mutagens (bacterial mutagenicity positive without S9)[a]
Micronucleus test (blood or bone marrow)	• For directly acting mutagens (bacterial mutagenicity positive without S9) and compounds known to be clastogenic[a]
Rat liver UDS test	• In particular for bacterial mutagenicity positive with S9 only • Responsible liver metabolite known • To be generated in test species used • To induce bulky adducts
Comet assay	• Justification needed (chemical class specific mode of action to form alkaline labile sites or single-strand breaks as preceding DNA damage that can potentially lead to mutations) • Justify selection of assay tissue/organ
Others	• With convincing justification

Note: UDS, unscheduled DNA synthesis

[a] For indirectly acting mutagens (requiring metabolic activation), justification is needed for sufficient exposure to metabolites.

Note 4: Example of linear extrapolation from TD50:

It is possible to calculate a compound-specific acceptable intake based on rodent carcinogenicity potency data such as TD50 values (doses giving a 50% tumor incidence equivalent to a cancer risk probability level of 1:2). Linear extrapolation to a probability of 1 in 105 (i.e., the accepted lifetime risk level used) is achieved by simply dividing TD50 by 50,000. This procedure is similar to that employed for the derivation of TTC.

Calculation example: ethylene oxide

TD50 values for ethylene oxide according to the Carcinogenic Potency Database are 21.3 mg/kg body weight/day (rat) and 63.7 mg/kg body weight/day (mouse). For the calculation of an acceptable intake, the lower (i.e., more conservative) value of the rat is used.

To derive a dose to cause tumors in 1 in 10^5 animals, divide by 50,000:

$$21.3\ \text{mg/kg} \div 50{,}000 = 0.42\ \mu\text{g/kg}$$

To derive a total human daily dose,

$$0.42\ \mu g/kg/day \times 50\ kg\ Body\ Weight = 21.3\ \mu g/person/day$$

Hence, a daily lifelong intake of 21.3 μg ethylene oxide would correspond to a theoretical cancer risk of 10^{-5} and therefore be an acceptable intake when present as an impurity in a drug substance.

As an alternative to using the most conservative TD50 value from rodent carcinogenicity studies irrespective of its relevance to humans, an in-depth toxicological expert assessment of the available carcinogenicity data can be done to initially identify the findings (species, organs, etc.) with the highest relevance to human risk assessment as a basis for deriving a reference point for linear extrapolation. Also, to better take into account directly the shape of the dose–response curve, a benchmark dose (BMD) such as a benchmark dose lower confidence limit 10% (BMDL10) (an estimate of the lowest dose that is 95% certain to cause no more than a 10% cancer incidence in rodents) may be used instead of TD50 values as a numerical index for carcinogenic potency. Linear extrapolation to a probability of 1 in 10^5 (i.e., the accepted lifetime risk level used) is then achieved by simply dividing BMDL10 by 10,000. Compound-specific acceptable intakes can also be derived from published recommended values from internationally recognized bodies such as WHO (IPCS Cancer Risk Assessment Program) and others using the appropriate 10^{-5} lifetime risk level. In general, a regulatory limit that is applied should be based on the most current and scientifically supported data and/or methodology.

Note 5: A compound-specific calculation of acceptable intakes for mutagenic impurities may be applied for mutagenic impurities (without carcinogenicity data) that are structurally similar to a chemically defined class of known carcinogens. For example, factors that are associated with the carcinogenic potency of alkyl halides have been identified and can be used to modify the safe acceptable intake of monofunctional alkyl halides, a group of alkyl halides commonly used in drug synthesis. Compared to multifunctional alkyl halides, the monofunctional compounds are much less potent carcinogens with TD50 values ranging from 36 to 1810 mg/kg/day ($n = 15$; epichlorohydrin with two distinctly different functional groups is excluded). A TD50 value of 36 mg/kg/day can thus be used as a still very conservative class-specific potency reference point for the calculation of acceptable intakes for monofunctional alkyl halides. This potency level is at least ten-fold lower than the TD50 of 1.25 mg/kg/day corresponding to the default lifetime TTC (1.5 μg/day) and therefore justifies lifetime and LTL daily intakes for monofunctional alkyl halides 10 times the default ones.

Note 6: Some published data give reliable experimental evidence for (practical) thresholds in the dose response for compounds that are positive for bacterial mutagenicity. This includes examples of thresholds in error-free repair capacity of the mutagenic DNA-ethylating agent ethyl methanesulfonate or similarly for methylating agents. Thresholds involving metabolic detoxification processes also appear to exist for 1,3-butadiene. Further, a threshold for oxidative DNA damage associated with the buildup of hemosiderin has been shown for *p*-chloroaniline hydrochloride. Aside from mechanistic considerations supporting an experimentally observed threshold, it is important that a proper statistical analysis supports this assumption as well.

Note 7: See Figure 1.5

Establishing LTL acceptable intakes for mutagenic impurities in pharmaceuticals has precedent in the establishment of staged TTC limits for clinical development. The

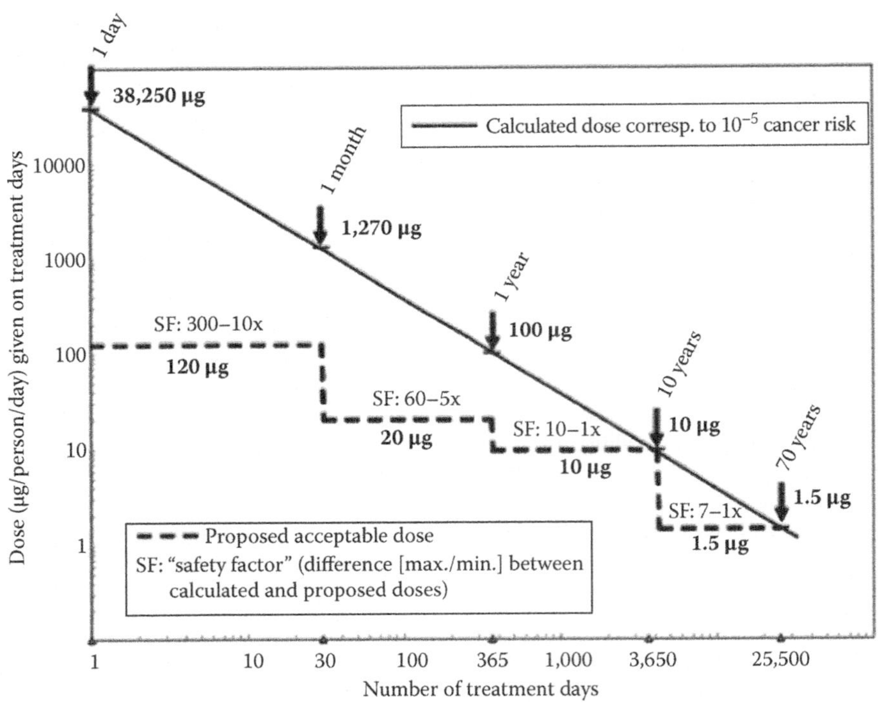

FIGURE 1.5 Illustration of calculated daily dose of a mutagenic impurity corresponding to a theoretical 1:100,000 cancer risk as a function of duration of treatment in comparison with the acceptable intake levels, ICH M7 draft guideline.

calculation of LTL acceptable intakes is predicated on the principle of Haber's rule, a fundamental concept in toxicology where concentration (C) × time (T) = a constant (k). Therefore, the carcinogenic effect is based on both dose and duration of exposure.

The solid line in Figure 1.5 represents the linear relationship between the amount of daily intake of a mutagenic impurity corresponding to a 10^{-5} cancer risk and the number of treatment days. The calculation is based on the TTC level as applied in this guideline for lifelong treatment, that is, 1.5 µg/person/day using the following formula:

$$\text{LTL AI} = \frac{1.5\ \mu\text{g} \times (365\ \text{days}\ \times 70\ \text{Years Lifetime} = 25{,}550)}{\text{Total Number of Treatment Days}}$$

where AI refers to acceptable intake.

Thus, the calculated daily intake levels would be 1.5 µg for a treatment duration of 70 years, 10 µg for 10 years, 100 µg for 1 year, 1270 µg for 1 month, and approximately 38.3 mg as a single dose, all resulting in the same cumulative intake and therefore theoretically in the same cancer risk (1 in 10^5). The dashed step-shaped curve in Figure 1.5 represents the actual daily intake levels adjusted to LTL exposure for products in clinical development and marketed products. These proposed levels are in general significantly lower than the calculated values, thus providing safety factors that increase with shorter treatment durations. The proposed accepted daily intakes are also in compliance with a 10^{-6} cancer risk level if treatment durations are not longer than 6 months* and are therefore applicable in early clinical trials with volunteers/patients where benefit has not yet been established. In this case, the safety factors as shown in the upper graph in Figure 1.5 would be reduced by a factor of 10.

1.8.6 Glossary

acceptable intake: in the context of the ICH M7 guideline, an intake level that is without appreciable cancer risk.

acceptable limit: maximum acceptable concentration of an impurity in a drug substance or drug product derived from the acceptable intake and the daily dose of the drug.

acceptance criterion: numerical limits, ranges, or other suitable measures for the acceptance of the results of analytical procedures.

BMDL10: the lower 95% confidence interval of a BMD representing a 10% response (e.g., tumor response upon lifetime exposure), that is, the lower 95% confidence interval of a BMD10. BMD10 is the BMD associated with a 10% response adjusted for background.

control strategy: a planned set of controls derived from current product and process understanding that ensures process performance and product quality. The controls can include parameters and attributes related to drug substance and drug product materials and components, facility and equipment operating conditions, in-process controls, finished product specifications, and the associated methods and frequency of monitoring and control.

cumulative intake: the total intake of a substance that a person is exposed to over time.

degradant: degradation product as defined in ICH Q3B.

DNA reactive: substances that have a potential to induce direct DNA damage through chemical reactions with DNA.

expert knowledge: in the context of this guideline, expert knowledge can be generalized as a review of preexisting data and the use of any other relevant information to evaluate the accuracy of an in silico model prediction for mutagenicity.

genotoxicity: a broad term that refers to any deleterious change in genetic material regardless of the mechanism by which the change is induced.

in-process control: checks performed during production to monitor and, if appropriate, to adjust the process and/or to ensure that the intermediate or drug substance conforms to its specifications.

mutagenic impurity: an impurity that has been demonstrated to be mutagenic in an appropriate mutagenicity test model, for example, the bacterial mutagenicity assay.

---

* At 6 months, the calculated dose at a 10^{-6} risk level would be 20 μg, which is identical to the recommended accepted dose, that is, there is no extra safety factor; at longer durations, the theoretical 10^{-6} risk level would be exceeded.

NOEL: abbreviation for no-observed-effect dose (level). It is the highest dose of substance at which there are no biologically significant increases in frequency or severity of any effects in exposed humans or animals.

periodic (verification) testing: also known as periodic or skip testing in ICH Q6A.

purge factor: purge reflects the ability of a process to reduce the level of an impurity, and the purge factor is defined as the level of an impurity at an upstream point in a process divided by the level of an impurity at a downstream point in a process. Purge factors may be measured or predicted.

QSAR and SAR: in the context of this guideline, refers to the relationship between the molecular substructure of a compound and its mutagenic activity using QSARs derived from experimental data.

statistical process control: application of statistical methodology and procedures to analyze the inherent variability of a process.

structural alert: in the context of this guideline, a chemical grouping or molecular substructure that is associated with mutagenicity.

TD50: the dose rate in milligram per kilogram of body weight per day that if administered chronically for the standard life span of a species will halve the probability of remaining tumorless throughout that period.

threshold: categorically, a dose of a substance or exposure concentration below which a stated effect is not observed or expected to occur.

Table 1.10 illustrates scope scenarios for application of the ICH M7 Guideline.

TABLE 1.10
Scope Scenarios for the Application of the ICH M7 Draft Guideline

Scenario	Applies to Drug Substance	Applies to Drug Product	Comments
Registration of new drug substances and associated drug product	Yes	Yes	Primary intent of the M7 guideline
Clinical trial applications for new drug substances and associated drug product	Yes	Yes	Primary intent of the M7 guideline
Clinical trial applications for new drug substances for an anticancer drug per ICH S9	No	No	Out of scope of the M7 guideline
Clinical trial applications for new drug substances for an orphan drug	Yes	Yes	There may be exceptions on a case-by-case basis for higher impurity limits.
Clinical trial application for a new drug product using an existing drug substance where there are no changes to the drug substance manufacturing process	No	Yes	Retrospective application of the M7 guideline is not intended for marketed products unless there are changes made to the synthesis. Since no changes are made to the drug substance synthesis, the drug

TABLE 1.10
Scope Scenarios for the Application of the ICH M7 Draft Guideline

Scenario	Applies to Drug Substance	Applies to Drug Product	Comments
			substance would not require reevaluation. Since the drug product is new, applications of this guideline is expected.
A new formulation of an approved drug substance is filed	No	Yes	See text on postapproval changes to drug product.
A product that is previously approved in a member region is filed for the first time in a different member region. The product is unchanged.	Yes	Yes	As there is no mutual recognition, an existing product in one member region filed for the first time in another member region would be considered a new product.
A new supplier or new site of the drug substance is registered. There are no changes to the manufacturing process used in this registered application.	No	No	As long as the synthesis of the drug substance is consistent with previously approved methods, reevaluation of mutagenic impurity risk is not necessary. The applicant would need to demonstrate that no changes have been made to a previously approved process/product.
An existing product (approved after the issuance of ICH M7 with higher limits based on ICH S9) associated with an advanced cancer indication is now registered for use in a nonlife-threatening indication.	Yes	Yes	Since the patient population and acceptable cancer risk have changed, the previously approved impurity control strategy and limits will require reevaluation.
New combination product is filed that contains one new drug substance and an existing drug substance (no changes to the manufacturing process).	Yes (new drug substance) No (existing drug substance)	Yes	The M7 guideline would apply to the new drug substance. For the existing drug substance, retrospective application of the M7 guideline to existing products is not intended. For the drug product, this would classify as a new drug product so the guideline would apply to any new or higher levels of degradants.

Source: *Assessment and Control of DNA Reactive (Mutagenic) Impurities in Pharmaceuticals to Limit Potential Carcinogenic Risk*, ICH M7, ICH Harmonised Tripartite Guideline, International Conference on Harmonisation of Technical Requirements for Registration of Pharmaceuticals for Human Use, 2013.

Case Examples to Illustrate Potential Control Approaches

CASE 1: EXAMPLE OF AN OPTION 3 CONTROL STRATEGY

Impurity A: intermediate X is introduced into the second to last step of the synthesis, and impurity A is routinely detected in the intermediate material X. The impurity A is a stable compound and carries over to the drug substance. A spike study of the impurity A with different concentration levels was performed. As a result of these studies, it was determined that up to 1.0% of the impurity A in the intermediate material X can be removed consistently to less than 30% of the TTC, 100 ppm in this case. This purge is consistent with the determined solubility of the impurity in the process solvents. This purge ability of the process has been confirmed by determination of any residue of impurity A in the drug substance in multiple pilot-scale batches, and results ranged from 16–29 ppm. Therefore, control of the impurity A in the intermediate material X with an acceptance limit of 1.0% is established. As the purge of impurity A is based on the solubility of the impurity in the process solvents and is determined to be scale independent, submission of data on initial commercial batches would not be expected.

CASE 2: EXAMPLE OF AN OPTION 3 CONTROL STRATEGY—BASED ON PREDICTED PURGE FROM A SPIKING STUDY USING STANDARD ANALYTICAL METHODS

Impurity B: a starting material Y is introduced in step 3 of a five-step synthesis, and an impurity B is routinely detected in the starting material Y at less than 0.1% using standard analytical methods. To determine if the 0.1% specification in the starting material is acceptable, a purge study was conducted at laboratory scale where impurity B was spiked into starting material Y with different concentration levels up to 10% and a purge factor above 500-fold was determined across the final three processing steps. This purge factor applied to a 0.1% specification in starting material Y would result in a predicted level of impurity B in the drug substance of less than 2 ppm. As this is below the TTC-based limit of 50 ppm for this impurity in the drug substance, the 0.1% specification of impurity B in starting material Y is justified without the need for testing in the drug substance on pilot-scale or commercial-scale batches.

CASE 3: EXAMPLE OF AN OPTION 2 AND 4 CONTROL STRATEGY—CONTROL OF STRUCTURALLY SIMILAR MUTAGENIC IMPURITIES

The step 1 intermediate of a five-step synthesis is a nitroaromatic compound that may contain low levels of impurity C, a positional isomer of the step 1 intermediate and also a nitroaromatic compound. The amount of impurity C in the step 1 intermediate has not been detected by ordinary analytical methods, but it may be present at lower levels. The step 1 intermediate is positive in the bacterial mutagenicity assay. The step 2 hydrogenation reaction results in a 99% conversion of the step 1 intermediate to the corresponding aromatic amine. This is confirmed via in-process testing. An assessment of purge of the remaining step 1 nitroaromatic intermediate was conducted, and a high purge factor was predicted based on purge points in the subsequent step 3 and 4 processing steps. Purge across the step 5 processing step was not expected, and a specification for the step 1 intermediate at TTC levels was established at the step 4 intermediate (option 2 control approach). The positional isomer impurity C would be expected to purge via the same purge points as the step 1 intermediate and therefore would always be much lower than the step 1 intermediate itself; therefore, no testing is required and an option 4 control strategy for impurity C can be supported without the need for any additional laboratory or pilot scale data.

CASE 4: EXAMPLE OF AN OPTION 4 CONTROL STRATEGY—A HIGHLY REACTIVE IMPURITY

Thionyl chloride is a highly reactive compound that is mutagenic. This reagent is introduced in step 1 of a five-step synthesis. At multiple points in the synthesis, significant amounts of water are used. Since thionyl chloride reacts instantaneously with water, there is no chance of any residual thionyl chloride being present in the drug substance. An option 4 control approach is suitable without the need for any laboratory- or pilot-scale data.

CASE 5: OPTION 1 CONTROL STRATEGY—APPLICATION OF PERIODIC VERIFICATION TESTING

A mutagenic reagent is used in the last step of synthesizing a drug substance. This reagent is a liquid at room temperature, is not used in excess, and is soluble in reaction and isolation solvents. A test and acceptance criteria for this reagent is contained in the drug substance specification due to the fact that the reagent is used in the final synthetic step. This impurity was tested for in the first 10 commercial batches, and all test results were less than 5% of the acceptance criteria. In this situation, periodic verification testing could be accepted.

1.9 GUIDANCE ON GENOTOXICITY TESTING AND DATA INTERPRETATION FOR PHARMACEUTICALS INTENDED FOR HUMAN USE, ICH S2(R1)

ICH S2(R1) [9] addresses the standards for follow-up testing and interpretation of positive findings in standard genetic toxicology battery for new "small molecules" being developed as human pharmaceuticals. The basic principle underlining these tests is that compounds that induce genetic damage to DNA and its fixation have the potential to be human carcinogens and/or mutagens.

1.9.1 Standard Test Battery for Genotoxicity

Tumorigenesis is a complex process, and no single test can detect all relevant mechanisms. Therefore, the standard test battery takes a multipronged approach to assessing risk by using a combination of in vitro and in vivo tests in mammalian and bacterial test systems.

Assessment of mutagenicity is performed using the bacterial reverse gene mutation assay, which has been shown to detect most, but not all, rodent and human carcinogens, highlighting the need for a comprehensive test battery. Three comparable in vitro mammalian cell systems are typically used: the in vitro metaphase chromosome aberration assay, in vitro micronucleus assay, and mouse-lymphoma L5178Y cell TK gene mutation assay. In vivo tests are included in the battery to detect those not picked up in in vitro assays to account for exposure. The most commonly used evaluations are the measurement of micronuclei in erythrocytes or chromosome aberrations in metaphase cells in bone marrow. Occasionally, cytogenic analysis in lymphocytes cultured from treated animals is performed, but the experience with this assay is considerably less than that with the others described. The combination of these tests for chromosome aberrations can detect diverse changes, including chromatid or chromosome breaks, aneuploidy, gene mutation, and chromosome damage and loss.

Positive in vivo or in vitro tests can be followed up with further testing to develop a weight-of-evidence assessment regarding genotoxic risk. Negative results in two appropriate in vivo assays are usually sufficient to demonstrate absence of a risk defined by positive results in the standard genotoxicity battery.

1.9.2 Description of Two Options for the Standard Battery

Two options are equally suitable for the standard test battery, but there is more historical experience with option 1. Regardless of the initial approach, when a positive in vitro result is obtained, negative results in two in vivo assays are sufficient evidence for a lack of genotoxic potential in vivo.

Option 1:

1. A test for gene mutation in bacteria.
2. A cytogenic test for chromosomal damage (either the in vitro metaphase chromosome aberration test or the in vitro micronucleus test), or an in vitro mouse-lymphoma assay.
3. An in vivo test for genotoxicity, generally a test for chromosomal damage using rodent hematopoietic cells, either for micronuclei or for chromosomal aberrations in metaphase cells.

Option 2:

1. A test for gene mutation in bacteria.
2. An in vivo assessment of genotoxicity in two different tissues, usually an assay for micronuclei using rodent hematopoietic cells and a second in vivo assay, typically a DNA strand breakage assay in the liver. Tissues other than the liver can be used, considering potential mechanism of action for the substance, metabolism, or distribution.

For the in vivo studies, the administration of the test substance may be either acute or repeated dose. When multiple assessments are performed it is best to include all of the end points in a single study.

Negative results in the standard battery provide assurance that the substance does not have genotoxic activity. Positive results in the standard battery are typically a trigger for further follow-up testing to better understand the genotoxic activity of the agent.

1.9.3 Modifications to the Test Battery

ICH S2(R1) outlines situations where modifications to the standard test battery might be appropriate, for example, exploratory clinical studies, testing compounds that are toxic to bacteria, compounds bearing structural alerts for genotoxic activity, and instances where there are limitations to the use of in vivo tests.

1.9.4 Detection of Germ Mutations

Based on historical experience, most germ cell mutagens are detected as genotoxic in somatic cell tests. Therefore, these assays are considered adequate to assess germ cell effects.

1.9.5 Recommendations for In Vitro Tests

The recommended test systems and study designs are sufficiently characterized and have sufficient adequate internal controls that repetition of a clear positive or negative result is not warranted. Statistical analysis may aid in interpretation, but biological interpretation is also critical. Some adequately performed test may result in an equivocal result and when the test is repeated, the result may be clearly positive or negative, describing the overall result as positive, or negative, respectively; or, the repeated assay may also provide equivocal results, causing the overall conclusion to be equivocal.

The maximal dose level is 5000 μg/plate (or 5 μL/plate for a liquid test substance) when not limited by solubility or cytotoxicity. For bacterial cultures, in the absence of toxicity the lowest precipitating dose should be used as the top dose. If dose-related cytotoxicity or mutagenicity occurs, then the top dose should be based on cytotoxicity as determined by number of revertants and/or clearing or diminution of the background lawn. Bacterial strains typically used to detect base substitution and frameshift mutations are (1) *Salmonella typhimurium* TA 98, TA100, TA1535, (2) either TA1537 or TA97 or TA97a, and (3) either *S. typhimurium* TA102 or *E. WP2 uvrA* or *E. coli WP2 uvrA* (pKM101).

1.9.6 Recommended Protocols for the Mammalian Cell Assays

The maximum top concentration recommended is 1 mM or 0.5 mg/mL, whichever is lower, when not limited by solubility in a solvent or culture medium or by cytotoxicity. When solubility is the limiting factor, the top concentration should be that which provides the minimal visible precipitate. For cytotoxicity as a dose-limiting criterion the top concentration should not cause more and a 50% reduction in cell growth (for assays evaluating metaphase chromosome aberrations or for micronuclei), and for the mouse-lymphoma assay the top dose should not exceed 80%–90% cytotoxicity as determined by relative total growth.

When conducting an in vitro cytogenetic assay, the test conditions should include scenarios both with and without metabolic activation, and positive and negative controls should always be used. In addition, colony sizing, with appropriate comparators, should be assessed in the mouse-lymphoma assay. Treatment in a metaphase aberration assay should be for 3 to 6 hours with periodic sampling to capture several cell cycles. Treatment in the mouse-lymphoma assay is typically 3 to 4 hours plus 24 hours without metabolic activation. For in vitro mammalian cell assays, similar confirmatory elements as those described should be utilized. Positive controls for in vitro mammalian tests may be limited to conditions where metabolic activation is used to demonstrate the capability of the metabolic system and the responsiveness of the assay.

1.9.7 Recommendations for In Vivo Tests

For the detection of clastogens in rats and mice, either the analysis of chromosomal aberrations or the measurement of micronucleated polychromatic erythrocytes in bone marrow cells is considered appropriate. Chromosome aberrations can also be analyzed in peripheral lymphocytes from rodents. Investigation of chromosome aberrations or of gene mutations in endogenous genes is not feasible with standard methods in most tissues. Mutations can be measured in transgenes in rodents after prolonged treatment, but a more common approach is to measure DNA damage as a surrogate end point. Typical assays for DNA damage include DNA strand breaks (e.g., the "comet" assay and the alkaline elution assay), transgenic mouse mutation assays, and covalent binding assays, and the unscheduled DNA synthesis assay.

Study designs usually evaluate three dose levels, and they can be short-term or long-term studies. Short-term studies administer the test substance one to three times at the maximal tolerated dose or the limit dose of 2000 mg/kg. When the genotoxicity assessment is integrated into a repeat dose toxicity study, the doses employed to support human clinical trials are used. When conducting a follow-up study to address an indication of genotoxicity, or when an in vitro mammalian cell assay is not available, several criteria are satisfactory for establishing the top dose to be used, including the maximal feasible dose, a limit dose of 1000 mg/kg for studies of 14 days or longer, maximal possible exposure, or a dose that is greater than or equal to 50% of the top dose used in an acute study.

1.9.8 Testing Compounds That Are Toxic for Blood or Bone Marrow

Sometimes, genotoxicity is only detectable at doses very close to cytotoxic doses for the red blood cell lineage, in which case doses should be spaced no more than twofold below the cytotoxic dose. Alternatively, one could conduct early blood sampling in a longer term study (e.g., collect after 3 to 4 days of dosing), perform an in vitro mammalian cell micronucleus assay, or perform an acute bone marrow micronucleus assay.

1.9.9 Demonstration of Target Tissue Exposure

When negative in vivo test results are obtained, measures must be employed to ensure that there was adequate exposure of the test compound in the target tissue. Little value is obtained from conducting the tests if adequate exposure is not achieved.

When an in vitro genotoxicity test is positive, or not done, the follow-up in vivo test should have data demonstrating adequate exposure at the highest dose, or the most relevant dose, tested. Demonstration of exposure can be by cytotoxicity measurements or systemic exposure. Cytotoxicity end points such as the proportion of immature erythrocytes among total erythrocytes, a reduction in the mitotic index, histopathology, or clinical chemistry may be used. Systemic exposure may be demonstrated by measuring drug-related material in plasma or blood or target tissues.

When in vitro genotoxicity tests are negative, adequate exposure in the in vivo test may be demonstrated by the measurement of systemic exposure or by using exposure results from other toxicity studies.

1.9.10 Additional Study Design Considerations

The appropriate sample time for in vivo assays depends on the test being conducted. When integrating micronucleus evaluations in a repeat dose study, sampling can be done on the day following the final administered dose. For assays looking at DNA damage/strand breaks, samples are usually collected 2 to 6 hours after the final dose when evaluated in a repeat dose study and a few hours and 24 hours after the administration of a single dose.

If the genotoxicity evaluation is conducted as part of a repeat dose toxicity study, it is sufficient to evaluate a random subset of the study animals.

For sex-specific drugs, the target sex to human use should be used in the genotoxicity assessments. When using an acute protocol, males are typically used unless there is a meaningful difference in exposure between the sexes, in which case both sexes should be used. When a repeat dose protocol is used, samples are typically collected from males and females; but if there are no significant differences in toxicity or metabolism, then only the males are evaluated.

Generally, the route of administration for in vivo genotoxicity assessments is the same as the intended route of administration in humans.

Once a laboratory demonstrates competence in the use of an assay, a positive control is not required to be administered concurrently with every assay.

1.9.11 Guidance on Evaluation of Test Results and on Follow-Up Test Strategies

Genotoxicity assays detect carcinogens acting primarily via a direct mechanism and not via nongenotoxic mechanisms. Both false negative and false positive results may be obtained in these assays. The assay battery is designed to reduce the risk of false negative results, but a false positive result does not necessarily mean that the test substance is a hazard to humans. Therefore, the biological significance for in vitro positive results must be established. Results that are not considered biologically meaningful could be obtained when changes are statistically significant compared to the negative or solvent controls, but the values are still within the confidence intervals of the historical controls, or when the changes are weak/equivocal and not reproducible.

1.9.12 Evaluation of Results Obtained in In Vitro and In Vivo Tests

Positive results in the bacterial mutation assay imply direct DNA reactivity. However, artifactual or system-specific changes that are not relevant for humans may occur, such as an increase in colonies that are not true revertants due to contamination with amino acids or when bacteria-specific metabolism occurs.

Similarly, results of questionable relevance to humans can also be obtained in mammalian cell assays due to conditions that do not occur in vivo (pH, osmolality, and precipitates), or when the effect occurs only at toxic concentrations.

There are instances when follow-up testing should be conducted, even when a negative in vitro result is obtained. For example, there should be follow-up testing if the test substance is known to be metabolized by routes not standardly used in the in vitro assays, or when the structure or activity of the compound indicates that the use of standard test systems is inappropriate.

In vivo test systems may be more relevant than in vitro systems for humans because in vivo systems account for absorption, distribution, and excretion and metabolism may be more relevant. When differences between in vivo and in vitro assays occur, an understanding of the differences in the test systems and relevance to humans should be considered. Similar to in vitro assays, misleading positive results can occur in in vivo assays. Examples include increases in micronuclei due to disturbances in erythropoiesis, evaluation of DNA adduct data without consideration of background levels, and indirect toxicity.

1.9.13 Follow-Up Strategies for Positive Results

When a negative result is obtained in an Ames bacterial mutation assay but a positive result is obtained in an in vitro mammalian cell test, a weight-of-evidence approach can be taken with additional in vitro studies in conjunction with a single in vivo assay or by carrying out two additional in vivo assays. In instances where the in vitro positive result was obtained only with S9 metabolic activation, additional work to establish the relevance of these in vitro conditions to in vivo should be conducted. When there is a positive result in an in vivo micronucleus assay, the contribution of nongenotoxic effects should be evaluated.

1.9.14 Follow-Up Genotoxicity Testing in Relation to Tumor Findings in a Carcinogenicity Bioassay

When a compound is negative in the standard test battery but test substance–related tumors develop in carcinogenicity studies and there is insufficient evidence to demonstrate that these tumors develop via nongenotoxic mechanisms, there are additional test strategies that can be employed. Further evaluation of metabolic activation in in vitro and in vivo test systems can be undertaken, such as measurement of genetic damage in the target tissues themselves or induction of transgenes or molecular characterization of genetic changes in tumor-related genes.

REFERENCES

1. *Impurities in New Drug Substances*, ICH Q3A(R2), ICH Harmonised Tripartite Guideline, International Conference on Harmonisation of Technical Requirements for Registration of Pharmaceuticals for Human Use, October 2006.
2. *Impurities in New Drug Products*, ICH Q3B(R2), ICH Harmonised Tripartite Guideline, International Conference on Harmonisation of Technical Requirements for Registration of Pharmaceuticals for Human Use, June 2006.
3. ICH official website, www.ich.org.
4. *Impurities: Guideline for Residual Solvents*, ICH Q3C (R5), ICH Harmonised Tripartite Guideline, International Conference on Harmonisation of Technical Requirements for Registration of Pharmaceuticals for Human Use, February 2011.
5. Committee for Medicinal Products for Human Use, *Guideline on the Limits of Genotoxic Impurities*, London: European Medicines Agency, January 2007.
6. Committee for Medicinal Products for Human Use, *Questions and Answers on the "Guideline on the Limits of Genotoxic Impurities,"* London: European Medicines Agency, September 2010 (Rev. 3).
7. Center for Drug Evaluation and Research, *Guidance for Industry, Genotoxic and Carcinogenic Impurities in Drug Substances and Products: Recommended Approaches*, Draft Guidance, Food and Drug Administration, December 2008.
8. *Assessment and Control of DNA Reactive (Mutagenic) Impurities in Pharmaceuticals to Limit Potential Carcinogenic Risk*, ICH M7, ICH Harmonised Tripartite Guideline, International Conference on Harmonisation of Technical Requirements for Registration of Pharmaceuticals for Human Use, February 2013.
9. *Guidance on Genotoxicity Testing and Data Interpretation for Pharmaceuticals Intended for Human Use*, ICH S2(R1), ICH Harmonised Tripartite Guideline, International Conference on Harmonisation of Technical Requirements for Registration of Pharmaceuticals for Human Use, November 2011.

2 Structural Alerts for Genotoxicity and Carcinogenicity

Heewon Lee

CONTENTS

The underlying principle of chemical carcinogenesis [1–3] is the concept that "chemical structure defines toxicity." Vast investigations and experimental studies are available in the literature on chemical carcinogenesis. This chapter provides a brief introduction and a few illustrations of chemical carcinogenesis to appreciate the concept and practice of structural alerts and their application. Well-known and representative structural alerts for genotoxicity, mutagenicity, and carcinogenicity in the literature are included.

The observations in the 1700s of nasal cancer being a consequence of excessive use of tobacco snuff and skin cancer of the scrotum among chimney sweeps were the beginning of knowledge on chemical carcinogenesis. A number of investigations to demonstrate the carcinogenic activity of soots and tars in experimental animals and to search for the active agents resulted in the identification of polycyclic aromatic hydrocarbons (PAHs) as the responsible carcinogens. Dibenz[*a*,*h*]anthracene was the first ever carcinogen synthesized, and other PAHs, including benzo[*a*]pyrene, 3-methylcholanthrene, and 7,12-dimethylbenz[*a*]anthracene, have been extensively studied. Some of the structures of chemical carcinogens discovered or identified prior to 1960 are shown in Figure 2.1.

As the number and variety of known chemical carcinogens increased, it became evident that these chemicals lacked a common structural feature. Some carcinogens, such as aromatic amine derivatives, caused tumors at distant sites (liver and urinary bladder) regardless of the route of administration, providing an early clue to the involvement of metabolism. The importance of metabolic activation in chemical carcinogenesis was stated by Miller: "The great majority of chemical carcinogens were active only after metabolism to ultimate carcinogens (i.e., the derivatives that actually initiate the neoplastic event). The known exceptions are the carcinogens that are alkylating or acylating agents per se. Further, the data then available suggested that the ultimate forms of chemical carcinogens might all be strong electrophilic reactants... . The known ultimate carcinogens contain relatively electron-deficient atoms

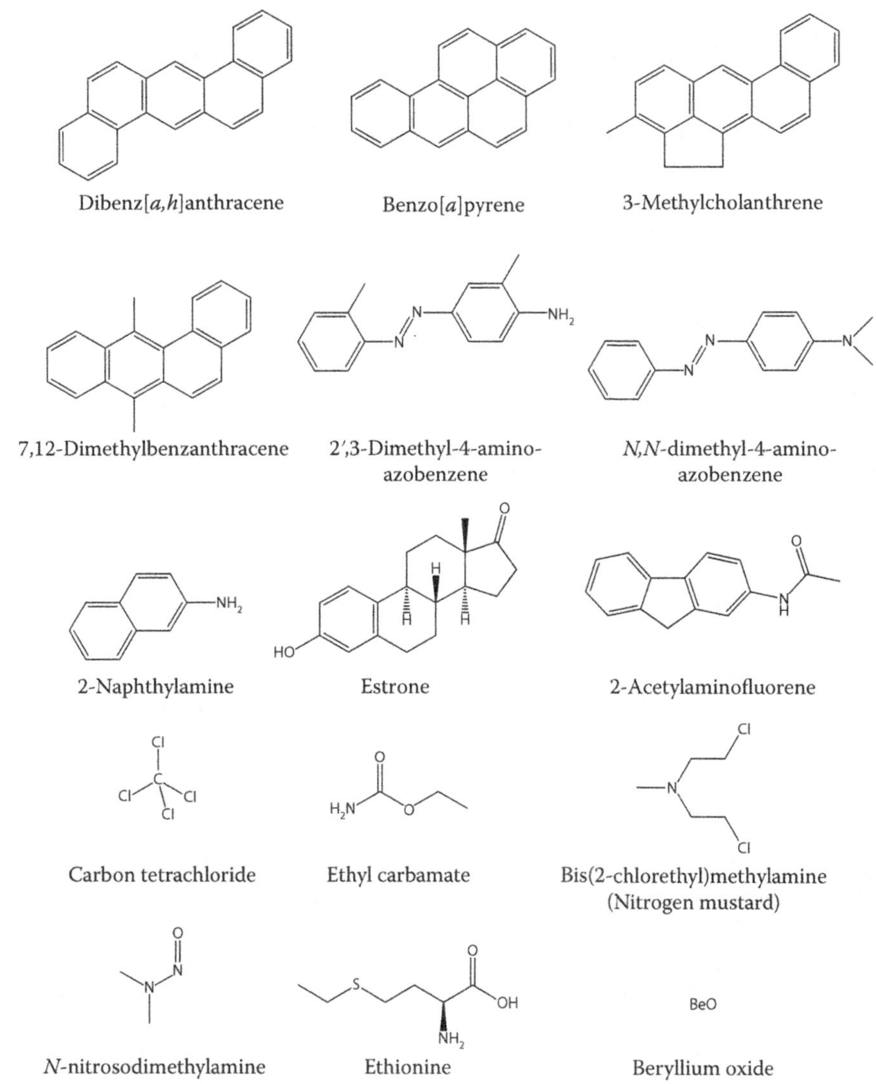

FIGURE 2.1 The structures of chemical carcinogens discovered prior to 1960. (Adapted from Miller, E.C., *Cancer Res.*, 38, 1479–1496, 1978.)

that seek to react with nucleophilic sites, i.e., atoms that have easily shared electrons. These nucleophilic sites are relatively abundant in DNA's, RNA's, and proteins" [1]. Accordingly, a common property of ultimate carcinogens is electrophilicity.

Precarcinogens can be metabolized to more than one ultimate carcinogen, and there are many nucleophilic sites in target macromolecules (DNAs, RNAs, and proteins); therefore, it is possible to have multiple macromolecule-bound derivatives of each precarcinogen. The possibility that the major adducts are not necessarily the most important ones in carcinogenesis, in addition to the multiplicity of adducts, gives rise to the complexity of metabolic activation in carcinogenesis. Figure 2.2

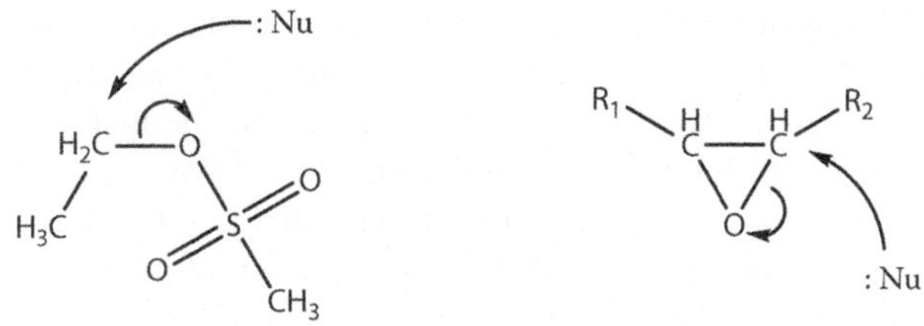

Ethyl methane sulfonate Epoxide

: Nu comprises DNAs, RNAs, and proteins.

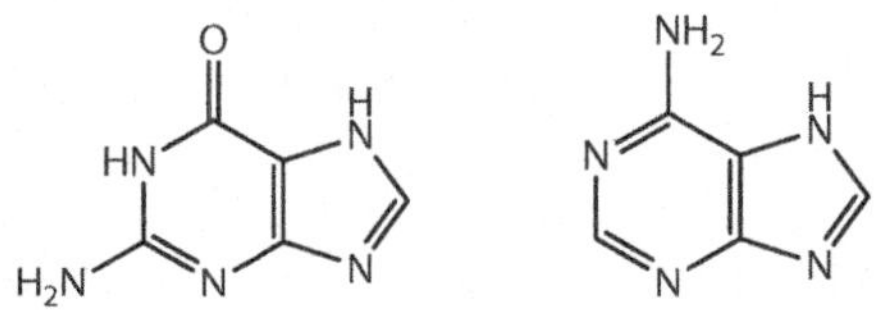

Guanine Adenine

FIGURE 2.2 Examples of strong electrophilic reactants and their reaction with nucleophiles (: Nu). (Adapted from Miller, E.C., *Cancer Res.*, 38, 1479–1496, 1978.)

shows examples of strong electrophilic reactants (positive ions or molecules with electron-deficient atoms) and their reactions with nucleophilic sites in DNA, RNA, and proteins (such as oxygen, nitrogen, and sulfur atoms).

A few example compounds with reported chemical carcinogenesis are shown with proposed mechanisms to illustrate the importance of metabolic bioactivation. Simply alkylating agents, such as alkyl halides, were not selected, as they proceed via direct alkylation on macromolecules (DNAs, RNAs, and proteins) due to their chemical reactivity. Rather, highly potent compounds inducing neoplasia (tumor) that require metabolic activation (in this sense, they are procarcinogens) are chosen. Four examples shown in the literature are benzo[*a*]pyrene, aflatoxin B1, aristolochic acid, and aromatic amines. These chemical carcinogens are, strictly speaking, precarcinogens that form the electrophilic ultimate carcinogens via metabolic activation.

Benzo[*a*]pyrene is a ubiquitous environmental pollutant produced during combustion processes and also found in tobacco and diet [4]. It is the major genotoxic component in cigarette smoke and was the cause of the high rate of cancer in Britain among chimney sweeps. The majority of human exposure occurs through smoking and consumption of charbroiled foods by ingestion, inhalation, and dermal absorption. Benzo[*a*]pyrene must be activated by enzymes such as cytochrome P450 (P450) to acquire its mutagenic and carcinogenic properties. The metabolic activation pathway is shown in Figure 2.3. The first step is the formation of the 7,8-epoxide, followed by hydrolysis by microsomal epoxide hydrolase to the 7,8-diol metabolite. The 7,8-diol metabolite is activated by P450 further and becomes the mutagenic diol epoxide. The diol epoxide is extremely reactive and binds to macromolecules, including DNAs, RNAs, and proteins.

Aflatoxins' name originated from *Aspergillus flavus* toxins, as they are naturally occurring mycotoxins produced by a species of fungi, *A. flavus*. Aflatoxin B1 is acutely toxic to most animal species and is a potent mutagen and a potent hepatocarcinogen [5]. A major source of human exposure to aflatoxins is dietary contamination with molded grain. The formation of a DNA adduct with aflatoxin B1 through metabolic activation is shown in Figure 2.4. Aflatoxin B1 is first converted to the reactive 2,3-exo-epoxide via microsomal conversion. The reactive 2,3-epoxide then reacts with nucleophilic sites of DNA, the N7 atom of guanine in the example

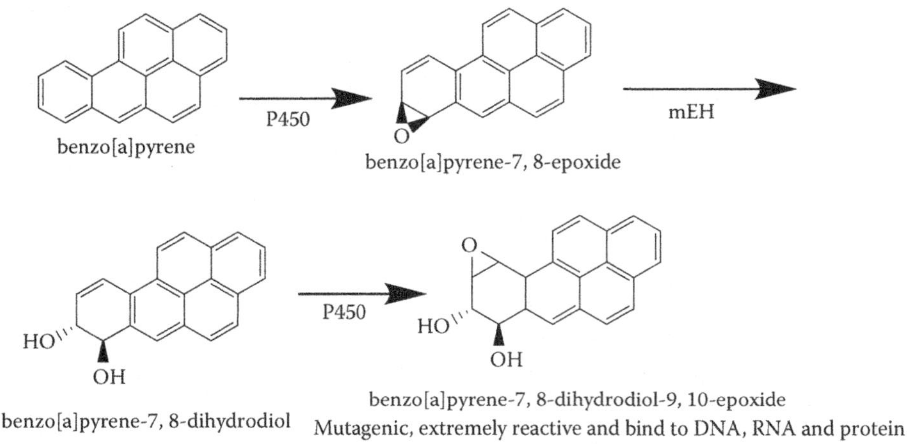

FIGURE 2.3 Metabolic activation of benzo[*a*]pyrene. mEH, microsomal epoxide hydrolase. (Adapted from Kim et al., *Carcinogenesis*, 19(10), 1847–1853, 1998.)

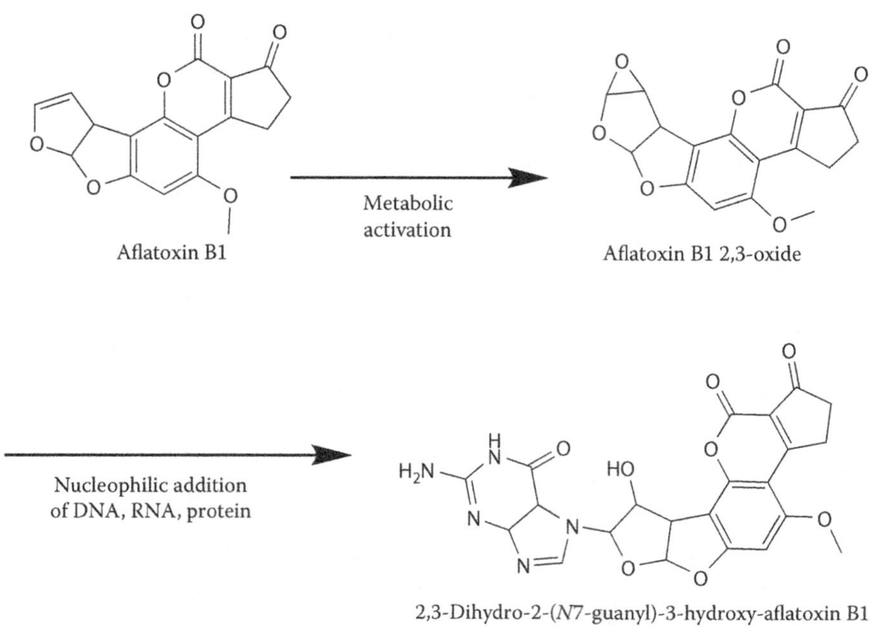

FIGURE 2.4 DNA adduct formation of aflatoxin B1 via metabolic activation. (Adapted from Croy et al., *Proc. Natl. Acad. Sci. USA*, 75(4), 1745–1749, 1978.)

shown in Figure 2.4, covalently binding to DNA. Aflatoxin is one of the three compounds, together with *N*-nitroso and azoxy compounds, named as high-potency genotoxic carcinogens in the European Medicines Agency [6] and Food and Drug Administration [7] guidelines on genotoxic impurities.

Aristolochic acids are a family of structurally related nephrotoxic and carcinogenic nitrophenanthrene compounds found in Aristolochia herbal plants. Some of the Aristolochia herbal plants have been used worldwide, especially in Asia, for medical purposes for over 2000 years [8]. Exposure to aristolochic acids has been linked to nephropathy (kidney disease) and urothelial cancer of the upper urinary tract. Aristolochic acids are enzymatically metabolized via the reduction of the nitro group to produce the reactive *N*-hydroxyaristolactam. The *N*-hydroxyaristolactams undergo the postulated nitrenium intermediates, which covalently bind to the exocyclic amino groups of the purine nucleobases, as shown in Figure 2.5.

Exposure to aromatic amines has been linked to human urinary bladder cancer. Aromatic amines, and any other structures that can be metabolized into aromatic amines, are readily converted into the extremely electrophilic nitrenium ion, which in turn forms covalent adducts with biological macromolecules such as DNA [9,10]. The major metabolic pathway of aniline is shown in Figure 2.6. The first step involves N-oxidation to *N*-hydroxyarylamines by P450, followed by esterification by enzymes such as *O*-acetyltransferase, phosphotransferase, and sulfotransferase. On leaving the corresponding ester group, a reactive electrophilic intermediate, nitrenium ion, is formed, which is the ultimate carcinogen. Subsequently, the nitrenium ion reacts with a macromolecular nucleophile such as DNA to form covalent adducts resulting in genetic mutation and/or cancer.

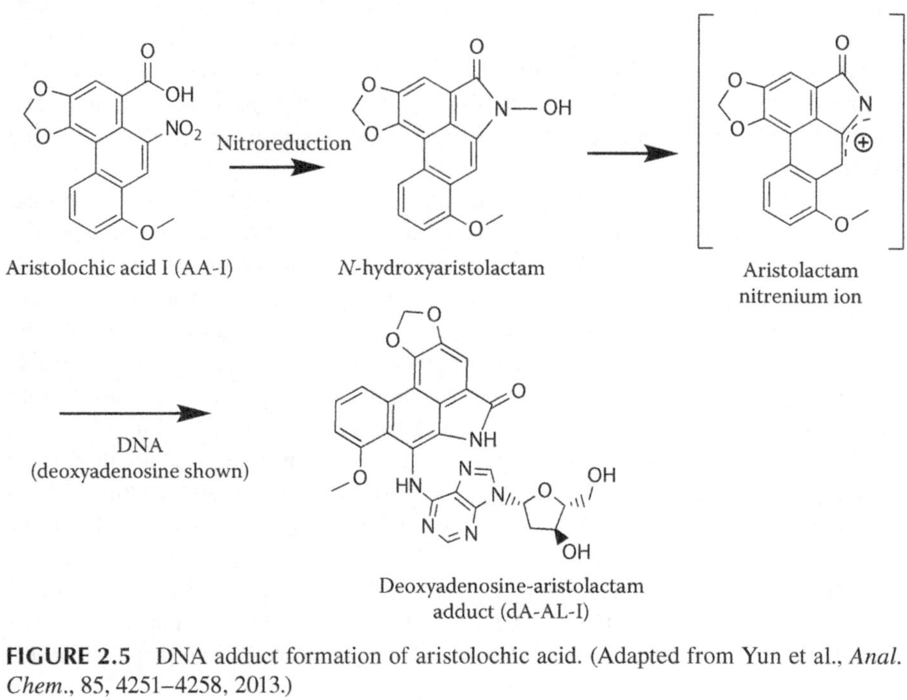

FIGURE 2.5 DNA adduct formation of aristolochic acid. (Adapted from Yun et al., *Anal. Chem.*, 85, 4251–4258, 2013.)

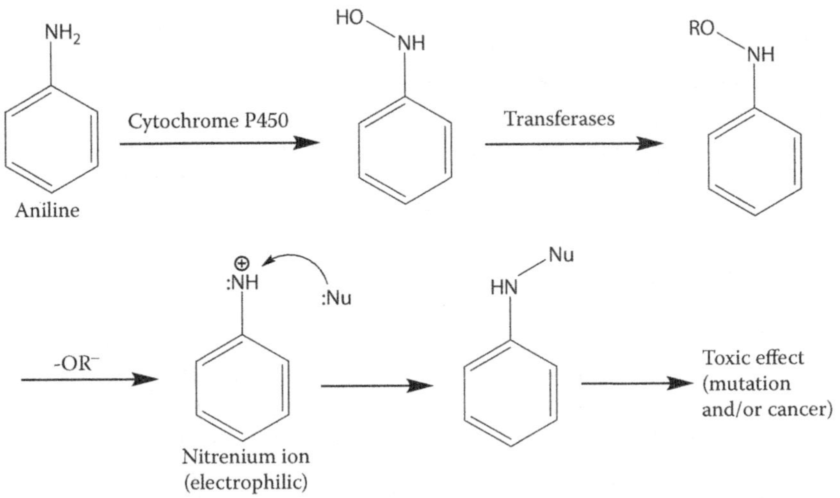

FIGURE 2.6 Metabolic activation of aniline (aromatic amines) to form an electrophilic nitrenium ion. (Adapted from Benigni et al., *Chem. Rev.*, 100, 3697–3714, 2000; Enoch and Cronin, *Crit. Rev. Tox.*, 40(8), 728–748, 2010.)

These four examples clearly illustrate the significance of metabolic activation in the induction of mutation and tumors by most chemical carcinogens.

Experimental determination of chemical carcinogenesis involves time-consuming, labor-intensive, and expensive animal testing, and experimental testing data on carcinogenicity are available for only a relatively small portion of chemicals in commerce today. Therefore, there is a strong demand to develop general prediction models that can be used to predict carcinogenicity for chemicals, for which no experimental data are available and possibly will not be available any time soon.

The relationship of the chemical structure of a molecule and its biological activity is called the "structure–activity relationship" (SAR). In cases where the biological activity is toxic effects, the toxic properties are often related to chemical structures, specifically, to particular substructures. These particular substructures are generally identified as toxicophores (or structural alerts). In 1985, Ashby [11] published a hypothetical compound structure with chemical functionalities (structural alerts) that may contribute to potential carcinogenicity, to demonstrate the relationship of chemical structures and biological activities. The structure, famously known as Ashby's polycarcinogen, is shown in Figure 2.7 adapted from the studies of Ashby and others [11–14] with the list of known structural alerts included in the polycarcinogen. This was the first graphical display of known structural alerts to carcinogenicity.

Cramer and others [15] developed a decision tree to estimate toxic hazard of chemicals based on information including SARs, metabolic mechanisms, chemical reactivity, and human exposure. Chemicals are categorized into three structural

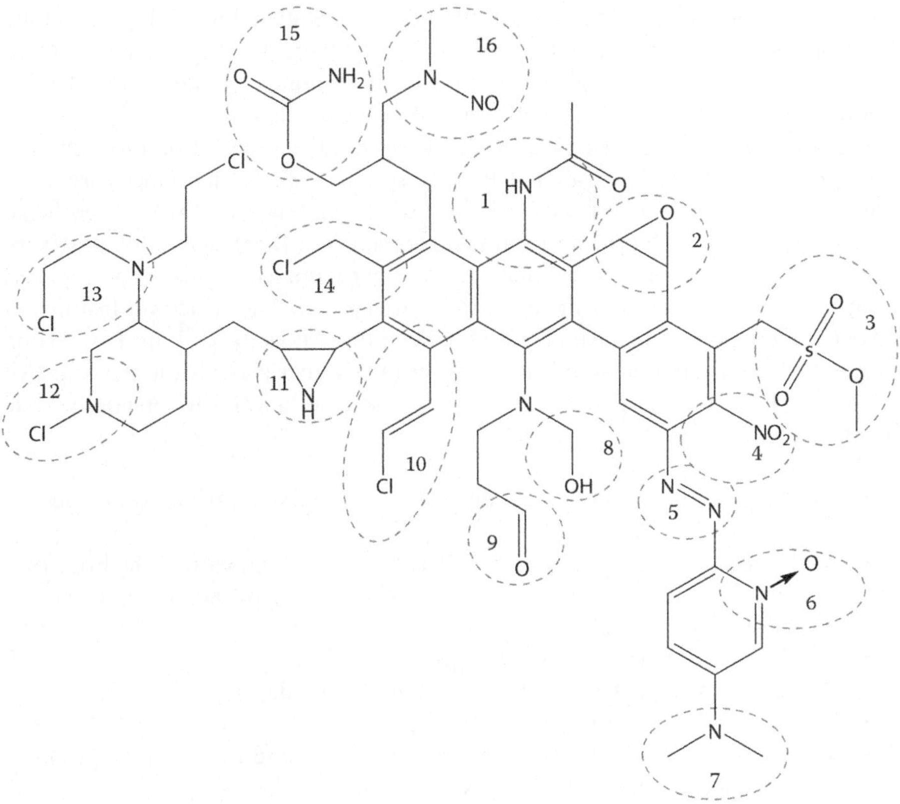

FIGURE 2.7 Polycarcinogen: structural alerts to potential carcinogenicity.

1. Aryl amines
2. Ring epoxides
3. Alkane sulfonate esters
4. Aryl nitro groups
5. Azo groups
6. Ring *N*-oxides
7. Dimethyl amines
8. Methylols
9. Aliphatic aldehydes
10. Ring vinyl groups
11. Aziridine groups
12. Chloramines
13. Nitrogen mustards
14. Benzyl halides
15. Alkyl urethanes
16. Alkyl nitrosamines

(From Ashby and others, see references 11 through 14.)

classes based on 33 questions. The 33 questions are tabulated in Table 2.1, and the decision flow occurs according to "yes" or "no" answers. Cramer emphasized that the decision tree and classification was intended to be used as a guide to the acquisition of toxicity/safety data, but not as a substitute for the data.

Class I (low-toxicity) substances are those with a low order of oral toxicity and are a low priority for investigation. Class III (serious-toxicity) substances are those that permit no strong initial presumptions of safety, or may even suggest significant toxicity, deserving high priority for investigation. Class II (moderate-toxicity) substances are simply intermediates that are less clearly innocuous than those of class I substances but do not offer the basis to be in class III. The structural features of class III substances indicative of a high potential for toxicity [16] are listed here. Metabolic bioactivation is an inherent part of the Cramer classification as many of these class III structures undergo metabolic bioactivation to form potentially toxic chemical entities:

- Aliphatic secondary amino-, cyano-, *N*-nitroso-, diazo-, triazeno-, quaternary nitrogen
- Unionized substituents containing elements other than carbon, hydrogen, oxygen, nitrogen, or sulfur (divalent), for example, halogeno-compounds
- Safrole-like compounds
- Fused lactone or α,β-unsaturated lactone
- Three-membered heterocyclics, for example, epoxides
- Unsubstituted heteroaromatic compounds
- Three or more different functional groups (excluding the methoxy group and considering acids and esters as one group)
- Unsubstituted aromatic hydrocarbons
- Compounds without a strongly anionic group for every 20 (or fewer) carbon (C) atoms (for compounds not classified at earlier steps)

Another effort in the literature to predict the genotoxicity of chemicals by structure classification was demonstrated by the general toxicophore concept for mutagenicity. Mutagenicity is the ability of a compound to cause mutations in DNA. A compound can be mutagenic because of its direct reactivity toward DNA, reactive metabolite formation by the metabolic activation of a nonreactive compound, or intercalation. Intercalation is a reversible, noncovalent process in which a compound (such as PAH) inserts itself between and parallel to base pairs of the DNA double helix, forming stable π-stacking interactions. Kazius and others [17] published a list of eight general toxicophores for mutagenicity prediction by a substructure search of a data set. A data set of over 4000 molecular structures with corresponding Ames test data (both mutagens and nonmutagens) was evaluated by a substructure search. It showed that most mutagens were detected by applying only eight general toxicophores with a satisfactory prediction rate. The names, substructures, and examples of the eight toxicophores are shown in Table 2.2. A more specific final set 29 toxicophores was proposed in the report that could be applied to risk assessment processes.

Structural alerts were described by Benigni and others [18] as molecular substructures or reactive groups that are related to carcinogenic and mutagenic properties

TABLE 2.1
Questions Asked in the Decision Tree Approach for Estimation of Toxic Hazard

No.	Question	Comment	If "No," Proceed to	If "Yes," Proceed to
1	Is the substance a normal constituent of the body (F) or an optical isomer of such?	This question throws into class I all normal constituents of body tissues and fluids, including normal metabolites. Hormones are excluded, as are, by implication, the metabolites of environmental and food contaminants or those resulting from disease states.	2	I
2	Does the substance contain any of the following functional groups: an aliphatic (A) secondary amine or a salt thereof, cyano, *N*-nitroso, diazo, triazeno ($RN{=}NNH_2$), or quaternary nitrogen, except in any of the following forms: $>C{=}N^+R_2$, $>C{=}N^+H_2$, or the hydrochloride or sulfate salt of a primary or tertiary amine?	Questions 2, 3, and 4 are a means of placing in class III those structures that contain elements or valency states often associated with enhanced toxicity. Halo-, nitro-, *N*-nitroso-, and diazo-compounds; organophosphates; quaternary nitrogen compounds (but note modifying comments); and similar xenobiotic structures should cause "yes" answers to questions 2 and 3 and a "no" answer to question 4.	3	**III**
3	Does the structure contain elements other than carbon, hydrogen, oxygen, nitrogen, or divalent sulfur?		5	4
4	Do all elements not listed in question 3 occur only as (a) a sodium, potassium, calcium, magnesium, or ammonium salt of a carboxylic acid; or (b) a sulfate or hydrochloride of an amine; or (c) a sodium, potassium, or calcium sulfonate, sulfamate, or sulfate? (If the answer is "yes," treat as the free acid, amine, unsulfonated, or unsulfated compound, except for the purposes of questions 24 and 33, and proceed.)	This is intended to let through, for further consideration, certain acid, amine, sulfonate, and sulfate salts. Sulfamate salts are treated as such because they are not readily hydrolyzed.	**III**	7
5	Is it a simply branched (I) acyclic aliphatic (A) hydrocarbon or a common carbohydrate?	This drops out the generally innocuous hydrocarbons and carbohydrates.	6	**I**

(Continued)

TABLE 2.1 (*Continued*)
Questions Asked in the Decision Tree Approach for Estimation of Toxic Hazard

No.	Question	Comment	If "No," Proceed to	If "Yes," Proceed to
6	Is the substance a benzene derivative bearing substituents consisting only of (a) hydrocarbon chains or 1′-hydroxy or hydroxy ester-substituted hydrocarbon chains and (b) one or more alkoxy groups, one of which must be para to the hydrocarbon chain in (a)?	This places in class III safrole, myristicin, and related substances.	7	**III**
7	Is the substance heterocyclic?		16	8
8	Is it a lactone or a cyclic diester?	This question separates the lactones and cyclic diesters from other heterocyclic compounds.	10	9
9	Is it a lactone fused to another ring, or a five- or six-membered α,β-unsaturated lactone?	This places certain lactones known or suspected to be of unusual toxicity in class III.[a] If it is a lactone, from this point on treat the structure as if it were the hydroxy acid in the form of its more stable tautomer and proceed to question 20 if it is open chain, to 10 if it is heterocyclic, and to 23 if it is carbocyclic; if it is a cyclic diester, treat as the separate components.	[a]	**III**
10	Is it a three-membered heterocycle?	This places substances such as epoxides and ethylenimine in class III.	11	**III**

11	Disregarding only the heteroatoms in any one ring, does that heterocyclic ring contain or bear substituents other than simply branched (I) hydrocarbons (including bridged chains and monocyclic aryl or alkyl structures), alkyl alcohols, aldehydes, acetals, ketones, ketals, acids, esters (including cyclic esters other than lactones), mercaptans, sulfides, methyl ethers, or hydroxy or single rings (hetero or aryl) with no substituents other than those just listed?	Questions 11–15 separate out various categories of heteroaromatic substances. Under 11, set aside and do not consider the atoms, usually oxygen, nitrogen, or sulfur, making the ring heterocyclic. If there is more than one hetero ring regard each ring separately, with the remainder of the structure as substituents of that hetero ring. Other than the heterocyclic atoms, does the ring carry anything besides the simple groups listed? If so, the answer is "yes" and the next question is 33. If not, then classify further by question 12 et seq. Bridged-chain derivatives may be represented by structures like the bicyclic ether 1,4-cineole, whereas monocyclic aryl derivatives may be represented by compounds like benzaldehyde, propylene glycol, acetal, or 3-phenyl-2-furancarboxaldehyde.	12	33
12	Is it heteroaromatic (B)?	This question separates the aromatic heterocyclics for the purpose of considering whether they are polynuclear (question 14) or unsubstituted (question 13).	22	13
13	Does the ring bear any substituents?		**III**	14
14	Does the structure contain more than one aromatic (B) ring?		22	15
15	Is it readily hydrolyzed (H) to mononuclear residues? (If yes, treat the mononuclear heterocyclic residues by question 22 and any carbocyclic residue by question 16.)		33	22
16	Is it a common terpene (D)-hydrocarbon, -alcohol, -aldehyde, or –carboxylic acid (not a ketone)?	Questions 16 and 17 deal with terpenes. A hydrocarbon terpene that is a common terpene (D) and has not already been put in class I by question 5 would go into class I by question 16.	17	**I**

(Continued)

TABLE 2.1 (*Continued*)
Questions Asked in the Decision Tree Approach for Estimation of Toxic Hazard

No.	Question	Comment	If "No," Proceed to	If "Yes," Proceed to
17	Is the substance readily hydrolyzed (H) to a common terpene (D), -alcohol, -aldehyde, or –carboxylic acid? (If the answer is "yes," treat the hydrolyzed residues separately and proceed to 18 for the terpene moiety and to 19 for any nonterpenoid moiety.)	Since there may be substances that are hydrolyzed to two or more residues one of which is a terpene, treat the residues separately from question 18 onward to conclusion.	19	18
18	Is the substance one of the following: (a) a vicinal diketone, or a ketone or ketal of a ketone attached to a terminal vinyl group (b) a secondary alcohol or ester of a secondary alcohol attached to a terminal vinyl group (c) an allyl alcohol or its acetal, ketal, or ester derivative (d) an allyl mercaptan, allyl sulfide, allyl thioester, or allyl amine (e) a crolein, a methacrolein, or their acetals (f) an acrylic or a methacrylic acid (g) an acetylenic compound (h) an acyclic aliphatic (A) ketone, ketal, or ketoalcohol with no other functional groups and with four or more carbons on either side of the keto group (i) a substance in which the functional groups (E) are all sterically hindered (J)	Question 18 examines the terpenes (and later the open-chain and mononuclear substances by reference) to determine whether they contain certain structural features generally thought to be associated with some enhanced toxicity.	I	II

19	Is the substance open chain (G)?	Questions 19–21 deal with open-chain substances.	23	20
20	Is the structure a linear or simply branched (I) aliphatic (A) compound containing any one or a combination of only the following functional groups (E): (a) four or less, each, of alcohol, aldehyde, carboxylic acid, or esters and/or (b) one each of one or more of acetal, either ketone or ketal but not both; mercaptan; sulfide (mono or poly); thioester; polyoxyethylene [(–OCH_2CH_2–)$_x$, with x not greater than 4]; or primary or tertiary amine?	This question should be answered "yes" if the structure contains one or any possible combination of alcoholic, aldehydic, or carboxylic acid or ester groups, provided there are no more than four of any one kind. It should be answered "yes" if the structure contains in addition to, or instead of, those just listed any assortment of no more than one each of the following: acetal, either ketone or ketal but not both; mercaptan; mono- or polysulfide; thioester; polyoxyethylene; or primary or tertiary amine. Answer the question "no" if the structure contains more than four of any of the first set of groups, more than one of the second set, or any substituent not listed.	22	21
21	Does the structure contain three or more different types of functional groups (exclude methoxy and consider acids and esters as one functional type)?	Aliphatic (A) compounds containing three or more different functional groups (excluding methoxy) are too complex to permit a satisfactory prediction of toxicity. They should go, therefore, into class III. However, we do not wish to put into class III polyesters and similar substances, so these and the methoxy compounds get passed along to question 18.	18	**III**
22	Is the substance a common component of food (C) or structurally closely related (K) to a common component of food (C)?	This question places in class II the natural, nature-identical, and nearly nature-identical substances not already put into class I by physiological occurrence or structural criteria. An artificial (i.e., non-nature-identical) substance, or one not closely related, goes to question 33.	33	**II**
23	Is the substance aromatic (B)?	Questions 23–26 deal with alicyclic substances.	24	27
24	Is the substance monocarbocyclic (excluding cyclopropane or cyclobutane and their derivatives) with ring or aliphatic (A) side chains, unsubstituted or containing only alcohol; aldehyde; side-chain ketone; acid; ester; or sodium, potassium, or calcium sulfonate or sulfamate; or acyclic acetal or ketal?		25	18

(Continued)

TABLE 2.1 (*Continued*)
Questions Asked in the Decision Tree Approach for Estimation of Toxic Hazard

No.	Question	Comment	If "No," Proceed to	If "Yes," Proceed to
25	Is the substance (a) a cyclopropane or cyclobutane with only the substituents mentioned in question 24 or (b) a mono- or bicyclic sulfide or mercaptan?		26	**II**
26	Does the structure contain no functional groups other than those listed in question 24, and is it either a monocycloalkanone or a bicyclic compound with or without a ring ketone?		22	**II**
27	Do the rings have any substituents?	Questions 27–31 deal with aromatic compounds.	**III**	28
28	Does the structure contain more than one aromatic (B) ring?		30	29
29	Is it readily hydrolyzed (H) to mononuclear residues? (If yes, treat the individual aromatic mononuclear residues by question 30 and any other residue by question 19.)		33	30
30	Disregarding ring hydroxy or methoxy, does the ring bear substituents other than 1-5-carbon aliphatic (A) groups, either hydrocarbon or containing alcohol, ketone, aldehyde, carboxyl, or simple esters that may be hydrolyzed to ring substituents of five or less carbons (if it is a simple ester that may be hydrolyzed, treat the aromatic portion by question 18 and the residue portion by question 19.)	This should be answered "no" if the ring bears only aliphatic groups of five carbons or less, which either are hydrocarbon in nature or contain the groups listed. If the ring bears any other substituents than those listed, the question should be answered "yes" and one should proceed to question 31.	18	31
31	Is the substance an acyclic acetal, -ketal, or –ester of any of the aforementioned substances (see question 30)? (If yes, assume hydrolysis and treat the nonaromatic residues by question 19 and the aromatic residue by question 18.)	This question is simply designed to see whether the substance would fit within the definition of question 30 if it were not an acetal, a ketal, or an ester. In other words, would the substance carry only the groups listed in question 30?	32	18

32	Does the substance contain only the functional groups (E) listed in question 30, or their derivatives listed in question 31, but with any or all of the following: (a) a single fused nonaromatic carbocyclic ring, (b) aliphatic (A) substituent chains longer than five carbon atoms, or (c) a polyoxyethylene [$(-OCH_2CH_2-)_x$ with x not greater than 4] chain either on the aromatic ring or on an aliphatic (A) side chain?	Part (a) is intended to allow simple derivatives of tetralin into class II while putting polycyclic compounds such as the steroids ultimately into class III except those that may be normal food components. Part (b) allows compounds with permitted functional groups but longer side chains into class II instead of sending them eventually to class III. Part (c) puts short-chain polyoxyethylene derivatives of aryl compounds into class II rather than class III.	22	**II**
33	Does the substance bear on every major structural component at least one sodium, potassium, or calcium sulfonate or sulfamate for every 20 or fewer carbon atoms without any free primary amines except those adjacent to the sulfonate or sulfamate?	Sodium, potassium, and calcium sulfonate and sulfamate salts have a strong tendency to decrease toxicity by promoting solubility and rapid excretion. This is particularly noticeable, e.g., with some of the food colorings. It is important that the substance bears sufficient sulfonate groups, including one on each of the major structural fragments into which the original compound might be metabolized. This question serves to steer sulfonated compounds except those with amines nonadjacent to the sulfonate into a presumptively less toxic classification than what the compounds would occupy if unsulfonated.	**III**	**I**

Source: Cramer and Ford, *Food Cosmet. Toxicol.*, 16, 255–276, 1978.

Note: This approach categorizes compounds into class I (low), class II (moderate), and class III (serious) toxicity.

(A) Aliphatic includes olefinic and polyolefinic, but not acetylenic or alicyclic, compounds.

(B) Aromatic means that the substance has at least one benzene, furan, thiophene, pyridine, or pyrrole ring, however substituted and whether or not fused to another ring.

(C) Common component of food. In something as diverse, changing, and occasionally uncertain as natural occurrence, it is only possible to define a guideline, not a firm rule. For this decision tree, the term common component of food denotes a substance that has been reported in the recognized literature as occurring in significant quantity (~50 ppm or more) in at least one major food, or in trace quantities at the parts-per-million level or less in several foods, including minor or less frequently consumed foods. The latter include spices, herbs, and ethnic specialties. This definition excludes natural or man-made contaminants, and hormones.

(Continued)

TABLE 2.1 (*Continued*)
Questions Asked in the Decision Tree Approach for Estimation of Toxic Hazard

(D) Common terpene means an isoprenoid compound (the carbon skeleton made up of two or more 5-carbon isoprene units) reported in the literature as a more than trace constituent of two or more generally consumed foods, either raw or as ordinarily prepared for consumption, without added ingredients.

(E) Functional group is a portion (sometimes called a radical) of an organic molecule consisting of a combination of atoms of two or more elements (at least one of which is not hydrogen or carbon) and causing the molecule to exhibit a characteristic set of reactions. For the purpose of this classification, this definition excludes C=C double bonds and aromatic rings.

(F) Normal constituent of the body means any systemic constituent present at a normal physiological level, whether free or combined, except hormones. This includes essential nutrients and major food constituents and the physiologically normal metabolites of each. It excludes transitory substances present only as a result of (a) trace constituents of food, (b) gut contents, or (c) products of the actions of the gut flora.

(G) Open chain means the absence of any ring structure.

(H) Readily hydrolyzed means known to be or, in the absence of any contraindication based on structure, assumed to be hydrolyzed either during food preparation or by physiological processes after consumption.

(I) Simply branched means branched at C–C bonds, with branches of two or more carbon atoms, at no more than two points along the main chain, with no secondary branching. Multiple branching, consisting only 1-carbon moieties, falls within this definition of simply branched.

(J) Sterically hindered means posing steric hindrance to a functional group equivalent to or greater than that exhibited by *o*-*tert*-butyl or 2,6-disubstitution on an aromatic ring.

(K) Structurally closely related means (a) a member, not more than two carbon atoms removed, in a homologous series, except ethoxy and higher homologues of a methoxy compound; (b) a primary alcohol and its aldehyde, or either of these and the corresponding carboxylic acid; (c) a secondary alcohol and the corresponding ketone; (d) structural isomers unlikely to involve steric hindrance of a functional group; (e) compounds with the same functional groups, the remainder of the compounds being known to be easily and harmlessly metabolized; and (f) non–sterically hindered esters, thioesters, acetals, ketals, and their components.

TABLE 2.2
Eight General Toxicophores for Mutagenicity Prediction

Toxicophore Name	Substructure	Example
Aromatic nitro	Ar—NO_2	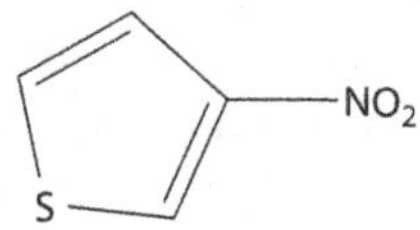
Aromatic amine	Ar—NH_2	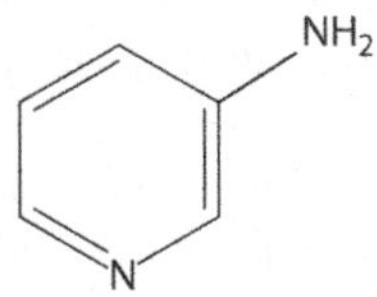
Three-membered heterocycle	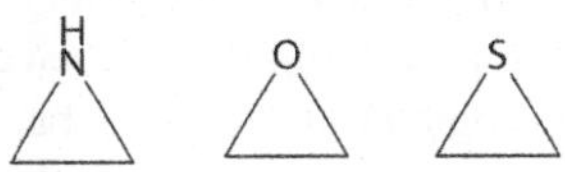	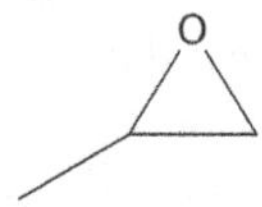
Nitroso	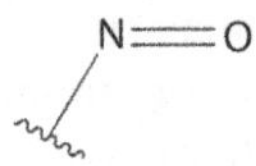	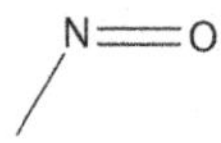
Unsubstituted heteroatom-bonded heteroatom	NH_2, OH | N, O	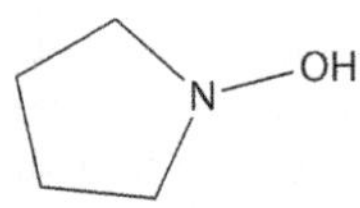
Azo type	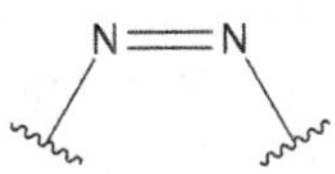	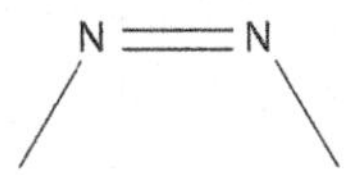
Aliphatic halide	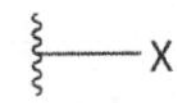	
Polycyclic aromatic system	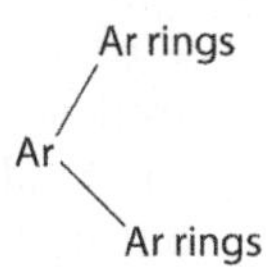	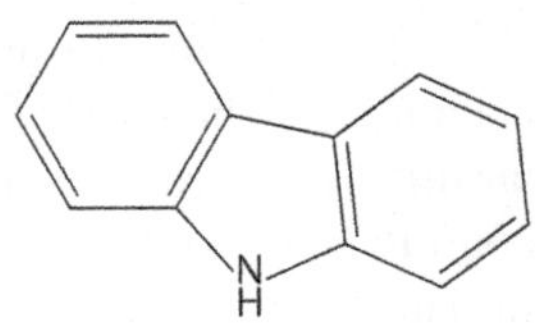

Source: Kazius et al., *J. Med. Chem.*, 48, 312–320, 2005.
Note: Ar, aromatic; Ar rings, part of multiple aromatic rings; X, halide (Cl, Br, or I).

of the chemicals, and as a representation of "codification" aimed at highlighting the mechanism of action of the mutagenic and carcinogenic chemicals. The identification of the structural alerts has facilitated the understanding of mechanisms and assessment of the risk posed by chemicals. The carcinogens are classified into either genotoxic carcinogens or epigenetic carcinogens from the point of view of the mechanism of action. Genotoxic carcinogens cause damage directly to DNA, and many known mutagens are in this category with mutation being one of the first steps in cancer development. In contrast, epigenetic carcinogens do not bind covalently to

DNA, do not cause direct DNA damage, and are usually negative in standard mutagenicity assays. Genotoxic carcinogens have a unifying feature of electrophilicity per se or via metabolic activation. With this electrophilicity hypothesis [1,2], several chemical functional groups and substructures (structural alerts) were identified for genotoxic carcinogens based on mechanisms of action and metabolic fate. However, the recognition of structural alerts for nongenotoxic carcinogens is not well established due to the lack of a unifying theory.

Structural alerts for mutagenicity and carcinogenicity, known as the Benigni/Bossa rulebase, with corresponding structures are listed in Table 2.3 [18–20]. Each of the structural alerts is a code for a well-characterized chemical class, with its own specific mechanism of action.

The chemical category principle states that similar chemicals should have similar toxicological profiles enabling placement into a chemical category. Such categories can be used to predict a range of toxicological end points for chemicals for which no toxicological data exist. The key step is the ability to select similar chemicals, and one of the most powerful methodologies to this end is based on a common mechanism of action. The term mechanism of action refers to the chemical mechanism by which a chemical forms a covalent adduct with a biological macromolecule, such as DNA, RNA, or proteins. The information on mechanisms of action needs to be defined through chemical structures. This is the underlying principle of chemical carcinogenesis: "chemical structure defines toxicity."

Covalent adduct formation has been defined as the molecular initiating event, an event that is the first step in a series that can ultimately lead to a toxic response [9]. Although a number of biological steps are required for a toxic response, the molecular initiating event must occur to initiate the remaining steps in the adverse outcome pathway. The chemistry controlling covalent adduct formation could be utilized to construct chemical categories. The knowledge of well-established mechanistic organic chemistry can be used to define the structural requirements of an exogenous chemical to form a covalent bond with a biological macromolecule. In this case, the chemical mechanism refers to the type of covalent bond–forming reaction (molecular initiating event) between the exogenous chemical and a biological nucleophile. In contrast, the biological mechanism refers to the entire pathway after the molecular initiating event to a toxic response. This type of mechanistic chemistry analysis has been used in the development of chemical categories.

It is important to understand that a chemical in a mechanistic category with a reactive structural feature, based on the ability to undergo a common molecular initiating event, does not necessarily mean that the chemical is toxic. Other factors including toxicokinetic or toxicodynamic profiles could prevent the completion of the adverse outcome pathway. An example is that aromatic amines with a sulfate substituent attached to the aromatic ring are not mutagenic. This is related to the increased detoxification of the chemicals due to increased solubility and hence lower availability in a cell or in vivo. This so-called mitigating factor [9] is an additional structural feature present within a molecule that removes the toxic effect. In the example of the aromatic amine, the sulfate group is the mitigating factor as its presence within the molecule removes the mutagenic activity.

TABLE 2.3
Structural Alerts for the Rodent In Vivo Micronucleus Assay

Structural Alert	Generic Structure
Acyl halide	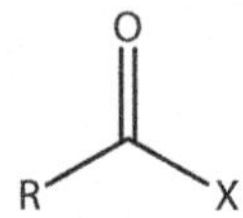
Alkyl (C<5) or benzyl ester of sulfonic or phosphonic acid	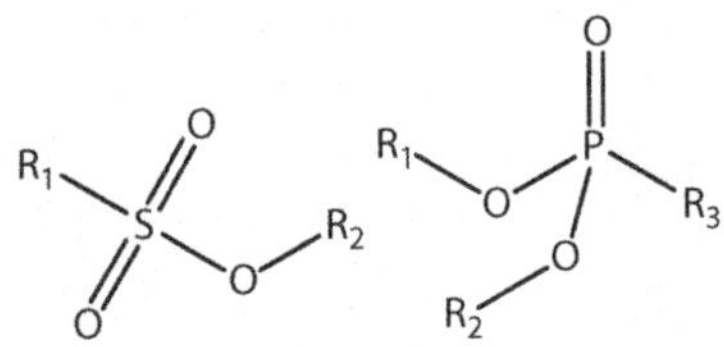
N-methylol derivative	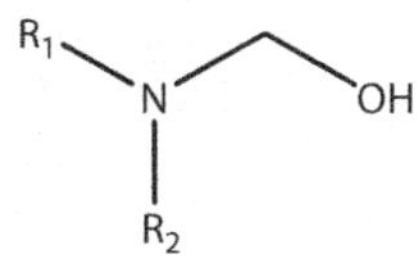
Monohaloalkene	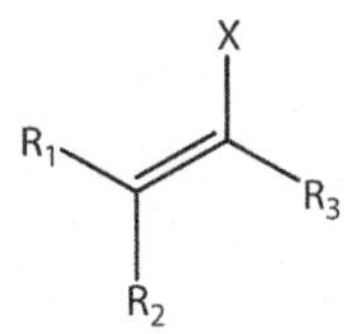
S- or *N*-mustard	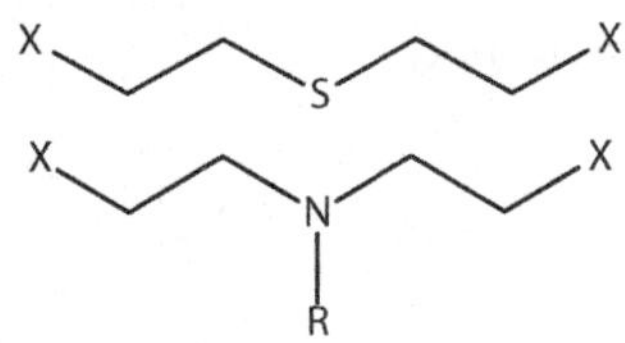
Propiolactone or propiosultone	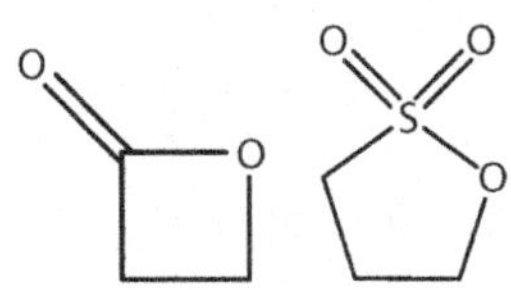
Epoxides and aziridine	
Aliphatic halogens	R—X
Alkyl nitrite	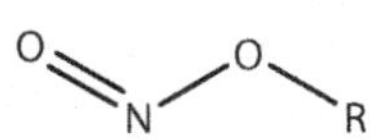

(*Continued*)

TABLE 2.3 (*Continued*)
Structural Alerts for the Rodent In Vivo Micronucleus Assay

Structural Alert	Generic Structure
α,β-Unsaturated carbonyl	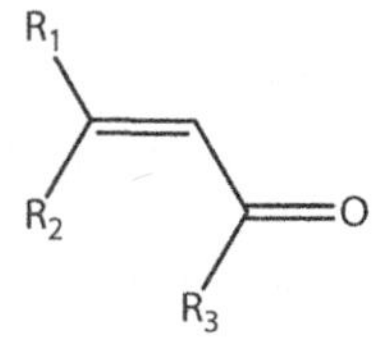
Simple aldehyde	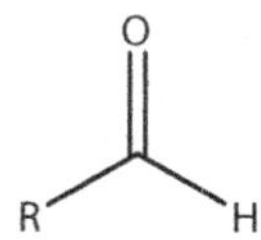
Quinone	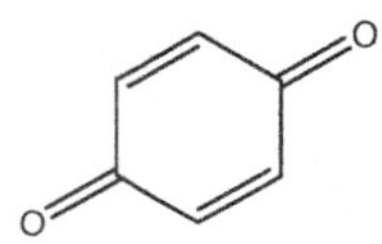
Hydrazine	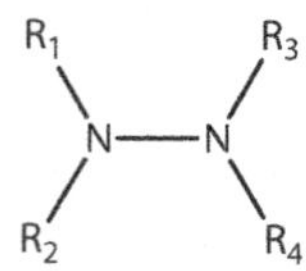
Aliphatic azo and azoxy	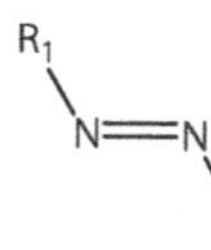 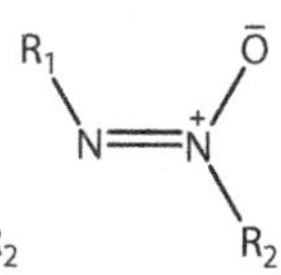
Isocyanate and isothiocyanate	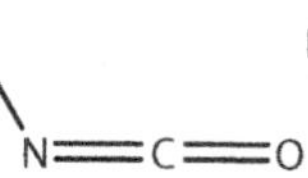 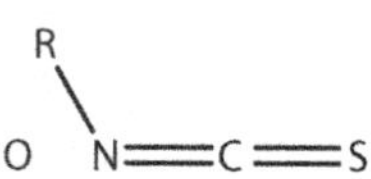
Alkyl carbamate and thiocarbamate	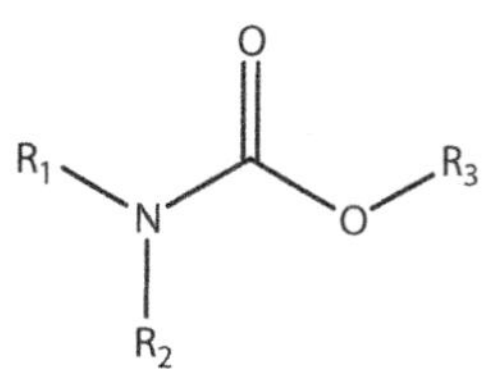 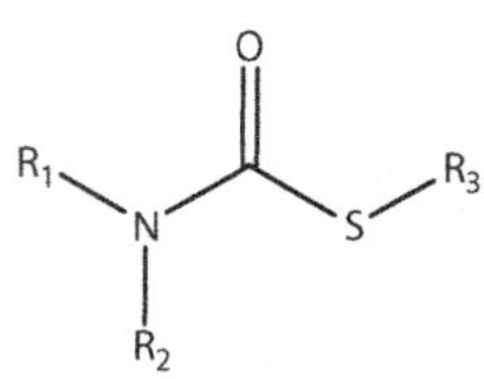 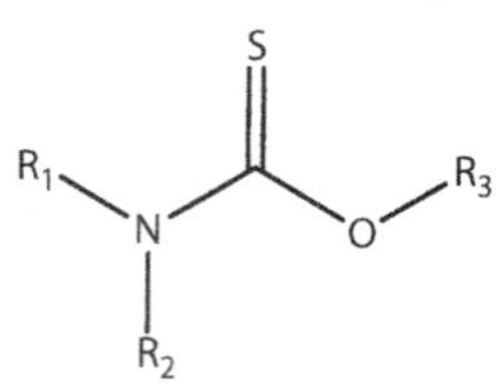

TABLE 2.3 (*Continued*)
Structural Alerts for the Rodent In Vivo Micronucleus Assay

Structural Alert	Generic Structure
Polycyclic aromatic hydrocarbon	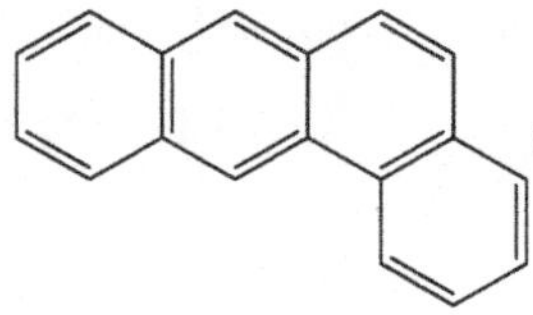 Example
Heterocyclic polycyclic aromatic hydrocarbon	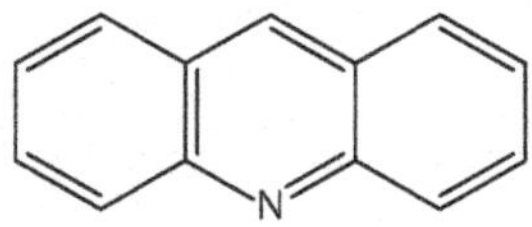 Example
Alkyl and aryl *N*-nitroso	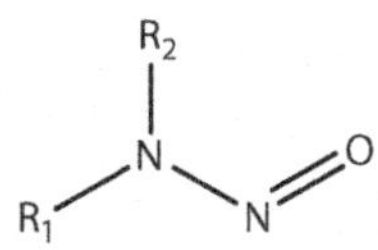
Azide and triazene	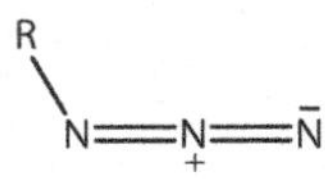 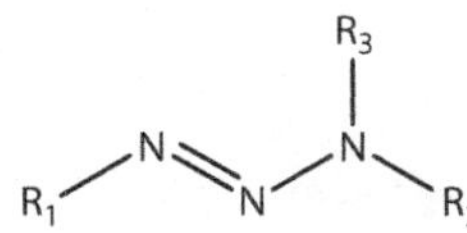
Aliphatic *N*-nitro	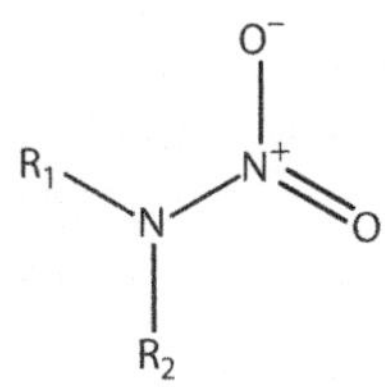
α,β-Unsaturated aliphatic alkoxy	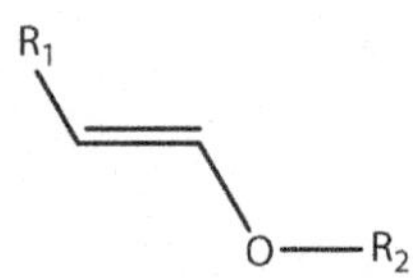
Aromatic nitroso	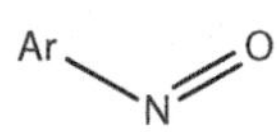
Aromatic ring *N*-oxide	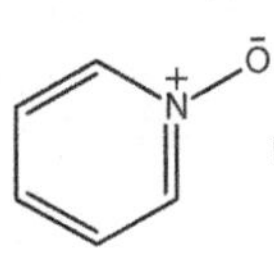 Example
Nitroaromatic	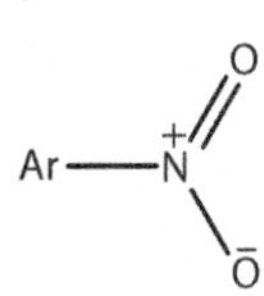

(*Continued*)

TABLE 2.3 (*Continued*)
Structural Alerts for the Rodent In Vivo Micronucleus Assay

Structural Alert	Generic Structure
Primary aromatic amine, hydroxy amine, and their derived esters	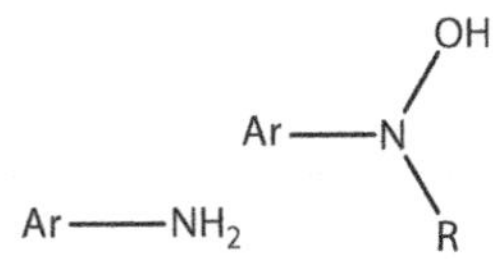
Aromatic mono- and dialkylamine	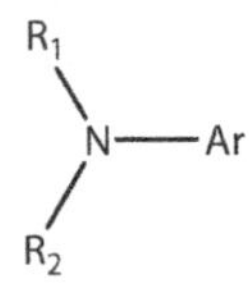
Aromatic *N*-acyl amine	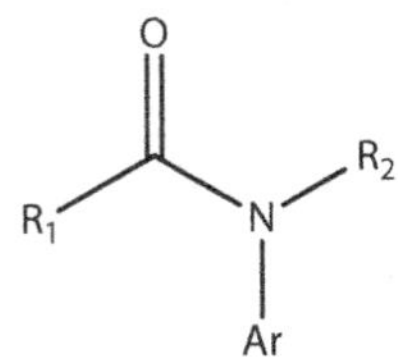
Aromatic diazo	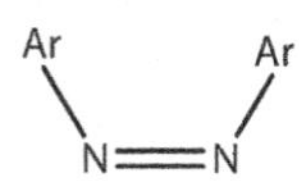
Coumarin and furocoumarin	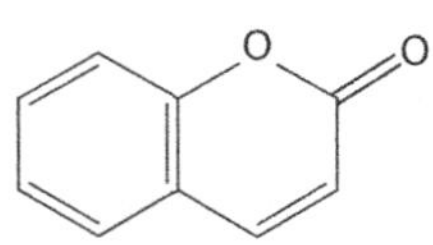 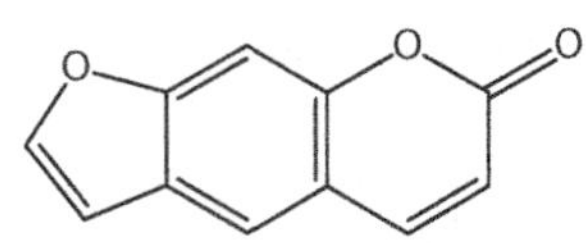
1,3-Dialkoxy benzene	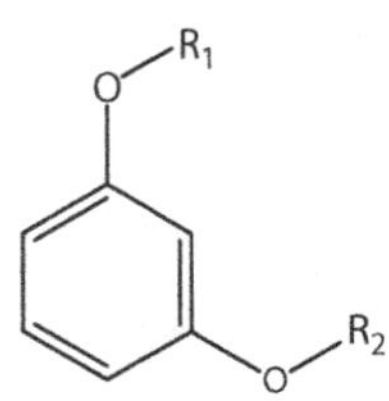
1-Phenoxy benzene	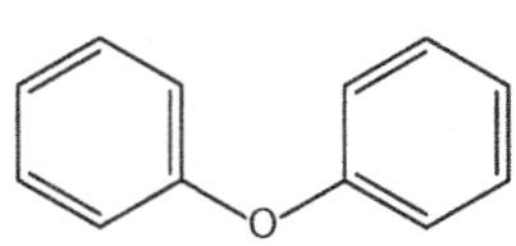
H acceptor-path 3 -H acceptor (two hydrogen acceptors at a distance of 3)	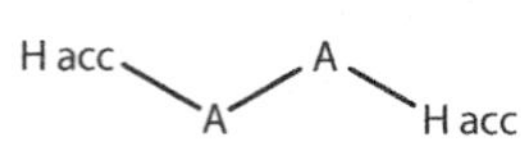

TABLE 2.3 (*Continued*)
Structural Alerts for the Rodent In Vivo Micronucleus Assay

Structural Alert	Generic Structure
Oxolane	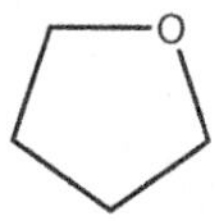
Carbodiimide	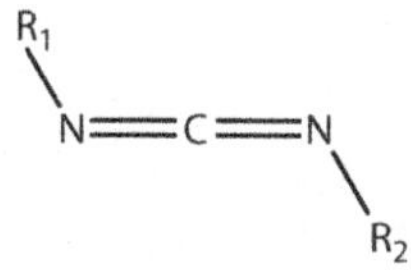

Source: Benigni and others, see references 14 through 16.
Note: R, H, alkyl or generic group; X, halogen; Ar, aromatic ring; H-acc, hydrogen bond acceptor; A, any atom except hydrogen.

The terms "mechanism of action" and "mode of action" were defined by Aptula and others [21] with respect to a defined domain of applicability, that is, identification of the range of compounds for which a quantitative structure–activity relationship (QSAR) can confidently be applied for the purposes of toxicity prediction. In this context, the chemical mechanism of action was defined as "what the toxic chemical and the biological chemicals do together in vivo" and the biological mechanism of action comprises the events that occur at the cellular or subcellular level subsequently. The mode of action was defined as "what happens to the organism." In the example of skin sensitization, the reaction of electrophiles with nucleophilic groups on proteins (e.g., formation of antigenic modified proteins and Michael adducts) is the chemical mechanism of action. The biological mechanism of action is the expression of antigenic modified proteins by Langerhans cells followed by T cell proliferation in lymph nodes. The mode of action is the sensitization state of the individual. The underlying principles are as follows: (1) the compound and the target organism together obey the laws of chemistry (including those that we may not know about), (2) the laws of chemistry determine the chemical mechanism of action, (3) the chemical mechanism of action determines the biological mechanism of action, and (4) the biological mechanism of action determines the mode of action. In this example, the distinction between mode and mechanism is made: skin sensitization is a single mode of action phenomenon with several mechanisms of action according to how the electrophiles bind to protein.

Seven reaction mechanistic domains were reported [9,21] that categorize the mechanistic chemistry of molecular initiating events, which occur when an exogenous chemical binds covalently to biological macromolecules. These seven reaction mechanistic domains are shown in Figure 2.8. These classifications are based on organic reaction mechanistic principles and were developed with the purpose

Michael addition (MA) X = electron-withdrawing substituent	
S_NAr (SNAr) (aromatic nucleophilic substitution) X = halogen or pseudohalogen; Y = (at least two) NO_2, CN, CHO, CF_3, halogen	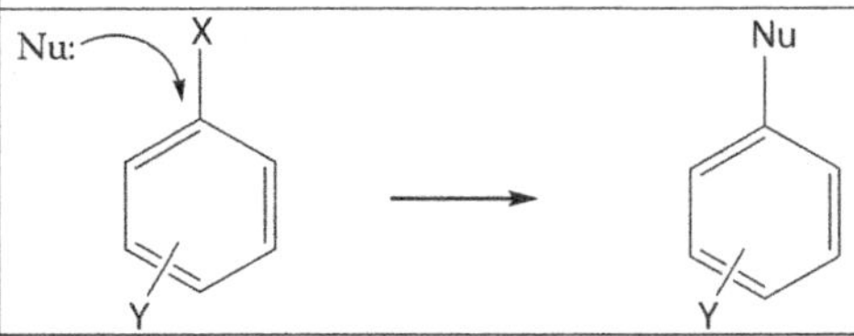
S_N2 (SN2) (bimolecular aliphatic nucleophilic substitution) X = halogen or other electronegative leaving group	
Schiff base formers (SB) Reactive carbonyl species such as aliphatic aldehyde or diketones	
Acylating agents (AC) X = halogen or electronegative leaving group	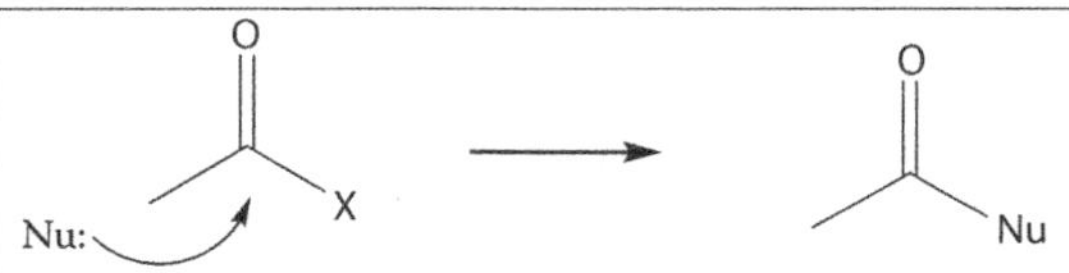
S_N1 (SN1) (unimolecular aliphatic nucleophilic substitution) Ability to form a stabilized positive charge, typically on a carbon or nitrogen	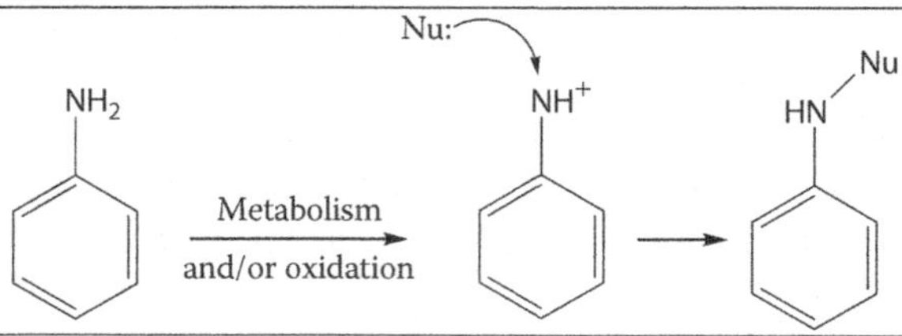
Radical (Rad)	No general scheme

FIGURE 2.8 General principles defining the seven reaction mechanistic domains. (Adapted from Aptula and Roberts, *Chem. Res. Toxicol.*, 19, 1097–1105, 2006; Enoch and Cronin, *Crit. Rev. Tox.*, 40(8), 728–748, 2010.)

of applying mechanism-based QSARs to broader ranges of compounds for toxicity prediction. Each of the structure alerts were assigned to one of the seven mechanistic domains. The mechanistic domains and structural alerts enable in silico toxicity risk assessments of a chemical to act as a mutagen or genotoxic carcinogen (genotoxicity) based on the knowledge of the structure.

An important application of the structural alerts concept was the establishment of the threshold of toxicological concern (TTC) with the aim of reducing extensive toxicological evaluation. The concept that there are levels of exposure that do not cause adverse effects is inherent in setting acceptable daily intakes for chemicals with known toxicological profiles. It should be recognized that it is neither possible nor necessary to thoroughly test the toxicity of all chemicals in use or present in the environments or food supply. The TTC approach establishes a generic human exposure threshold value below which there would be no appreciable risk to human health. The underlying principle is that TTC can be identified for many chemicals, including those of unknown toxicity, taking into consideration their chemical structures and the known toxicity of chemicals that share similar structural characteristics [18,16].

Munro and others [3] proposed thresholds of acceptable human exposure based on the Cramer decision tree [15] of structural classes I, II, and III. Kroes and others [16] reported that the TTC principle can be applied for low concentrations in food of chemicals that lack toxicity data, provided that there is a sound intake estimate. They also proposed a decision tree to apply the TTC principle, shown in Table 2.4. Proteins, heavy metals, and polyhalogenated dibenzodioxins and related compounds were excluded from this approach due to the following reasons. The most potent allergens are proteins foreign to the host that are capable of inducing immunologically mediated allergic reactions. It is very difficult to predict and quantify such toxic effects. Many heavy metals are known to accumulate (e.g., cadmium in the kidney), and accumulation needs special consideration when there are major species differences in clearance that exceed the usual uncertainty factor for species differences. Steroids and polyhalogenated dibenzo-*p*-dioxins and dibenzofurans are considered to be nongenotoxic

TABLE 2.4
Decision Tree for Low-Molecular-Weight Compounds with Limited Toxicity Data

Step	Question	If "No"	If "Yes"
1	Is the substance a nonessential metal or metal-containing compound, or is it a polyhalogenated dibenzodioxin, dibenzofuran, or biphenyl?	Go to 2	Risk assessment requires compound-specific toxicity data
2	Are there structural alerts that raise concern for potential genotoxicity?	Go to 5	Go to 3
3	Is the chemical an aflatoxin-like azoxy or *N*-nitroso compound?	Go to 4	Risk assessment requires compound-specific toxicity data
4	Does estimated intake exceed a TTC of 0.15 μg/day?[a]	Negligible risk (low probability of a lifetime cancer risk greater than 1 in 10^6)[a]	Risk assessment requires compound-specific toxicity data
5	Does estimated intake exceed a TTC of 1.5 μg/day?	Substance would not be expected to be a safety concern	Go to 6
6	Is the compound an organophosphate?	Go to 8	Go to 7

(Continued)

TABLE 2.4 (*Continued*)
Decision Tree for Low-Molecular-Weight Compounds with Limited Toxicity Data

Step	Question	If "No"	If "Yes"
7	Does estimated intake exceed a TTC of 18 μg/day?	Substance would not be expected to be a safety concern	Risk assessment requires compound-specific toxicity data
8	Is the compound in Cramer structural class III?	Go to 10	Go to 9
9	Does estimated intake exceed 90 μg/day?	Substance would not be expected to be a safety concern	Risk assessment requires compound-specific toxicity data
10	Is the compound in Cramer structural class II?	Go to 12	Go to 11
11	Does estimated intake exceed 540 μg/day?	Substance would not be expected to be a safety concern	Risk assessment requires compound-specific toxicity data
12	Does estimated intake exceed 1800 μg/day?	Substance would not be expected to be a safety concern	Risk assessment requires compound-specific toxicity data

Source: Kroes et al., *Food Chem.* Tox., 42, 65–83, 2004.

[a] These TTC values are developed for food contact substances present at low levels in the diet. A TTC value of 1.5 μg/day intake of a genotoxic impurity is considered to be associated with an acceptable risk for most pharmaceuticals (1 in 10^5 over a lifetime) considering pharmaceutical benefits.

carcinogens, which would show a threshold in their dose–response relationships; therefore, it is irrelevant to the linearized models used to develop the TTC.

The concepts discussed in this chapter include chemical carcinogenesis, electrophilic reactivity of ultimate carcinogens, reaction mechanistic domains, metabolic activation, mechanism of action, mode of action, toxicophores, structural alerts, and the TTC. The goal of this chapter is to succinctly introduce the interconnected concepts to interested readers with a few well-known examples. These topics have been investigated extensively over the decades by numerous researchers, and the disciplines are still evolving with significant advancements being made constantly. All of these subjects are critically important in the field of genotoxicity.

REFERENCES

1. E. C. Miller, "Some current perspectives on chemical carcinogenesis in humans and experimental animals: Presidential address," *Cancer Res*, 38, 1479–1496 (1978).
2. J. A. Miller, "Carcinogenesis by chemicals: An overview—G. H. A. Clowes memorial lecture," *Cancer Res*, 30, 559–576 (1970).
3. I. C. Munro, R. A. Ford, E. Kennepohl, J. G. Sprenger, "Correlation of structural class with no-observed-effect levels: A proposal for establishing a threshold of concern," *Food Chem Tox*, 34, 829–867 (1996).

4. J. H. Kim, K. H. Stansbury, N. J. Walker, M. A. Trush, P. T. Strickland, T. R. Sutter, "Metabolism of benzo[*a*]pyrene and benzo[*a*]pyrene-7,8-diol by human cytochrome P450 1B1," *Carcinogenesis*, 19(10), 1847–1853 (1998).
5. R. G. Croy, J. M. Essigmann, V. N. Reinhold, G. N. Wogan, "Identification of the principal aflatoxin B1-DNA adduct form in vivo in rat liver," *Proc Natl Acad Sci USA*, 75(4), 1745–1749 (1978).
6. Committee for Medicinal Products for Human Use, *Guideline on the Limits of Genotoxic Impurities*, London: European Medicines Agency, January (2007).
7. Center for Drug Evaluation and Research, *Guidance for Industry, Genotoxic and Carcinogenic Impurities in Drug Substances and Products: Recommended Approaches*, Draft Guidance, Food and Drug Administration, December (2008).
8. B. H. Yun, T. A. Rosenquist, J. Nikolic, D. Dragicevic, K. Tomic, B. Jelakovic, K. G. Dickman, A. P. Grollman, R. J. Turesky, "Human formalin-fixed paraffin-embedded tissues: An untapped specimen for biomonitoring of carcinogen DNA adducts by mass spectrometry," *Anal Chem*, 85, 4251–4258 (2013).
9. S. J. Enoch, M. T. D. Cronin, "A review of the electrophilic reaction chemistry involved in covalent DNA binding," *Crit Rev Tox*, 40(8), 728–748 (2010).
10. R. Benigni, A. Giuliani, R. Franke, A. Gruska, "Quantitative structure-activity relationships of mutagenic and carcinogenic aromatic amines," *Chem Rev*, 100, 3697–3714 (2000).
11. J. Ashby, "Fundamental structural alerts to potential carcinogenicity or noncarcinogenicity," *Environ Mutagenesis*, 7, 919–921 (1985).
12. J. Ashby, R. W. Tennant, "Chemical structure, *Salmonella* mutagenicity and extent of carcinogenicity as indicators of genotoxic carcinogenesis among 222 chemicals tested in rodents by the U.S. NCT/NTP," *Mutat Res*, 204, 17–115 (1988).
13. J. Ashby, R. W. Tennant, E. Zeiger, S. Stasiewicz, "Classification according to chemical structure, mutagenicity to *Salmonella* and level of carcinogenicity of a further 42 chemicals tested for carcinogenicity by the U.S. National Toxicology Program," *Mutat Res*, 223, 73–103 (1989).
14. R. W. Tennant, J. Ashby, "Classification according to chemical structure, mutagenicity to *Salmonella* and level of carcinogenicity of a further 39 chemicals tested for carcinogenicity by the U.S. National Toxicology Program," *Mutat Res*, 257, 209–227 (1991).
15. G. M. Cramer, R. A. Ford, "Estimation of toxic hazard—A decision tree approach," *Food Cosmet Toxicol*, 16, 255–276 (1978).
16. R. Kroes, A. G. Renwick, M. Cheeseman, J. Kleiner, I. Mangelsdorf, A. Piersma, B. Schilter, et al., "Structure-based thresholds of toxicological concern (TTC): Guidance for application to substances present at low levels in the diet," *Food Chem Tox*, 42, 65–83 (2004).
17. J. Kazius, R. McGuire, R. Bursi, "Derivation and validation of toxicophores for mutagenicity prediction," *J Med Chem*, 48, 312–320 (2005).
18. R. Benigni, C. Bossa, "Structural alerts of mutagens and carcinogens," *Curr Computer-Aided Drug Design*, 2(2), 169–176 (2006).
19. R. Benigni, C. Bossa, A. Worth, "Structural analysis and predictive value of the rodent in vivo micronucleus assay results," *Mutagenesis*, 25(4), 335–341 (2010).
20. R. Benigni, C. Bossa, O. Tcheremenskaia, A. Worth, *Development of structural alerts for the in vivo micronucleus assay in rodents*, JRC Scientific and Technical Reports, European Commission: Institute for Health and Consumer Protection (2009).
21. A. O. Aptula, D. W. Roberts, "Mechanistic applicability domains for non-animal-based prediction of toxicological end points: General principles and application to reactive toxicity," *Chem Res Toxicol*, 19, 1097–1105 (2006).

3 Acceptable Exposure Calculations for Impurities and Selected Compounds from the Carcinogenic Potency Database

Heewon Lee

CONTENTS

Several regulatory guidelines [1–10] recommend quantitative exposure limits of impurities or toxic chemicals. This chapter provides a collection of tables with the recommended and calculated exposure limits that are relevant to impurities and genotoxic impurities, to serve as a systematic reference. Different terminologies are used to describe the quantitative exposure limits by regulatory authorities, and some of the terms are listed in Table 3.1 for comparison.

The International Conference on Harmonisation of Technical Requirements for Registration of Pharmaceuticals for Human Use (ICH) Q3A guideline *Impurities in New Drug Substances* [1] recommends reporting, identification, and qualification thresholds in drug substances, as shown in Table 3.2. The threshold values depend on the maximum daily dose (MDD). In Table 3.3, calculated values of the impurity thresholds are shown based on the MDD, in percentage or in milligrams as appropriate.

The ICH Q3B guideline *Impurities in New Drug Products* [2] recommends reporting, identification, and qualification degradation product thresholds in drug products, as shown in Table 3.4. The threshold values depend on the MDD. In Table 3.5, calculated values of the degradation product thresholds are shown based on the MDD, in percentage or in milligrams as appropriate.

The ICH Q3C guideline *Impurities: Guideline for Residual Solvents* [3] classifies solvents by risk assessment and recommends concentration limits based on permitted daily exposure (PDE) and daily dose. The concentration limits recommended

TABLE 3.1
Terminologies Used to Describe Quantitative Exposure Limits by Regulatory Authorities

Guideline (References)	Term	Comment
ICH Q3A [1]	Threshold	Reporting, identification, qualification
ICH Q3B [2]	Threshold	Reporting, identification, qualification
ICH Q3C [3]	Permitted daily exposure (PDE)	Pharmaceutical acceptable intake of residual solvents
		Term PDE is used to avoid confusion of differing values for ADIs of the same substance used by IPCS and WHO
		Calculations of PDE are based on the NOEL
		Based on 50 kg body weight
EMA genotoxic impurities [4]	Threshold of toxicological concern (TTC)	1.5 μg/person/day Based on 60 kg body weight
EMA genotoxic impurities Q&A [5]	Staged TTC (sTTC); Allowable daily intake (ADI)	Depending on the duration of exposure during clinical development
FDA genotoxic impurities [6]	Staged TTC (sTTC); Acceptable daily intake (ADI)	Depending on the duration of clinical trial exposure
ICH M7 [7]	Acceptable daily intake (ADI)	Depending on the duration of treatment
EMA metal [8]	Permitted daily exposure (PDE)	
USP 232 [9]	Permissible daily exposure (PDE)	Based on 50 kg body weight
ICH Q3D [10]	Permitted daily exposure (PDE)	
IPCS	Tolerable daily intake (TDI)	International Program on Chemical Safety
WHO	Acceptable daily intake (ADI)	World Health Organization

TABLE 3.2
ICH Q3A: Thresholds of Impurities in New Drug Substances

MDD (g/day)	Reporting Threshold	Identification Threshold	Qualification Threshold
≤2	0.05%	0.10% or 1.0 mg/day Intake (whichever is lower)	0.15% or 1.0 mg/day Intake (whichever is lower)
>2	0.03%	0.05%	0.05%

Source: International Conference on Harmonisation of Technical Requirements for Registration of Pharmaceuticals for Human Use, *ICH Harmonised Tripartite Guideline, Impurities in New Drug Substances*, ICH Q3A(R2), October 2006.

TABLE 3.3
ICH Q3A: Impurity Thresholds Based on Maximum Daily Dose

MDD (mg)	Reporting Threshold	Identification Threshold	Qualification Threshold
<667	0.05%	0.10%	0.15%
667	0.05%	0.10% (0.67 mg)	0.15% (1.0 mg)
700	0.05%	0.10% (0.7 mg)	0.14% (1.0 mg)
800	0.05%	0.10% (0.8 mg)	0.13% (1.0 mg)
900	0.05%	0.10% (0.9 mg)	0.11% (1.0 mg)
1000	0.05%	0.10% (1.0 mg)	0.10% (1.0 mg)
1100	0.05%	0.09% (1.0 mg)	0.09% (1.0 mg)
1200	0.05%	0.08% (1.0 mg)	0.08% (1.0 mg)
1300	0.05%	0.08% (1.0 mg)	0.08% (1.0 mg)
1400	0.05%	0.07% (1.0 mg)	0.07% (1.0 mg)
1500	0.05%	0.07% (1.0 mg)	0.07% (1.0 mg)
1600	0.05%	0.06% (1.0 mg)	0.06% (1.0 mg)
1700	0.05%	0.06% (1.0 mg)	0.06% (1.0 mg)
1800	0.05%	0.06% (1.0 mg)	0.06% (1.0 mg)
1900	0.05%	0.05% (1.0 mg)	0.05% (1.0 mg)
2000	0.05%	0.05% (1.0 mg)	0.05% (1.0 mg)
>2000	0.03%	0.05%	0.05%

TABLE 3.4
ICH Q3B: Thresholds for Degradation Products in New Drug Products

MDD	Reporting Threshold	Identification Threshold	Qualification Threshold
<1 mg	0.1%	1.0% or 5 μg TDI, whichever is lower	1.0% or 50 μg TDI, whichever is lower
1–10 mg	0.1%	0.5% or 20 μg TDI, whichever is lower	1.0% or 50 μg TDI, whichever is lower
10–100 mg	0.1%	0.2% or 2 mg TDI, whichever is lower	0.5% or 200 μg TDI, whichever is lower
100 mg to 1 g	0.1%	0.2% or 2 mg TDI, whichever is lower	0.2% or 3 mg TDI, whichever is lower
>1–2 g	0.05%	0.2% or 2 mg TDI, whichever is lower	0.2% or 3 mg TDI, whichever is lower
>2 g	0.05%	0.10%	0.15%

Source: International Conference on Harmonisation of Technical Requirements for Registration of Pharmaceuticals for Human Use, *ICH Harmonised Tripartite Guideline, Impurities in New Drug Products*, ICH Q3B(R2), June 2006.

TABLE 3.5
ICH Q3B: Degradation Product Thresholds Based on MDD

MDD (mg)	Reporting Threshold	Identification Threshold	Qualification Threshold
<0.5	0.1%	1.0%	1.0%
0.5	0.1%	1.0% (5 μg)	1.0% (5 μg)
0.6	0.1%	0.8% (5 μg)	1.0% (6 μg)
0.7	0.1%	0.7% (5 μg)	1.0% (7 μg)
0.8	0.1%	0.6% (5 μg)	1.0% (8 μg)
0.9	0.1%	0.6% (5 μg)	1.0% (9 μg)
1	0.1%	0.5% (5 μg)	1.0% (10 μg)
2	0.1%	0.5% (10 μg)	1.0% (20 μg)
3	0.1%	0.5% (15 μg)	1.0% (30 μg)
4	0.1%	0.5% (20 μg)	1.0% (40 μg)
5	0.1%	0.4% (20 μg)	1.0% (50 μg)
6	0.1%	0.3% (20 μg)	0.8% (50 μg)
7	0.1%	0.3% (20 μg)	0.7% (50 μg)
8	0.1%	0.3% (20 μg)	0.6% (50 μg)
9	0.1%	0.2% (20 μg)	0.6% (50 μg)
10	0.1%	0.2% (20 μg)	0.5% (50 μg)
20	0.1%	0.2% (40 μg)	0.5% (100 μg)
30	0.1%	0.2% (60 μg)	0.5% (150 μg)
40	0.1%	0.2% (80 μg)	0.5% (200 μg)
50	0.1%	0.2% (100 μg)	0.4% (200 μg)
60	0.1%	0.2% (120 μg)	0.3% (200 μg)
70	0.1%	0.2% (140 μg)	0.3% (200 μg)
80	0.1%	0.2% (160 μg)	0.3% (200 μg)
90	0.1%	0.2% (180 μg)	0.2% (200 μg)
100	0.1%	0.2% (200 μg)	0.2% (200 μg)
200	0.1%	0.2% (400 μg)	0.2% (400 μg)
300	0.1%	0.2% (600 μg)	0.2% (600 μg)
400	0.1%	0.2% (800 μg)	0.2% (800 μg)
500	0.1%	0.2% (1 mg)	0.2% (1 mg)
600	0.1%	0.2% (1 mg)	0.2% (1 mg)
700	0.1%	0.2% (1 mg)	0.2% (1 mg)
800	0.1%	0.2% (2 mg)	0.2% (2 mg)
900	0.1%	0.2% (2 mg)	0.2% (2 mg)
1000	0.1%	0.2% (2 mg)	0.2% (2 mg)
1100	0.05%	0.18% (2 mg)	0.2% (2.2 mg)
1200	0.05%	0.17% (2 mg)	0.2% (2.4 mg)
1300	0.05%	0.15% (2 mg)	0.2% (2.6 mg)
1400	0.05%	0.14% (2 mg)	0.2% (2.8 mg)
1500	0.05%	0.13% (2 mg)	0.2% (3 mg)
1600	0.05%	0.13% (2 mg)	0.19% (3 mg)
1700	0.05%	0.12% (2 mg)	0.18% (3 mg)
1800	0.05%	0.11% (2 mg)	0.17% (3 mg)
1900	0.05%	0.11% (2 mg)	0.16% (3 mg)
2000	0.05%	0.10% (2 mg)	0.15% (3 mg)
>2000	0.05%	0.10%	0.15%

are shown in the ICH Q3C section of Chapter 1. Two options are described for setting limits for class 2 solvents. The concentration limit is calculated from PDE and administered daily dose (equation follows), and option 1 is based on a 10 g daily dose:

$$\text{Concentration (ppm)} = \frac{1000 \times \text{PDE (mg/day)}}{\text{Dose (g/day)}}$$

No further calculation is necessary to use option 1 concentration limits, provided that the daily dose does not exceed 10 g. Products that are administered in doses greater than 10 g/day should be considered under option 2. Using option 2, PDE (in milligrams per day) can be used with the known MDD and the aforementioned equation to determine the concentration of residual solvent allowed in a drug product. Such limits are considered acceptable provided that it has been demonstrated that the residual solvent has been reduced to the practical minimum. Concentration limits for class 2 solvents, shown in Table 3.6, were calculated using option 2 with PDE and MDD values.

Inorganic impurities can result from the manufacturing process, and they are normally known and identified. Inorganic impurities include reagents, ligands, catalysts, heavy metals, and inorganic salts. The *United States Pharmacopeia* (USP) General Chapter 232 recommends permissible daily exposure (PDE) for elemental impurities [9], as shown in Table 3.7. The European Medicines Agency (EMA) guidelines on the specification limits for residues of metal catalysts [8] classify metals into one of the following three classes:

Class 1 metals: metals of significant safety concern.
  Known or suspect human carcinogens, or possible causative agents of other significant toxicity.
Class 2 metals: metals with low safety concern.
  Metals with lower toxic potential to man. They are generally well tolerated up to exposures that are relevant to the context of the guidelines. They may be trace metals required for nutritional purposes, or they are often present in foodstuff or readily available nutritional supplements.
Class 3 metals: metals with minimal safety concern.
  Metals with no significant toxicity. Their safety profile is well established. They are generally well tolerated up to doses that are well beyond the doses relevant to the context of the guidelines. Typically, they are ubiquitous in the environment or the plant and animal kingdoms.

Table 3.8 shows the recommendation of maximum acceptable limits of metal residues as described in the EMA guideline. It should be noted that an adult body weight of 50 kg was used in the PDE considerations of this EMA guideline.

The USP and EMA guidelines include the concentration limits of metals based on a 10 g MDD. Both the USP and EMA guidelines state that the option 2 approach (as described earlier for class 2 solvents in ICH Q3C) is acceptable, which determines the concentration limit based on PDE and MDD values. As can be seen in

TABLE 3.6
Concentration Limits (in Parts per Million) for Class 2 Solvents Based on Option 2, ICH Q3C, [PDE (mg/day)]

MDD (mg)	Acetonitrile [4.1]	Chlorobenzene [3.6]	Chloroform [0.6]	Cumene [0.7]
1	n/a	n/a	600,000	700,000
2	n/a	n/a	300,000	350,000
3	n/a	n/a	200,000	233,333
4	n/a	900,000	150,000	175,000
5	820,000	720,000	120,000	140,000
6	683,333	600,000	100,000	116,667
7	585,714	514,286	85,714	100,000
8	512,500	450,000	75,000	87,500
9	455,556	400,000	66,667	77,778
10	410,000	360,000	60,000	70,000
20	205,000	180,000	30,000	35,000
30	136,667	120,000	20,000	23,333
40	102,500	90,000	15,000	17,500
50	82,000	72,000	12,000	14,000
60	68,333	60,000	10,000	11,667
70	58,571	51,429	8,571	10,000
80	51,250	45,000	7,500	8,750
90	45,556	40,000	6,667	7,778
100	41,000	36,000	6,000	7,000
200	20,500	18,000	3,000	3,500
300	13,667	12,000	2,000	2,333
400	10,250	9,000	1,500	1,750
500	8,200	7,200	1,200	1,400
600	6,833	6,000	1,000	1,167
700	5,857	5,143	857	1,000
800	5,125	4,500	750	875
900	4,556	4,000	667	778
1,000	4,100	3,600	600	700
1,100	3,727	3,273	545	636
1,200	3,417	3,000	500	583
1,300	3,154	2,769	462	538
1,400	2,929	2,571	429	500
1,500	2,733	2,400	400	467
1,600	2,563	2,250	375	438
1,700	2,412	2,118	353	412
1,800	2,278	2,000	333	389
1,900	2,158	1,895	316	368
2,000	2,050	1,800	300	350
3,000	1,367	1,200	200	233
4,000	1,025	900	150	175
5,000	820	720	120	140
6,000	683	600	100	117
7,000	586	514	86	100
8,000	513	450	75	88
9,000	456	400	67	78
10,000	410	360	60	70

TABLE 3.6 (*Continued*)
Concentration Limits (in Parts per Million) for Class 2 Solvents Based on Option 2, ICH Q3C, [PDE (mg/day)]

MDD (mg)	Cyclohexane [38.8]	1,2-Dichloroethene [18.7]	Dichloro methane [6.0]	1,2-Dimethoxyethane [1.0]
1	n/a	n/a	n/a	n/a
2	n/a	n/a	n/a	500,000
3	n/a	n/a	n/a	333,333
4	n/a	n/a	n/a	250,000
5	n/a	n/a	n/a	200,000
6	n/a	n/a	n/a	166,667
7	n/a	n/a	857,143	142,857
8	n/a	n/a	750,000	125,000
9	n/a	n/a	666,667	111,111
10	n/a	n/a	600,000	100,000
20	n/a	935,000	300,000	50,000
30	n/a	623,333	200,000	33,333
40	970,000	467,500	150,000	25,000
50	776,000	374,000	120,000	20,000
60	646,667	311,667	100,000	16,667
70	554,286	267,143	85,714	14,286
80	485,000	233,750	75,000	12,500
90	431,111	207,778	66,667	11,111
100	388,000	187,000	60,000	10,000
200	194,000	93,500	30,000	5,000
300	129,333	62,333	20,000	3,333
400	97,000	46,750	15,000	2,500
500	77,600	37,400	12,000	2,000
600	64,667	31,167	10,000	1,667
700	55,429	26,714	8,571	1,429
800	48,500	23,375	7,500	1,250
900	43,111	20,778	6,667	1,111
1,000	38,800	18,700	6,000	1,000
1,100	35,273	17,000	5,455	909
1,200	32,333	15,583	5,000	833
1,300	29,846	14,385	4,615	769
1,400	27,714	13,357	4,286	714
1,500	25,867	12,467	4,000	667
1,600	24,250	11,688	3,750	625
1,700	22,824	11,000	3,529	588
1,800	21,556	10,389	3,333	556
1,900	20,421	9,842	3,158	526
2,000	19,400	9,350	3,000	500
3,000	12,933	6,233	2,000	333
4,000	9,700	4,675	1,500	250
5,000	7,760	3,740	1,200	200
6,000	6,467	3,117	1,000	167
7,000	5,543	2,671	857	143
8,000	4,850	2,338	750	125
9,000	4,311	2,078	667	111
10,000	3,880	1,870	600	100

(*Continued*)

TABLE 3.6 (*Continued*)
Concentration Limits (in Parts per Million) for Class 2 Solvents Based on Option 2, ICH Q3C, [PDE (mg/day)]

MDD (mg)	*N,N*-Dimethylacetamide [10.9]	*N,N*-Dimethyl-formamide [8.8]	1,4-Dioxane [3.8]	2-Ethoxyethanol [1.6]
1	n/a	n/a	n/a	n/a
2	n/a	n/a	n/a	800,000
3	n/a	n/a	n/a	533,333
4	n/a	n/a	950,000	400,000
5	n/a	n/a	760,000	320,000
6	n/a	n/a	633,333	266,667
7	n/a	n/a	542,857	228,571
8	n/a	n/a	475,000	200,000
9	n/a	977,778	422,222	177,778
10	n/a	880,000	380,000	160,000
20	545,000	440,000	190,000	80,000
30	363,333	293,333	126,667	53,333
40	272,500	220,000	95,000	40,000
50	218,000	176,000	76,000	32,000
60	181,667	146,667	63,333	26,667
70	155,714	125,714	54,286	22,857
80	136,250	110,000	47,500	20,000
90	121,111	97,778	42,222	17,778
100	109,000	88,000	38,000	16,000
200	54,500	44,000	19,000	8,000
300	36,333	29,333	12,667	5,333
400	27,250	22,000	9,500	4,000
500	21,800	17,600	7,600	3,200
600	18,167	14,667	6,333	2,667
700	15,571	12,571	5,429	2,286
800	13,625	11,000	4,750	2,000
900	12,111	9,778	4,222	1,778
1,000	10,900	8,800	3,800	1,600
1,100	9,909	8,000	3,455	1,455
1,200	9,083	7,333	3,167	1,333
1,300	8,385	6,769	2,923	1,231
1,400	7,786	6,286	2,714	1,143
1,500	7,267	5,867	2,533	1,067
1,600	6,813	5,500	2,375	1,000
1,700	6,412	5,176	2,235	941
1,800	6,056	4,889	2,111	889
1,900	5,737	4,632	2,000	842
2,000	5,450	4,400	1,900	800
3,000	3,633	2,933	1,267	533
4,000	2,725	2,200	950	400
5,000	2,180	1,760	760	320
6,000	1,817	1,467	633	267
7,000	1,557	1,257	543	229
8,000	1,363	1,100	475	200
9,000	1,211	978	422	178
10,000	1,090	880	380	160

TABLE 3.6 (*Continued*)
Concentration Limits (in Parts per Million) for Class 2 Solvents Based on Option 2, ICH Q3C, [PDE (mg/day)]

MDD (mg)	Ethyleneglycol [6.2]	Formamide [2.2]	Hexane [2.9]	Methanol [30.0]
1	n/a	n/a	n/a	n/a
2	n/a	n/a	n/a	n/a
3	n/a	733,333	966,667	n/a
4	n/a	550,000	725,000	n/a
5	n/a	440,000	580,000	n/a
6	n/a	366,667	483,333	n/a
7	885,714	314,286	414,286	n/a
8	775,000	275,000	362,500	n/a
9	688,889	244,444	322,222	n/a
10	620,000	220,000	290,000	n/a
20	310,000	110,000	145,000	n/a
30	206,667	73,333	96,667	n/a
40	155,000	55,000	72,500	750,000
50	124,000	44,000	58,000	600,000
60	103,333	36,667	48,333	500,000
70	88,571	31,429	41,429	428,571
80	77,500	27,500	36,250	375,000
90	68,889	24,444	32,222	333,333
100	62,000	22,000	29,000	300,000
200	31,000	11,000	14,500	150,000
300	20,667	7,333	9,667	100,000
400	15,500	5,500	7,250	75,000
500	12,400	4,400	5,800	60,000
600	10,333	3,667	4,833	50,000
700	8,857	3,143	4,143	42,857
800	7,750	2,750	3,625	37,500
900	6,889	2,444	3,222	33,333
1,000	6,200	2,200	2,900	30,000
1,100	5,636	2,000	2,636	27,273
1,200	5,167	1,833	2,417	25,000
1,300	4,769	1,692	2,231	23,077
1,400	4,429	1,571	2,071	21,429
1,500	4,133	1,467	1,933	20,000
1,600	3,875	1,375	1,813	18,750
1,700	3,647	1,294	1,706	17,647
1,800	3,444	1,222	1,611	16,667
1,900	3,263	1,158	1,526	15,789
2,000	3,100	1,100	1,450	15,000
3,000	2,067	733	967	10,000
4,000	1,550	550	725	7,500
5,000	1,240	440	580	6,000
6,000	1,033	367	483	5,000
7,000	886	314	414	4,286
8,000	775	275	363	3,750
9,000	689	244	322	3,333
10,000	620	220	290	3,000

(*Continued*)

TABLE 3.6 (*Continued*)
Concentration Limits (in Parts per Million) for Class 2 Solvents Based on Option 2, ICH Q3C, [PDE (mg/day)]

MDD (mg)	2-Methoxyethanol [0.5]	Methylbuty ketone [0.5]	Methylcyclo-hexane [11.8]	N-Methylpyrrolidone [5.3]
1	500,000	500,000	n/a	n/a
2	250,000	250,000	n/a	n/a
3	166,667	166,667	n/a	n/a
4	125,000	125,000	n/a	n/a
5	100,000	100,000	n/a	n/a
6	83,333	83,333	n/a	883,333
7	71,429	71,429	n/a	757,143
8	62,500	62,500	n/a	662,500
9	55,556	55,556	n/a	588,889
10	50,000	50,000	n/a	530,000
20	25,000	25,000	590,000	265,000
30	16,667	16,667	393,333	176,667
40	12,500	12,500	295,000	132,500
50	10,000	10,000	236,000	106,000
60	8,333	8,333	196,667	88,333
70	7,143	7,143	168,571	75,714
80	6,250	6,250	147,500	66,250
90	5,556	5,556	131,111	58,889
100	5,000	5,000	118,000	53,000
200	2,500	2,500	59,000	26,500
300	1,667	1,667	39,333	17,667
400	1,250	1,250	29,500	13,250
500	1,000	1,000	23,600	10,600
600	833	833	19,667	8,833
700	714	714	16,857	7,571
800	625	625	14,750	6,625
900	556	556	13,111	5,889
1,000	500	500	11,800	5,300
1,100	455	455	10,727	4,818
1,200	417	417	9,833	4,417
1,300	385	385	9,077	4,077
1,400	357	357	8,429	3,786
1,500	333	333	7,867	3,533
1,600	313	313	7,375	3,313
1,700	294	294	6,941	3,118
1,800	278	278	6,556	2,944
1,900	263	263	6,211	2,789
2,000	250	250	5,900	2,650
3,000	167	167	3,933	1,767
4,000	125	125	2,950	1,325
5,000	100	100	2,360	1,060
6,000	83	83	1,967	883
7,000	71	71	1,686	757
8,000	63	63	1,475	663
9,000	56	56	1,311	589
10,000	50	50	1,180	530

TABLE 3.6 (*Continued*)
Concentration Limits (in Parts per Million) for Class 2 Solvents Based on Option 2, ICH Q3C, [PDE (mg/day)]

MDD (mg)	Nitromethane [0.5]	Pyridine [2.0]	Sulfolane [1.6]	Tetrahydrofuran [7.2]
1	500,000	n/a	n/a	n/a
2	250,000	n/a	800,000	n/a
3	166,667	666,667	533,333	n/a
4	125,000	500,000	400,000	n/a
5	100,000	400,000	320,000	n/a
6	83,333	333,333	266,667	n/a
7	71,429	285,714	228,571	n/a
8	62,500	250,000	200,000	900,000
9	55,556	222,222	177,778	800,000
10	50,000	200,000	160,000	720,000
20	25,000	100,000	80,000	360,000
30	16,667	66,667	53,333	240,000
40	12,500	50,000	40,000	180,000
50	10,000	40,000	32,000	144,000
60	8,333	33,333	26,667	120,000
70	7,143	28,571	22,857	102,857
80	6,250	25,000	20,000	90,000
90	5,556	22,222	17,778	80,000
100	5,000	20,000	16,000	72,000
200	2,500	10,000	8,000	36,000
300	1,667	6,667	5,333	24,000
400	1,250	5,000	4,000	18,000
500	1,000	4,000	3,200	14,400
600	833	3,333	2,667	12,000
700	714	2,857	2,286	10,286
800	625	2,500	2,000	9,000
900	556	2,222	1,778	8,000
1,000	500	2,000	1,600	7,200
1,100	455	1,818	1,455	6,545
1,200	417	1,667	1,333	6,000
1,300	385	1,538	1,231	5,538
1,400	357	1,429	1,143	5,143
1,500	333	1,333	1,067	4,800
1,600	313	1,250	1,000	4,500
1,700	294	1,176	941	4,235
1,800	278	1,111	889	4,000
1,900	263	1,053	842	3,789
2,000	250	1,000	800	3,600
3,000	167	667	533	2,400
4,000	125	500	400	1,800
5,000	100	400	320	1,440
6,000	83	333	267	1,200
7,000	71	286	229	1,029
8,000	63	250	200	900
9,000	56	222	178	800
10,000	50	200	160	720

Note: n/a, not applicable.

Tables 3.7 and 3.8, the metals included in the two guidelines do not match and some of the recommended PDE values differ between the two guidelines (e.g., the oral daily dose PDE for nickel is 500 μg/day according to the USP, whereas the oral PDE is 300 μg/day according to the EMA). Concentration limits (in parts per million) based on MDD values were calculated using PDE values recommended by either USP or EMA guidelines and are shown in Table 3.9. International efforts to harmonize the guidances for elemental impurities are in progress, and the step 2 version of the ICH Q3D guideline for elemental impurities is currently available. The permitted daily exposures recommended in the ICH Q3D draft guideline are shown in Table 3.10 as reference.

A threshold of toxicological concern (TTC) was developed to establish a common exposure level for any unstudied chemical that will not pose a risk of significant carcinogenicity or other toxic effects [4,11]. This TTC value was estimated to be 1.5 μg/person/day. The TTC, originally developed as a "threshold of regulation" at the Food and Drug Administration (FDA) for food-contact materials, was established based on the analysis of 343 carcinogens from a carcinogenic potency database (CPDB) and was repeatedly confirmed by evaluations expanding the database to more than 700 carcinogens. The probability distribution of carcinogenic potencies has been used to derive an estimate of a daily exposure level of most carcinogens, which would give rise to less than a 1 in 10^6 (1×10^6) upper-bound lifetime risk of cancer ("virtually safe dose"). Further analysis of subsets of high-potency carcinogens led

TABLE 3.7
Elemental Impurities for Drug Products, UPS 232

Element	Oral Daily Dose PDE (μg/day)	Parenteral Daily Dose PDE (μg/day)	Inhalational Daily Dose PDE (μg/day)
Cadmium	2.5	2.5	1.5
Lead	5	5	5
Inorganic arsenic	1.5	1.5	1.5
Inorganic mercury	15	1.5	1.5
Iridium	100	10	1.5
Osmium	100	10	1.5
Palladium	100	10	1.5
Platinum	100	10	1.5
Rhodium	100	10	1.5
Ruthenium	100	10	1.5
Chromium	Not a safety concern	Not a safety concern	25
Molybdenum	100	10	10
Nickel	500	50	1.5
Vanadium	100	10	30
Copper	1000	100	100

Source: United States Pharmacopeial Forum, General Chapter 232, Elemental Impurities—Limits, In: *United States Pharmacopeia*, Rockland, MD, February 2013.

Note: PDE, permissible daily exposure based on a 50 kg person.

TABLE 3.8
Class Exposure and Concentration Limits for Individual Metal Catalysts and Metal Reagents, EMA

Classification	Oral Exposure		Parenteral Exposure		Inhalation Exposure
	PDE (μg/day)	Concentration (ppm)	PDE (μg/day)	Concentration (ppm)	PDE (μg/day)
Class 1A: Pt, Pd	100	10	10[a]	1[a]	Pt: 70[a]
Class 1B: Ir, Rh, Ru, Os	100[b]	10[b]	10[b]	1[b]	
Class 1C: Mo, Ni, Cr, V	300	30	30	3	Ni: 100, Cr(VI): 10
Class 2: Cu, Mn	2,500	250	250	25	
Class 3: Fe, Zn	13,000	1,300	1,300	130	

Source: European Medicines Agency, *Guideline on the Specification Limits for Residues of Metal Catalysts or Metal Reagents*, Committee for Medicinal Products for Human Use, 2008.

[a] Specific limits have been set for inhalation exposure to Pt, Cr(VI), and Ni; Pt as hexachloroplatinic acid.

[b] Subclass limit: the total amount of listed metals should not exceed the indicated limit.

to the suggestion of a 10-fold lower TTC (0.15 μg/day) for chemicals with structural alerts that raise concern for potential genotoxicity. However, for the application of a TTC in the assessment of acceptable limits of genotoxic impurities in drug substances, a value of 1.5 μg/day, corresponding to a 10^{-5} lifetime risk of cancer, can be justified as for pharmaceuticals a benefit exists. It should be recognized that the methods on which the TTC value is based are generally considered very conservative since they involved a simple linear extrapolation (LE) from the dose giving a 50% tumor incidence (TD50) to a 1 in 10^6 incidence, using TD50 data for the most sensitive species and the most sensitive site (several worst-case assumptions).

Acceptable genotoxic impurity levels during clinical development are recommended in the FDA draft guideline [6]. The most pragmatic approach to calculate acceptable short-term exposures to known genotoxic carcinogens is to linearly extrapolate the short-term exposure from the acceptable lifetime exposure or virtually safe dose. The acceptable daily intakes (ADIs) (also termed "acceptable qualification thresholds" in the FDA guideline) of genotoxic impurities during clinical development are shown in Table 3.11, based on the LE approach. The impurity threshold exposures for exposure durations of up to 12 months are based on a 10^{-6} cancer risk level (0.15 μg/day for lifetime exposure), since these trials often include healthy subjects for whom there is no expected health benefit and the efficacy of the drug may still be uncertain. The values are derived from a LE from the

TABLE 3.9
Metals Concentration Limit (in Parts per Million) Based on MDD and PDE, UPS and EMA

MDD (mg)	PDE (μg/day)															
	1.5	2.5	5	10	15	25	30	50	70	100	250	500	1,000	1,300	2,500	13,000
1	1,500	2,500	5,000	10,000	15,000	25,000	30,000	50,000	70,000	100,000	250,000	500,000	n/a	n/a	n/a	n/a
2	750	1,250	2,500	5,000	7,500	12,500	15,000	25,000	35,000	50,000	125,000	250,000	500,000	650,000	n/a	n/a
3	500	833	1,667	3,333	5,000	8,333	10,000	16,667	23,333	33,333	83,333	166,667	333,333	433,333	833,333	n/a
4	375	625	1,250	2,500	3,750	6,250	7,500	12,500	17,500	25,000	62,500	125,000	250,000	325,000	625,000	n/a
5	300	500	1,000	2,000	3,000	5,000	6,000	10,000	14,000	20,000	50,000	100,000	200,000	260,000	500,000	n/a
6	250	417	833	1,667	2,500	4,167	5,000	8,333	11,667	16,667	41,667	83,333	166,667	216,667	416,667	n/a
7	214	357	714	1,429	2,143	3,571	4,286	7,143	10,000	14,286	35,714	71,429	142,857	185,714	357,143	n/a
8	188	313	625	1,250	1,875	3,125	3,750	6,250	8,750	12,500	31,250	62,500	125,000	162,500	312,500	n/a
9	167	278	556	1,111	1,667	2,778	3,333	5,556	7,778	11,111	27,778	55,556	111,111	144,444	277,778	n/a
10	150	250	500	1,000	1,500	2,500	3,000	5,000	7,000	10,000	25,000	50,000	100,000	130,000	250,000	n/a
20	75	125	250	500	750	1,250	1,500	2,500	3,500	5,000	12,500	25,000	50,000	65,000	125,000	650,000
30	50	83	167	333	500	833	1,000	1,667	2,333	3,333	8,333	16,667	33,333	43,333	83,333	433,333
40	38	63	125	250	375	625	750	1,250	1,750	2,500	6,250	12,500	25,000	32,500	62,500	325,000
50	30	50	100	200	300	500	600	1,000	1,400	2,000	5,000	10,000	20,000	26,000	50,000	260,000
60	25	42	83	167	250	417	500	833	1,167	1,667	4,167	8,333	16,667	21,667	41,667	216,667
70	21	36	71	143	214	357	429	714	1,000	1,429	3,571	7,143	14,286	18,571	35,714	185,714
80	19	31	63	125	188	313	375	625	875	1,250	3,125	6,250	12,500	16,250	31,250	162,500
90	17	28	56	111	167	278	333	556	778	1,111	2,778	5,556	11,111	14,444	27,778	144,444
100	15	25	50	100	150	250	300	500	700	1,000	2,500	5,000	10,000	13,000	25,000	130,000
200	7.50	13	25	50	75	125	150	250	350	500	1,250	2,500	5,000	6,500	12,500	65,000
300	5.00	8.33	17	33	50	83	100	167	233	333	833	1,667	3,333	4,333	8,333	43,333
400	3.75	6.25	13	25	38	63	75	125	175	250	625	1,250	2,500	3,250	6,250	32,500
500	3.00	5.00	10	20	30	50	60	100	140	200	500	1,000	2,000	2,600	5,000	26,000

600	2.50	4.17	8.33	17	25	42	50	83	117	167	417	833	1,667	2,167	4,167	21,667
700	2.14	3.57	7.14	14	21	36	43	71	100	143	357	714	1,429	1,857	3,571	18,571
800	1.88	3.13	6.25	13	19	31	38	63	88	125	313	625	1,250	1,625	3,125	16,250
900	1.67	2.78	5.56	11	17	28	33	56	78	111	278	556	1,111	1,444	2,778	14,444
1,000	1.50	2.50	5.00	10	15	25	30	50	70	100	250	500	1,000	1,300	2,500	13,000
1,100	1.36	2.27	4.55	9.09	14	23	27	45	64	91	227	455	909	1,182	2,273	11,818
1,200	1.25	2.08	4.17	8.33	13	21	25	42	58	83	208	417	833	1,083	2,083	10,833
1,300	1.15	1.92	3.85	7.69	12	19	23	38	54	77	192	385	769	1,000	1,923	10,000
1,400	1.07	1.79	3.57	7.14	11	18	21	36	50	71	179	357	714	929	1,786	9,286
1,500	1.00	1.67	3.33	6.67	10	17	20	33	47	67	167	333	667	867	1,667	8,667
1,600	0.94	1.56	3.13	6.25	9.38	16	19	31	44	63	156	313	625	813	1,563	8,125
1,700	0.88	1.47	2.94	5.88	8.82	15	18	29	41	59	147	294	588	765	1,471	7,647
1,800	0.83	1.39	2.78	5.56	8.33	14	17	28	39	56	139	278	556	722	1,389	7,222
1,900	0.79	1.32	2.63	5.26	7.89	13	16	26	37	53	132	263	526	684	1,316	6,842
2,000	0.75	1.25	2.50	5.00	7.50	13	15	25	35	50	125	250	500	650	1,250	6,500
3,000	0.50	0.83	1.67	3.33	5.00	8.33	10	17	23	33	83	167	333	433	833	4,333
4,000	0.38	0.63	1.25	2.50	3.75	6.25	7.50	13	18	25	63	125	250	325	625	3,250
5,000	0.30	0.50	1.00	2.00	3.00	5.00	6.00	10	14	20	50	100	200	260	500	2,600
6,000	0.25	0.42	0.83	1.67	2.50	4.17	5.00	8.33	12	17	42	83	167	217	417	2,167
7,000	0.21	0.36	0.71	1.43	2.14	3.57	4.29	7.14	10	14	36	71	143	186	357	1,857
8,000	0.19	0.31	0.63	1.25	1.88	3.13	3.75	6.25	8.75	13	31	63	125	163	313	1,625
9,000	0.17	0.28	0.56	1.11	1.67	2.78	3.33	5.56	7.78	11	28	56	111	144	278	1,444
10,000	0.15	0.25	0.5	1.0	1.5	2.5	3.0	5.0	7.0	10	25	50	100	130	250	1,300

Note: n/a, not applicable.

TABLE 3.10
Permitted Daily Exposures for Elemental Impurities,[a] ICH Q3D Draft Guideline

Element	Class[b]	Oral PDE (µg/day)	Parenteral PDE (µg/day)	Inhalation PDE (µg/day)
As	1	15	15	1.9
Cd	1	5.0	6.0	3.4
Hg	1	40	4.0	1.2
Pb	1	5.0	5.0	5.0
Co	2A	50	5.0	2.9
Mo	2A	180	180	7.6
Se	2A	170	85	140
V	2A	120	12	1.2
Ag	2B	170	35	6.9
Au	2B	130	130	1.3
Ir[c]	2B	1,000	10	1.4
Os[c]	2B	1,000	10	1.4
Pd	2B	100	10	1.0
Pt	2B	1,000	10	1.4
Rh[c]	2B	1,000	10	1.4
Ru[c]	2B	1,000	10	1.4
Tl	2B	8.0	8.0	69
Ba	3	13,000	1,300	340
Cr	3	11,000	1,100	2.9
Cu	3	1,300	130	13
Li	3	780	390	25
Ni	3	600	60	6.0
Sb	3	1,200	600	22
Sn	3	6,400	640	64

Source: International Conference on Harmonisation of Technical Requirements for Registration of Pharmaceuticals for Human Use, *ICH Harmonised Tripartite Guideline, Guideline for Elemental Impurities*, ICH Q3D Draft Guideline, July 2013.

[a] PDEs reported in this table are rounded to two significant figures (micrograms per day).

[b] Classification as defined in Section 4 of the ICH Q3D draft guideline.

[c] Insufficient data to establish an appropriate PDE; the PDE was established based on platinum PDE.

qualification threshold using the maximum duration of dosing for each time period. In addition, these values incorporate an uncertainty factor of 2 to allow for deviations from the LE model. For trials of duration greater than 1 year, the threshold value is identical to the threshold for a marketing application and is based on a 10^{-5} cancer risk level (1.5 µg/day derived from lifetime exposure); subjects in these trials generally have the condition or disease being studied and are more certain to derive benefit from the treatment than subjects in early trials.

The EMA Committee for Medicinal Products for Human Use agrees with the use of a staged TTC (sTTC) concept during clinical development, as described

TABLE 3.11
Comparison of the sTTC Values in FDA and EMA Guidelines

FDA	Duration of Clinical Trial Exposure					
	<14 Days	14 Days to 1 Month	1–3 Months	3–6 Months	6–12 Months	>12 Months
Genotoxic and carcinogenic impurity threshold (μg/day)	120	60	20	10	5	1.5
EMA	**Duration of Exposure**					
	Single Dose	**≤1 Month**	**≤3 Months**	**≤6 Months**	**≤12 Months**	**>12 Months**
Allowable daily intake	120 μg	60 μg	20 μg	10 μg	5 μg	1.5 μg

Sources: Food and Drug Administration, *Guidance for Industry, Genotoxic and Carcinogenic Impurities in Drug Substances and Products: Recommended Approaches, Draft Guidance*, Center for Drug Evaluation and Research, December 2008; European Medicines Agency, Committee for Medicinal Products for Human Use, *Guideline on the Limits of Genotoxic Impurities*, January 2007; European Medicines Agency, *Questions and Answers on the 'Guideline on the Limits of Genotoxic Impurities'*, September 2010 (Rev. 3).

in the EMA Q&A document on genotoxic impurities [5]. The allowable daily intake (ADI) values are shown in Table 3.11 in comparison with the FDA guideline. The concentration limits (in parts per million) calculated using MDD and ADI ("acceptable daily intake" according to the FDA, and "allowable daily intake" according to the EMA) values for genotoxic and carcinogenic impurities are shown in Table 3.12.

TTC is a pragmatic risk assessment tool that can be used to assess the likelihood of toxic effects of a particular level of exposure to a chemical, in the absence of chemical-specific toxicity data [11]. The TTC concept reduces unnecessary extensive toxicity testing and safety evaluations, and it enables focused use of limited resources (time, animal use, cost, and expertise) on substances with greater potential risks to human health. Application of the TTC principle would set priorities for toxicity testing and could be used to indicate analytical data needs. It is considered a preliminary step in safety assessment. The TTC values were derived after the evaluation of over 700 compounds extracted from the Gold carcinogenic potency database (also known as the "Gold database") [12,13], as described in Chapters 1.5, 1.6, 1.7, and 1.8.

The CPDB was developed by Gold and others [12,13], and it is publicly available on the Internet. The database includes Ames test results and TD50 and target sites on 1547 compounds currently. TD50 is defined as "chronic dose-rate in mg/kg body wt/day which would induce tumors in half the test animals at the end of a standard lifespan for the species" or more precisely "dose-rate in mg/kg body wt/day which, if administered chronically for the standard lifespan of the species, will halve the probability of remaining tumorless throughout that period."

TABLE 3.12
Genotoxic Impurities Concentration Limits (in Parts per Million) Based on MDD and ADI

	ADI (µg/day)					
MDD (mg)	120	60	20	10	5	1.5
1	120,000	60,000	20,000	10,000	5,000	1,500
2	60,000	30,000	10,000	5,000	2,500	750
3	40,000	20,000	6,667	3,333	1,667	500
4	30,000	15,000	5,000	2,500	1,250	375
5	24,000	12,000	4,000	2,000	1,000	300
6	20,000	10,000	3,333	1,667	833	250
7	17,143	8,571	2,857	1,429	714	214
8	15,000	7,500	2,500	1,250	625	188
9	13,333	6,667	2,222	1,111	556	167
10	12,000	6,000	2,000	1,000	500	150
20	6,000	3,000	1,000	500	250	75
30	4,000	2,000	667	333	167	50
40	3,000	1,500	500	250	125	38
50	2,400	1,200	400	200	100	30
60	2,000	1,000	333	167	83	25
70	1,714	857	286	143	71	21
80	1,500	750	250	125	63	19
90	1,333	667	222	111	56	17
100	1,200	600	200	100	50	15
200	600	300	100	50	25	7.5
300	400	200	67	33	17	5.0
400	300	150	50	25	13	3.8
500	240	120	40	20	10	3.0
600	200	100	33	17	8.3	2.5
700	171	86	29	14	7.1	2.1
800	150	75	25	13	6.3	1.9
900	133	67	22	11	5.6	1.7
1,000	120	60	20	10	5.0	1.5
1,100	109	55	18	9.1	4.5	1.4
1,200	100	50	17	8.3	4.2	1.3
1,300	92	46	15	7.7	3.8	1.2
1,400	86	43	14	7.1	3.6	1.1
1,500	80	40	13	6.7	3.3	1.0
1,600	75	38	13	6.3	3.1	0.94
1,700	71	35	12	5.9	2.9	0.88
1,800	67	33	11	5.6	2.8	0.83
1,900	63	32	11	5.3	2.6	0.79
2,000	60	30	10	5.0	2.5	0.75
3,000	40	20	6.7	3.3	1.7	0.50
4,000	30	15	5.0	2.5	1.3	0.38
5,000	24	12	4.0	2.0	1.0	0.30
6,000	20	10	3.3	1.7	0.83	0.25
7,000	17	8.6	2.9	1.4	0.71	0.21
8,000	15	7.5	2.5	1.3	0.63	0.19
9,000	13	6.7	2.2	1.1	0.56	0.17
10,000	12	6.0	2.0	1.0	0.50	0.15

Selected compounds from the CPDB [12] that are pharmaceutically relevant are presented in Table 3.13. These compounds include possible raw materials, reagents, catalysts, solvents, by-products, intermediates in drug substance synthesis, pharmaceutical agents, or compounds that would provide guidance on genotoxicity as a representative of chemical functionality. When a compound contains HCl in the salt form, HCl is omitted from the structure for simplicity of presentation in Table 3.13. When there are several compounds with similar structures (such as positional isomers) in the CPDB, the compounds with the most potent TD50 values are presented in Table 3.13. Generally, compounds that have either positive mutagenicity results or TD50 values (in rats and/or mice) are chosen. Some pharmaceutically relevant compounds that show no mutagenicity are included to serve as references for risk evaluation purposes.

Three structural classes were mentioned as high-potency carcinogens (cohort of concern) in EMA and FDA guidelines, that is, aflatoxin-like, *N*-nitroso, and azoxy compounds. The compounds that belong to the three classes are noted as "high-potency carcinogens" in Table 3.13.

For mutagenic compounds with TD50 data, a compound-specific acceptable exposure was calculated based on the rodent carcinogenicity potency data (TD50) in Table 3.14, using the simple LE procedure employed to derive the TTC. At TD50, one in two animals will develop cancer over its lifetime. The cancer risk of 1 in 10^5 is then calculated by dividing TD50 by 50,000, corresponding to the LE from a 1 in 2 to a 1 in 100,000 risk of cancer. This is the LE method used to derive the TTC from the TD50 values of the compounds in the CPDB [11]. An example of the LE calculations for ethylene oxide is shown in Note 4 of the ICH M7 step 2 document [7] and is repeated here. It should be noted that the ICH M7 draft guideline is under development and is subject to change.

ICH M7 Step 2 Note 4: Example of linear extrapolation from the TD50

It is possible to calculate a compound-specific acceptable intake based on rodent carcinogenicity potency data such as TD50 values (doses giving a 50% tumor incidence equivalent to a cancer risk probability level of 1:2). Linear extrapolation to a probability of 1 in 100,000 (i.e., the accepted lifetime risk level used) is achieved by simply dividing the TD50 by 50,000. This procedure is similar to that employed for derivation of the TTC.

Calculation example: Ethylene oxide

TD50 values for ethylene oxide according to the Carcinogenic Potency database [12] are 21.3 mg/kg body weight/day (rat) and 63.7 mg/kg body weight/day (mouse). For the calculation of an acceptable intake, the lower (i.e., more conservative) value of the rat is used.

To derive a dose to cause tumors in 1 in 100,000 animals, divide by 50,000:

$$21.3 \text{ mg/kg} \div 50{,}000 = 0.42\ \mu\text{g/kg}$$

To derive a total human daily dose,

$$0.42\ \mu\text{g/kg/day} \times 50 \text{ kg Body Weight} = 21.3\ \mu\text{g/person/day}$$

TABLE 3.13
Selected Compounds from the CPDB

Chemical Name	CAS	*Salmonella* (Ames)	Harmonic Mean of TD50 (mg/kg/day)		Structure	Note
			Rat	Mouse		
Acetaldehyde	75-07-0	neg	153	—	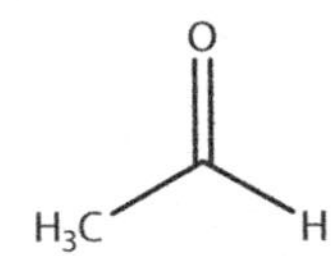	—
Acetamide	60-35-5	neg	180	3010	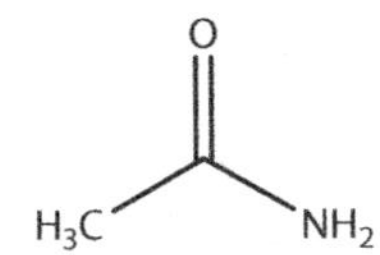	—
Acetaminophen	103-90-2	neg	495	1620	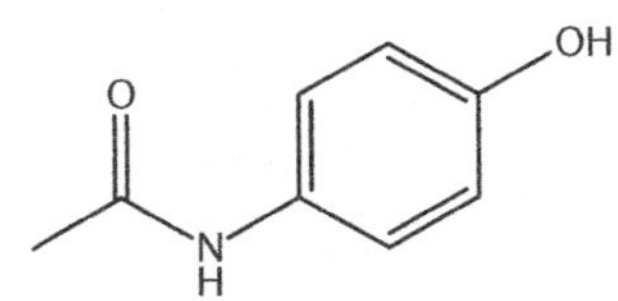	—
Acetonitrile	75-05-8	neg	—	—		Solvent, ICH Q3C class 2 (PDE 4.1 mg/day, 410 ppm)

(*Continued*)

1-Acetyl-2-phenylhydrazine	114-83-0	pos	—	51.2	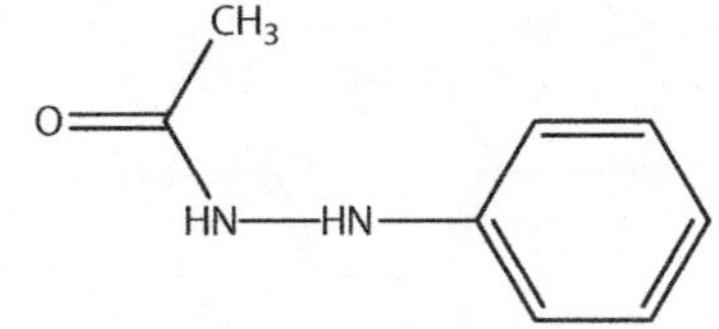	—
Acrylamide	79-06-1	neg	3.75	—	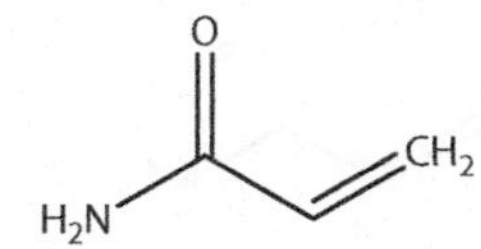	—
Aflatoxin B1	1162-65-8	pos	0.0032	—	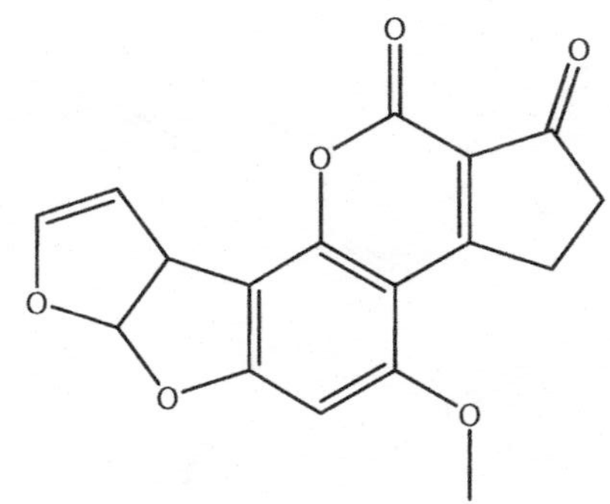	High-potency carcinogen, chemical carcinogenesis mechanism reported
Allyl chloride	107-05-1	pos	—	—	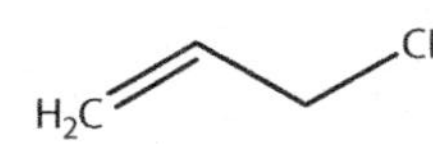	—

(*Continued*)

TABLE 3.13 (*Continued*)
Selected Compounds from the CPDB

Chemical Name	CAS	*Salmonella* (Ames)	Harmonic Mean of TD50 (mg/kg/day) Rat	Harmonic Mean of TD50 (mg/kg/day) Mouse	Structure	Note
Allyl isothiocyanate	57-06-7	pos	96	—	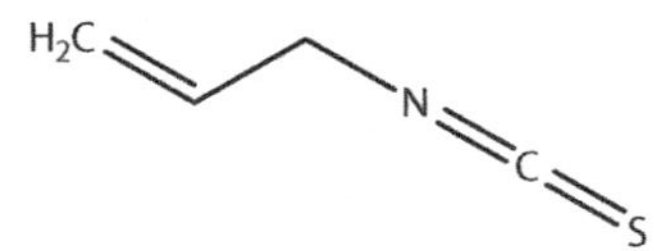	—
1-Allyl-1-nitrosourea	760-56-5	NE	0.341	—	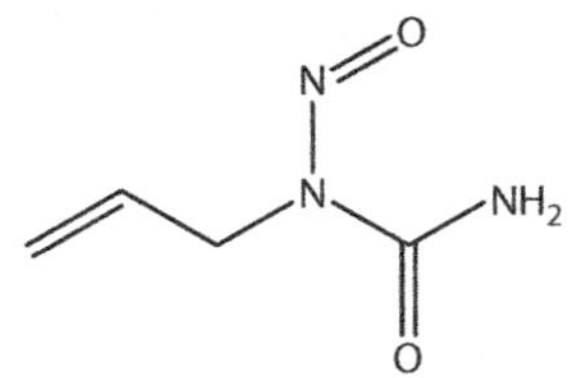	High-potency carcinogen
Allylhydrazine HCl	52207-83-7	NE	—	34.2	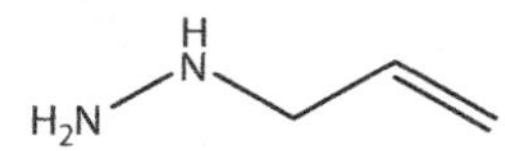	—
2-Amino-4-nitrophenol	99-57-0	pos	839	—	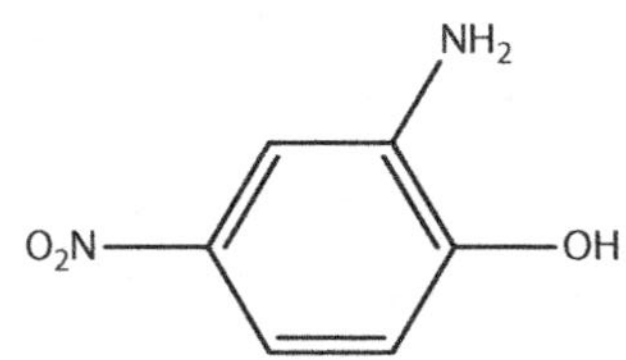	—

2-Amino-5-nitrophenol	121-88-0	pos	111	—	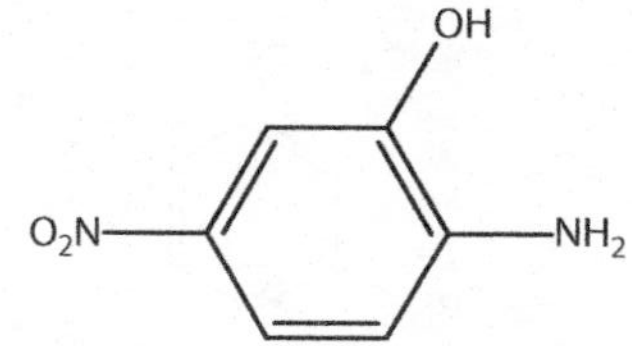	—
4-Amino-2-nitropohenol	119-34-6	pos	309	—	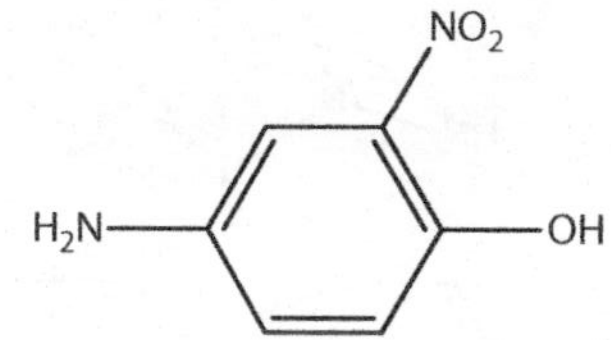	—
2-Amino-5-nitrothiazole	121-66-4	pos	44.6	—	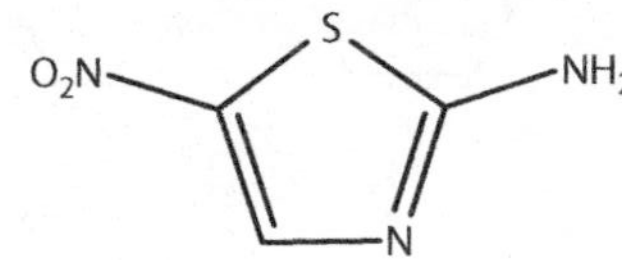	—
2-Aminoanthraquinone	117-79-3	pos	101	1190	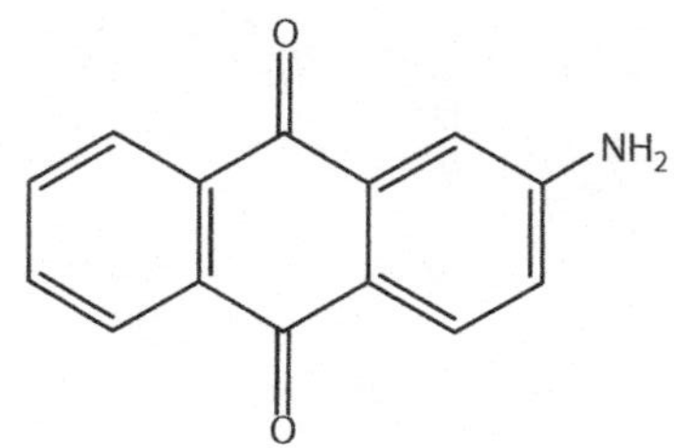	—

(Continued)

TABLE 3.13 (*Continued*)
Selected Compounds from the CPDB

Chemical Name	CAS	*Salmonella* (Ames)	Harmonic Mean of TD50 (mg/kg/day)		Structure	Note
			Rat	Mouse		
o-Aminoazotoluene	97-56-3	pos	4.04	—	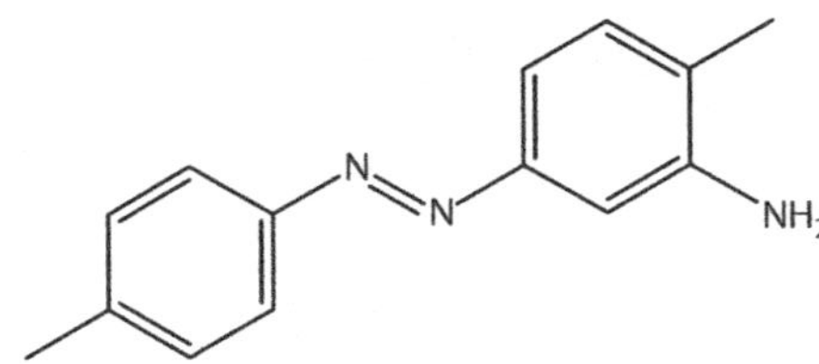	—
4-Aminobiphenyl	92-67-1	pos	—	2.1	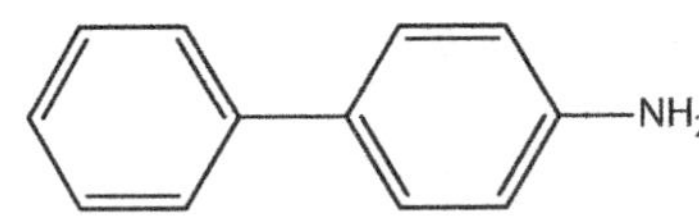	—
3-Aminotriazole	61-82-5	neg	9.94	25.3	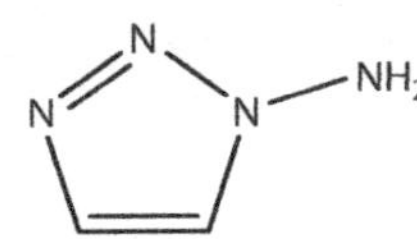	—
Aniline HCl	142-04-1	neg	269	—	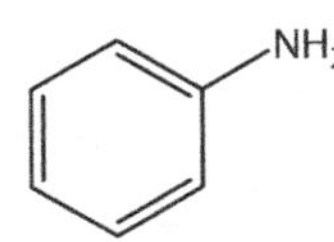	—

o-Anisidine HCl	134-29-2	pos	29.7	966	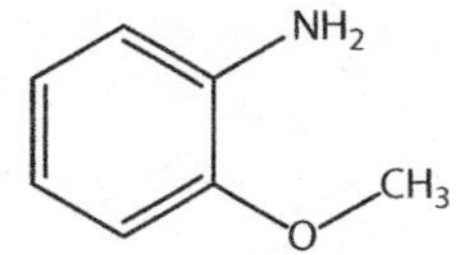	—
p-Anisidine HCl	20265-97-8	pos	—	—	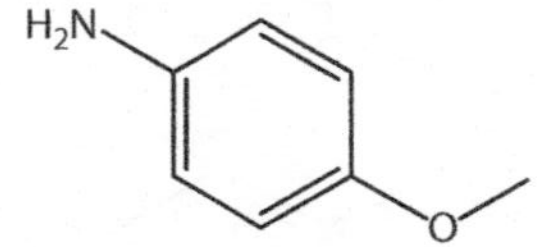	—
Aristolochic acid, sodium salt	10190-99-5	pos	0.0141	—	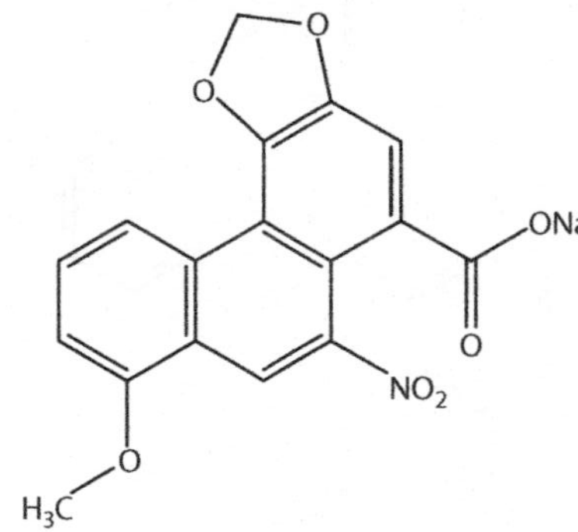	Found in some herbal plants, chemical carcinogenesis mechanism reported
Aspirin	50-78-2	neg	—	—	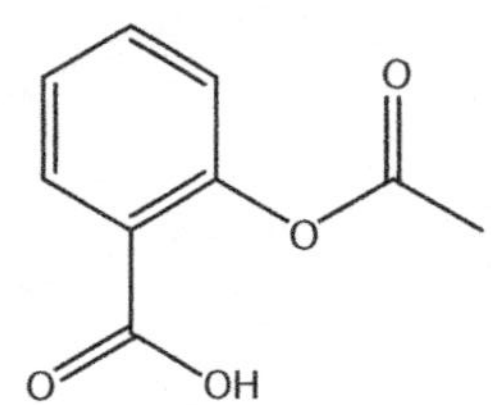	—

(Continued)

TABLE 3.13 (*Continued*)
Selected Compounds from the CPDB

Chemical Name	CAS	*Salmonella* (Ames)	Harmonic Mean of TD50 (mg/kg/day)		Structure	Note
			Rat	Mouse		
Azide, sodium	26628-22-8	pos	—	—	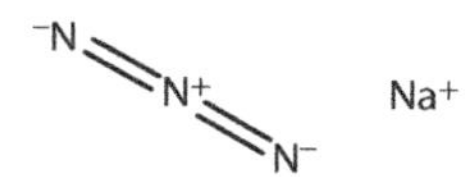	—
Azobenzene	103-33-3	pos	24.1	—	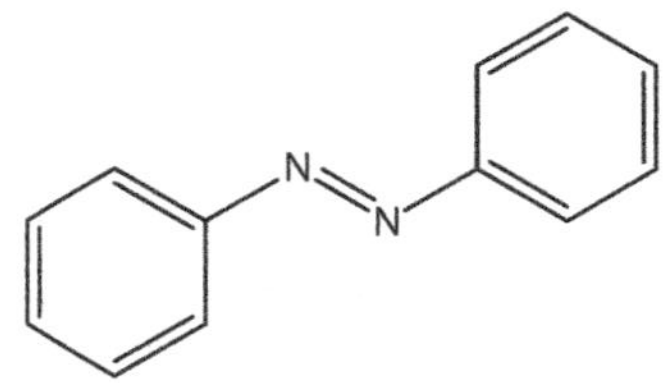	—
Azoxymethane	25843-45-2	pos	0.0466	—	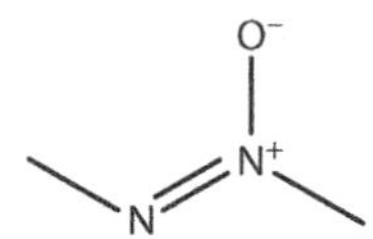	High-potency carcinogen
1-Azoxypropane	17697-55-1	NE	0.000241	—	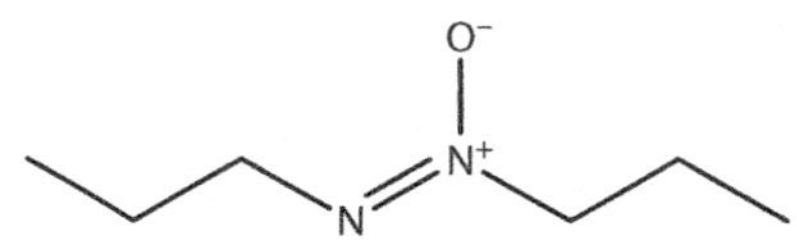	High-potency carcinogen

AZT (azidothymidine)	30516-87-1	pos	11,600	296	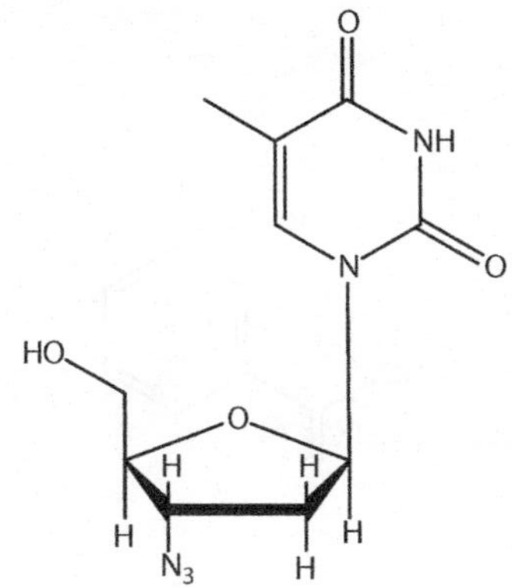	HIV treatment
Benzaldehyde	100-52-7	neg	—	1,490	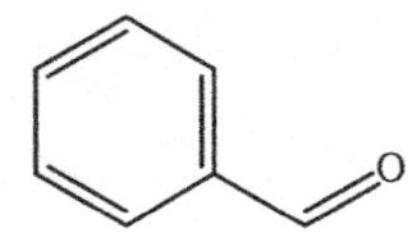	—
Benzene	71-43-2	neg	169	77.5		Solvent, ICH Q3C class 1 (2 ppm)
Benzidine	92-87-5	pos	1.73	19.9	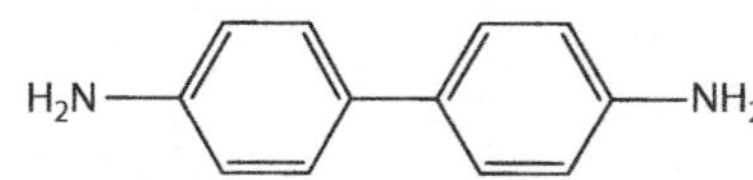	—

(*Continued*)

TABLE 3.13 (*Continued*)
Selected Compounds from the CPDB

Chemical Name	CAS	*Salmonella* (Ames)	Harmonic Mean of TD50 (mg/kg/day)		Structure	Note
			Rat	Mouse		
Benzo(*a*)pyrene	50-32-8	pos	0.956	3.47	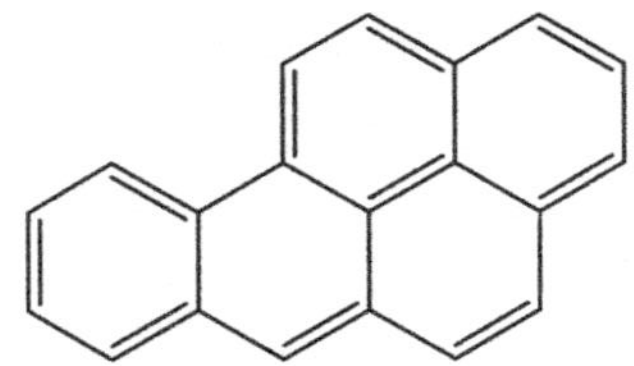	In tobacco smoke; chemical carcinogenesis mechanism via epoxide upon metabolism
Benzofuran	271-89-6	neg	424	25.1	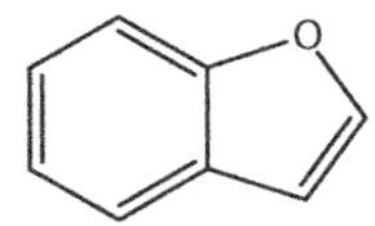	—
Benzoic acid	65-85-0	neg	—	—	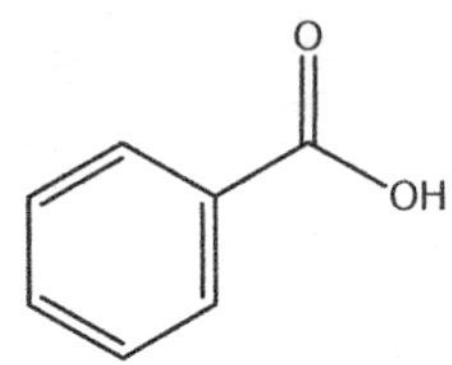	—

Benzoyl hydrazine	613-94-5	NE	—	9.59	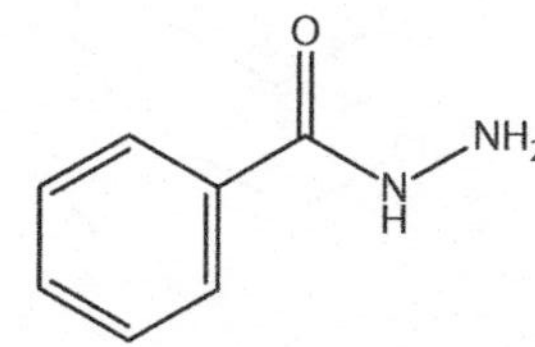	—
Benzyl acetate	140-11-4	neg	—	1,440	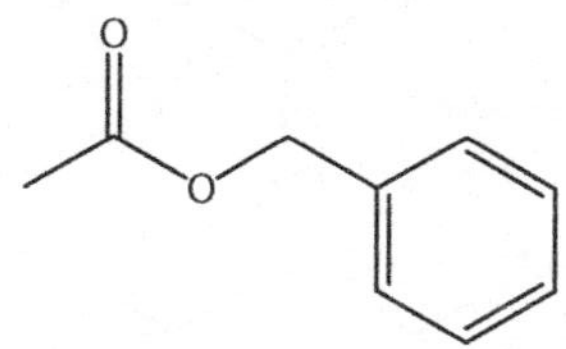	—
Benzyl alcohol	100-51-6	neg	—	—	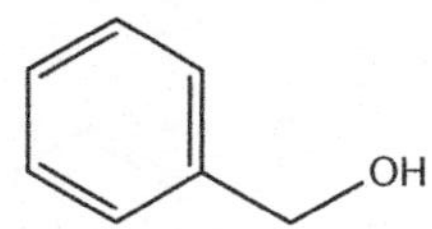	—
Benzyl chloride	100-44-7	pos	—	61.5	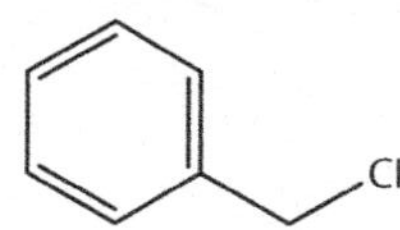	—

(*Continued*)

TABLE 3.13 (*Continued*)
Selected Compounds from the CPDB

Chemical Name	CAS	*Salmonella* (Ames)	Harmonic Mean of TD50 (mg/kg/day)		Structure	Note
			Rat	Mouse		
2-Biphenylamine HCl	2185-92-4	pos	—	1,120	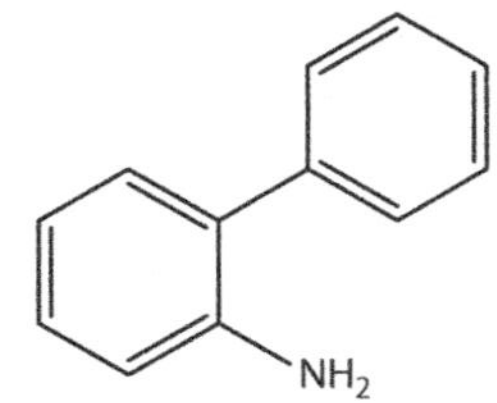	—
Bis-2-chloroethylether	111-44-4	pos	—	11.7	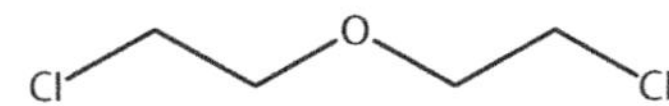	—
Bis-(chloromethyl)ether	542-88-1	NE	0.00357	0.182	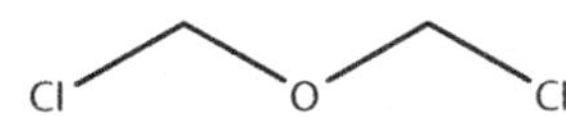	—
Boric acid	10043-35-3	neg	—	—	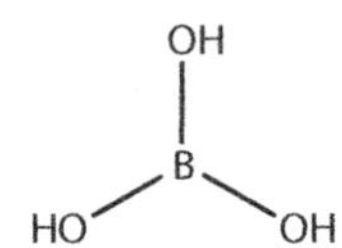	—
Bromodichloromethane	75-27-4	pos	72.5	47.7	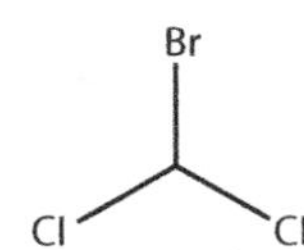	—
Bromoethane	74-96-4	pos	149	535	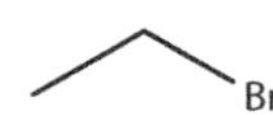	—

Bromoethanol	540-51-2	pos	—	76.1		—
1,3-Butadiene	106-99-0	pos	261	13.9		—
t-Butyl alcohol	75-65-0	neg	64.6	21,900	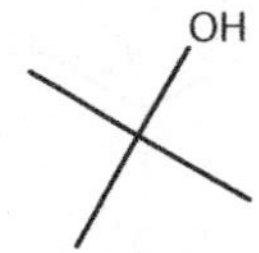	—
N-*n*-butyl-*N*-nitrosourea	869-01-2	pos	0.517	—	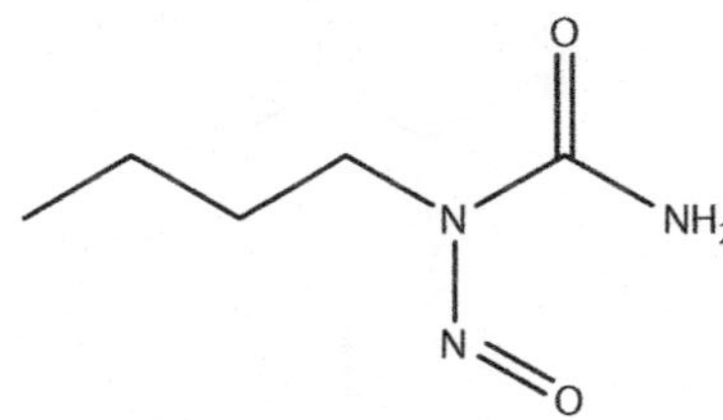	High-potency carcinogen
Butylated hydroxytoluene	128-37-0	neg	—	653	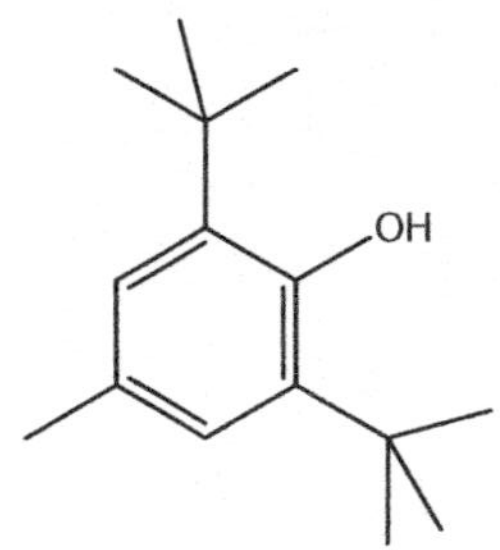	—

(*Continued*)

TABLE 3.13 (*Continued*)
Selected Compounds from the CPDB

Chemical Name	CAS	*Salmonella* (Ames)	Harmonic Mean of TD50 (mg/kg/day)		Structure	Note
			Rat	Mouse		
Caffeine	58-08-2	neg	—	—	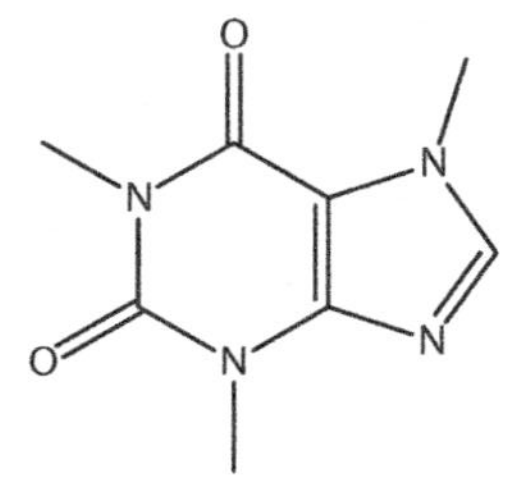	—
Carbamyl hydrazine HCl	563-41-7	neg	—	223	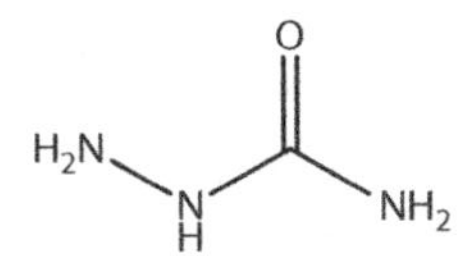	—
Carbon tetrachloride	56-23-5	neg	27.8	33.6	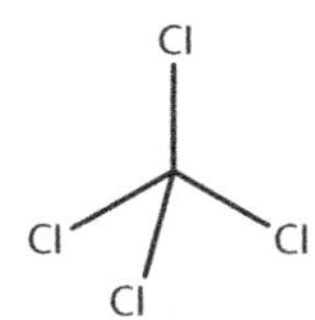	Solvent, ICH Q3C class 1 (4 ppm)

Catechol	120-80-9	neg	71.5	244	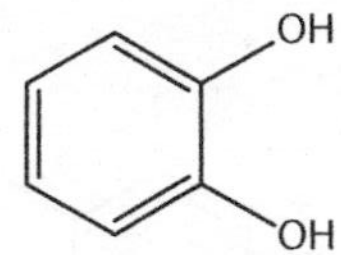	—
Chlorambucil	305-03-3	pos	0.896	0.133	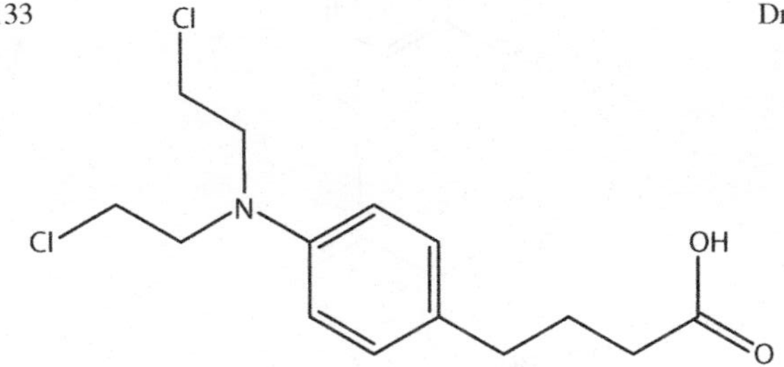	Drug, chemotherapy
Chlorine	7782-50-5	neg	—	—		—
1-Chloro-2-nitrobenzene	88-73-3	pos	—	157	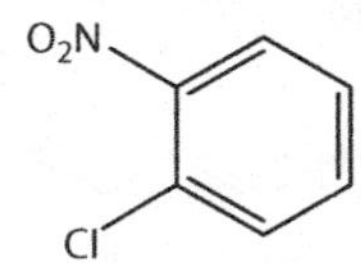	—
1-Chloro-4-nitrobenzene	100-00-5	pos	—	473	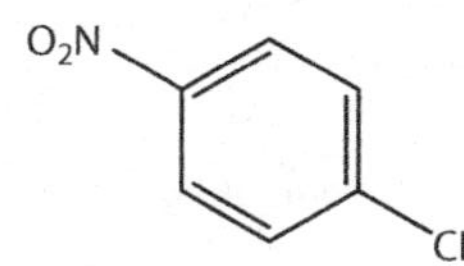	—

(Continued)

TABLE 3.13 (*Continued*)
Selected Compounds from the CPDB

Chemical Name	CAS	*Salmonella* (Ames)	Harmonic Mean of TD50 (mg/kg/day)		Structure	Note
			Rat	Mouse		
4-Chloro-*m*-phenylenediamine	5131-60-2	pos	315	1,230	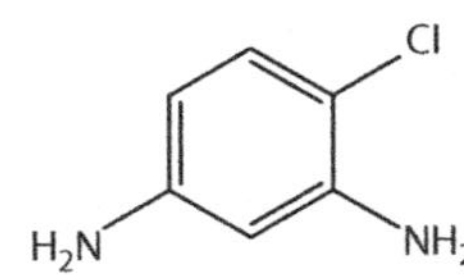	—
4-Chloro-*o*-phenylenediamine	95-83-0	pos	214	1,340	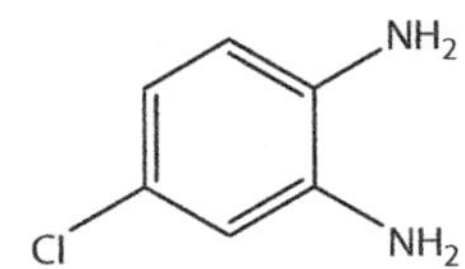	—
Chloroacetaldehyde	107-20-0	pos	—	36.1	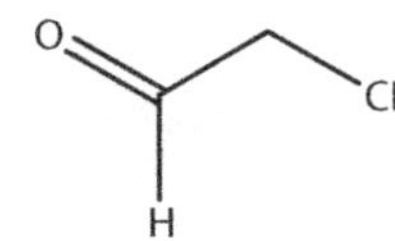	—
p-Chloroaniline HCl	20265-96-7	pos	7.62	89.5	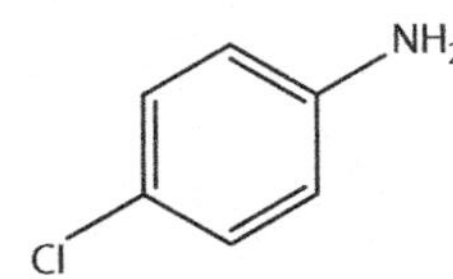	—

Chlorobenzene	108-90-7	neg	247	—	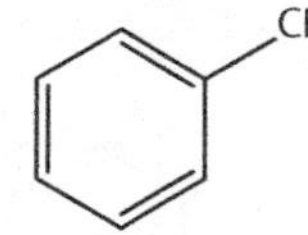	Solvent, ICH Q3C class 2 (PDE 3.6 mg/day, 360 ppm)
Chloroethane	75-00-3	pos	—	1,810		—
Chloroform	67-66-3	neg	262	111	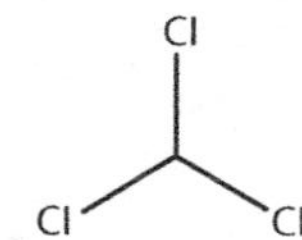	Solvent, ICH Q3C class 2 (PDE 0.6 mg/day, 60 ppm)
3-(Chloromethyl)pyridine HCl	6959-48-4	pos	433	229	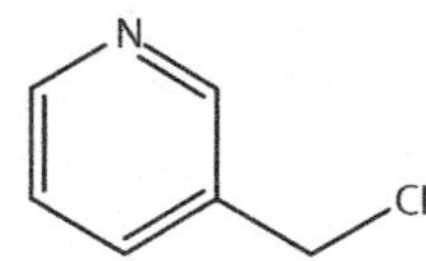	—
Citric acid	77-92-9	NE	—	—	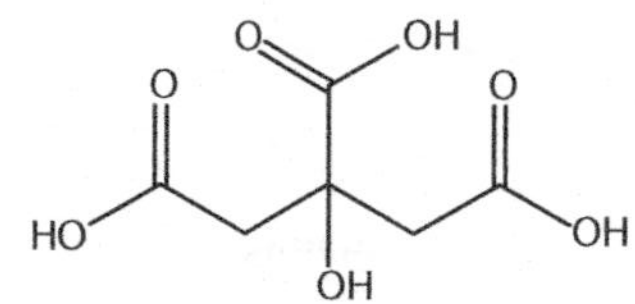	—

(*Continued*)

TABLE 3.13 (*Continued*)
Selected Compounds from the CPDB

Chemical Name	CAS	*Salmonella* (Ames)	Harmonic Mean of TD50 (mg/kg/day)		Structure	Note
			Rat	Mouse		
Coumarin	91-64-5	pos	39.2	103	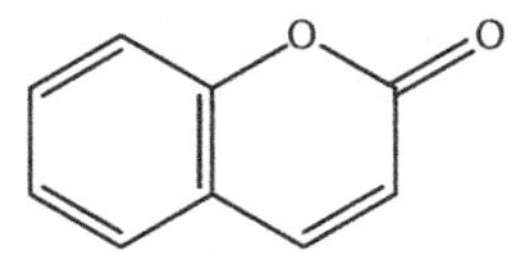	—
m-Cresidine	102-50-1	pos	470	—	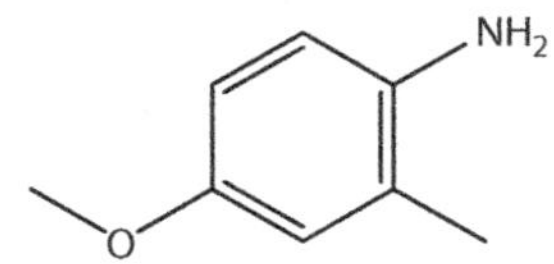	—
p-Cresidine	120-71-8	pos	98	54.3	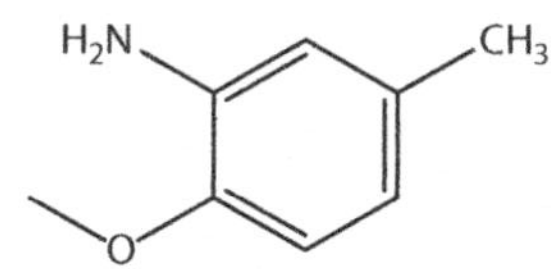	—
Cyclohexylamine HCl	4998-76-9	neg	—	—	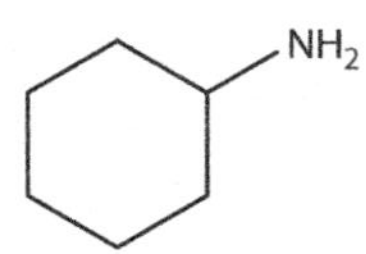	—

Cyclophosphamide	50-18-0	pos	2.21	5.96	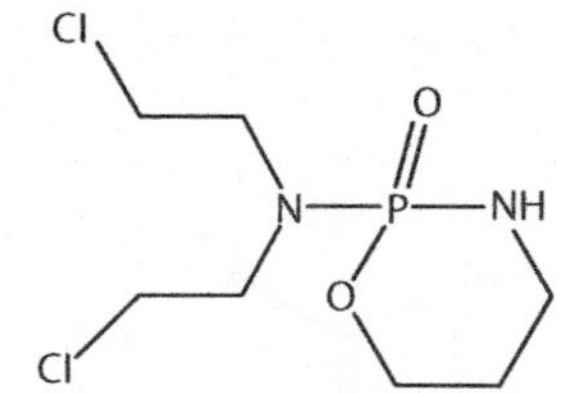	Cancer and immune disease treatment
2,4-Diaminophenol 2HCl	137-09-7	pos	—	143	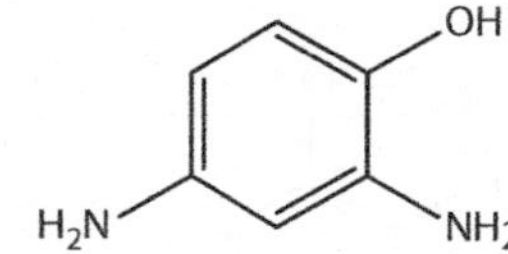	—
2,4-Diaminotoluene	95-80-7	pos	2.47	26.7	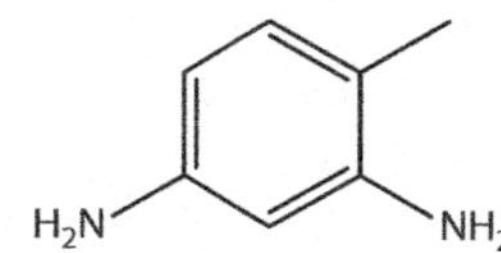	—
1,2-Dibromo-3-chloropropane	96-12-8	pos	0.259	2.72	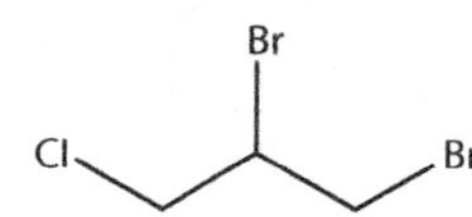	—
1,2-Dibromoethane	106-93-4	pos	1.52	7.45		—

(*Continued*)

TABLE 3.13 (*Continued*)
Selected Compounds from the CPDB

Chemical Name	CAS	*Salmonella* (Ames)	Harmonic Mean of TD50 (mg/kg/day)		Structure	Note
			Rat	Mouse		
Dibromomannitol	488-41-5	pos	27.6	14.9	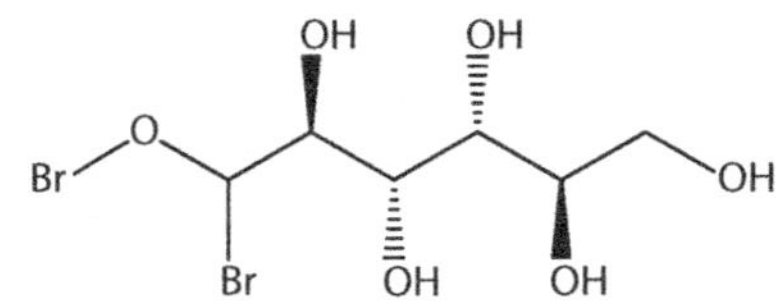	—
2,6-Dichloro-4-nitroaniline	99-30-9	pos	—	—	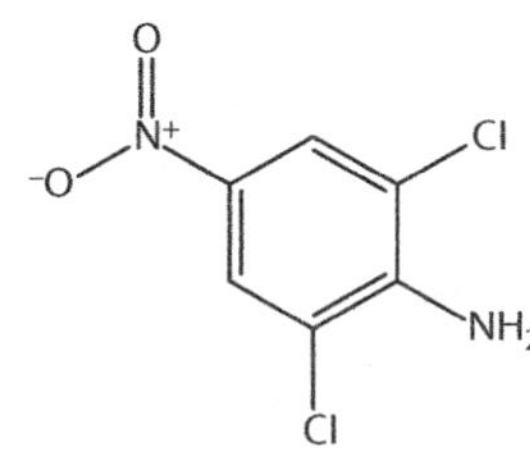	—
Dichloroacetic acid	79-43-6	pos	161	119	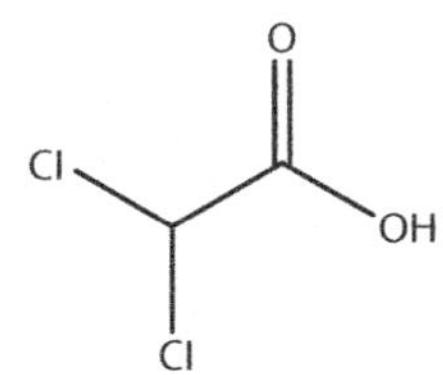	—

Dichloroacetylene	7572-29-4	NE	3.58	0.574		—
1,4-Dichlorobenzene	106-46-7	neg	644	323	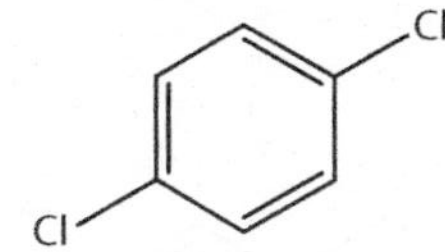	—
3,3′-Dichlorobenzidine	91-94-1	pos	28.1	—	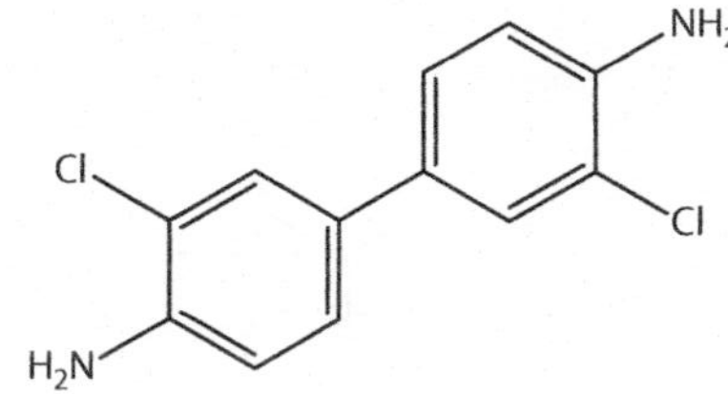	—
1,1-Dichloroethane	75-34-3	neg	—	—	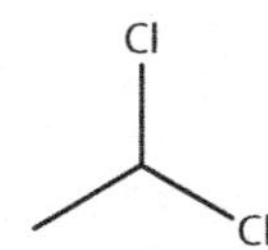	Solvent, ICH Q3C class 1 (8 ppm)
1,2-Dichloroethane	107-06-2	pos	14.6	138		Solvent, ICH Q3C class 1 (5 ppm)
2,4-Dimethoxyaniline HCl	54150-69-5	pos	—	—	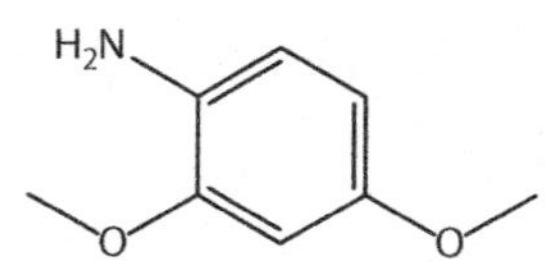	—

(*Continued*)

TABLE 3.13 (*Continued*)
Selected Compounds from the CPDB

Chemical Name	CAS	*Salmonella* (Ames)	Harmonic Mean of TD50 (mg/kg/day)		Structure	Note
			Rat	Mouse		
3,3′-Dimethoxybenzidine 2HCl	20325-40-0	pos	1.04	73.8	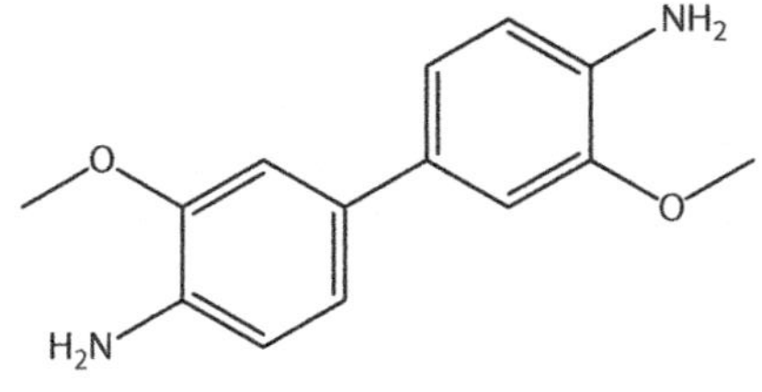	—
Dimethylacetamide	127-19-5	neg	—	—	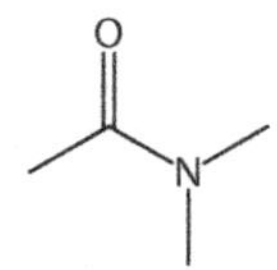	Solvent, ICH Q3C class 2 (PDE 10.9 mg/day, 1,090 ppm)
Dimethylformamide	68-12-2	neg	—	—	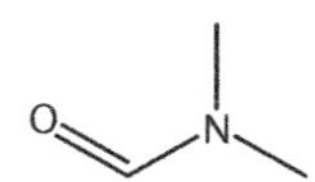	Solvent, ICH Q3C class 2 (PDE 8.8 mg/day, 880 ppm)

1,2-Dimethyl-5-nitroimidazole	551-92-8	pos	17	—	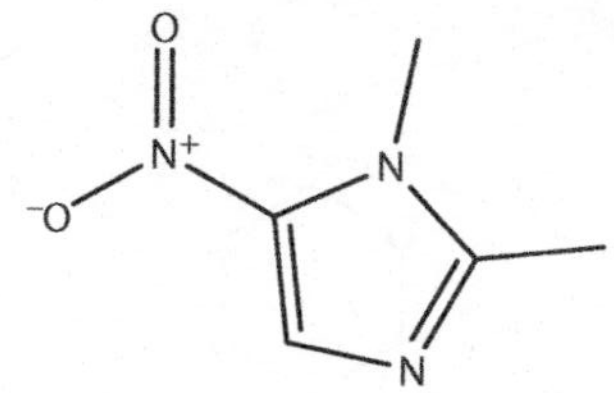	—
2,6-Dimethylaniline	87-62-7	pos	—	—	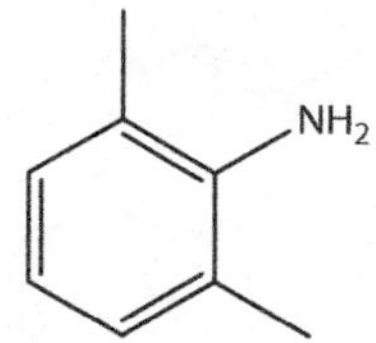	—
3,3′-Dimethylbenzidine 2HCl	612-82-8	pos	0.629	28.6	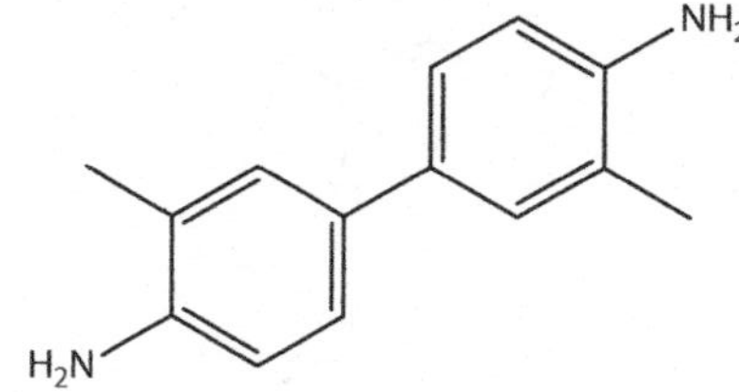	—
Dimethylcarbamyl chloride	79-44-7	pos	—	5.37	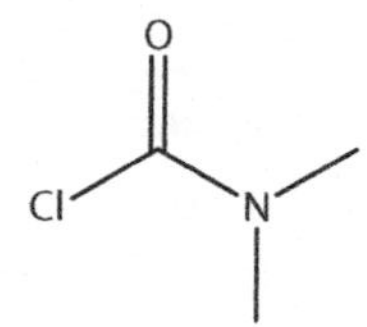	—

(Continued)

TABLE 3.13 (*Continued*)
Selected Compounds from the CPDB

Chemical Name	CAS	*Salmonella* (Ames)	Harmonic Mean of TD50 (mg/kg/day)		Structure	Note
			Rat	Mouse		
1,2-Dimethylhydrazine 2HCl	306-37-6	pos	—	0.114	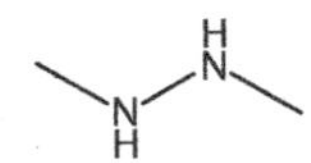	—
Dimethylvinyl chloride	513-37-1	pos	31.8	14.9	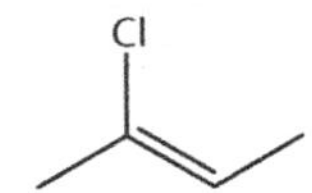	—
2,4-Dinitrotoluene	121-14-2	pos	6.21	29.4	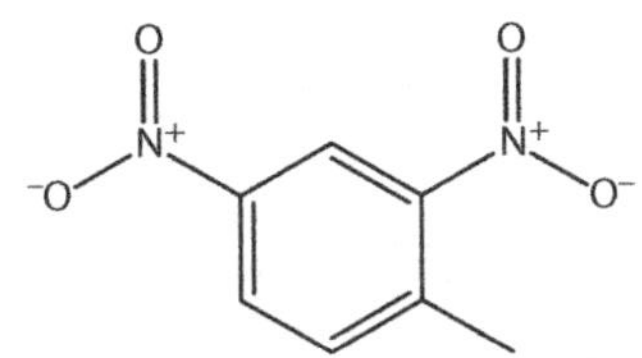	—
1,4-Dioxane	123-91-1	neg	267	204	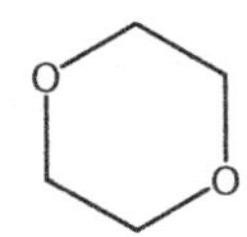	Solvent, ICH Q3C class 2 (PDE 3.8 mg/day, 380 ppm)

Dopamine HCl	62-31-7	pos	—	—	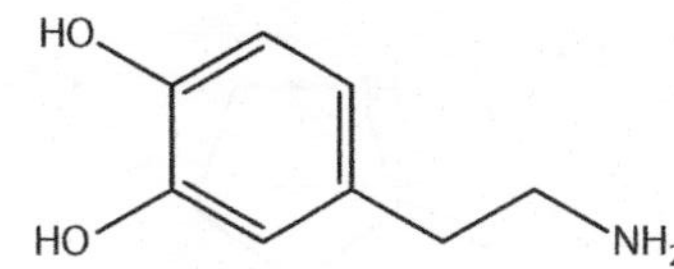	—
EDTA (ethylenediaminetet-raacetic acid), trisodium salt trihydrate	150-38-9	neg	—	—	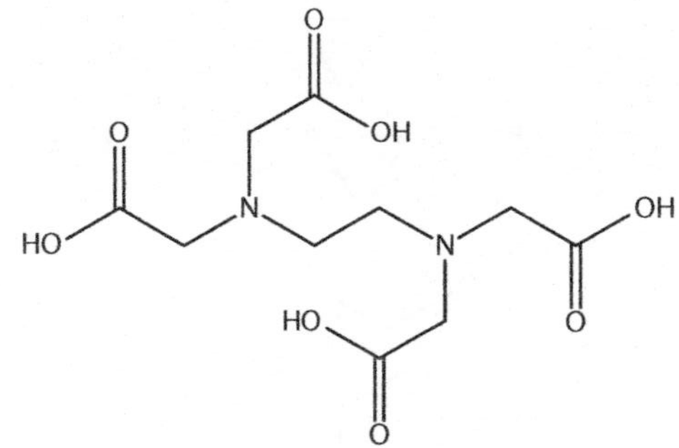	—
Epichlorohydrin	106-89-8	pos	2.96	—	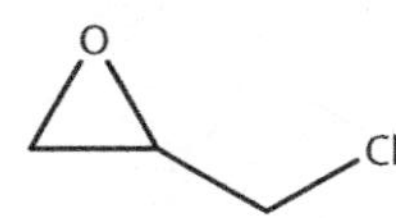	—
1,2-Epoxybutane	106-88-7	pos	220	—	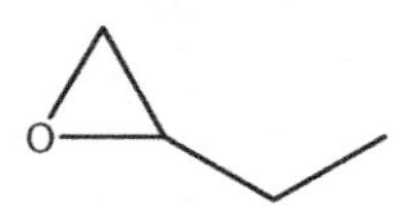	—
Ethyl alcohol	64-17-5	neg	9,110	—		Solvent, ICH Q3C class 3 (PDE 50 mg/day, 5,000 ppm)

(Continued)

TABLE 3.13 (*Continued*)
Selected Compounds from the CPDB

Chemical Name	CAS	*Salmonella* (Ames)	Harmonic Mean of TD50 (mg/kg/day)		Structure	Note
			Rat	Mouse		
Ethyl bromoacetate	105-36-2	pos	—	—	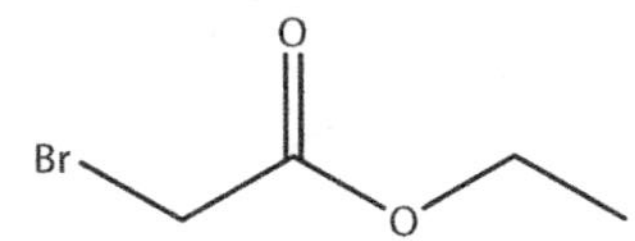	—
Ethylbenzene	100-41-4	neg	72.6	1,600	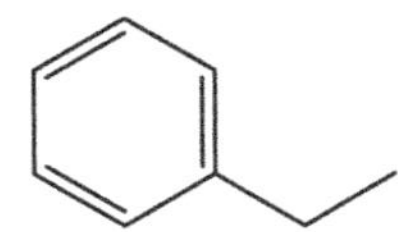	—
Ethylene glycol	107-21-1	neg	—	—		Solvent, ICH Q3C class 2 (PDE 6.2 mg/day, 620 ppm)
Ethylene imine	151-56-4	pos	—	0.377		—
Ethylene oxide	75-21-8	pos	21.3	63.7	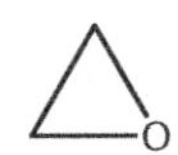	—
Ethylene thiourea	96-45-7	pos	8.13	23.5	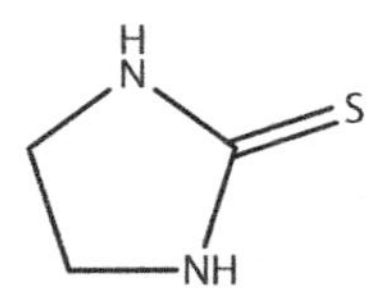	—

Ethylhydrazine HCl	18413-14-4	NE	—	6.56	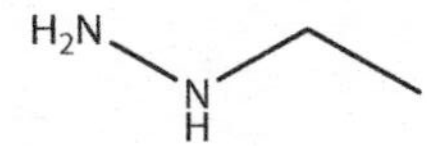	—
Formaldehyde	50-00-0	pos	1.35	43.9	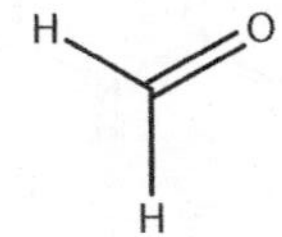	—
Furan	110-00-9	neg	0.396	2.72	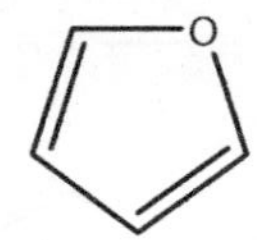	—
Furfural	98-01-1	pos	683	197	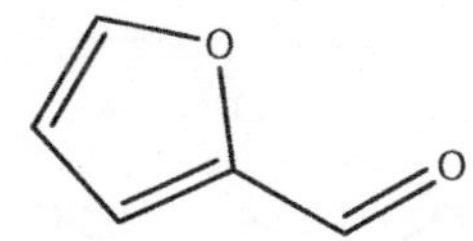	—
Furfuryl alcohol	98-00-0	neg	9.6	263	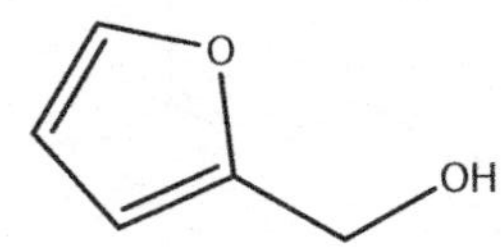	—
Glycidol	556-52-5	pos	4.28	34.7	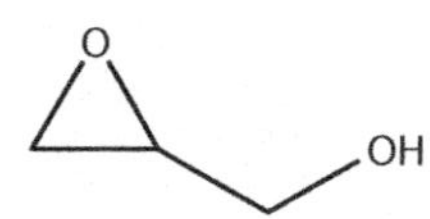	—

(*Continued*)

TABLE 3.13 (*Continued*)
Selected Compounds from the CPDB

Chemical Name	CAS	*Salmonella* (Ames)	Harmonic Mean of TD50 (mg/kg/day)		Structure	Note
			Rat	Mouse		
Hexachlorobenzene	118-74-1	neg	3.86	65.1	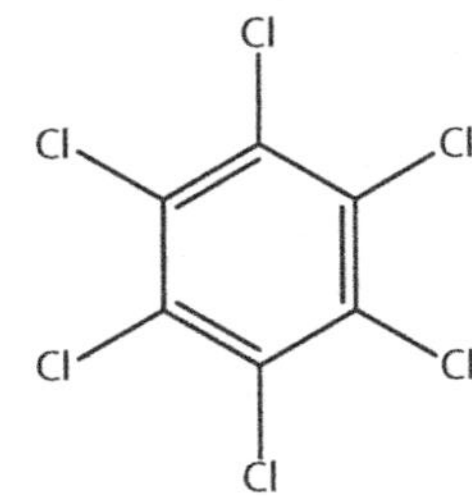	—
Hexane	110-54-3	neg	—	30,300		Solvent, ICH Q3C class 2 (PDE 2.9 mg/day, 290 ppm)
Hydrazine	302-01-2	pos	0.613	2.93	H_2N-NH_2	—
Hydrazobenzene	122-66-7	pos	5.59	26	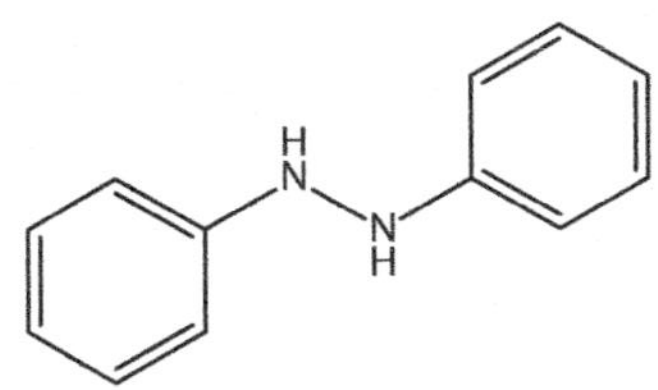	—

Hydrochloric acid	7647-01-0	NE	—	—	H—Cl	—
Hydrogen peroxide	7722-84-1	pos	—	7,540	HO—OH	—
Hydroquinone	123-31-9	neg	82.8	225	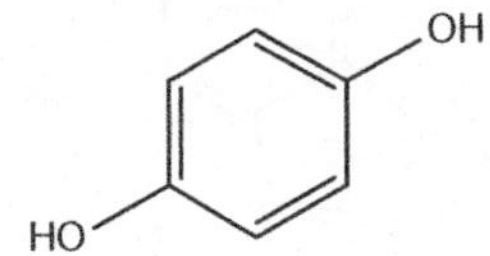	—
2-Hydroxyethylhydrazine	109-84-2	pos	—	0.397	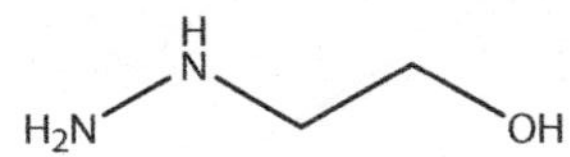	—
Indole	120-72-9	NE	—	—	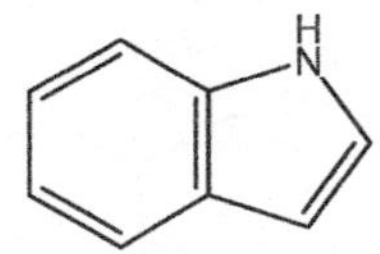	—
IQ (3-methyl-3*H*-imidazo [4,5-*F*]quinolin-2-amine)	76180-96-6	pos	0.812	19.6	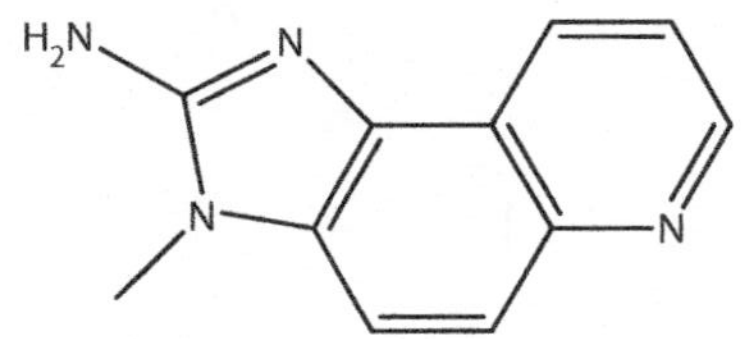	Found in cooked meat
Isobutyl nitrite	542-56-3	pos	54.1	310	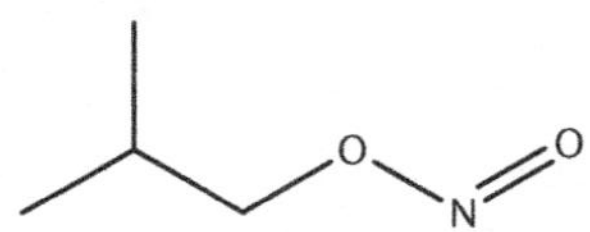	—

(*Continued*)

TABLE 3.13 (*Continued*)
Selected Compounds from the CPDB

Chemical Name	CAS	*Salmonella* (Ames)	Harmonic Mean of TD50 (mg/kg/day)		Structure	Note
			Rat	Mouse		
Isophosphamide	3778-73-2	pos	0.739	5.06	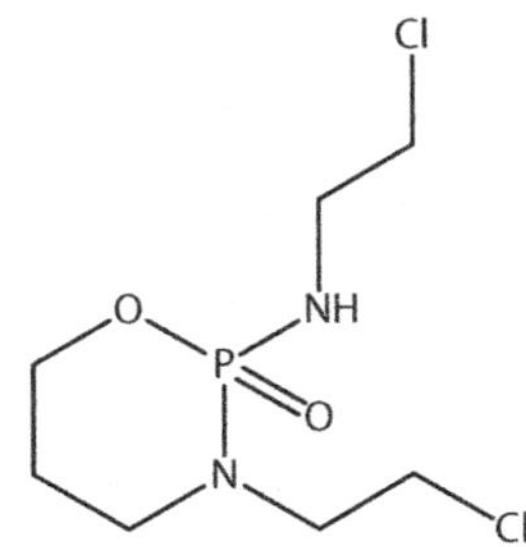	Cancer treatment
Isopropanol	67-63-0	neg	—	—	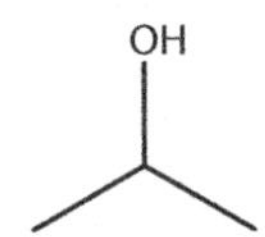	Solvent, ICH Q3C class 3 (PDE 50 mg/day, 5,000 ppm)
MeIQ (2-amino-3,4-dimethyl-3*H*-imidazo[4,5-*F*] quinoline)	77094-11-2	pos	—	15.5	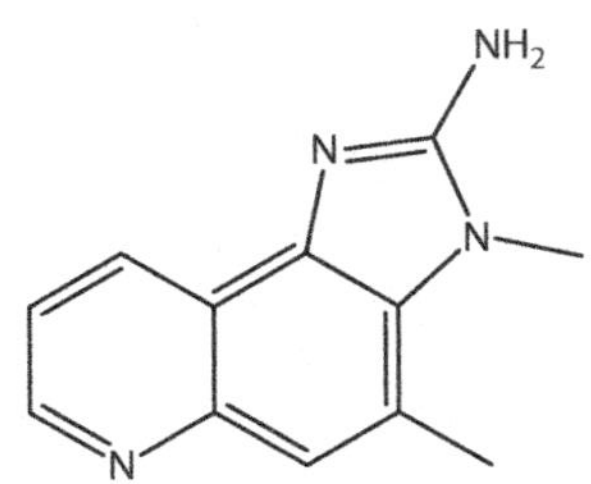	Found in cooked meat

MeIQx (2-amino-3,8-dimethylimidazo[4,5-*F*] quinoxaline)	77500-04-0	pos	1.66	24.3	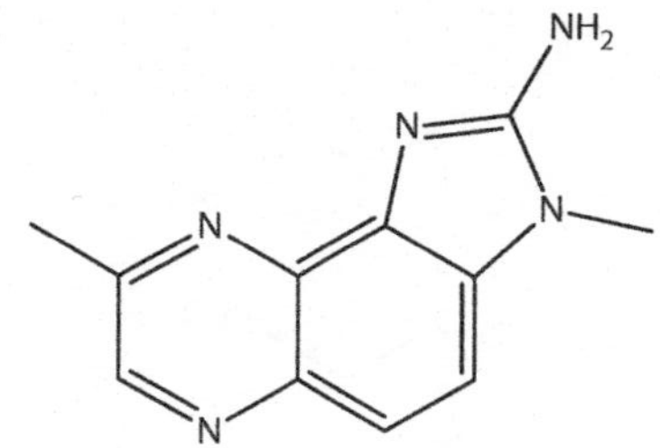	Found in cooked meat
4-Methoxyphenol	150-76-5	neg	658	—	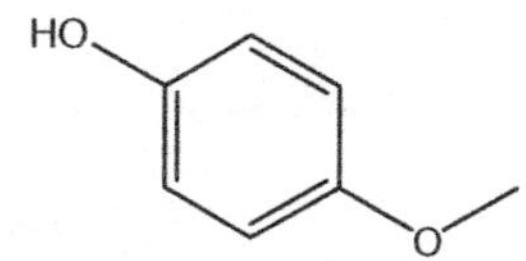	—
Methyl bromide	74-83-9	pos	—	—		—
Methyl *tert*-butyl ether	1634-04-4	neg	702	6,550	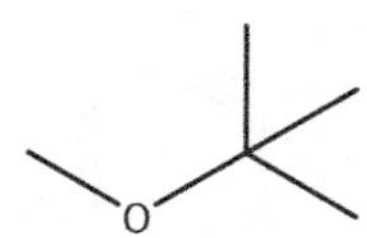	Solvent, ICH Q3C class 3 (PDE 50 mg/day, 5,000 ppm)
Methyl carbamate	598-55-0	neg	56.6	—	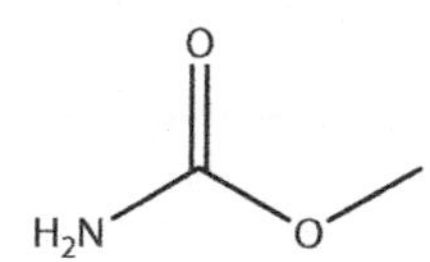	—

(*Continued*)

TABLE 3.13 (*Continued*)
Selected Compounds from the CPDB

Chemical Name	CAS	*Salmonella* (Ames)	Harmonic Mean of TD50 (mg/kg/day)		Structure	Note
			Rat	Mouse		
4-Methylcatechol	452-86-8	neg	248	—	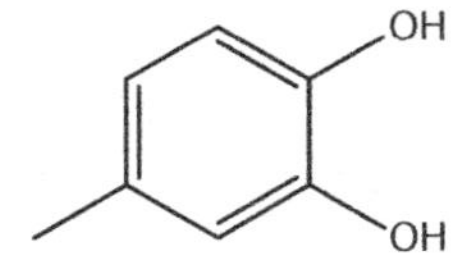	—
Methyl methanesulfonate	66-27-3	pos	—	31.8	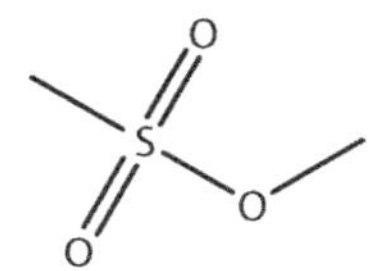	—
N-methyl-2-pyrrolidone	872-50-4	neg	—	2,050	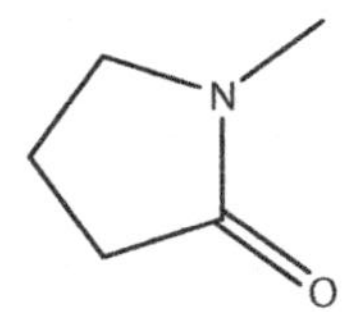	Solvent, ICH Q3C class 2 (PDE 5.3 mg/day, 530 ppm)
Methylene chloride	75-09-2	pos	724	1,100		Solvent, ICH Q3C class 2 (PDE 6.0 mg/day, 600 ppm)
Methylhydrazine	60-34-4	pos	—	7.55	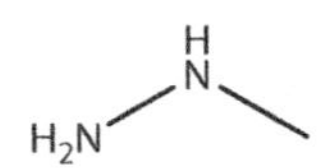	—

2-Methylimidazole	693-98-1	neg	868	782	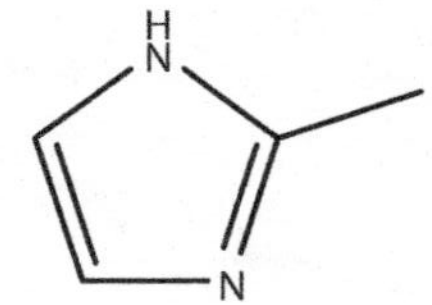	—
Methylnitramine	598-57-2	NE	17.4	—	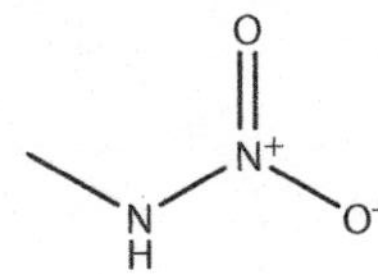	—
6-Methylquinoline	91-62-3	pos	—	—	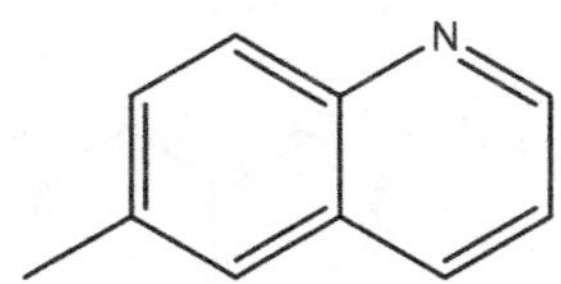	—
Mitomycin-C	50-07-7	pos	0.00102	—	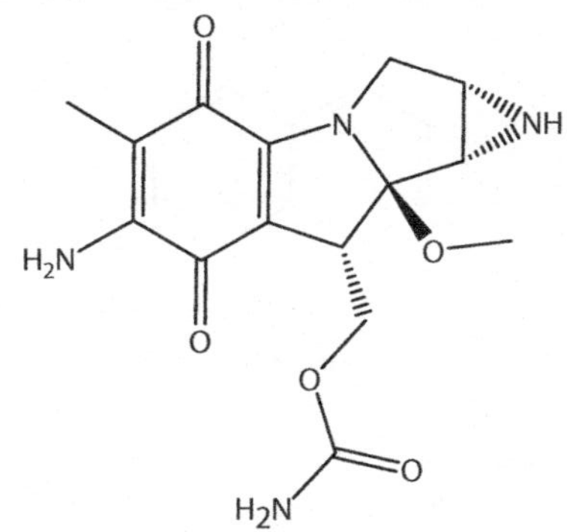	Cancer treatment, DNA cross-linking

(*Continued*)

TABLE 3.13 (*Continued*)
Selected Compounds from the CPDB

Chemical Name	CAS	*Salmonella* (Ames)	Harmonic Mean of TD50 (mg/kg/day)		Structure	Note
			Rat	Mouse		
Naphthalene	91-20-3	neg	22.1	163	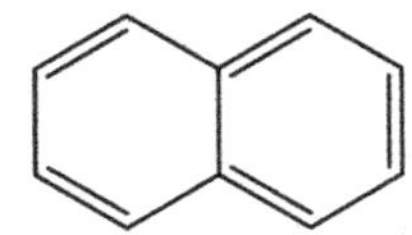	—
2-Naphthylamine	91-59-8	pos	61.6	39.4	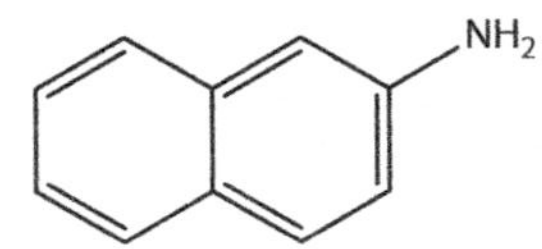	—
Nitrate, sodium	7631-99-4	NE	—	—	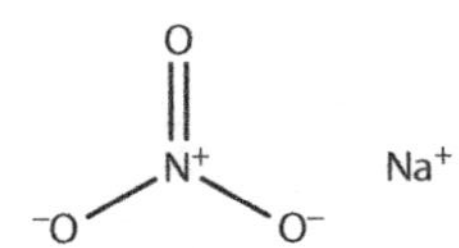	—
Nitrite, sodium	7632-00-0	pos	167	—	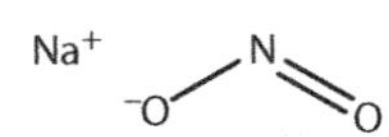	—

5-Nitro-*o*-anisidine	99-59-2	pos	53.9	3,720	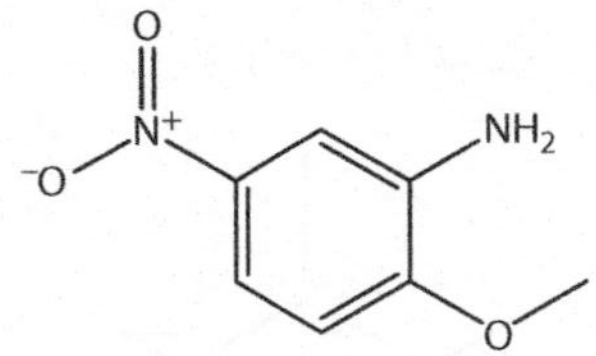	—
p-Nitroaniline	100-01-6	pos	—	—	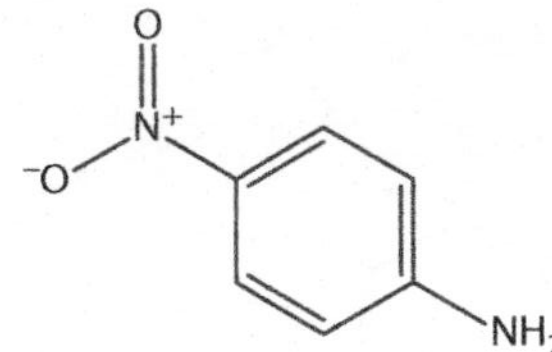	—
o-Nitroanisole	91-23-6	pos	15.6	178	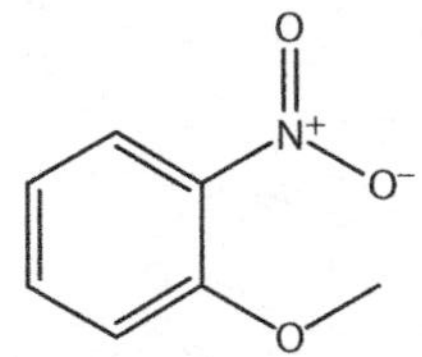	—
Nitrobenzene	98-95-3	neg	25.5	296	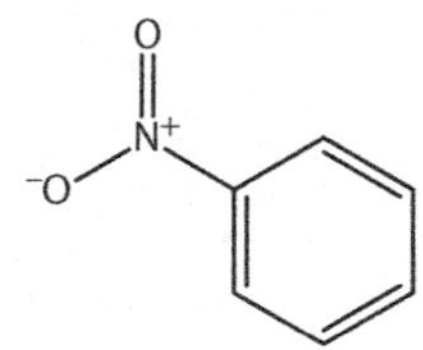	—

(*Continued*)

TABLE 3.13 (*Continued*)
Selected Compounds from the CPDB

Chemical Name	CAS	*Salmonella* (Ames)	Harmonic Mean of TD50 (mg/kg/day)		Structure	Note
			Rat	Mouse		
6-Nitrobenzimidazole	94-52-0	pos	—	372	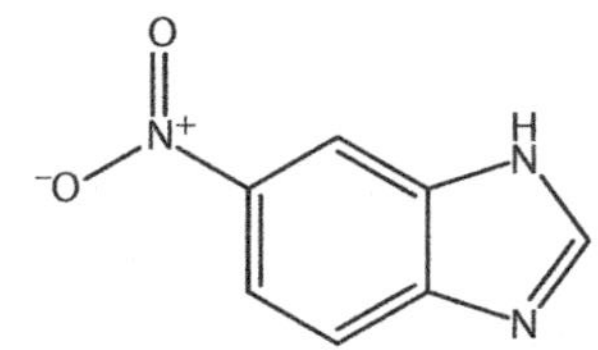	—
p-Nitrobenzoic acid	62-23-7	pos	287	—	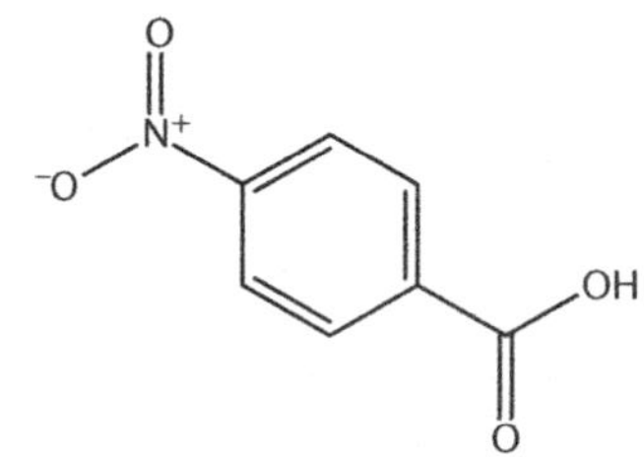	—
Nitrogen mustard	51-75-2	pos	0.0114	—	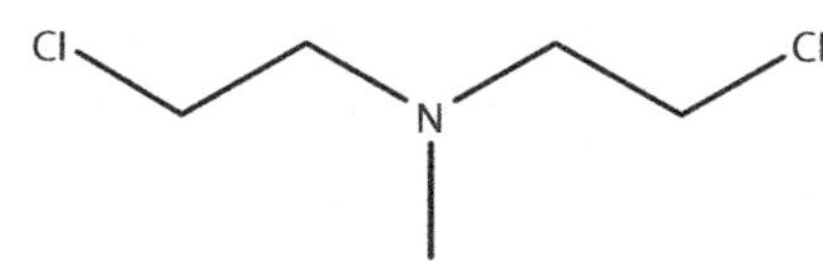	—

Nitrogen mustard *N*-oxide	126-85-2	NE	0.764	—	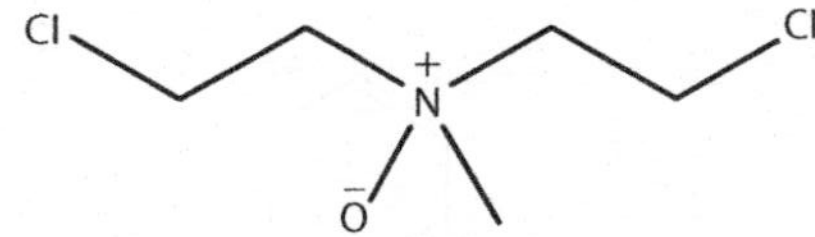	—
Nitromethane	75-52-5	neg	40.4	469	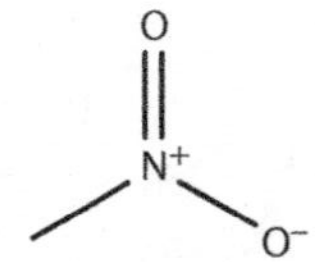	Solvent, ICH Q3C class 2 (PDE 0.5 mg/day, 50 ppm)
1-Nitronaphthalene	86-57-7	pos	—	—	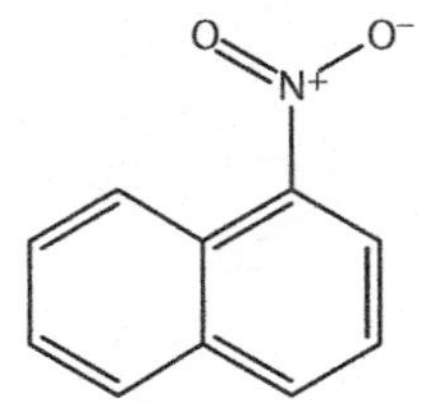	—
8-Nitroquinoline	607-35-2	pos	9.82	—	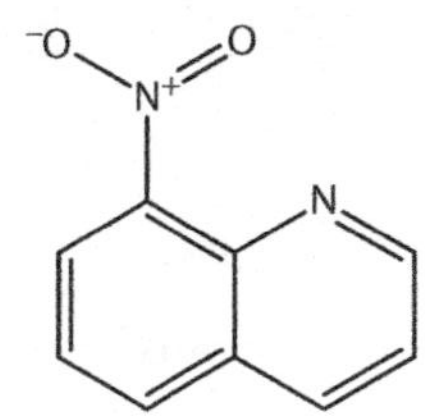	—

(*Continued*)

TABLE 3.13 (*Continued*)
Selected Compounds from the CPDB

Chemical Name	CAS	*Salmonella* (Ames)	Harmonic Mean of TD50 (mg/kg/day)		Structure	Note
			Rat	Mouse		
Omeprazole	73590-58-6	NE	119	—	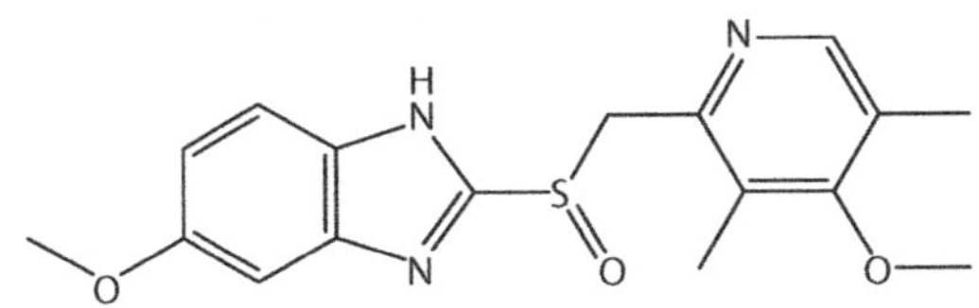	Drug, proton pump inhibitor
Ozone	10028-15-6	pos	—	1.88	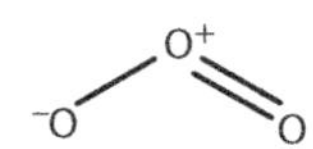	—
Phenobarbital	50-06-6	pos	—	7.37	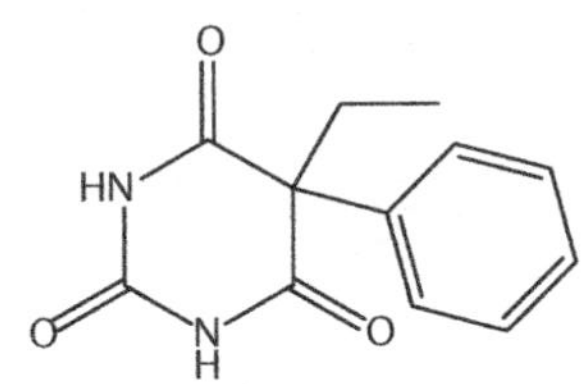	Drug, anticonvulsant
Phenol	108-95-2	neg	—	—	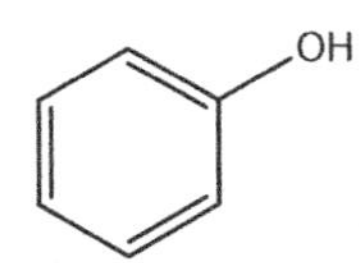	—

o-Phenylenediamine 2HCl	615-28-1	pos	248	735	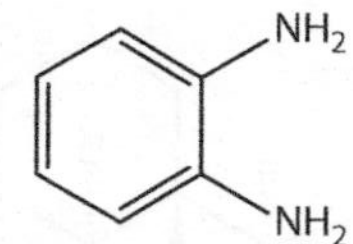	—
Phenylhydrazine HCl	59-88-1	pos	—	71.3	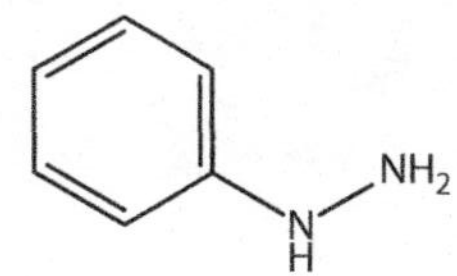	—
o-Phenylphenol	90-43-7	pos	232	—	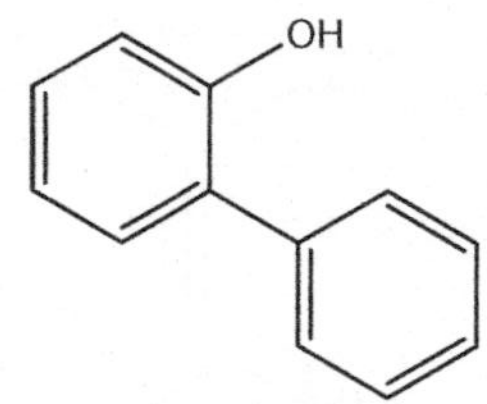	—
Piperazine	110-85-0	neg	—	—	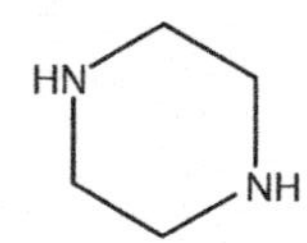	—

(Continued)

TABLE 3.13 (*Continued*)
Selected Compounds from the CPDB

Chemical Name	CAS	*Salmonella* (Ames)	Harmonic Mean of TD50 (mg/kg/day)		Structure	Note
			Rat	Mouse		
Piperidine	110-89-4	NE	—	—	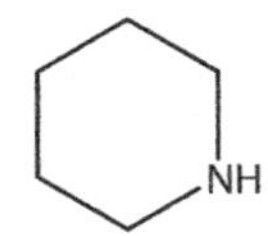	—
Potassium bicarbonate	198-14-6	NE	13,000	—	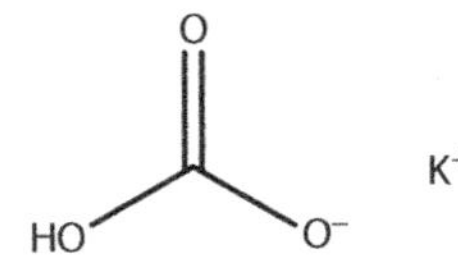	—
Potassium chloride	7447-40-7	neg	—	—	K^+ Cl^-	—
Potassium iodide	7681-11-0	NE	440	—	K^+ I^-	—
Prednisolone	50-24-8	neg	1.53	—	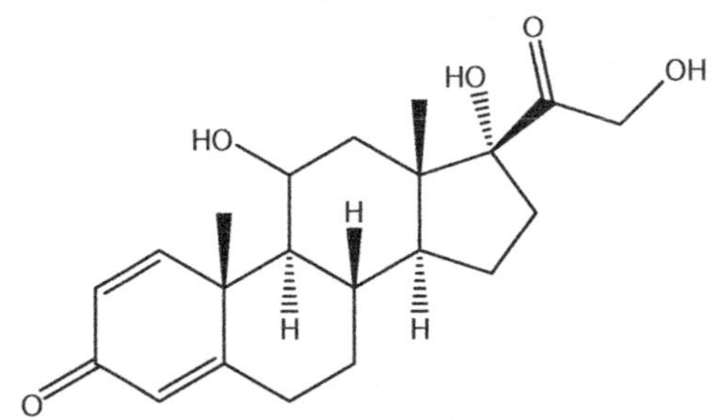	Drug, anti-inflammation

Propranolol HCl	318-98-9	NE	—	—	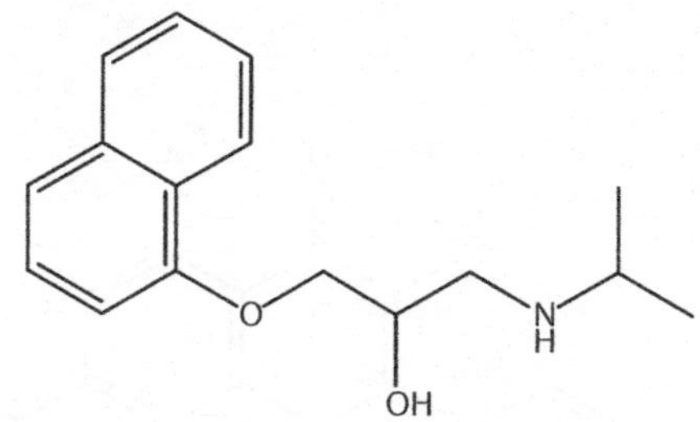	Drug, β-blocker
Propylene	115-07-1	pos	—	—	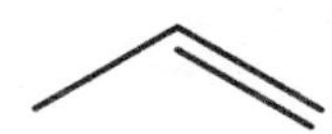	—
Propylene glycol	57-55-6	neg	—	—	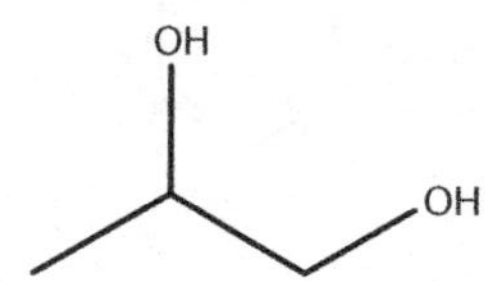	—
1,2-Propylene oxide	75-56-9	pos	74.4	912	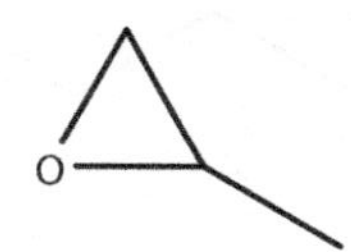	—

(*Continued*)

TABLE 3.13 (*Continued*)
Selected Compounds from the CPDB

Chemical Name	CAS	*Salmonella* (Ames)	Harmonic Mean of TD50 (mg/kg/day)		Structure	Note
			Rat	Mouse		
Propylhydrazine HCl	56795-66-5	NE	—	45.5	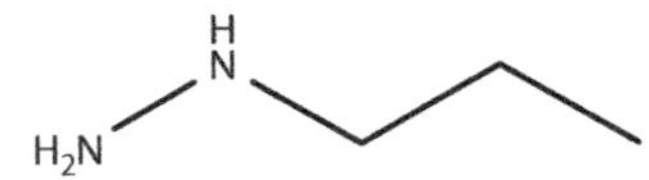	—
Pyridine	110-86-1	neg	67.3	24.4	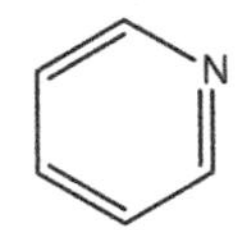	Solvent, ICH Q3C class 2 (PDE 2.0 mg/day, 200 ppm)
Quercetin	117-39-5	pos	10.1	—	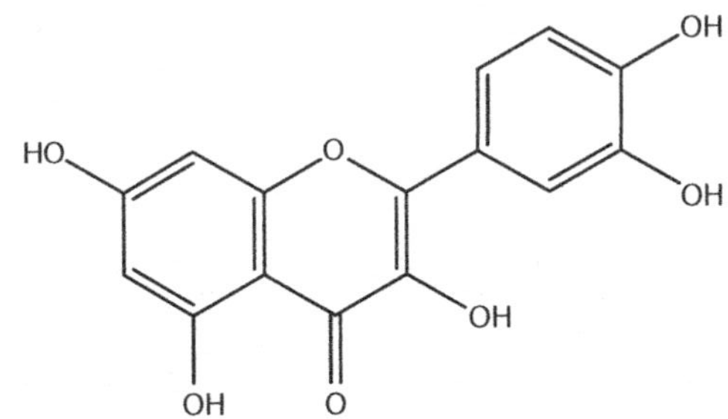	—

Reserpine	50-55-5	neg	0.306	5.02	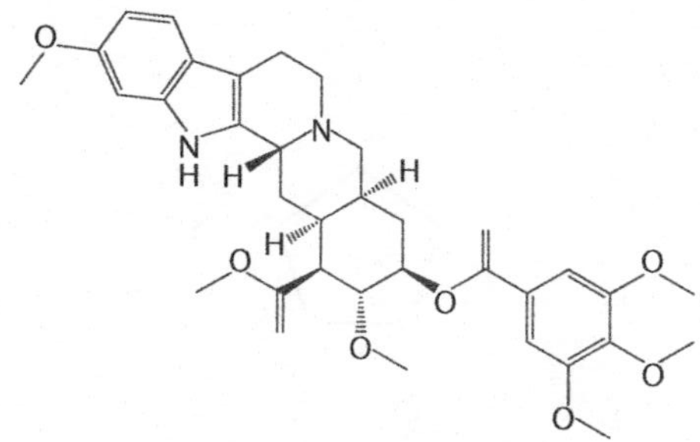	High blood pressure treatment
Safrole	94-59-7	neg	441	51.3	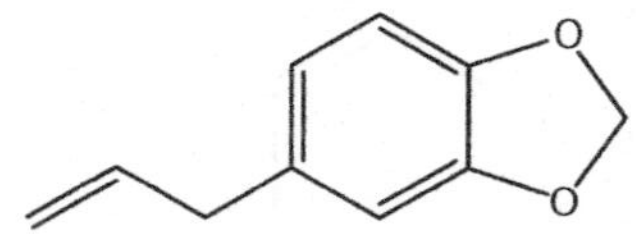	Naturally occurring in herbs and spices
Styrene	100-42-5	pos	23.3	210	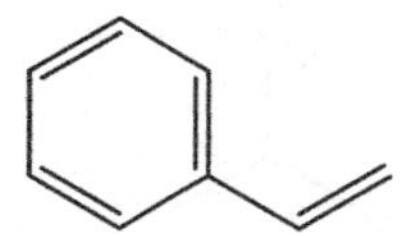	—
Styrene oxide	96-09-3	pos	55.4	118	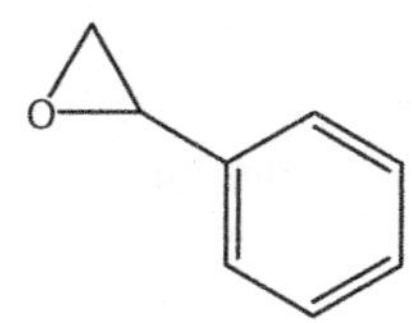	—

(*Continued*)

TABLE 3.13 (*Continued*)
Selected Compounds from the CPDB

Chemical Name	CAS	*Salmonella* (Ames)	Harmonic Mean of TD50 (mg/kg/day)		Structure	Note
			Rat	Mouse		
1,1,2,2-Tetrachloroethane	79-34-5	neg	—	38.3	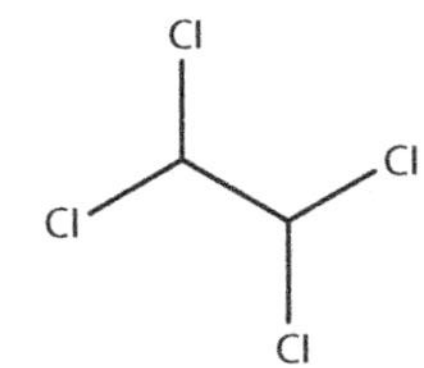	—
Tetrachloroethylene	127-18-4	neg	145	158	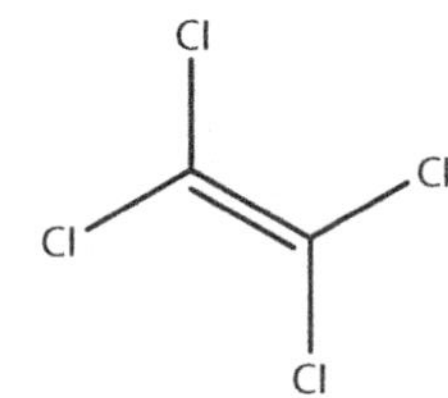	—
Tetrahydrofuran	109-99-9	neg	407	1,300		Solvent, ICH Q3C class 2 (PDE 7.2 mg/day, 720 ppm)

Thioacetamide	62-55-5	neg	11.5	8.81	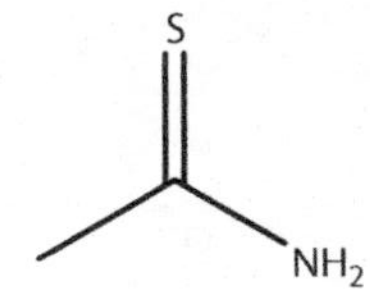	—
4,4′-Thiodianiline	139-65-1	pos	3.71	33.2	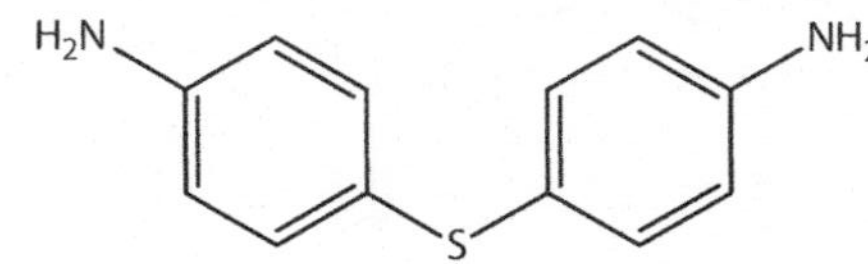	—
Thiourea	62-56-6	neg	98.5	—	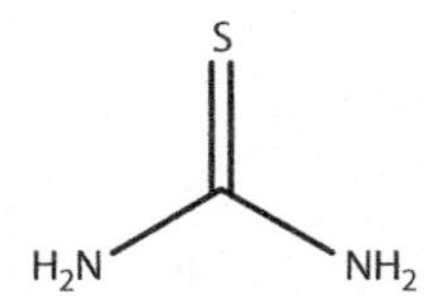	—
Toluene	108-88-3	neg	3,060	—	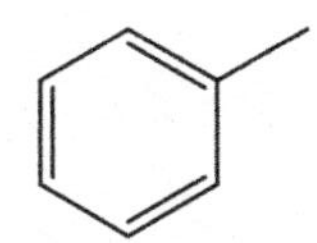	Solvent, ICH Q3C class 2 (PDE 8.9 mg/day, 890 ppm)
o-Toluenesulfonamide	88-19-7	neg	3,960	—	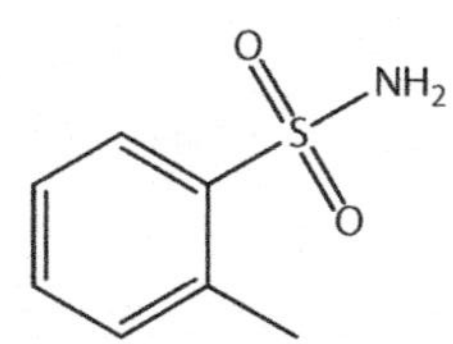	—

(*Continued*)

TABLE 3.13 (*Continued*)
Selected Compounds from the CPDB

Chemical Name	CAS	*Salmonella* (Ames)	Harmonic Mean of TD50 (mg/kg/day)		Structure	Note
			Rat	Mouse		
o-Toluidine HCl	636-21-5	pos	43.6	840	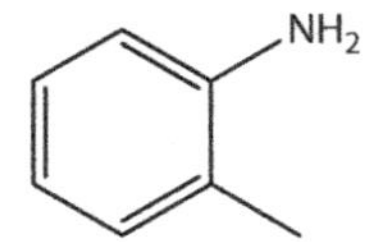	—
p-Tolylurea	622-51-5	neg	—	206	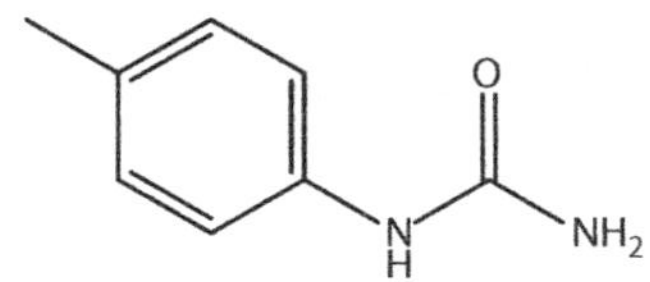	—
Tributyl phosphate	126-73-8	neg	191	1,120	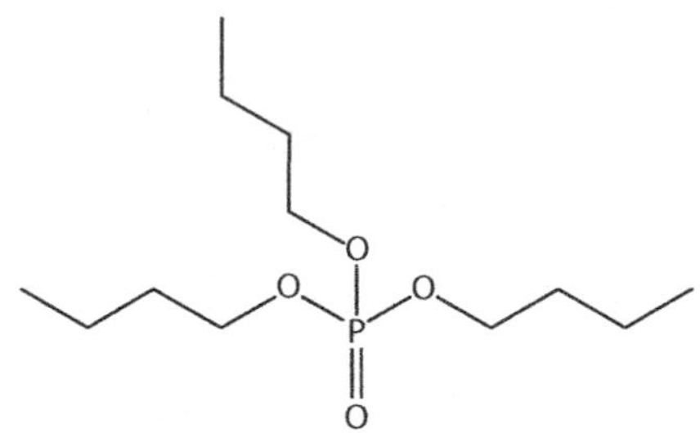	—

Trichloroacetic acid	76-03-9	neg	—	584	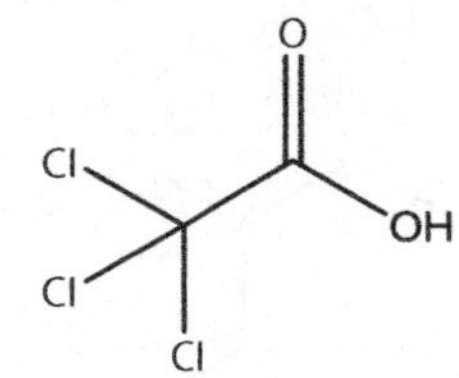	—
2,4,6-Trichloroaniline	634-93-5	neg	—	259	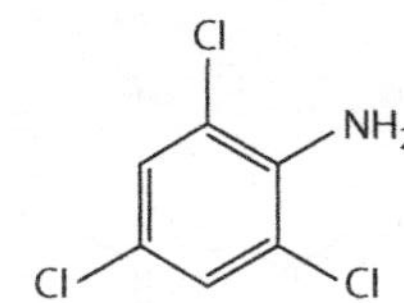	—
1,1,2-Trichloroethane	79-00-5	neg	—	55	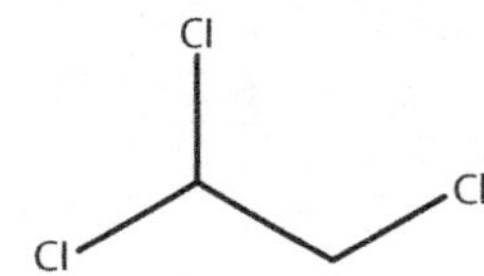	—
1,1,1-Trichloroethane	71-55-6	pos	—	—	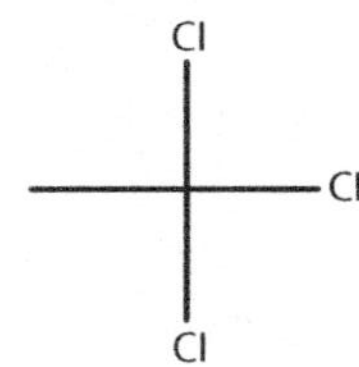	Solvent, ICH Q3C class 1 (1,500 ppm)

(Continued)

TABLE 3.13 (*Continued*)
Selected Compounds from the CPDB

Chemical Name	CAS	*Salmonella* (Ames)	Harmonic Mean of TD50 (mg/kg/day)		Structure	Note
			Rat	Mouse		
Trichloroethylene	79-01-6	neg	668	1,580	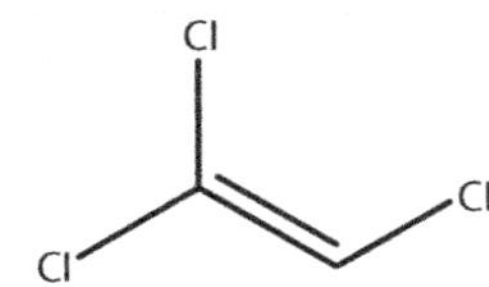	Solvent, ICH Q3C class 2 (PDE 0.8 mg/day, 80 ppm)
2,4,6-Trichlorophenol	88-06-2	neg	405	1,070	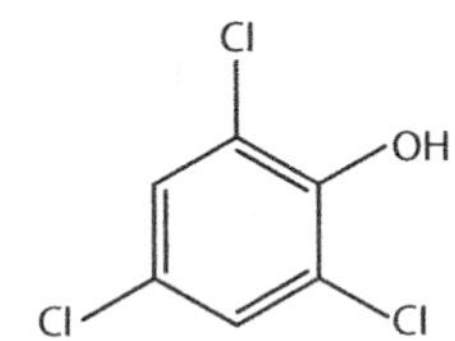	—
Triethanolamine	102-71-6	neg	—	100	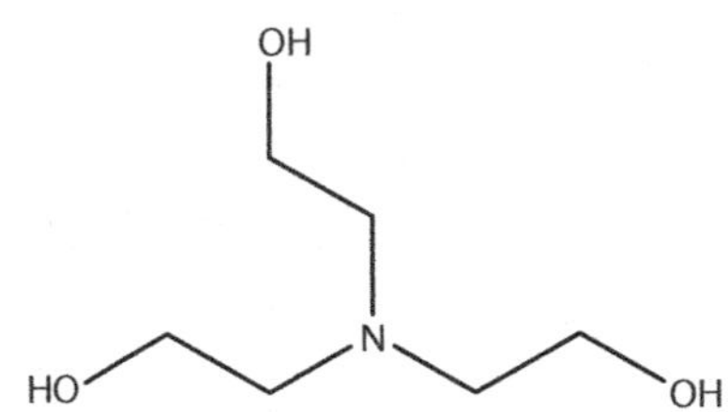	—

2,4,6-Trimethylaniline HCl	6334-11-8	NE	5.17	24.8	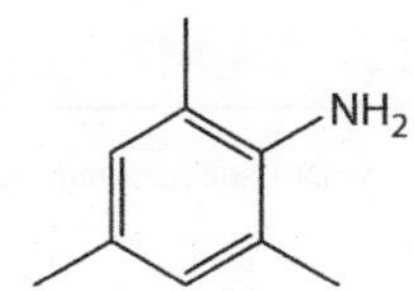	—
Trimethylphosphate	512-56-1	pos	—	335	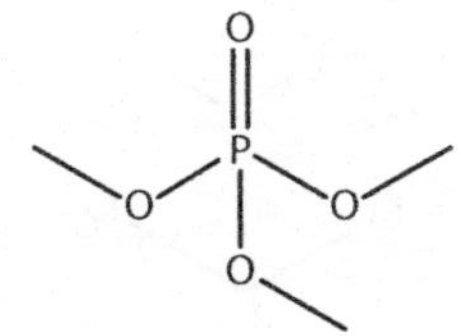	—
Trimethylthiourea	2489-77-2	neg	25.8	—	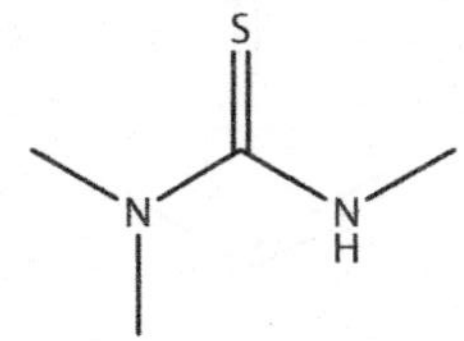	—
Urea	57-13-6	neg	—	—	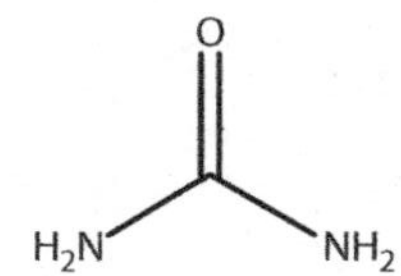	—
Vinyl acetate	108-05-4	neg	341	3,920	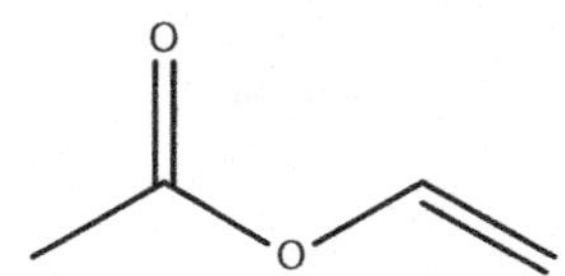	—

(*Continued*)

TABLE 3.13 (*Continued*)
Selected Compounds from the CPDB

Chemical Name	CAS.	*Salmonella* (Ames)	Harmonic Mean of TD50 (mg/kg/day)		Structure	Note
			Rat	Mouse		
Vinyl carbamate	15805-73-9	NE	—	0.124	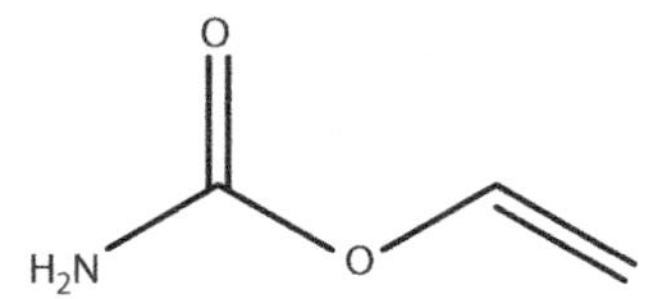	—
Vinyl chloride	75-01-4	pos	6.11	21.8	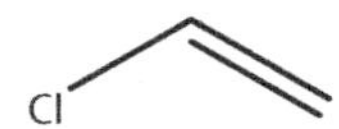	—
Vinylidene chloride	75-35-4	pos	—	34.6	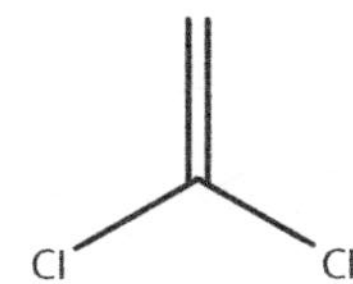	—
Xylene mixture	1330-20-7	neg	3,110	—	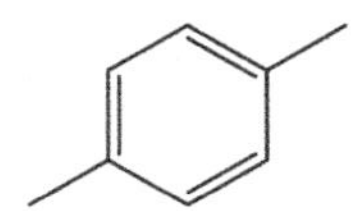	Solvent, ICH Q3C class 2 (PDE 21.7 mg/day, 2,170 ppm)

Note: For mutagenicity, pos, neg, and NE indicate positive, negative, and not evaluated, respectively. CAS, chemical abstract service registry number.

TABLE 3.14
Linear Extrapolation (LE) Calculation from TD50 for Selected Mutagenic Compounds in CPDB

			Harmonic Mean of TD50 (mg/kg/day)		LE Calculation, 50 kg Body Wt. Used (μg/day)		Most Sensitive Species (μg/day)
Chemical Name	**CAS**	***Salmonella* (Ames)**	**Rat**	**Mouse**	**Rat**	**Mouse**	**Lower Value**
1-Acetyl-2-phenylhydrazine	114-83-0	pos	—	51.2	—	51.2	51.2
Aflatoxin B1	1162-65-8	pos	0.0032	—	0.0032	—	0.0032
Allyl chloride	107-05-1	pos	—	—	—	—	
Allyl isothiocyanate	57-06-7	pos	96	—	96	—	96
2-Amino-4-nitrophenol	99-57-0	pos	839	—	839	—	839
2-Amino-5-nitrophenol	121-88-0	pos	111	—	111	—	111
4-Amino-2-nitropohenol	119-34-6	pos	309	—	309	—	309
2-Amino-5-nitrothiazole	121-66-4	pos	44.6	—	44.6	—	44.6
2-Aminoanthraquinone	117-79-3	pos	101	1,190	101	1,190	101
o-Aminoazotoluene	97-56-3	pos	4.04	—	4.04	—	4.04
4-Aminobiphenyl	92-67-1	pos	—	2.1	—	2.1	2.1
o-Anisidine HCl	134-29-2	pos	29.7	966	29.7	966	29.7
p-Anisidine HCl	20265-97-8	pos	—	—	—	—	—
Aristolochic acid, sodium salt	10190-99-5	pos	0.0141	—	0.0141	—	0.0141
Azide, sodium	26628-22-8	pos	—	—	—	—	—
Azobenzene	103-33-3	pos	24.1	—	24.1	—	24.1
Azoxymethane	25843-45-2	pos	0.0466	—	0.0466	—	0.0466
AZT	30516-87-1	pos	11,600	296	11,600	296	296
Benzidine	92-87-5	pos	1.73	19.9	1.73	19.9	1.73
Benzo(*a*)pyrene	50-32-8	pos	0.956	3.47	0.956	3.47	0.956

(Continued)

TABLE 3.14 ***(Continued)***
Linear Extrapolation (LE) Calculation from TD50 for Selected Mutagenic Compounds in CPDB

			Harmonic Mean of TD50 (mg/kg/day)		LE Calculation, 50 kg Body Wt. Used (μg/day)		Most Sensitive Species (μg/day)
Chemical Name	**CAS**	***Salmonella* (Ames)**	**Rat**	**Mouse**	**Rat**	**Mouse**	**Lower Value**
2-Biphenylamine HCl	2185-92-4	pos	—	1120	—	1,120	1120
Bis-2-chloroethylether	111-44-4	pos	—	11.7	—	11.7	11.7
Bromodichloromethane	75-27-4	pos	72.5	47.7	72.5	47.7	47.7
Bromoethane	74-96-4	pos	149	535	149	535	149
Bromoethanol	540-51-2	pos	—	76.1	—	76.1	76.1
1,3-Butadiene	106-99-0	pos	261	13.9	261	13.9	13.9
N-n-butyl-*N*-nitrosourea	869-01-2	pos	0.517	—	0.517	—	0.517
Chlorambucil	305-03-3	pos	0.896	0.133	0.896	0.133	0.133
4-Chloro-*m*-phenylenediamine	5131-60-2	pos	315	1,230	315	1,230	315
4-Chloro-*o*-phenylenediamine	95-83-0	pos	214	1,340	214	1,340	214
Chloroacetaldehyde	107-20-0	pos	—	36.1	—	36.1	36.1
3-(Chloromethyl)pyridine HCl	6959-48-4	pos	433	229	433	229	229
Coumarin	91-64-5	pos	39.2	103	39.2	103	39.2
m-Cresidine	102-50-1	pos	470	—	470	—	470
Cyclophosphamide	50-18-0	pos	2.21	5.96	2.21	5.96	2.21
2,4-Diaminophenol 2HCl	137-09-7	pos	—	143	—	143	143
2,4-Diaminotoluene	95-80-7	pos	2.47	26.7	2.47	26.7	2.47
1,2-Dibromo-3-chloropropane	96-12-8	pos	0.259	2.72	0.259	2.72	0.259
1,2-Dibromoethane	106-93-4	pos	1.52	7.45	1.52	7.45	1.52
Dibromomannitol	488-41-5	pos	27.6	14.9	27.6	14.9	14.9

2,6-Dichloro-4-nitroaniline	99-30-9	pos	—	—	—	—	—
Dichloroacetic acid	79-43-6	pos	161	119	161	119	119
3,3′-Dichlorobenzidine	91-94-1	pos	28.1	—	28.1	—	28.1
1,2-Dichloroethane	107-06-2	pos	14.6	138	14.6	138	14.6
2,4-Dimethoxyaniline HCl	54150-69-5	pos	—	—	—	—	—
3,3′-Dimethoxybenzidine 2HCl	20325-40-0	pos	1.04	73.8	1.04	73.8	1.04
1,2-Dimethyl-5-nitroimidazole	551-92-8	pos	17	—	17	—	17
2,6-Dimethylaniline	87-62-7	pos	—	—	—	—	—
3,3′-Dimethylbenzidine 2HCl	612-82-8	pos	0.629	28.6	0.629	28.6	0.629
1,2-Dimethylhydrazine 2HCl	306-37-6	pos	—	0.114	—	0.114	0.114
Dimethylvinyl chloride	513-37-1	pos	31.8	14.9	31.8	14.9	14.9
2,4-Dinitrotoluene	121-14-2	pos	6.21	29.4	6.21	29.4	6.21
Dopamine HCl	62-31-7	pos	—	—	—	—	—
1,2-Epoxybutane	106-88-7	pos	220	—	220	—	220
Ethyl bromoacetate	105-36-2	pos	—	—	—	—	—
Ethylene imine	151-56-4	pos	—	0.377	—	0.377	0.377
Ethylene oxide	75-21-8	pos	21.3	63.7	21.3	63.7	21.3
Ethylene thiourea	96-45-7	pos	8.13	23.5	8.13	23.5	8.13
Furfural	98-01-1	pos	683	197	683	197	197
Glycidol	556-52-5	pos	4.28	34.7	4.28	34.7	4.28
Hydrazobenzene	122-66-7	pos	5.59	26	5.59	26	5.59
2-Hydroxyethylhydrazine	109-84-2	pos	—	0.397	—	0.397	0.397
IQ (3-methyl-3*H*-imidazo[4,5-*F*] quinolin-2-amine)	76180-96-6	pos	0.812	19.6	0.812	19.6	0.812
Isobutyl nitrite	542-56-3	pos	54.1	310	54.1	310	54.1
Isophosphamide	3778-73-2	pos	0.739	5.06	0.739	5.06	0.739
MeIQ (2-amino-3,4-dimethyl-3*H*-imidazo[4,5-*F*]quinoline)	77094-11-2	pos	—	15.5	—	15.5	15.5

(*Continued*)

TABLE 3.14 (*Continued*)
Linear Extrapolation (LE) Calculation from TD50 for Selected Mutagenic Compounds in CPDB

			Harmonic Mean of TD50 (mg/kg/day)		LE Calculation, 50 kg Body Wt. Used (μg/day)		Most Sensitive Species (μg/day)
Chemical Name	**CAS**	***Salmonella* (Ames)**	**Rat**	**Mouse**	**Rat**	**Mouse**	**Lower Value**
MeIQx (2-amino-3,8-dimethylimidazo[4,5-*F*] quinoxaline)	77500-04-0	pos	1.66	24.3	1.66	24.3	1.66
Methyl bromide	74-83-9	pos	—	—	—	—	—
Methyl methanesulfonate	66-27-3	pos	—	31.8	—	31.8	31.8
Methylene chloride	75-09-2	pos	724	1,100	724	1,100	724
Methylhydrazine	60-34-4	pos	—	7.55	—	7.55	7.55
6-Methylquinoline	91-62-3	pos	—	—	—	—	—
Mitomycin-C	50-07-7	pos	0.00102	—	0.00102	—	0.00102
2-Naphthylamine	91-59-8	pos	61.6	39.4	61.6	39.4	39.4
Nitrite, sodium	7632-00-0	pos	167	—	167	—	167
5-Nitro-*o*-anisidine	99-59-2	pos	53.9	3720	53.9	3720	53.9
p-Nitroaniline	100-01-6	pos	—	—	—	—	
o-Nitroanisole	91-23-6	pos	15.6	178	15.6	178	15.6
6-Nitrobenzimidazole	94-52-0	pos	—	372	—	372	372
p-Nitrobenzoic acid	62-23-7	pos	287	—	287	—	287
Nitrogen mustard	51-75-2	pos	0.0114	—	0.0114	—	0.0114
1-Nitronaphthalene	86-57-7	pos	—	—	—	—	—
8-Nitroquinoline	607-35-2	pos	9.82	—	9.82	—	9.82

Ozone	10028-15-6	pos	—	1.88	—	1.88	1.88
Phenobarbital	50-06-6	pos	—	7.37	—	7.37	7.37
o-Phenylenediamine 2HCl	615-28-1	pos	248	735	248	735	248
Phenylhydrazine HCl	59-88-1	pos	—	71.3	—	71.3	71.3
o-Phenylphenol	90-43-7	pos	232	—	232	—	232
Propylene	115-07-1	pos	—	—	—	—	—
1,2-Propylene oxide	75-56-9	pos	74.4	912	74.4	912	74.4
Quercetin	117-39-5	pos	10.1	—	10.1	—	10.1
Styrene	100-42-5	pos	23.3	210	23.3	210	23.3
Styrene oxide	96-09-3	pos	55.4	118	55.4	118	55.4
4,4′-Thiodianiline	139-65-1	pos	3.71	33.2	3.71	33.2	3.71
o-Toluidine HCl	636-21-5	pos	43.6	840	43.6	840	43.6
1,1,1-Trichloroethane	71-55-6	pos	—	—	—	—	—
Trimethylphosphate	512-56-1	pos	—	335	—	335	335
Vinyl chloride	75-01-4	pos	6.11	21.8	6.11	21.8	6.11
Vinylidene chloride	75-35-4	pos	—	34.6	—	34.6	34.6

Note: For mutagenicity, pos, neg, and NE indicate positive, negative, and not evaluated, respectively.

Hence, a daily lifelong intake of 21.3 μg ethylene oxide would correspond to a theoretical cancer risk of 10^{-5} and therefore be an acceptable intake when present as an impurity in a drug substance.

The human body weight of 50 kg was used in the calculation to yield more conservative exposure limits, even though 60 kg body weight was used to derive the TTC of 1.5 μg/day [11]. For reference, the TTC value of 1.5 μg/day corresponds to 0.025 μg/kg/day using 60 kg body weight (60 kg body weight was used to derive the TTC), which then corresponds to a TD50 of 1.25 mg/kg/day. The step-by-step calculations are as follows:

$$(1.5\ \mu g/day)/60\ kg = 0.025\ \mu g/kg/day = 0.000025\ mg/kg/day$$

To convert 1 in 10^5 to 1 in 2, multiply by 50,000:

$$(0.000025\ mg/kg/day) \times 50{,}000 = 1.25\ mg/kg/day$$

Therefore, the 1.5 μg/day TTC corresponds to the TD50 value of 1.25 mg/kg/day.

This LE method is very conservative [11], as several conservative assumptions were made to derive the TTC value from TD50 values of compounds in the CPDB. These conservative assumptions include the following:

- Establishment of the dose giving a 50% tumor incidence (TD50) using data for the most sensitive species and the most sensitive site.
- Based on a selected subset of the database containing 730 carcinogenic substances, which had adequate estimates of the TD50 following oral dosage.
- Simple LE from the TD50 to a 1 in 10^6 incidence.
- The approach assumes that all biological processes involved in the generation of tumors at high dosages are linear over a 500,000-fold range of extrapolation.
- Simple linear low-dose extrapolation is conservative because the possible effects of cytoprotective, DNA repair, apoptotic, and cell cycle control processes on the shape of the dose–response relationship are not taken into account.
- All of the compounds were analyzed assuming that there is no threshold in the dose–response relationship.

It should be noted that the 10^{-6} lifetime risk of cancer was used for the derivation of TTC, which was originally developed for food-contact materials (in the diet). Considering the existence of pharmaceutical benefits, a 10^{-5} lifetime risk of cancer is justified for pharmaceuticals.

Due to the nature of the databases used to derive the TTC values, the TTC approach would not normally be applied to the following cases [11]:

- Heavy metals, such as arsenic, cadmium, lead, and mercury, because of the concern of accumulation

- Compounds with extremely long half-lives that show very large species differences in bioaccumulation, such as 2,3,7,8-tetrachlorodibenzo-p-dioxin (TCDD) and structural analogs
- Proteins

Table 3.14 illustrates the calculation of LE based on TD50 for selected compounds from the CPDB. The calculated LE values shown in Table 3.14 are for the 10^{-5} acceptable lifetime cancer risk. In other words, these values are related to daily exposure throughout a person's lifetime (70 years). In cases where the treatment duration is less than lifetime, including clinical development, higher exposures could be justified [7] by the analogy of the sTTC concept [5,6].

The LE values presented in Table 3.14 could be used for the preliminary compound-specific risk assessment purpose and for the analysis and control of genotoxic and carcinogenic impurities in pharmaceutical development.

The solvents that are included in both the ICH Q3C guideline [3] and the CPDB [12] are compared in Table 3.15. For class 2 and class 3 solvents, the permitted daily exposure values from ICH Q3C are presented. ICH Q3C does not explicitly state the PDE for class 1 solvents. The PDE values of class 1 solvents were back-calculated from the concentration limits based on a 10 g daily dose in Table 3.15 for comparison purposes only. The simple LE values from TD50 were calculated using the same procedure as in Table 3.14. It is clear that the vast majority of the solvents show lower LE values (calculated from TD50) than PDE (calculated from the no-observed-effect level [NOEL] in ICH Q3C). This illustrates the conservativeness of the LE method.

Table 3.16 shows LE values and PDE values for metals included in EMA [8] and USP [9] guidelines and in the CPDB [12], for comparison purposes only. The values were calculated using the same procedure of simple LE from TD50 as in Table 3.14. PDE values were obtained from either USP or EMA guidelines. For the majority of the metals that have both PDE and TD50 values, the LE values calculated from TD50s are lower than PDEs. The only two exceptions are lead (as lead acetate in the CPDB) and arsenic (as trimethylarsine oxide in the CPDB). The TTC approach was considered to be not applicable to heavy metals, such as arsenic, cadmium, lead, and mercury, due to the concern of accumulation [11]. Again, this comparison demonstrates that the LE method from TD50 is very conservative.

There are several regulatory guidelines with recommended quantitative exposure limits. This chapter summarizes and compares the exposure limits that are relevant to impurities and genotoxic impurities in drug development. Calculation procedures of acceptable exposures are discussed, and the calculated values are presented in a tabular format to be used as an essential and versatile reference.

TABLE 3.15
Solvents in the ICH Q3C Guideline and the CPDB

			Harmonic Mean of TD50 (mg/kg/day)		LE Calculation (μg/day) 50 kg Body Wt. Used		Most Sensitive Species (μg/day)	PDE from ICH Q3C (μg/day)	ICH Q3C Concentration Limit (ppm)	ICH Q3C Class
Chemical Name	CAS	*Salmonella* (Ames)	Rat	Mouse	Rat	Mouse	Lower Value	(50 kg Body Wt. Used)	(10 g Daily Dose Used)	
Acetonitrile	75-05-8	neg	—	—	—	—	—	4,100	410	2
Benzene	71-43-2	neg	169	77.5	169	77.5	77.5	20[a]	2	1
Carbon tetrachloride	56-23-5	neg	27.8	33.6	27.8	33.6	27.8	40[a]	4	1
Chlorobenzene	108-90-7	neg	247	—	247	—	247	3,600	360	2
Chloroform	67-66-3	neg	262	111	262	111	111	600	60	2
1,1-Dichloroethane	75-34-3	neg	—	—	—	—	—	80[a]	8	1
1,2-Dichloroethane	107-06-2	pos	14.6	138	14.6	138	14.6	50[a]	5	1
Dimethylacetamide	127-19-5	neg	—	—	—	—	—	10,900	1,090	2
Dimethylformamide	68-12-2	neg	—	—	—	—	—	8,800	880	2
1,4-Dioxane	123-91-1	neg	267	204	267	204	204	3,800	380	2
Ethyl alcohol	64-17-5	neg	9,110	—	9,110	—	9,110	50,000	5,000	3
Ethylene glycol	107-21-1	neg	—	—	—	—	—	6,200	620	2
Hexane	110-54-3	neg	—	30,300	—	30,300	30,300	2,900	290	2
Isopropanol	67-63-0	neg	—	—	—	—	—	50,000	5,000	3
N-methyl-2-pyrrolidone	872-50-4	neg	—	2,050	—	2,050	2,050	5,300	530	2
Methyl *tert*-butyl ether	1634-04-4	neg	702	6,550	702	6,550	702	50,000	5,000	3
Methylene chloride	75-09-2	pos	724	1,100	724	1,100	724	6,000	600	2

Nitromethane	75-52-5	neg	40.4	469	40.4	469	40.4	500	50	2
Pyridine	110-86-1	neg	67.3	24.4	67.3	24.4	24.4	2,000	200	2
Tetrahydrofuran	109-99-9	neg	407	1,300	407	1,300	407	7,200	720	2
Toluene	108-88-3	neg	3,060	—	3,060	—	3,060	8,900	890	2
1,1,1-Trichloroethane	71-55-6	pos	—	—	—	—	—	15,000[a]	1,500	1
Trichloroethylene	79-01-6	neg	668	1,580	668	1,580	668	800	80	2
Xylene mixture	1330-20-7	neg	3,110	—	3,110	—	3,110	21,700	2,170	2

Note: For mutagenicity, pos, neg, and NE indicate positive, negative, and not evaluated, respectively.

[a] ICH Q3C does not show PDE values explicitly for class 1 solvents. The PDE values of class 1 solvents were calculated from the concentration limits based on a 10 g daily dose.

TABLE 3.16
Metals in USP and EMA Guidelines and the Carcinogenic Potency Database

Chemical Name	CAS	*Salmonella* (Ames)	Harmonic Mean of TD50 (mg/kg/day)		LE Calculation (μg/day) 50 kg Body Wt. Used		Most Sensitive Species (μg/day)	PDE (μg/day)	Concentration Limit (ppm)	Metal Class
			Rat	Mouse	Rat	Mouse	Lower Value	(50 kg Body Wt. Used)	(10 g Daily Dose Used)	
Aluminum potassium sulfate	10043-67-1	NE								
Arsenate, sodium	7631-89-2	NE						15	1.5	USP 1
Arsenite, sodium	7784-46-5	NE								
Beryllium sulfate	13510-49-1	neg								
Bismuth dimethyldithiocarbamate	21260-46-8	NE								
Bismuth oxychloride	7787-59-9	NE								
Cadmium acetate	543-90-8	NE						5	0.5	USP 1
Cadmium chloride	10108-64-2	neg	0.0136		0.0136		0.0136			
Cadmium chloride monohydrate	35658-65-2	NE								
Cadmium diethyldithiocarbamate	14239-68-0	NE								
Cadmium sulfate (1:1)	10124-36-4	NE	0.0217		0.0217		0.0217			
Cadmium sulfate (1:1) hydrate (3:8)	7790-84-3	NE								
Chromic oxide pigment	1308-38-9	NE						250	25	USP 2, EMA 1C

Chromium (III) acetate	1066-30-4	NE								
Cobalt sulfate heptahydrate	10026-24-1	pos	0.137	0.756	0.137	0.756	0.137			
Copper dimethyldithiocarbamate	137-29-1	NE						2500	250	USP 2, EMA 2
Copper-8-hydroxyquinoline	10380-28-6	NE								
Dimethylarsinic acid	75-60-5	neg	11.4		11.4		11.4	15	1.5	USP 1
Ethyl tellurac	20941-65-5	neg								
Ferric chloride	7705-08-0	neg						13,000	1,300	EMA 3
Ferric citrate tetrahydrate	5/8/2338	NE								
Ferric dimethyldithiocarbamate	14484-64-1	NE								
Germanate, sodium	12025-19-3	NE								
Lead acetate	301-04-2	neg	46.6		46.6		46.6	10	1	USP 1
Lead acetate, basic	1335-32-6	NE	181	472	181	472	181			
Lead dimethyldithiocarbamate	19010-66-3	pos								
Magnesium chloride hexahydrate	7791-18-6	NE						2,500	250	EMA 2
Manganese ethylenebisthiocarbamate	12427-38-2	NE	157		157		157	2,500	250	USP 2, EMA 2
Manganese (II) sulfate monohydrate	10034-96-5	neg								
Mercuric chloride	7487-94-7	neg	3.12		3.12		3.12	15	1.5	USP 1
Mercurous chloride	7546-30-7	NE								
Mercurymethyl chloride	115-09-3	neg								
Molybdenum trioxide	1313-27-5	neg		6.13		6.13	6.13	250	25	USP 2, EMA 1C

(Continued)

TABLE 3.16 (*Continued*)
Metals in USP and EMA Guidelines and the Carcinogenic Potency Database

Chemical Name	CAS	*Salmonella* (Ames)	Harmonic Mean of TD50 (mg/kg/day)		LE Calculation (µg/day) 50 kg Body Wt. Used		Most Sensitive Species (µg/day)	PDE (µg/day)	Concentration Limit (ppm)	Metal Class
			Rat	Mouse	Rat	Mouse	Lower Value	(50 kg Body Wt. Used)	(10 g Daily Dose Used)	
Monomethylarsonic acid	124-58-3	NE						15	1.5	USP 1
Nickel	7440-02-0	NE						250	25	USP 2, EMA 1C
Nickel (II) acetate	373-02-4	NE								
Nickel dibutyldithiocarbamate	13927-77-0	NE								
Nickel (II) sulfate hexahydrate	10101-97-0	neg								
Niobate, sodium	12034-09-2	NE								
Phenylmercuric acetate	62-38-4	neg								
Potassium bicarbonate	298-14-6	NE	13,000		13,000		13,000			
Potassium chloride	7447-40-7	neg								
Potassium iodide	7681-11-0	NE	440		440		440			
Selenium	7782-49-2	NE								
Selenium diethyldithiocarbamate	5456-28-0	NE		1.49		1.49	1.49			
Selenium dimethyldithiocarbamate	144-34-3	NE								

Selenium dioxide	7446-08-4	NE								
Selenium sulfide	7446-34-6	pos	8.01	69.3	8.01	69.3	8.01			
Sodium copper chlorophyllin	28302-36-5	NE						2,500	250	EMA 2
Tin (II) chloride	7772-99-8	neg								
Titanium dioxide	13463-67-7	neg								
Titanium oxalate, postassium	14481-26-6	NE								
Titanocene dichloride	1271-19-8	pos								
Trimethylarsine oxide	4964-14-1	NE	24.1		24.1		24.1	15	1.5	USP 1
Tungstate, sodium	13472-45-2	NE								
Vanadyl sulfate	27774-13-6	NE						250	25	USP 2, EMA 1C
Zinc (II) acetate dihydrate	5970-45-6	NE						13,000	1,300	EMA 3
Zinc dibutyldithiocarbamate	136-23-2	NE								
Zinc diethyldithiocarbamate	14324-55-1	NE								
Zinc dimethyldithiocarbamate	137-30-4	pos	40.7		40.7		40.7			
Zinc ethylenebisthiocarbamate	12122-67-7	neg	255		255		255			
Zirconium (IV) sulfate	14644-61-2	NE								
Metals not included in CPDB										
Palladium								100	10	USP 2, EMA 1A
Platinum								100	10	USP 2, EMA 1A

(Continued)

TABLE 3.16 (*Continued*)
Metals in USP and EMA Guidelines and the Carcinogenic Potency Database

Chemical Name	CAS	*Salmonella* (Ames)	Harmonic Mean of TD50 (mg/kg/day)		LE Calculation (μg/day) 50 kg Body Wt. Used		Most Sensitive Species (μg/day)	PDE (μg/day)	Concentration Limit (ppm)	Metal Class
			Rat	Mouse	Rat	Mouse	Lower Value	(50 kg Body Wt. Used)	(10 g Daily Dose Used)	
Metals not included in CPDB										
Osmium								100 (Combination not to exceed)	10 (Combination not to exceed)	USP 2, EMA 1B
Rhodium										
Ruthenium										
Iridium										

Note: Daily exposure values are shown in micrograms per day. For mutagenicity, pos, neg, and NE indicate positive, negative, and not evaluated, respectively. USP *and EMA guidelines apply to metal in general, not to the specific metal compounds.*

REFERENCES

1. International Conference on Harmonisation of Technical Requirements for Registration of Pharmaceuticals for Human Use, *ICH Harmonised Tripartite Guideline, Impurities in New Drug Substances*, ICH Q3A(R2), October 2006.
2. International Conference on Harmonisation of Technical Requirements for Registration of Pharmaceuticals for Human Use, *ICH Harmonised Tripartite Guideline, Impurities in New Drug Products*, ICH Q3B(R2), June 2006.
3. International Conference on Harmonisation of Technical Requirements for Registration of Pharmaceuticals for Human Use, *ICH Harmonised Tripartite Guideline, Impurities: Guideline for Residual Solvents*, ICH Q3C (R5), February 2011.
4. European Medicines Agency, *Guideline on the Limits of Genotoxic Impurities*, Committee for Medicinal Products for Human Use, January 2007.
5. European Medicines Agency, *Questions and Answers on the 'Guideline on the Limits of Genotoxic Impurities'*, Committee for Medicinal Products for Human Use, September 2010 (Rev. 3).
6. Food and Drug Administration, *Guidance for Industry, Genotoxic and Carcinogenic Impurities in Drug Substances and Products: Recommended Approaches, Draft Guidance*, Center for Drug Evaluation and Research, December 2008.
7. International Conference on Harmonisation of Technical Requirements for Registration of Pharmaceuticals for Human Use, *ICH Harmonised Tripartite Guideline, Assessment and Control of DNA Reactive (Mutagenic) Impurities in Pharmaceuticals to Limit Potential carcinogenic Risk*, ICH M7 Draft Guideline, February 2013.
8. European Medicines Agency, *Guideline on the Specification Limits for Residues of Metal Catalysts or Metal Reagents*, Committee for Medicinal Products for Human Use, 2008.
9. United States Pharmacopeial Forum, General Chapter 232, Elemental Impurities—Limits, In: *United States Pharmacopeia*, Rockville, MD, February 2013.
10. International Conference on Harmonisation of Technical Requirements for Registration of Pharmaceuticals for Human Use, *ICH Harmonised Tripartite Guideline, Guideline for Elemental Impurities*, ICH Q3D Draft Guideline, July 2013.
11. R. Kroes, A. G. Renwick, M. Cheeseman, J. Kleiner, I. Mangelsdorf, A. Piersma, B. Schilter, et al., "Structure-based thresholds of toxicological concern (TTC): Guidance for application to substances present at low levels in the diet," *Food and Chemical Toxicology*, 42, 65–83 (2004).
12. L. Gold, *Carcinogenic Potency Database (CPDB)*, University of California, Berkeley, CA. Accessed at www.toxnet.nlm.nih.gov.
13. L. W. Gold, T. H. Slone, N. B. Manley, G. B. Garfinkel, E. S. Hudes, L. Rohrback, B. N. Ames, "The carcinogenic potency database: Analyses of 4000 chronic animal cancer experiments published in the general literature and by the U.S. National Cancer Institute/National Toxicology Program," *Environmental Health Perspectives*, 96, 11–15 (1991).

4 Genotoxic Impurities from Toxicology Perspectives Including Cancer Risk Assessment, In Silico, In Vitro/In Vivo Testings, and Regulatory Aspects

Esther Vock

CONTENTS

4.1 CANCER RISK ASSESSMENT

Cancer risk assessment consists of hazard identification, dose–response, exposure assessment, and finally risk characterization. In terms of genotoxic impurities in pharmaceuticals, hazard identification is done using in silico analysis for the prediction of mutagenicity or an Ames test. A hazard is given in case of a positive result. A positive in silico prediction can be overruled by a negative Ames test, that is, a negative result is accepted such that the test article is classified as not mutagenic (or presently nongenotoxic with respect to impurities). A positive Ames test can be overruled by a negative outcome in a relevant in vivo mutagenicity follow-up study. The exposure assessment for genotoxic impurities in pharmaceuticals is simple; since 1997, impurities should be identified at 0.05% and qualified above 1 mg/day according to the International Conference on Harmonisation of Technical Requirements for Registration of Pharmaceuticals for Human Use (ICH) quality guidelines on impurities in drug substances and drug products [1,2]. Since 2007, genotoxic impurities should be controlled to the threshold of toxicological concern (TTC) limit of 1.5 μg/day at the cancer risk level of 1 in 100,000 according to the European *Guideline on genotoxic impurities* [3,4] and the respective Food and Drug Administration (FDA) draft guidance [5]. The risk characterization for genotoxic impurities is predefined to be a mutagenic, very potent carcinogen unless data can prove otherwise and, therefore, they are classified as unusually toxic.

Generally speaking, genotoxicity means damage to the DNA, the macromolecule that contains the genetic code transcribed in genes and when condensed to chromosomes allows the transfer of genetic information to daughter cells or the next generation. Genotoxicity as it is understood presently includes mutations in the DNA (base pair or frameshift/deletion mutations, i.e., the genetic information has a different sequence compared to the origin), clastogenicity (damage to chromosomes), and aneugenicity (numerical change in the number of chromosomes). Genotoxic impurities that must be controlled to 1.5 μg/day are by definition those that induce mutations, that is, mutagenic impurities. Mutagenic compounds react with DNA, causing DNA damage, which can be fixed as mutation once the DNA is duplicated. Mutations are most frequently detected in the bacterial reverse mutation assay, the Ames test. Clastogenicity and aneugenicity can be detected in the chromosomal aberration test and/or the in vitro and in vivo micronucleus test. The chromosomal aberration test is part of the genotoxicity package for impurities above the qualification threshold in ICH quality guidelines. Nevertheless, it is only mutagenicity that is of concern, since the dose response for DNA reactivity and mutagenicity can span several orders of magnitude while clastogenic or aneugenic events have a threshold effect and therefore do not require extrapolation to the very low dose of 1.5 μg/day. On the other hand, the rate of fixation is dependent on the rate of DNA damage formation, DNA repair, and kinetics of cell replication; therefore, a threshold may also exist for mutagenic compounds, but this conclusion needs to be supported by data.

Accumulation of DNA damage in specific genes in a cell can lead to cancer, a process that develops in the course of several years [6]. Genotoxicity studies are short-term studies in which DNA damage or mutations are investigated. They are part of the clinical phase I–enabling preclinical safety package for the active pharmaceutical ingredient (API).

Carcinogenicity studies are performed in mice and rats for a treatment period of 18–24 months, that is, the average lifetime of these rodents. These studies are performed late in development mainly in parallel to clinical phase III trials with the results being available at the time of submission. A few dog and monkey studies were also performed with a few animals and 5–15 years of exposure. All species are regarded to be as sensitive as humans, and if available the most sensitive species that responds at a lower dose is taken as the basis for risk assessment. The doses are chosen based on the maximum tolerated dose studies in rats and mice of generally 13-week duration. The top dose should cause some toxicity like body weight gain reduction by maximally 20% for chemicals or a sufficient high systemic exposure of 25- to 50-fold for an API compared to human exposure established in the clinical phase II trials. Two lower doses are included to see whether there is a dose–response relationship. In general, there are 50 males and 50 females/group. The majority of all 18- or 24-month carcinogenicity studies are performed with dose levels in the milligram per kilogram range or at a limit dose of 1500 mg/kg for an API. In contrast, the human exposure of a mutagenic impurity would be in the range of 25 ng/kg for a 60 kg patient at the TTC limit. In a carcinogenicity study, the so-called ED01 megamouse study with 2-aminofluorene, 24,192 mice were tested in seven dose groups (and different durations of treatment) with 1,728 mice each in the lowest-dose groups with serial sacrifice following 18 and 24 months and natural death. This high number of animals was required for statistical reasons to adjust for the low incidence. This study showed a rather linear dose–response relationship for liver tumors but clearly not for bladder tumors [7]. The DNA adducts study using nine dose groups also showed a linear dose–response relationship in the liver and bladder [8], indicating that additional tissue-specific factors influenced the nonlinearity for bladder tumors but not for liver tumors. In the absence of such carcinogenicity data, it is assumed that there is a linear dose–response relationship for all tumor types from the milligram per kilogram range tested in carcinogenicity studies down to the nanogram per kilogram, which is acceptable for humans. The correct point of departure for linear extrapolation from the high doses in the carcinogenicity studies down to the low levels that people are exposed to is a matter of wide debate, that is, whether the TD50 value (the chronic daily dose that induces tumors in 50% of the animals), the TD10, the TD25, or other points of departure for linear extrapolation can be used. The TD50 value could be either the harmonic mean of several studies as given in the Carcinogenic Potency Database (http://potency.berkeley.edu/) or the lowest TD50 value in one species, strain, and gender [9]. Acceptable levels, also called virtual safe doses (VSDs) are calculated by linear extrapolation from the rodent TD50 value (risk level of 0.5) by division with 50,000 down to the cancer risk level of 1 in 10^5 (0.00001) for mutagenic and carcinogenic impurities in pharmaceuticals. A VSD of 1.5 μg/day of TTC corresponds to a highly carcinogenic potent compound with a TD50 value of 1.25 mg/kg/day (1.5 μg/day times 50,000/60 kg patient). The risk level of 1 in 10^5 is considered appropriate for pharmaceuticals since there is a benefit for patients taking the drug. The risk level for the general population exposed to contaminants that are without a benefit is 1 in 10^6. It has to be noted that animals also form tumors spontaneously, and the same is true for humans. It is estimated that at least 10% of all people will have a tumor by the end of their lifetime, that is, the overall spontaneous risk level for cancer in humans is 0.1.

There is human experience with chemicals, primarily those used in working places like polycyclic aromatic hydrocarbons (PAHs), aromatic amines, reactive chemicals, and alkylating anticancer drugs, which are known to be mutagenic, human carcinogens. The discovery by Percivall Pott in the eighteenth century that chimney sweeps had higher occurrences of testicular cancer than any other group and the recognition that these tumors were due to the extensive exposure of this group to soot initiated the research on human carcinogens. The PAHs present in soot were the first identified human carcinogens. PAHs are in general not used as intermediates in the synthesis of APIs and, therefore, mutagenic impurities in APIs do not belong to the class of PAHs. However, aromatic amines and alkylating agents are frequently used in the synthesis of APIs. Workers exposed to aromatic amines, for example, 2-naphthylamine, benzidine, and 4-aminobiphenyl, have a higher risk of developing bladder cancer [10,11]. One study suggests an association between exposure to benzidine/β-naphthylamine and cancers of the liver, gallbladder, bile duct, large intestines, and lungs in humans [12], although the confounding factors like food and smoking habits are very difficult to assess. The three human carcinogens (naphthylamine, benzidine, and 4-aminobiphenyl) as well as o-toluidine were either banned or strict control of workplace exposures was introduced. Epidemiology data from workers exposed to alkylating agents are limited to α-chlorinated toluenes and benzoyl chloride, and the cohorts are small. Nevertheless, the excess lung cancer risk is estimated to be approximately threefold compared to control [13]. Intermediates can be impurities, and since the most potent carcinogens are banned from workplaces it is less likely that highly potent carcinogenic impurities are introduced in the synthesis of pharmaceuticals.

4.2 IN SILICO TESTING

In principle, electrophilic chemicals can react with DNA. Relevant DNA-reactive chemical groups were described [14,15], including those that require first metabolic activation to DNA-reactive metabolites, as described for aromatic amines. The initial in silico systems for the prediction of mutagenicity were based on the Ashby and Tennant alerts. All in silico systems for the prediction of mutagenicity are either based on knowledge or based on quantitative, statistical structural activity relationships. These systems predict mutagenicity as seen in the bacterial reverse mutation assay (the Ames test). They are continuously updated with the goal to reduce the false negative predictivity without increasing the false positive predictivity. The false negative rate needs to be low since such impurities are currently controlled at ICH Q3A/B levels (666-fold higher than the TTC with a calculated risk of $666 \times 10^5 = 7$ in 10^3 under the assumption that the impurity is mutagenic and highly carcinogenic). The negative predictivity for the Ames test is in a good case around 90%, that is, if all in silico negative compounds would be tested in the Ames test, 90% are expected to be negative. The predictivity depends on the variety of chemical spaces. The better the prediction, the closer the chemical space and the higher the coverage, that is, more Ames data are available to create and/or optimize a rule for structural alert. Nevertheless, some classes are difficult to predict, like aromatic amines. The false positive predictivity is in the range of 30%–60%. The false positive predictivity triggers an unnecessary burden on chemistry and toxicology since efforts are undertaken to reduce potential genotoxic impurity to the TTC level or the impurity is synthesized and tested in the Ames test. All predictivity models are limited by the fact that the Ames test can be tested only up to bacteriotoxic and/or insoluble concentrations (or 5000 μg/plate), that is, insoluble, bacteriotoxic compounds do not exert their mutagenic properties under the test conditions.

4.3 IN VITRO TESTING

There are numerous in vitro mutagenicity assays; however, only the following tests are described in the Organisation for Economic Co-operation and Development (OECD) guidelines and are regulatorily well accepted. In general, only high-quality, reproducible, and well-described mutagenicity tests should be evaluated (Table 4.1). In the published literature, at least a detailed description of the model used, test article identity (CAS number), batch, and purity should be provided as basic information. The maximum concentrations should be up to 10 mmol/L for chemicals (1 mmol/L for pharmaceuticals), that is, concentrations that are in general far above the systemic exposures achieved in patients. It is suggested to use a conversion factor and calculate the concentrations for the base if a salt form of the test article is used. The salt form of an intermediate or impurity may influence the solubility, and solubility is a limiting factor in all in vitro tests. The solvents used in in vitro tests are water, phosphate buffer, and dimethylsulfoxide. Other solvents like ethylene glycol dimethyl ether, formamide, dimethylformamide, 0.01 M NaOH, acetone, or chloroform need to be tested first for bacterio-/cytotoxicity. It is assumed that the impurities are stable under the test conditions (usually for 1–6 hours at room temperature). However, in specific cases the stability in the vehicle needs to be assessed. All in vitro tests are performed in absence and presence of a metabolic activation system (S9 mix). S9 is prepared from rats treated with Aroclor 1254® or phenobarbital/β-naphthoflavone that induce enzymes like cytochromes P450 in the liver. S9 is an organ (usually the liver) homogenate isolated as supernatant following centrifugation at 9000g for 20 minutes containing cytosol and microsomes. The S9 mix is the organ fraction and an NAPDH-generating system. The metabolic activation system is in a first tier with rat liver S9 mix. In a second tier, S9 mix from other species like hamsters or humans or other enzyme inducers may be used. In general, the cytochromes in human livers

TABLE 4.1
List of In Vitro Mutagenicity Tests Relevant for Impurities in Pharmaceuticals

	System	Strains	Test Article Demand	Duration	OECD Guideline
Bacterial reverse mutation assay (Ames test)	Bacteria	*S. typhimurium* [TA1535, TA1537 or TA97 or TA97a, TA98, TA100, TA102 or *E. coli* WP2 *uvrA*, or *E. coli* WP2 *uvrA* (pKM101)]	1 g	1 month	471
ML-TK gene mutation assay	Mammalian cells	L5178Y $TK^{+/-}$	2 g[a]	3 months	476

[a] The highest test concentration for pharmaceuticals is 1 mmol/L but 10 mmol/L for chemicals and, therefore, the test article demand is 3 g.

vary and are in general lower than those in induced animals. Rat S9 mix apparently does not permit cytochrome P450 2E1–mediated metabolic activation as required for urethane-like structures. Therefore, this chemical class is usually negative in the Ames test. Otherwise, there is a very good relationship between rat S9 mix and all important human primary metabolic pathways [16].

Good descriptions for the Ames and mouse lymphoma assays are given in various books and publications and in *Redbook 2000* IV.C.1a–d from the FDA. The target molecule, DNA, is ubiquitously present in all species and, therefore, even the data from bacteria are representative of events in mammalian systems.

4.3.1 Ames Test

The bacterial reverse mutation assay Ames test was named after Bruce Ames, who created this test in the early 1970s. The Ames test is performed according to OECD Guideline No. 471, and it is recommended to perform it under good laboratory practice conditions to reach regulatory acceptability. At least five different bacterial strains of *Salmonella typhimurium* and/or specific strains of *Escherichia coli* from the list in the OECD guideline should be tested with and without a metabolic activation system (S9 mix). The bacterial strains are mutated and therefore not able to synthesize histidine (tryptophan in the case of *E. coli*) by themselves, and thus bacterial growth is not possible. The bacterial strains, which are reverse mutated by the test article to the wild type or with restored histidine/tryptophan-producing pathways, can grow to visible colonies within 2 to 3 days. The Ames test is positive if the number of revertants in treated cultures increased twofold to threefold above those of the negative control. The judgment of an Ames positive response is based on a concentration-dependent, reproducible increase in one of the tester strains with or without S9 mix compared to control. Equivocal responses are sometimes seen with test articles that induce only weak (less than twofold to threefold) but reproducible increases. Whenever possible the isolated impurity should be tested, and for critical classes only test articles of the highest purity (> 99.9%, e.g., for aromatic amines) should be used or a repurification should be taken into consideration. The maximum recommended concentration in the Ames test is either 5000 µg/plate or a bacteriotoxic or insoluble concentration. Under specific circumstances, an impurity like a degradation product may be tested in the presence of the drug substance/product as long as the impurity levels reach 250 µg/plate. In this case, a chemical rationale should be provided that it is technically not feasible to synthesize the impurity. Besides the major challenges of purity and stability of the test article, it is recognized that histidine-like structures may be responsible for false positive responses.

A positive genotoxic outcome indicates a high probability for a positive outcome in the carcinogenicity study and, therefore, there is a tendency in drug development to use Ames data like a switch from negative = GO to positive = avoid in view of the higher risk for cancer. The sensitivity of the Ames test to correctly identify rodent carcinogens is 58.8%, and the specificity to correctly recognize rodent noncarcinogens is 73.9% [17]. This means that at least 26.1% of the noncarcinogens are Ames positive. For example, *p*-anisidine hydrochloride, a potential impurity, is

Ames positive but negative in the carcinogenicity studies in mice and rats. Therefore, in specific cases with low exposure levels it is worthwhile to assess the biological relevance of the Ames test in further relevant in vivo genotoxicity studies.

4.3.2 Mammalian Gene Mutation Assay

In general, mutagenic impurities below the ICH Q3A/B qualification threshold do not need to be tested in the mouse lymphoma thymidine kinase (ML-TK) assay. However, if mouse lymphoma data are already available, for example, for worker safety reasons according to in-house environmental health protection rules, then the risk assessment should be done on the mutagenic outcome, that is, an increase in mutation frequency of the large clones. Various cell lines are listed in OECD Guideline No. 476, but the L5178Y$^{+/-}$ cell line is used for pharmaceuticals according to the ICH S2(R1) guideline [18]. The cell line is heterozygotic for the mutated thymidine kinase gene, that is, only one additional mutation is necessary to deactivate the thymidine kinase from the second chromosome. Cells with inactivated thymidine kinase can grow in the presence of trifluorothymidine, whereas the nonmutated cells die. The reason is that thymidine is synthesized by the mutated cells via a de novo synthesis route and not via thymidine kinase. Thymidine is one of the nucleotides of DNA and is necessary for replicating/growing cells. Cells are treated with the test article for 3 to 4 hours with and without a metabolic activation system. The 24-hour incubation experiment is not necessary for impurities since aneugens and clastogens are primarily detected under these conditions. Four analyzable test concentrations should be used, the highest concentration reaching a survival of 10%–20% or up to 1 mmol/L for the API and large-molecular-weight impurities. Small-molecular-weight impurities could be tested up to 10 mmol/L as recommended by the OECD guideline for testing of chemicals, whereas API-like impurities could be limited to 1 mmol/L. Only the mutant frequencies for large clones need to be evaluated since large clones are due to mutations and small clones due to clastogenic events. The sensitivity of the ML-TK assay (both mutagenic and clastogenic events) is 80.8%, and the specificity is 47.6% [17].

Impurities for which genetic toxicity data from the chromosomal aberration assay, in vitro micronucleus test, or mouse lymphoma assay (small clones and 24-hour incubation data) are available do not need to be controlled at the TTC level since they do not follow the linear extrapolation for cancer risk but have an apparent threshold effect. In these cases, the risk assessment follows a different approach, which is not yet defined. One approach could be to calculate permitted daily exposure levels using safety factors, as described in the ICH Q3C guideline [19].

4.4 IN VIVO TESTING

The biological relevance of a positive result in the Ames test or the ML-TK may be assessed in a relevant in vivo follow-up study (Table 4.2). Several questions need to be addressed like the sensitivity of the in vivo model to assess the DNA reactivity of a specific chemical compound. Furthermore, it needs to be considered whether the impurity should be spiked to the drug substance/product or the "pure" isolated

TABLE 4.2
List of In Vivo Mutagenicity Tests

Study Title	Guidelines	Species	Treatment Duration	Time from Initiation to Report	Test Article Demand[a]
Comet/micronucleus	Comet: draft OECD; micronucleus: OECD No. 474 ICH S2(R1) for API	Rats or mice ($n = 5$ to 6/dose group, one gender, negative and positive control, 2 to 3 dose levels) 23–30 animals[b]	3 days	3 to 4 months	20 g
UDS test with mammalian liver cells in vivo	OECD No. 486 ICH S2(R1) for API	Rats (or mice) ($n = 3$ to 4/dose group, one gender, negative and positive control, 2 dose levels) 18 animals[b]	Single dose	3 to 4 months	5 g
Transgenic mice (Muta mouse, Big Blue mouse)	OECD No. 488	Mice ($n = 5$–10/dose group, one gender, negative control, 2 to 3 dose levels) 30 animals[b]	7 days dose range finder, 28 days treatment	6 to 7 months	40 g
Pig-a gene mutation assay	Validation phase	Rats or other species (details of protocol to be developed; $n = 5$/dose group, one gender, negative and positive control, 2 to 3 dose levels) 23–28 animals[b]	7 days dose range finder, 28 days treatment	6 to 7 months	170 g
DNA adduct determination (CBI, ^{32}P-postlabeling, HPLC-MS)	None	Rats, mice, or other species ($n = 3$–5/dose group, one gender, negative control, 2 to 3 dose levels) 15–26 animals[b]	1 to 2 days	3 to 4 months	12 g
Mouse spot test (no longer in use)	OECD No. 484				

Note: CBI, covalent binding index; HPLC-MS, high-performance liquid chromatography mass spectrometry.

[a] Test article demand calculations are based on the dose levels of 100, 500, and 2000 mg/kg and body weights of 250 g for rats and 30 g for mice.

[b] The number of animals includes the dose range finding and the main study.

impurity and finally which doses should be tested. It is recommended to test the pure impurity to fully assess the mutagenic potential since the dose should be several multiples of 1.25 mg/kg (see Chapter 1, VSD of 1.5 μg/kg). Currently, the top dose for chemicals should be the maximum tolerated dose, that is, about 50%–66% of the lethal dose that induced death in 50% of the animals (LD50) but not higher than 2000 mg/kg for acute dosing or 1000 mg/kg for repeated daily treatment for 4 weeks.

4.4.1 Comet/Micronucleus

Alkaline single cell gel electrophoresis (SCGE), or comet assay, is a sensitive and reliable method for detecting DNA damage expressed as single- and double-strand breaks, alkaline-labile sites, DNA–DNA, and DNA–protein cross-links including incomplete excision repair sites in single cells of different organs [20]. Rats or mice are treated with the test article and organs are isolated at the time of suspected highest systemic exposure (3–6 hours). The liver is a default organ with high metabolic activity. The stomach, duodenum, and jejunum (first site of contact or organ with highest absorption) are taken in case the Ames positive result was obtained in the absence of the metabolic activation system. However, in principle, any known target organ of toxicity can be included. Single cells are prepared and embedded in agarose, the DNA unwinded at pH ≥ 13, and the single-strand fragments migrated by electrophoresis. After neutralization, the DNA is stained with the fluorescent dye ethidium bromide and evaluated by microscopy. The resulting images, which were subsequently named for their appearance as "comets," are measured to determine the extent of DNA damage. The light intensity of the cell nucleus "head" (normal DNA or large fragments) in relation to the "tail" (fragmented DNA) is quantified. The comet assay is easily combined with the micronucleus tests since only the bone marrow or peripheral blood needs to be used. Furthermore, the integration of further mutagenic end points, that is, gene expression profile, would be possible. The treatment duration should be extended to three administrations of the test article in the combined comet/micronucleus assay since it takes more time to express micronuclei in the bone marrow. The administration schedule is that the first two administrations are given within 24 hours and the last 3–6 hours prior to necropsy. The micronucleus assay is designed to detect clastogens and aneugens; nevertheless, it was shown that it also detects mutagenic carcinogens when the doses were sufficiently high.

For aromatic nitro intermediates and impurities for which the bacterial nitroreductase renders this chemical class to be oversensitive in the Ames test, a combined in vivo comet/micronucleus test is considered suitable. 2,6-Dinitrotoluene is Ames positive and has a high carcinogenic potency with a TD50 value of 0.29 mg/kg. The comet assay in rat liver was positive when animals were treated with 25 mg/kg [21], thereby confirming the validity and sensitivity of this test to detect mutagenic carcinogens.

Furthermore, the combined comet/micronucleus test is considered suitable if, for example, the impurity has the properties to react with the N7/C8 of the deoxyguanosine-forming alkaline-labile sites. Deoxyguanosine is one of four nucleotides of the DNA and is the major acceptor for many electrophilic compounds [22].

4.4.2 Unscheduled DNA Synthesis in Liver

Rodents are given a single dose of the test article. Viable liver cells are harvested at about 2–4 hours postadministration (p.a.) and 12–16 hours p.a. by in situ collagenase perfusion. The established hepatocyte cultures are incubated in the presence of ^{3}H-thymidine for 18 hours. The radiolabeled nucleotide is incorporated into the DNA in the course of DNA repair synthesis if the test article had caused DNA damage in vivo. Quantification of the unscheduled DNA synthesis (UDS) response is done by silver grain counting (net nuclear grain count, i.e., grains over the nucleus minus the background grains in plasma) using autoradiography. The grains over the nucleus are of high density and the grains over the cytoplasma are of low density for positive compounds.

The UDS test in liver has a low sensitivity for carcinogens [23]. It was positive only for one-third of the Ames positive carcinogens. Strong mutagenic liver carcinogens of the aryl amine class or nitro aromatic compounds were positive or equivocal in this test. Therefore, this test may be used for specific chemical classes known to be metabolically activated in the liver and to have an Ames positive result in the presence of the rat liver metabolic activation system.

4.4.3 In Vivo Mutation Assay

Two different transgenic mouse strains are available for in vivo mutagenicity studies, the Big Blue® and the Muta™ mouse [24,25]. These animals are transgenic for bacterial genes (*λlacI* and *λlacZ*, respectively) incorporated in all their organs and therefore suitable to detect mutations in any organ of interest. Animals are treated daily for at least 28 days, and organs are sampled 3 days after the final treatment. Treatment duration depends on the cell replication rate, that is, the rate of DNA synthesis. It can be shorter for organs with high cell replication, for example, the hematopoietic system, skin, the gastrointestinal tract, and testes, and longer for organs with low cell replication, for example, the heart and kidneys. Genomic DNA is carefully extracted from the organ of interest. The DNA is packed into bacteriophages. These packed bacteriophages infect a special bacterial strain of *E. coli* and lyse the bacterial lawn forming plaques. The bacterial gene product of *lacZ* is β-galactosidase, which cleaves X-gal (5-bromo-4-chloro-3-indolyl-*b*-galactopyranoside) to galactose and an insoluble blue precipitate. If a mutation occurred in the bacterial *lacZ* gene in the Muta mouse model, the gene product may be reduced or inactivated and, therefore, X-gal is not cleaved, that is, the mutant clear plaques are counted in a background of blue plaques. The *lacI* gene product in the Big Blue model is a Lac repressor protein and, therefore, a mutation allows the subsequent *lacZ* gene product to become active. The total number of mutants is determined by counting the blue (mutant) plaques among a background of clear plaques in the Big Blue model.

In principle, all types of DNA-reactive chemicals that induce mutations in vivo should be detectable in these models. It is possible to detect the type of mutagenic events, that is, whether transversion or transition occurred since DNA sequencing could be performed, and thus assign a specific molecular signature for specific chemical classes. The spontaneous rate of mutations is age and organ dependent. The isolation of high-molecular-weight and pure DNA from organs is a prerequisite for good-quality results. Necrosis or iron deposits or any other DNA–protein cross-links may interfere with the isolation process. Therefore, the maximum dose should not cause severe morphological changes.

4.4.4 Pig-a Mutation Test

The phosphatidylinositol glycan class A (*pig-a*) gene is located on the X chromosome and, therefore, a single mutation inactivates the gene product in all daughter cells. The gene product, a catalytic subunit of *N*-acetylglucosaminyltransferase, is required for glycosylphosphatidyl inositol (GPI) anchor synthesis [26]. Among all cell surface proteins, CD59 is anchored by GPI to the cell surface and can be easily detected in the blood using, for example, flow cytometric methods, that is, a decrease in CD59 is indicative for mutations in the *pig-a* gene.

The method can be used in a variety of animal species including humans. Furthermore, all types of DNA-reactive chemicals that induce mutations in vivo should be detectable in this model. However, the detected mutation events are restricted to organs with high cell replication like the hematopoietic cell lineage. Therefore, impurities that are metabolically activated, for example, in the liver or highly reactive first site of contact chemicals, are less likely to be mutagenic in the pig-a test. A thorough validation will show whether the pig-a test can be used for all different types of chemical classes. Furthermore, it requires repeated dosing to accumulate a sufficiently high number of mutagenic events and, therefore, the dose might be lower than what it is after acute or short-term treatment due to cumulative toxicity or intolerability.

4.4.5 Emerging Models

Gene expression profiles may be used to elucidate the mode of action of test articles, including genotoxicity. Toxicogenomics (mRNA transcript profiling) allows the identification of genes that respond to toxic effects. Investigations are ongoing whether it is possible to discriminate the gene expression profile of genotoxic carcinogens from that of nongenotoxic carcinogens and noncarcinogens [27].

4.5 REGULATORY ASPECTS

In 1959, health authorities became aware that residues of the carcinogenic pesticide aminotriazole were present in cranberries. They ensured that the complete cranberry harvest was analyzed just prior to Thanksgiving. As a consequence, the U.S. health authorities released the "Delaney Clause," setting the zero cancer risk standard. The Delaney Clause reads that "the FDA shall not approve for use in food any chemical additive found to induce cancer in man, or, after tests, found to induce cancer in animals."

Health authorities have the task to protect public health. This includes identifying the major sources for cancer-causing agents and enforcing the avoidance or reduction of exposure to hazard-causing agents. The International Agency for Cancer Registry (IARC) in France have identified repetitively that food, drinking, and lifestyle are the major risk factors that induce cancer. It was acknowledged that many food components are damaging the DNA themselves. However, it is the mixture of nutrition, anticancer food components, and DNA-damaging components that needs to be balanced. Furthermore, it was recognized that it is impossible to avoid carcinogens in food. For pragmatic reasons, TTC concepts for different toxicological end points were established by the FDA for indirect food additives [9]. The carcinogenic potency data from 709 Ames positive and therefore mutagenic rodent carcinogens were taken as the basis. VSDs were calculated down to the risk level of 1 in 10^6. The logarithmic VSD values form a normal distribution curve in a semilogarithmic scale with the most likely potency corresponding to 0.5 ppb. This value corresponds to an exposure of 0.15 μg/day assuming a daily consumption of 1500 g of food and 1500 g of fluids. Therefore, if the daily lifetime exposure to a mutagen and presumptive carcinogen is kept at 0.15 μg/day, the cancer risk is limited to 1 excess cancer case per 1,000,000 lives. In 2004, the TTC concept was published to set human exposure threshold values also for all chemicals with no toxicity data based on established toxicity profiles [28].

In the pharmaceutical industry, the toxicological qualification of impurities in drug substances and drug products is addressed in the ICH Q3A and Q3B guidelines [1,2]. These guidelines ascertain that the exposure to a putative genotoxic impurity is not greater than 1 mg/day. In the Safety Working Party of the European Medicines Agency (EMA), discussions on the limits of genotoxic impurities started in 2002. The EMA *Guideline on the limits of genotoxic impurities* [3] came into effect in January 2007 along with subsequent questions and answers on this guideline [4]. In this guideline, a TTC value of 1.5 μg/day intake of a genotoxic impurity is considered to be associated with an acceptable risk (excess cancer risk of 1 in 100,000 over a lifetime) for most pharmaceuticals. The FDA released a draft guidance [5] in December 2008. The ICH M7 guideline *Assessment and control of DNA reactive (mutagenic) impurities in pharmaceuticals to limit potential carcinogenic risk* is in preparation. Until finalization, the respective local guidelines are active.

REFERENCES

1. *Q3A(R2) Impurities in new drug substances*, International Conference on Harmonisation of Technical Requirements for Registration of Pharmaceuticals for Human Use, 2006.
2. *Q3B(R2) Impurities in new drug products*, International Conference on Harmonisation of Technical Requirements for Registration of Pharmaceuticals for Human Use, 2006.
3. *Guideline on the limits of genotoxic impurities*, European Medicines Agency, CPMP/SWP/5199/02; EMEA/CHMP/QWP/251344/2006, 2006.
4. *Questions and answers on the 'guideline on the limits of genotoxic impurities'*, European Medicines Agency, EMA/CHMP/SWP/431994/2007 Rev. 3, 2010.
5. *Genotoxic and carcinogenic impurities in drug substances and products: Recommended approaches*, Food and Drug Administration draft guidance, 2008.
6. Cohen SM, Arnold LL. Chemical carcinogenesis. *Toxicological Sciences* 2011 120(S1): S76–S92.
7. Littlefield NA, Farmer JH, Gaylor DW, Sheldon WG. Effects of dose and time in a long term low dose carcinogenic study. *Journal of Environmental Pathology and Toxicology* 1980 3: 17–34.
8. Poirier MC, Fullerton NF, Kinouchi T, Smith BA, Beland FA. Comparison between DNA adduct formation and tumorigenesis in livers and bladders of mice chronically fed 2-acetlyaminofluorene. *Carcinogenesis* 1991 12(5): 895–900.

9. Cheeseman MA, Machuga EJ, Baily AB. A tiered approach to threshold of regulation. *Food and Chemical Toxicology* 1999 37: 387–412.
10. Vineis P, Pirastu R. Aromatic amines and cancer. *Cancer Causes and Control* 1997 8(3): 346–355.
11. Pira E, Piolatto G, Negri E, Romano C, Boffetta P, Lipworth L, McLaughlin JK, La Vecchia C. Bladder cancer mortality of workers exposed to aromatic amines: A 58-year follow-up. *Journal of the National Cancer Institute* 2010 102(14): 1096–1099.
12. Morinaga K, Oshima A, Hara I. Multiple primary cancers following exposure to benzidine and beta-naphthylamine. *American Journal of Industrial Medicine* 1982 3: 243–246.
13. Wong O. A cohort mortality study of employees exposed to chlorinated chemicals. *American Journal of Industrial Medicine* 1988 14: 417–431.
14. Ashby J, Tennant RW. Chemical structure, *Salmonella* mutagenicity and extent of carcinogenicity as indicators of genotoxic carcinogenesis among 222 chemicals tested in rodents by the U.S. NCI/NTP. *Mutation Research* 1988 204(1): 7–115.
15. Enoch SJ, Cronin MTD. A review of the electrophilic reaction chemistry involved in covalent DNA binding. *Critical Reviews in Toxicology* 2010 40(8): 728–748.
16. Hakura A, Shimada H, Nakajima M, Sui H, Kitamoto S, Suzuki S, Satoh T. *Salmonella*/human S9 mutagenicity test: A collaborative study with 58 compounds. *Mutagenesis* 2005 20(3): 217–228.
17. Kirkland D, Aardema M, Henderson L, Müller L. Evaluation of the ability of a battery of three in vitro genotoxicity tests to discriminate rodent carcinogens and non-carcinogens I. Sensitivity, specificity and relative predictivity. *Mutation Research* 2005 584: 1–256.
18. *S2(R1) Genotoxicity testing and data interpretation for pharmaceuticals intended for human use*, International Conference on Harmonisation of Technical Requirements for Registration of Pharmaceuticals for Human Use, 2011.
19. *Q3C(R4) Impurities: Guideline for residual solvents*, International Conference on Harmonisation of Technical Requirements for Registration of Pharmaceuticals for Human Use, February 2009.
20. Brendler-Schwaab S, Hartmann A, Pfuhler S, Speit G. The in vivo comet assay: Use and status in genotoxicity testing. *Mutagenesis* 2005 20(4): 245–254.
21. Suzuki H, Imamura T, Koeda A, Morimoto K, Wakasa Y, Takei Y, Amemiya R, Hatakeyama H, Satoh H, Sato S. Genotoxicity studies of 2,6-dinitrotoluene (2,6-DNT). *Journal of Toxicology Science* 2011 36(4): 499–505.
22. Sing R, Farmer PB. Liquid chromatography-electrospray ionization-mass spectrometry: The future of DNA adduct detection. *Carcinogenesis* 2006 27(2): 178–196.
23. Kirkland D, Speit G. Evaluation of the ability of a battery of three in vitro genotoxicity tests to discriminate rodent carcinogens and non-carcinogens III. Appropriate follow-up testing in vivo. *Mutation Research* 2008 654: 114–132.
24. Wahnschaffe U, Bitsch A, Kielhorn J, Mangelsdorf I. Mutagenicity testing with transgenic mice. Part I: Comparison with the mouse bone marrow micronucleus test. *Journal of Carcinogenesis* 2005 4(3): 1–14.
25. Wahnschaffe U, Bitsch A, Kielhorn J, Mangelsdorf I. Mutagenicity testing with transgenic mice. Part II: Comparison with the mouse spot test. *Journal of Carcinogenesis* 2005 4(4): 1–8.
26. Dertinger SD, Heflich RH. In vivo assessment of *pig-a* gene mutation—recent developments and assay validation. *Environmental and Molecular Mutagenesis* 2011 52(9): 681–684.
27. Mahadevan B, Snyder RD, Waters MD, Benz DR, Kemper RA, Tice RR, Richard AM. Genetic toxicology in the 21st century: Reflections and future directions. *Environmental Molecular Mutagenesis* 2011 25(5): 339–354.
28. Kroes R, Kleiner J, Renwick A. The threshold of toxicological concern concept in risk assessment. *Toxicological Sciences* 2004 86(2): 226–230.

5 In Silico/Computational Assessment for the Evaluation of Genotoxic Impurities

Alexander Amberg, Andreas Czich, and Véronique Thybaud

CONTENTS

5.1 INTRODUCTION

The ability to predict mutagenic activity of chemicals based on their structure and potential reactivity toward DNA has been used for a few decades. Mainly two types of in silico systems are currently in use based on either rules developed from scientific knowledge (i.e., substructures known to be responsible for interaction with DNA) or the fragment-based quantitative structure–activity relationship (QSAR) paradigm relying on experimental data sets (e.g., results obtained in the Ames test). Such in silico systems are key elements of discovery and occupational safety processes in industry for the selection of new candidates, and for prioritization of genotoxicity testing. More recently, they were identified as potential powerful tools for the identification of potential genotoxic impurities (GTIs) that might result from chemical synthesis and compound degradation. The purpose of this chapter is to exemplify, using in-house experience, how in silico systems could be integrated, alone or in association, in the risk assessment of potential GTIs and whether expert knowledge could contribute to the interpretation and validation of the in silico predictions.

5.2 IN SILICO/COMPUTATIONAL TOOLS USED FOR THE ASSESSMENT OF GENOTOXIC IMPURITIES

The in silico/computational assessment of potential GTIs generally combines the use of various databases and prediction systems together with expert knowledge. The databases organize toxicity data (e.g., Ames results) together with their chemical structure from the public literature, and also from in-house laboratory results or archives. Important to note is that new in-house data are permanently produced in the laboratory every year and, therefore, in-house data sets are continuously updated. These public and in-house databases can directly be used for in silico predictions, that is, the search for genotoxicity data and use of structure–activity relationship (SAR) data for the development of new rules. They can also be considered as the training data sets for the fragment-based prediction systems that statistically correlate substructure fragments with mutagenic activity. It is important for these systems to comply with the Organisation for Economic Co-operation and Development (OECD) principles for QSAR validation [1,2]. Results found in databases, predictions made by in silico systems, and review of these data and predictions by experts (often named "expert knowledge") are the three key elements of the structure-related safety assessment of GTIs. An overview of our in-house process is summarized in Figure 5.1. More details on this process are described in Sections 5.1 through 5.3.

5.2.1 DATABASES

The first step in the in silico assessment of potential GTIs is to perform a search in public or internal databases to determine if experimental genotoxicity data are already available. These are primarily mutagenicity data like that from the Ames test or mouse lymphoma assay (MLA) because compounds are classified as GTIs based on their DNA-reactive (mutagenic) activity. Public sources of genotoxicity

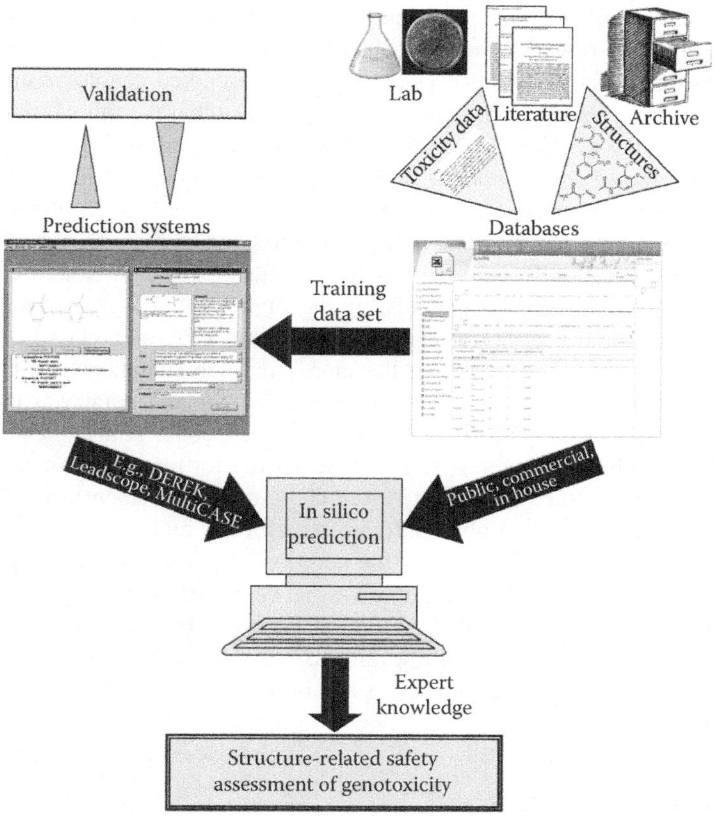

FIGURE 5.1 Different tools used for the structure-related safety assessment of genotoxicity.

data can be toxicological journals, data safety sheets, and public databases that summarize genotoxicity results. These databases can be searched by CAS number, compound name (should consider all synonyms), and chemical structure. Table 5.1 summarizes the most commonly used databases compiling genotoxicity results and includes their abbreviations and links for access. Examples are databases provided by several regulatory authorities or organizations such as the International Uniform Chemical Information Database (IUCLID) (European Union [EU]), Informatics and Computational Safety Analysis Staff (ICSAS) (Food and Drug Administration [FDA]), Integrated Risk Information System (IRIS) (Environmental Protection Agency [EPA]), National Toxicology Program (NTP), Toxicology Data Network (TOXNET), IPS INCHEM, and Japan Existing Chemical Data Base and Registry of Toxic Effects of Chemical Substances databases. Among toxicity results, data from carcinogenicity studies can help in the in silico genotoxicity assessment of GTIs. But the robustness of a carcinogenicity assay has to be evaluated before using it (e.g., route of administration, number of animals sufficient, appropriate controls included, single compound or mixture tested, duration of study, and top dose sufficient). For instance, negative carcinogenicity studies and a clear understanding of nongenotoxic versus genotoxic mechanisms of carcinogenicity can possibly help in overruling positive genotoxicity results obtained with potential GTIs. Similarly, positive carcinogenicity studies especially with an accepted genotoxic mechanism of carcinogenicity can help in confirming a GTI. Databases with carcinogenicity results include the Carcinogenic Potency Database, IUCLID, NTP, and TOXNET databases. Some of the regulatory authorities also provide classification systems that classify the compounds into categories with respect to their genotoxic and carcinogenic potential. Classification systems are provided by the European Chemical Substances Information System (EU) with classification systems for carcinogenic, mutagenic, and reprotoxic compounds, by the International Agency for Research on Cancer with classification for carcinogenicity and EPA with classifications for mutagenicity and carcinogenicity. Furthermore, commercial databases that compile the genotoxicity data from the different public sources and databases mentioned earlier are available. Among them are VITIC, Leadscope, PharmaPendium, and SciFinder. In addition to the possibility of searching for results by an exact structure search, substructure or similarity searches in the aforementioned databases can help in identifying data obtained with structurally related compounds. Some of these commercial databases allow the addition of internal in-house data to the public data. Finally, data-sharing initiatives between different companies are currently ongoing for GTIs (e.g., the Lhasa data-sharing initiative, which is organized via VITIC). The proper use of all these databases and literature information are also very important for the expert knowledge step in the GTI assessment described later in Section 5.2.3.

5.2.2 Prediction Systems

Most prediction systems provide information on many toxicological end points; but for the purposes of the GTI assessment described in this chapter, only mutagenicity alerts are considered that are based on Ames mutagenicity DNA reactivity [3,4].

5.2.2.1 Knowledge Rule-Based Systems

5.2.2.1.1 Deductive Estimation of Risk from Existing Knowledge

Deductive Estimation of Risk from Existing Knowledge (DEREK) is a knowledge rule-based expert computer system for the prediction of toxicity (www.lhasalimited.org/derek_nexus/) [5]. The existing toxicological knowledge is stored as rules/alerts in the computer system (e.g., 77 mutagenicity alerts in DEREK version 13 and 82 in version 14, respectively). The prediction of toxicity is based on the SAR analysis of a chemical, and the rules are based on the relationship of structural features to a toxicological activity or end point. When the software analyzes a compound, the rules identify features (also called scaffolds, moieties, and substructures) within the structure that were shown to be responsible for toxicological activity when they were present in other chemicals. DEREK highlights these substructures and gives structural alerts for this compound, including background information, example compounds, and references for each alert. These toxicological active substructures within the structures are also called toxicophores (www.lhasalimited.org). Since the

TABLE 5.1
Summary of Different Databases with Genotoxicity Results Including Their Link for Access

Name	Genotoxicity/Toxicity Data Available from	Link
ESIS	European Chemical Substances Information System, incl. EU classification for CMR compounds (carcinogenic, mutagenic, and reprotoxic)	esis.jrc.ec.europa.eu
IARC	International Agency for Research on Cancer, incl. classification for carcinogenicity	www.iarc.fr
IUCLID	International Uniform Chemical Information Database	iuclid.eu
ICSAS	Informatics and Computational Safety Analysis Staff databases from the FDA	www.fda.gov/AboutFDA/CentersOffices/OfficeofMedicalProductsandTobacco/CDER/ucm092217.htm
IRIS	Integrated Risk Information System from the EPA, incl. classification for carcinogenicity and mutagenicity	www.epa.gov/iris
NTP	National Toxicology Program	ntp.niehs.nih.gov
TOXNET	Toxicology Data Network	toxnet.nlm.nih.gov
IPS INCHEM	International Programme on Chemical Safety	www.inchem.org/
JECDB	Japan Existing Chemical Data Base	dra4.nihs.go.jp/mhlw_data/jsp/SearchPageENG.jsp
RTECS	Registry of Toxic Effects of Chemical Substances	www.cdc.gov/niosh/rtecs/default.html
CPDB	Carcinogenic Potency Database	potency.berkeley.edu
PharmaPendium	Database with toxicity data from FDA and EMA approval documents	Commercial (for information, see www.pharmapendium.com)
Leadscope	Database with toxicity data extracted from various public databases/sources and toxicological journals	Commercial (for information, see www.leadscope.com)
VITIC	Database with toxicity data extracted from various public databases/sources and toxicological journals	Commercial (for information, see www.lhasalimited.org/vitic_nexus)
Other sources from toxicological journals, databases, data safety sheets, etc.		

Note: EMA, European Medicines Agency.

same structures can exist in a variety of molecules, the rules are not chemical specific, but rather they serve as broad generalizations with respect to the chemical structure (e.g., alkylating agent and acid-/halogen-containing molecule) [6].

The knowledge is based on researching literature data or internal company data with an emphasis on the understanding of mechanisms of toxicity and metabolism required for the activation of a compound to a toxicological intermediate [7]. It covers a wide variety of toxicological end points, which include genotoxicity, carcinogenicity, irritation, skin and respiratory sensitization, hepatotoxicity, hERG channel inhibition, reproductive/developmental toxicity, and other miscellaneous end points (www.lhasalimited.org). Its main strengths lie in the prediction of carcinogenicity, mutagenicity, and skin sensitization [8]. But for the GTI assessment described in this chapter, carcinogenicity alerts should not be considered because genotoxic carcinogens will be alerted as mutagens and nongenotoxic carcinogens are not in the scope of the GTI assessment. Also, mutagenicity alerts are the most developed and validated compared to carcinogenicity alerts.

It is important to note the possibility of implementing custom internal in-house rules into the systems based on knowledge extrapolated from in-house data. However, such rules would have to be transparently described and reported if used for GTI genotoxicity assessment.

The DEREK software also incorporates a reasoning engine to predict the likelihood of a chemical to express its potential toxicity regarding a specific end point in the selected species. This reasoning engine combines both numerical and nonnumerical statements (like selected species, log P, molecular mass, end point, and toxicophores) to reach a conclusion about a given event. It is based on the mathematical framework of the logic of argumentation. The result of the reasoning as the likelihood of toxicity is then expressed in one of the following terms: certain (exact structure with Ames data is present in the training set), probable, plausible, equivocal, doubted, improbable, impossible, open, and contradicted. Additionally, DEREK provides validation data from up to seven different validation data sets (public and proprietary sources) for each alert. This validation data can help in the assessment of reliability of an alert.

The DEREK rules are written and maintained by experts from the nonprofit organization Lhasa Limited. They are regularly updated by Lhasa Limited and new versions of the computer software are available every year. Regular collaborative user group meetings are organized with representatives from pharmaceutical, agrochemicals, and regulatory organizations to discuss changes in computer software and knowledge base developments and to get feedback from users. This unique system encourages the sharing of toxicological information and knowledge for the benefit of all, without organizations compromising the confidentiality of their proprietary data [7,9].

It has to be mentioned that DEREK does not yield negative predictions since an absence of an alert can also mean an absence of knowledge (out of chemical space). Currently, Lhasa is working on the implementation of confidence metrics and negative predictions into future DEREK versions. For this purpose, the knowledge base will be exploited to define the predictive space and the model reliability domain to describe coverage of each alert. It will be defined based on predictions generated for large data sets. The structures or features with high similarity to compounds from the reference data set that are well predicted will be assigned as the model reliability domain. Thus, a "nothing to report" could be turned into a negative prediction [10].

5.2.2.1.2 Other Open Source Knowledge-Based Systems

A variety of open source knowledge rule-based expert computer systems are available for the prediction of genotoxicity. Examples include ToxTree (toxtree.sourceforge.net), OpenTox-ToxPredict (www.opentox.org/toxicity-prediction), and Bioclipse-DS (www.bioclipse.net/decision-support). ToxTree and OpenTox-Predict estimate toxic hazard by applying decision tree approaches. For both systems, several plugins are available for the prediction of mutagenicity, like the Cramer rules [11,12], the Benigni/Bossa rulebase [13], or Kazius-Bursi Salmonella models. Bioclipse is using for its rule-based prediction two approaches, the Smiles Arbitrary Target Specification and atom signatures. In validation studies that were published, ToxTree showed a reasonable performance but a lower predictivity for mutagenicity compared to DEREK [14]. Also, internal validation studies showed an acceptable predictivity for mutagenicity mainly for the Benigni/Bossa and Bioclipse rulebases.

5.2.2.2 Fragment-Based Quantitative Structure–Activity Relationship Systems

5.2.2.2.1 Leadscope

The Leadscope Enterprise software is an expert computer system using a fragment-based QSAR paradigm (www.leadscope.com) (Leadscope, Inc., Columbus, Ohio) [15]. The system consists of computer software to perform prediction and different training databases (models) for the prediction of respective toxicity end points. The fragments used for prediction are predefined in a hierarchically organized dictionary that is closely related to common organic/medicinal chemistry blocks. For binary classification problems, such as the Ames test, the algorithm identifies toxicity-modulating fragments using a χ^2 test. Furthermore, the software is able to build superstructures from smaller fragments if they improve predictivity. Additionally, eight global molecular properties are calculated (atom count, hydrogen bond acceptors, hydrogen bond donors, Lipinski score, log P, molecular weight, polar surface area, and rotatable bonds). These global molecular properties together with the set of fragments are then used as a descriptor set in a partial least squares logistic regression model of the activity class. Therefore, the predictions from this algorithm are continuous probabilities of class membership rather than binary outputs, given as the likelihood value between 0 (nontoxic) and 1 (toxic). All probabilities greater than 0.5 are considered as "active" predictions and probabilities smaller than 0.5 as "inactive" predictions. The higher the probability, the greater the chance of the test chemical being toxic in a particular end point. The program also assesses the applicability domain by measuring the distance to training set molecules by using two parameters: (1) having at least one feature defined in the model and (2) having at least one chemical in a training neighborhood with at least 30% similarity. Compounds that are annotated as "out of domain" or "missing descriptors" are counted as "not predicted" [16].

Currently, Leadscope offers QSAR models for the prediction of eight different toxicity end points. All these QSARs were constructed at the FDA by the ICSAS group. The training data sets were compiled by ICSAS, and the models were built within the Leadscope software using default settings [17].

The first group of toxicity end points includes QSAR models that predict the effects of compounds based on human clinical data, including adverse cardiological effects, adverse hepatobiliary effects, and adverse urinary tract effects. The second group includes models predicting toxicities of compounds based on the results of in vivo animal toxicity and in vitro studies. They include carcinogenicity in rodents, genetic toxicity (i.e., mutagenicity, clastogenicity, and DNA damage), reproductive toxicity in male and female rodents, developmental toxicity (i.e., dysmorphogenesis, fetal development, and survival of the rodent fetus), and neurotoxicity in newborn rodents.

Each toxicity end point has many different QSAR models. For some end points, submodels are constructed to improve the predictive performance, which depends highly on the ratio of active (toxic) to inactive (nontoxic) chemicals in a training set. A training set was divided into subsets to maintain the optimal active to inactive ratio between 0.30 and 0.35 to ensure high specificity. The rationale behind these QSAR models is that predicting true negatives must be maximized while false negatives must be minimized in product safety analyses within regulatory agencies. Leadscope runs each of the submodels behind the scenes, and the overall prediction results are based on averaging the probabilities (likelihood of being positive) from appropriate submodels (Leadscope FDA Model Applier Documentation 2008).

For the prediction of mutagenicity, Leadscope offers a public *Salmonella* gene mutation QSAR model trained with 3579 compounds. In addition to these public models, it is possible to use proprietary data for the construction of in-house QSAR models using the same Leadscope platform.

5.2.2.2.2 Multiple Computer Automated Structure Evaluation

Multiple computer automated structure evaluation (MultiCASE) is a fragment-based expert computer system for the prediction of toxicity (www.multicase.com, MultiCASE, Beachwood, Ohio) [18]. Like Leadscope, it consists of a computer system to perform the prediction and different training databases (modules) for the prediction of respective toxicity end points.

Each database contains a series of diverse noncongeneric chemical structures and their observed activity (quantitative or qualitative) for specific toxicological end points, including toxicologically active and inactive compounds. Some authors also classify MultiCASE as a hybrid QSAR and artificial expert structure-based program. The QSAR portion of the program is based on two-dimensional chemical descriptors that utilize a proprietary statistical analysis developed by Klopman [19,20]. The artificial expert structure–based program is based on the identification of atom fragments that are present in active and inactive molecules and that have a high probability of being relevant or responsible for the observed toxicological activity [21,22].

For the prediction of toxicity, in the first step each molecule of a database is broken down by MultiCASE into all possible fragments from 2 to 10 heavy (nonhydrogen) atoms also including overlapping fragments. These are then statistically classified as "biophores," fragments associated with toxicity, and "biophobes," fragments not associated with toxicity. In addition to utilizing molecular fragments, MultiCASE also identifies relevant two-dimensional distances between atoms within a chemical structure. MultiCASE then creates organized dictionaries of these biophores and biophobes and develops ad hoc local QSAR correlations that can be used to predict the activity of unknown molecules. The results of this first prediction step are saved, and identified biophores are visible to the users (www.multicase.com).

In the second step of the prediction, a new molecule is entered into MultiCASE; then, the program evaluates this molecule against the organized dictionary and the appropriate QSARs it has created and makes a prediction of the toxicological activity of the molecule for the corresponding end point. To do this, MultiCASE identifies all relevant biophores and biophobes of the unknown molecule, combines these into an equation, and calculates the toxicological activity expressed in computer automated structure evaluation (CASE) units with the help of the following equation [23]:

$$\text{CASE units} = \text{constant} + a\,(\text{fragment 1}) + b\,(\text{fragment 2}) + \ldots$$

The scale of CASE units has a linear range, and normally chemicals with an assigned value of 10–19 are inactive; 20–29 have marginal activity; and 30–99 are moderately active, active, very active, and extremely active, respectively. The system is also able to identify fragments that act as modifiers to the activity of each biophore class [8].

MultiCASE covers different toxicological end points like genotoxicity, carcinogenicity, irritation, and developmental toxicity/teratogenicity and adverse effects in humans for hepatobiliary, renal/urinary tract, and cardiac end points and other miscellaneous endpoints. For each of these end points, one or more databases (modules) containing active and inactive molecules are separately available and most of the modules were constructed at the FDA by the ICSAS group. The number of compounds varies from 70 to 6000 per module depending on the end points (www.multicase.com).

For the prediction of mutagenicity, MultiCASE offers the Ames Salmonella module (AZ2) trained with 7731 compounds. Like for Leadscope, it is possible in addition to these public models to build in-house modules using the same MultiCASE Leadscope platform but with proprietary data.

5.2.2.2.3 Other Open Source Fragment-Based Quantitative Structure–Activity Relationship Systems

Publicly available fragment-based QSAR expert computer systems also exist for the prediction of genotoxicity. Such models are offered, for example, by ToxTree (toxtree.sourceforge.net) for distinct chemical classes, such as aromatic amines and α-/β-unsaturated aldehydes [10]. Also, Bioclipse-DS (www.bioclipse.net/decision-support) offers QSAR models that are developed from the Kazius-Bursi Salmonella training data set. Bioclipse Modeling (www.genettasoft.com/) allows also the possibility of developing new prediction models from internal training data sets based on regression and classification methods. In internal validation studies, these publicly available fragment-based QSAR models showed an acceptable predictivity for mutagenicity.

5.2.3 Expert Knowledge

As mentioned in Sections 5.1 and 5.2, not only genotoxicity data but also physicochemical properties, metabolic activation, and mechanism of action for a given chemical and for structurally related compounds can help in improving mutagenicity prediction. Because this analysis is made on a case-by-case basis and could be compound specific, it is important to transparently report the elements and rationale that contribute to the analysis.

5.3 VALIDATION OF VARIOUS SYSTEMS FOR THE ASSESSMENT OF GENOTOXIC IMPURITIES

Because the absence of in silico alerts for mutagenicity during the in silico assessment of potential GTIs generally results in no further testing, an accurate and reliable prediction process that might consist of one or more systems/approaches is crucial. The goal of the present validation exercise was to (1) investigate the predictivity for GTI mutagenicity assessment of various prediction systems alone and in combination, (2) investigate the use of expert knowledge (mainly defined by the search of existing experimental genotoxicity results in public and internal databases), and (3) evaluate whether combinations of prediction systems and expert knowledge (i.e., expert assessment of all the available data) can improve the prediction systems.

5.3.1 Test Data Set Used for Validation

The goal of our validation was to investigate the predictivity of various in silico systems alone and in combination for GTI assessment using an in-house data set. For this purpose, we collected a test data set of all potential GTIs that were tested in house at Sanofi in the Ames assay mainly for occupational safety and GTI purposes in the years between 2009 and 2011. There are in total 269 compounds from which 39 (15%) were found positive and 230 (85%) were tested negative in Ames. It has to be noted that this is an unbalanced test data set in the way that it is dominated by negative compounds, but our main focus was to use a realistic data set. The imbalance was kept in mind for the later interpretation of the validation results. Regarding chemical diversity, Table 5.2 shows the number of compounds per chemical class with their results in the Ames assay.

5.3.2 Expert Systems Used for Validation

The expert prediction systems that were used for validation include the knowledge rule-based system DEREK (version 13.0) and the two fragment-based QSAR systems Leadscope (version 3.0) and MultiCASE (version 2.0). For mutagenicity prediction, we used the public Leadscope model FDA Salmonella 2010 and the public MultiCASE module AZ2 that were trained with 3600 and 7731 public Ames results, respectively. The analysis was completed by the use of an in-house Leadscope model and a MultiCASE module, both trained with 4200 proprietary Ames results (all compounds tested at Sanofi between 1990 and 2008).

TABLE 5.2
Number of Compounds per Chemical Class from the Test Data Set Used for Validation and Their Results in the Ames Assay

Chemical Class	Ames Positive	Ames Negative	Total
Alkylating agents	17	18	35
Aromatic nitro comp	9	9	18
Aromatic amine/amide	7	25	32
Alkyl aldehyde	2	11	13
Haloamine	2	0	2
Arylhydrazine	1	4	5
Geminal dihalide	1	2	3
Labile halogen	0	10	10
Carboxylic acid halide	0	7	7
Epoxide	0	2	2
Substituted vinyl ketone	0	2	2
Oxime	0	2	2
Quinoline	0	2	2
Others	0	136	136
Total	39	230	269

Furthermore, in addition to the expert prediction systems we also searched for available experimental mutagenicity data. To this end, we used all the various databases from public and commercial sources described in Table 5.1. Up to 10,000 Ames results are available from public sources and not always part of the training data sets used for building expert systems. Similarly, we looked for additional proprietary mutagenicity data (~8,000 Ames and other genotoxicity results). These results were also used in parts as training data sets for in-house Leadscope and MultiCASE models.

5.3.3 Validation Results

The validation results are described hereafter with typical validation parameters from a confusion matrix, that is, sensitivity, specificity, positive and negative predictivity, and concordance.

These parameters and their definitions are summarized in Figure 5.2. It must be noted that for GTI assessments a high confidence in negative predictions is very important since no further action (Ames testing or content controlling) is required in the absence of structural concerns from in silico/computational assessment for genotoxicity. In the context of GTI assessment, the expectation is a low number of false negative predictions, which results in a high sensitivity and a high negative predictivity. Also, these are parameters toward these expert systems are optimized.

5.3.3.1 Predictivity of Different Expert Systems When Used Alone

The predictivity of the three expert systems DEREK, Leadscope, and MultiCASE when they are used alone is summarized in Figure 5.3. For Leadscope and MultiCASE, individual validation parameters are shown by using both the public

		Prediction			
		Mutagenic	**Not mutegenic**		
Test result	**Positive**	True positives (TP)	False negative predictions (FN)	**Sensitivity** = TP/(TP + FN)	% correctly predicted positive compounds
ΣN = TP + FN + FP + TN	**Negative**	False positive predictions (FP)	True negatives (TN)	**Specificity** = TN/(FP + TN)	% correctly predicted negative compounds
		Positive predictivity = TP/(TP + FP)	**Negative predictivity** = TN/(FN + TN)	**Concordance** = TP + TN/(ΣN)	% correct overall predictions
		% correct positive predictions	% correct negative predictions		

FIGURE 5.2 Validation parameters and their definition from a confusion matrix.

FDA Ames and the in-house Sanofi Ames prediction model. For DEREK, Figure 5.3 shows the validation parameter by using the version with all the public alerts (i.e., no additional in-house rules). The Leadscope/MultiCASE predictions were classified as "mutagenic" for all positive predictions and as "not mutagenic" for all negative or not in domain predictions (predictions are out of the applicability/validity domain of the training data set). The DEREK predictions were classified as "mutagenic" for any DEREK mutagenicity alert and as "not mutagenic" for nothing to report. For each single Leadscope/MultiCASE Ames model, the percentage of not in domain predictions was between 30% and 44%. But by combining all public FDA and in-house Sanofi models from Leadscope and MultiCASE and by using public/in-house database search results, the not in domain predictions in all models were reduced to 3%. There were seven compounds in total that were all confirmed as no GTI later on by a negative Ames test. Due to this fact, we applied for all the following validation results the simplification of using the not in domain or nothing to report predictions as a not mutagenic prediction.

As single systems, Leadscope and MultiCASE showed comparable results for both FDA and Sanofi Ames models with negative predictivities between 88% and 89% but relatively low sensitivities between 26% and 36%. DEREK demonstrated a negative predictivity of 94% and a much higher sensitivity of 72% compared to the other in silico systems.

It must be mentioned that the validation results obtained for GTIs (mostly synthesis intermediates, reactants, and raw materials) showed better predictivity compared to in-house research and discovery compounds. A similar validation study for in-house research and discovery compounds tested in 2011 showed, for example, a sensitivity of only 36% and a negative predictivity of 84% for DEREK. The reason for the differences is that the public in silico systems have been developed from public data with chemicals very similar to GTIs. Therefore, these expert systems are better trained for GTIs than discovery compounds bearing newer and more innovative patented chemical structures. However, the chemical space of our proprietary in-house research and discovery compounds differs from the GTI chemical space and the chemical space used to build the public models, resulting in lower sensitivity, negative predictivity, and the fact that many of them are out of the domain of the training data set. Only Leadscope and MultiCASE models

Leadscope FDA Ames model

		Leadscope-prediction: Mutagenic	Leadscope-prediction: Not mutagenic	
Ames results	Pos. +	10	29	Sensitivity 26%
	Neg. −	24	206	Specificity 90%
		Pos. predictivity 29%	Neg. predictivity 206/235 = 88%	Concordance 80%

Leadscope Sanofi Ames model

		Leadscope-prediction: Mutagenic	Leadscope-prediction: Not mutagenic	
Ames results	Pos. +	14	25	Sensitivity 36%
	Neg. −	21	209	Specificity 91%
		Pos. predictivity 40%	Neg. predictivity 209/234 = 89%	Concordance 83%

MultiCASE FDA Ames model

		MultiCASE-prediction: Mutagenic	MultiCASE-prediction: Not mutagenic	
Ames results	Pos. +	14	25	Sensitivity 36%
	Neg. −	24	206	Specificity 90%
		Pos. predictivity 37%	Neg. predictivity 206/231 = 89%	Concordance 82%

MultiCASE Sanofi Ames model

		MultiCASE-prediction: Mutagenic	MultiCASE-prediction: Not mutagenic	
Ames results	Pos. +	11	28	Sensitivity 28%
	Neg. −	13	217	Specificity 94%
		Pos. predictivity 46%	Neg. predictivity 217/245 = 89%	Concordance 85%

DEREK

		DEREK prediction: Mutagenic	DEREK prediction: Not mutagenic	
Ames results	Pos. +	28	11	Sensitivity 72%
	Neg. −	69	161	Specificity 70%
		Pos. predictivity 29%	Neg. predictivity 161/172 = 94%	Concordance 70%

FIGURE 5.3 Validation parameters of different expert systems when they are used alone. For all validations, the same Sanofi test set of potential genotoxic impurities was used (269 compounds in total, from which 39 tested positive and 230 tested negative in Ames).

from in-house research and discovery compounds achieve increased sensitivity and negative predictivity (up to 82% sensitivity and 94% negative predictivity) for our in-house chemical space. In conclusion, the appropriateness of the chemical space used to train expert systems is a key element for validation and the development of a reliable prediction tool.

5.3.3.2 Predictivity of Different Expert Systems When Used in Combination

Since DEREK showed the best predictivity for GTIs, it was further investigated whether DEREK prediction could be improved by adding either another fragment-based QSAR system (Leadscope and MultiCASE) or by using a database search. DEREK was first combined with the public modules of Leadscope and MultiCASE, and then with both public and in-house models. The results of these different combinations are summarized in Figure 5.4.

When the knowledge rule-based system DEREK was used in combination with the public models of Leadscope and MultiCASE, predictivity could be improved and false negative predictions could be reduced to nine compounds. The combination of DEREK with Leadscope and MultiCASE increased the sensitivity to 77% in both cases, and negative predictivity increased to 94% for Leadscope and 95% for MultiCASE. When in-house Sanofi Ames Leadscope and MultiCASE models were

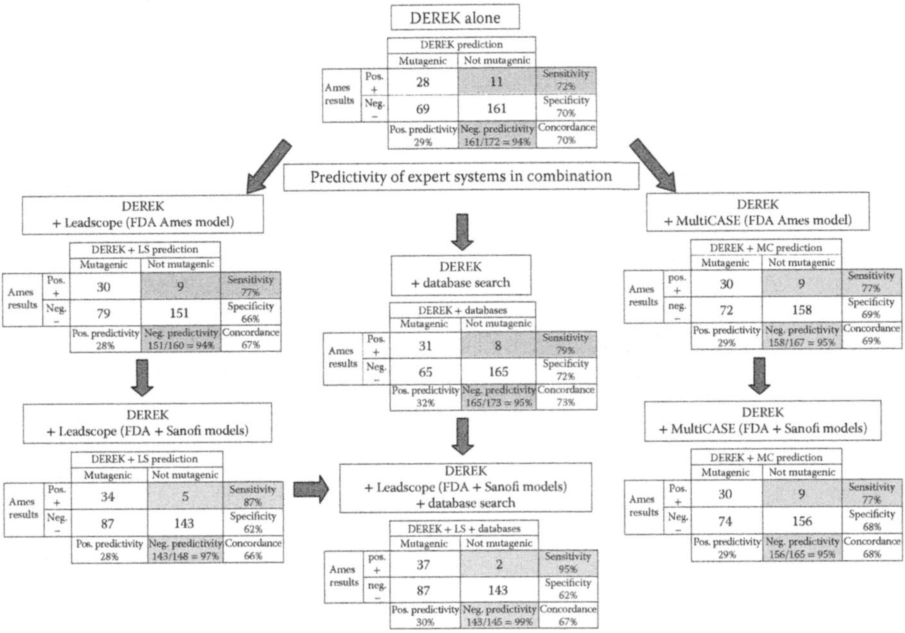

FIGURE 5.4 Validation parameters of the different systems when they are used in combination for a Sanofi test set of potential genotoxic impurities (GTIs). For all validations, the same Sanofi test set of potential GTIs was used (269 compounds in total, from which 39 tested positive and 230 tested negative in Ames).

added, the predictivity values were further improved only for Leadscope to only five false negative predictions, a sensitivity of 87%, and a negative predictivity of 97%.

Another approach to enhance the predictivity is searching the literature/databases for already existing Ames results, as described in Section 5.2.1. In these databases, we were able to find some Ames positive compounds that were not predicted as mutagens by DEREK. The combination of DEREK with database search reduced the false negative predictions to eight compounds, resulting in a sensitivity of 79% (vs. 72% with DEREK alone) and a negative predictivity of 95% (vs. 94% with DEREK alone).

The highest predictivity was achieved when the two expert systems DEREK and Leadscope based on both the FDA and Sanofi Ames models were combined with literature/database search. This combination reduced the false negative predictions to only two compounds and increased the sensitivity to 95% and negative predictivity to 99%. It is important to note that the addition of MultiCASE showed no added value. Moreover, even if the data set used for this study was unbalanced and showed only a small percentage of mutagenic compounds (39 out of 269), the combination of multiple approaches clearly improved the predictive values, moving from nine false negative compounds (leading to 72% sensitivity and 94% negative predictivity) for DEREK alone to only two false negative compounds (resulting in 95% sensitivity and 99% negative predictivity) when all approaches were combined.

5.3.4 Discussion/Comparison with Already Published Validation Data

This present validation study confirms and illustrates the already published data highlighting that the newest versions of the knowledge rule-based system DEREK are able to properly predict most of the Ames positive compounds (i.e., high negative predictive values) when used alone. Dobo and others [24] summarized an analysis conducted by eight pharmaceutical companies, which showed negative predictive average values of 94% for the in silico prediction of potential GTIs. In a complementary exercise by Sutter and others [10], our validation data presented in this chapter were published in parts together with the validation results from four pharmaceutical companies. In these studies, negative predictive values between 80% and 99% and sensitivity values between 44% and 97% were reported for a total number of 1449 potential GTIs evaluated in the Ames test. In both cases, complementary approaches have been shown to further improve the negative predictive value and/or the sensitivity. In Dobo's analysis, human interpretation of DEREK data increased the negative predictive value from 94% to 99%. In Sutter's analysis, DEREK data were complemented by database searches, a fragment-based QSAR system, and/or expert knowledge. These more sophisticated approaches resulted in a slight enhancement of negative predictivity (generally only a few percentages to reach up to 99%) and a much clearer increase in sensitivity (from 44% to 95%). Fragment-based QSAR systems are generally redundant and provide no added value when used in combinations. Several publications recently emphasized on the added value of combing multiple approaches in the evaluation of potential GTIs, and these considerations are taken into account in the preparation of the International Conference on Harmonisation of Technical Requirements for Registration of Pharmaceuticals for Human Use M7 guideline for the evaluation of potential GTIs.

Our validation study illustrated that in-house models only slightly improve the prediction parameters for the Leadscope system. The added value is much lower than that for discovery compounds, as mentioned in Section 5.3.3.1 and already published by Hillebrecht and others [16]. These findings confirm that public models have been trained with compounds having structures very similar to those of the potential GTIs that could result from drug synthesis.

It is important to note that the in silico assessment of GTIs should not be considered as a simple push button exercise. Even if it relies on the computational analysis of various expert prediction systems and databases, for a given potential GTI the results and data resulting from the different sources have to be evaluated individually with respect to their predictivity, validity, and applicability. Afterward, they have to be combined to allow an appropriate risk assessment. This data- and compound-specific evaluation can be summarized as expert knowledge.

The added value of such an evaluation is to identify some potential limitations. For example, the chemical space covered by the prediction systems or databases could be inappropriate for a given compound because it bears unknown substructures. In such cases, the applicability or validity of in silico predictions should be considered to be questionable and carefully handled to avoid any risk for the exposed populations. In those rare cases, follow-up testing or control to threshold of toxicological concern levels might be required.

5.3.5 Conclusion

In conclusion, the use of an appropriate in silico assessment can be regarded as an efficient and reliable approach for the evaluation of potential GTIs. Because the in silico systems and databases continuously evolve and improve, a dialog is needed to ensure that all stakeholders have the same level of understanding to efficiently and transparently collaborate on risk assessment.

CONFLICT OF INTEREST STATEMENT

The authors are employees of the Sanofi company, and the data presented in the analysis come from research funded by the Sanofi company.

ACKNOWLEDGMENT

The valuable contributions by Hans-Peter Spirkl and Salim Arslan in the performance of the different validation studies are gratefully acknowledged.

REFERENCES

1. Marchant, C.A.; Briggs, K.A.; Long, A. In silico tools for sharing data and knowledge on toxicity and metabolism: DEREK for Windows, Meteor, and VITIC. *Toxicol. Mech. Methods*. **2008**, *18*, 177–187.
2. Organisation for Economic Co-operation and Development. *Guidance Document on the Validation of (Quantitative) Structure-Activity Relationships [(Q)SAR] Models*. **2007**. http://search.oecd.org/officialdocuments/displaydocumentpdf/?doclanguage=en&cote=env/jm/mono(2007)2
3. Matthews, E.J.; Kruhlak, N.L.; Benz, R.D.; Contrera, J.F.; Marchant, C.A.; Yang, C. Combined use of MC4PC, MDL-QSAR, BioEpisteme, Leadscope PDM, and DEREK for Windows software to achieve high-performance, high-confidence, mode of action-based predictions of chemical carcinogenesis in rodents. *Toxicol. Mech. Methods*. **2008**, *18*, 189–206.
4. Snyder, R.D.; Pearl, G.S.; Mandakas, G.; Choy, W.N.; Goodsaid, F.; Rosenblum, I.Y. Assessment of the sensitivity of the computational programs DEREK, TOPKAT, and MCASE in the prediction of the genotoxicity of pharmaceutical molecules. *Environ. Mol. Mutagen*. **2004**, *43*, 143–158.
5. Sanderson, D.M.; Earnshaw, C.G. Computer prediction of possible toxic action from chemical structure; the DEREK system. *Hum. Exp. Toxicol*. **1991**, *10*, 261–273.
6. Cariello, N.F.; Wilson, J.D.; Britt, B.H.; Wedd, D.J.; Burlinson, B.; Gombar, V. Comparison of the computer programs DEREK and TOPKAT to predict bacterial mutagenicity. Deductive Estimate of Risk from Existing Knowledge. Toxicity Prediction by Komputer Assisted Technology. *Mutagenesis*. **2002**, *17*, 321–329.
7. Patlewicz, G.; Rodford, R.; Walker, J.D. Quantitative structure-activity relationships for predicting mutagenicity and carcinogenicity. *Environ. Toxicol. Chem*. **2003**, *22*, 1885–1893.
8. Greene, N. Computer systems for the prediction of toxicity: an update. *Adv. Drug Deliv. Rev*. **2002**, *54*, 417–431.

9. Greene, N.; Judson, P.N.; Langowski, J.J.; Marchant, C.A. Knowledge-based expert systems for toxicity and metabolism prediction: DEREK, StAR and METEOR. *SAR QSAR Environ. Res.* **1999**, *10*, 299–314.
10. Sutter, A.; Amberg, A.; Boyer, S.; Brigo, A.; Contrera, J.F.; Custer, L.L.; Dobo, K.L. et al. Use of in silico systems and expert knowledge for structure-based assessment of potentially mutagenic impurities. *Regul. Toxicol. Pharmacol.* **2013**, *67*(1), 39–52.
11. Cramer, G.M.; Ford, R.A.; Hall, R.L. Estimation of toxic hazard—a decision tree approach. *Food Cosmet. Toxicol.* **1978**, *16*, 255–276.
12. Patlewicz, G.; Jeliazkova, N.; Safford, R.J.; Worth, A.P.; Aleksiev, B. An evaluation of the implementation of the Cramer classification scheme in the ToxTree software. *SAR QSAR Environ. Res.* **2008**, *19*, 495–524.
13. Benigni, R.; Bossa, C. Structure alerts for carcinogenicity, and the Salmonella assay system: a novel insight through the chemical relational databases technology. *Mutat. Res.* **2008**, *659*, 248–261.
14. Devillers, J.; Mombelli, E. Evaluation of the OECD QSAR Application Toolbox and ToxTree for estimating the mutagenicity of chemicals. Part 1. Aromatic amines. *SAR QSAR Environ. Res.* **2010**, *21*, 753–769.
15. Valerio, L.G.; Yang, C.; Arvidson, K.B.; Kruhlak, N.L. A structural feature-based computational approach for toxicology predictions. *Expert Opin. Drug Metab. Toxicol.* **2010**, *6*, 505–518.
16. Hillebrecht, A.; Muster, W.; Brigo, A.; Kansy, M.; Weiser, T.; Singer, T. Comparative evaluation of in silico systems for Ames test mutagenicity prediction: scope and limitations. *Chem. Res. Toxicol.* **2011**, *24*, 843–854.
17. Yang, C.; Hasselgren, C.H.; Boyer, S.; Arvidson, K.; Aveston, S.; Dierkes, P.; Benigni, R. et al. Understanding genetic toxicity through data mining: the process of building knowledge by integrating multiple genetic toxicity databases. *Toxicol. Mech. Methods.* **2008**, *18*, 277–295.
18. Rosenkranz, H.S.; Cunningham, A.R.; Zhang, Y.P.; Claycamp, H.G.; Macina, O.T.; Sussman, N.B.; Grant, S.G.; Klopman, G. Development, characterization and application of predictive-toxicology models. *SAR QSAR Environ. Res.* **1999**, *10*, 277–298.
19. Klopman, G. Artificial intelligence approach to structure-activity studies: computer automated structure evaluation of biological activity of organic molecules. *J. Am. Chem. Soc.* **1984**, *106*, 7315–7321.
20. Klopman, G.; Rosenkranz, H.S. Approaches to SAR in carcinogenesis and mutagenesis. Prediction of carcinogenicity/mutagenicity using MultiCASE. *Mutat. Res.* **1994**, *305*, 33–46.
21. Durham, S.K.; Pearl, G.M. Computational methods to predict drug safety liabilities. *Curr. Opin. Drug Discov. Devel.* **2001**, *4*, 110–115.
22. Cunningham, A.R.; Klopman, G.; Rosenkranz, H.S. Identification of structural features and associated mechanisms of action for carcinogens in rats. *Mutat. Res.* **1998**, *405*, 9–27.
23. Dearden, J.C. In silico prediction of drug toxicity. *J. Comput. Aided Mol. Des.* **2003**, *17*, 119–127.
24. Dobo, K.L.; Greene, N.; Fred, C.; Glowienke, S.; Harvey, J.S.; Hasselgren, C.; Jolly, R. et al. In silico methods combined with expert knowledge rule out mutagenic potential of pharmaceutical impurities: an industry survey. *Regul. Toxicol. Pharmacol.* **2012**, *62*, 449–455.

6 Preclinical Assessment of Genotoxic Impurities

An Overview of Current Regulatory Guidance, Available Assays, and Data Interpretation

Todd J. Page, Vincent L. Reynolds, and J. Barry Phelps

CONTENTS

6.1 INTRODUCTION

Assessment of genotoxic impurities (GTIs) in pharmaceuticals is a complex issue; there are numerous guidelines that discuss the methods, rationales, and strategies for the assessment of genotoxicity in general as well as the assessment of GTIs specifically. In this chapter, we review the current regulatory guidance documents and discuss the impact that they have on strategies for GTI assessment as well as the various assays that can be used for the assessment of GTIs.

6.2 REGULATORY EXPECTATIONS AND GUIDANCES

6.2.1 Overall Genotoxicity Guidance

Although not specific for impurities, ICH S2(R1) (*Guidance on Genotoxicity Testing and Data Interpretation for Pharmaceuticals Intended for Human Use*) impacts the testing strategy as it forms the basis for the testing paradigm that is used to determine if an impurity is genotoxic and discusses how to evaluate the clinical risk that the potential GTI might pose.

Notably, ICH S2(R1) follows a *hazard identification* paradigm with the goal of detecting compounds that induce DNA damage. The aim of the battery of studies described in ICH S2(R1) is to determine if a chemical entity interacts with DNA and has the potential to be a mutagen. The tests in ICH S2(R1) are not designed to determine if a threshold to genotoxicity (or DNA reactivity) exists.

Because ICH S2(R1) was written to apply to drug substances, there is the implicit assumption that a standard 2-year carcinogenicity bioassay will be conducted prior to product registration that will clarify the carcinogenic risk posed by the hazard identified in the ICH S2(R1) test battery. This assumption is typically not valid for the assessment of impurities (except as present in the drug substance used in the 2-year bioassay). Although a 2-year bioassay could theoretically be conducted to determine the carcinogenic risk posed by an impurity per se, this would be a rare occurrence given the considerable time and resources required for these studies. For this reason, although the standard genotoxicity battery outlined in ICH S2(R1) is a starting point for assessing the DNA reactivity of impurities, additional tests may need to be conducted if a more thorough assessment of the potential human risk, rather than hazard identification, is a goal.

The principles behind the specific tests used to determine whether an impurity is genotoxic and the interpretation of these tests are discussed.

6.2.2 Impurity-Related Guidance Documents

Although ICH S2(R1) provides a description of the overall regulatory expectations for genotoxicity testing, there is currently no International Conference on Harmonisation of Technical Requirements for Registration of Pharmaceuticals for Human Use (ICH) guidance that defines the acceptable limits or a control strategy for GTIs in pharmaceuticals. ICH M7 (*Assessment and Control of DNA Reactive [Mutagenic] Impurities in Pharmaceuticals to Limit Potential Carcinogenic Risk*) is currently in development at ICH step 2. Documents at ICH step 2 are still subject to revision. Upon approval by the relevant competent authorities, it will have a significant impact on how potential GTIs are assessed and controlled; since this document has not been approved for implementation, this chapter focuses only on the currently published guidance documents pertaining to GTIs.

Currently, three guidance documents are available that directly discuss GTIs. These are as follows:

- *Guideline on the Limits of Genotoxic Impurities* (European Medicines Agency [EMA], 2006)
- *Genotoxic and Carcinogenic Impurities in Drug Substances and Products: Recommended Approaches* (Food and Drug Administration [FDA] draft guidance, 2008)
- *Questions and Answers on the "Guideline on the Limits of Genotoxic Impurities"* (EMA, 2010)

Although ICH Q3A, Q3B, and Q3C deal with impurities in pharmaceuticals, the scope of these documents is limited to marketed drugs rather than the drugs in clinical development. Specific guidances are needed to provide acceptable limits for GTIs during clinical development. Additionally, the identification and qualification thresholds for impurities that are defined in the Q3 documents are not stringent enough with respect to impurities that may be genotoxic to be appropriately protective for subjects or patients in clinical trials.

Determining an acceptable level of a GTI in a pharmaceutical product is challenging. It is commonly thought that there is no level of exposure to a GTI that is without increased cancer risk, however slight. This is based on the initiation/progression theory of carcinogenesis whereby any single mutation can lead to the development of neoplastic disease, provided it is in a sensitive region of the genome. It then follows from this theory that the likelihood of any single mutation leading to neoplastic disease is a function of the number of DNA base pairs in the human genome (~3 billion) and the number of mutations that occur [1]. While this is a simplistic model that does not account for endogenous DNA repair mechanisms, it is the basis for the idea that DNA-reactive chemicals do not have a threshold dose at which no damage or harm will occur.

6.2.3 A Fundamental Concept for Genotoxic Impurity Assessment: The Toxicological Threshold of Concern

The theory that there is no threshold at which DNA-reactive chemicals are safe necessitated the use of the toxicological threshold of concern (TTC) to define an acceptable limit for GTIs in pharmaceutical products. The TTC is based on an analysis of 384 known carcinogens [2] and confirmed by the evaluation of additional known carcinogens [3–5]. Based on these data, a daily intake of 1.5 μg/day of a GTI is predicted to result in 1:100,000 excess cancer risk, which is considered negligible. The FDA and EMA guidance documents both state that the TTC level of 1.5 μg/day is an acceptable daily intake (ADI) for a GTI. There are several assumptions that underlie the derivation of this value. They include the following:

- A linear extrapolation or carcinogenic potency from the TD50 (the calculated dose at which 50% of exposed animals will have a tumor in a 2-year bioassay) for the carcinogens included in the analysis.
- Patients will be exposed to the GTI for a lifetime (i.e., 70 years).
- There is no threshold for genotoxicity.

When considering these assumptions, it is clear that the TTC and the estimated excess cancer risk implied by the TTC level (1:100,000) are very conservative. It is based on the potency of roughly 700 carcinogens with the assumption that the dose–response relationship for the development of neoplasms is linear from the TD50.

Additionally, there is the assumption in the 1.5 μg/day ADI that patients will take the drug every day for 70 years (25,550 days). Even for chronically administered drugs, when patient compliance is considered it is highly unlikely that any patient would be exposed to a given drug for this time period.

Although the TTC is inherently conservative, it does represent a value that both global regulatory agencies and the pharmaceutical industry agree represents an appropriate level of risk. It is important to consider how the TTC ADI was derived and what assumptions underlie its value when evaluating risk.

6.2.4 Evaluation of Potential Impurities

In addition to setting a limit for GTIs, both the EMA and FDA guidances discuss the tests that are needed to determine if an impurity needs to be controlled as a GTI (i.e., to TTC levels). Both documents state that if an impurity contains a *structural alert* for genotoxicity it needs to be either tested in an in vitro mutation assay (e.g., a bacterial mutagenicity assay [or the Ames assay]) conducted to "regulatory acceptable standards" (EMA Q&A) or controlled as a GTI. The presence or absence of a structural alert can be evaluated using commercially available software packages. Typically, this evaluation would include two methods, one using an expert rules–based approach and the other using a statistics-based model. The absence of a structural alert is sufficient to conclude that an impurity is not genotoxic.

If an impurity contains a structural alert, the next step in the evaluation of its genotoxic potential is an Ames test. Typically, tests on the active pharmaceutical ingredient (API) containing the impurity in question will not be viewed sufficient to discharge the risk that an impurity is genotoxic. The EMA guidance states, "Moreover, negative carcinogenicity and genotoxicity data with the drug substance containing the impurity at low ppm levels do not provide sufficient assurance for setting acceptable limits for the impurity due to the lack of sensitivity of this testing approach." However, in the EMA Q&A document, it is stated that a structural alert for genotoxicity " … can be negated by carrying out an Ames test on the active ingredient containing the impurity as long as the impurity is present at a minimum concentration of 250 μg/plate." The 250 μg threshold is based on the work by Kenyon and others [6], which estimated the detection limit of the Ames assay for a variety of mutagens and estimated that 85% of known mutagens would be detected at the 250 μg threshold. Based on ICH S2(R1), the highest dose of API tested in the Ames test is 5000 μg/plate. To meet the 250 μg threshold, one would need an impurity concentration of 50,000 ppm or 5% (250 μg impurity/5000 μg API) in the API; it is unlikely that typical impurity concentrations would be this high and, thus, in the vast majority of cases the "neat" impurity should be tested.

6.2.5 Risk Characterization for DNA-Reactive Impurities

In the event that an impurity is Ames positive, the conservative TTC values (1.5 μg/day for market application) can be set as limit values to ensure adequate protection of patients. What can be done in cases where it is not possible to control an Ames positive impurity to TTC levels? Are there other considerations that can provide further perspective on the nature of the risk? The FDA and EMA guidances suggest that mechanism of action (MOA) should be considered and that those mechanisms (spindle apparatus disruption, topoisomerase inhibition, inhibition of DNA synthesis, etc.) clearly associated with a threshold dose response could represent the basis for a compound-specific assessment that results in a permissible daily exposure that is above the TTC value. However, in many cases the impurity in question will be DNA reactive and, as the FDA guidance states, " … at present it is extremely difficult to experimentally prove the existence of a threshold for the genotoxicity of a given mutagen."

The most well-publicized case for a DNA-reactive impurity having a threshold for genotoxicity (i.e., DNA reactivity) is ethyl methanesulfonate (EMS) (reviewed in detail in a special issue of *Toxicology Letters*) [7]. EMS is a genotoxic DNA-reactive carcinogen that, as a result of a production accident, was found as a contaminant in the protease inhibitor Viracept. As human exposures to the drug with the contaminant had already occurred, the impetus for the toxicology studies was risk characterization, rather than hazard identification, which is typically the goal of genotoxicity studies. To this end, the following studies were conducted with EMS:

- In vivo mouse bone marrow micronucleus test
- A 1-month repeated dose study to determine the dose–response relationship for induction of mutations in transgenic mice (Muta™ Mouse)
- A 1-month repeated dose mouse toxicity study
- Cross-species in vitro and in vivo evaluation of exposure to EMS

Prior to these investigations, it was widely assumed in the literature that as an alkylating agent EMS would not have a threshold for DNA reactivity and, thus, exposure to levels above the TTC would result in unacceptable risk to patients. The data generated in these studies demonstrated that EMS interacts with DNA in a threshold-dependent manner and that levels below 2 mg/kg/day do not represent a risk to patients. Contrast the 2 mg/kg/day threshold (120 mg/day, assuming a 60 kg human) with the TTC value of 1.5 μg/day.

These data raise some thought-provoking questions, that is, how many other DNA-reactive compounds that are assumed to act via a nonthreshold MOA actually have an experimentally definable threshold? A growing body of evidence suggests that many DNA-reactive compounds may in fact have a threshold for genotoxicity due primarily to the high efficiency of intracellular DNA repair mechanisms [8]. As mentioned in Section 6.2.1, the genotoxicity test battery outlined in ICH S2 (R1), and by extension the GTI guidances, is geared toward hazard identification (i.e., is a compound genotoxic?) rather than risk characterization (i.e., is there a dose of a genotoxic material that has an acceptable risk profile?). For APIs, this hazard identification paradigm has been appropriate as for most indications it would be difficult to justify an appropriate risk–benefit analysis (with some notable exceptions including oncolytic drugs).

For impurities that by their nature would have very low clinical exposure, the demonstration of a threshold for genotoxicity, rather than a generic TTC-based limit, would have a significant impact on the ADI. In the subsequent sections of this chapter, we discuss the various tests that can be used to determine if an impurity is genotoxic and how those data may be used to demonstrate a threshold MOA.

6.3 ASSAYS USED TO EVALUATE GENOTOXICITY

Although there are numerous end points to consider when evaluating the genotoxic potential of a test article, the three primary mutagenicity end points of concern are the point mutation (or "gene mutation"), structural chromosomal aberration (SCA), and numerical chromosomal aberration (NCA) (see Table 6.1). An adequate identification of the mutagenic potential of a test material must include the evaluation

TABLE 6.1
Primary Genetic Toxicology End Points

Gene mutation (also known as point mutation)	Any submicroscopic change in the DNA sequence • Includes single base-pair changes • May also include deletions, insertions, or inversions ranging from only a few base pairs to many thousands of base pairs
SCA	A change in the normal appearance of one or more chromosomes that can be visualized with proper staining under light microscopy • Sometimes referred to as clastogenic damage
NCA	An abnormal gain or loss of one or more entire chromosomes from a cell • Having a chromosome number that is not an exact multiple of the normal haploid number • Also known as aneuploidy

of all three end points. Because no single genetic toxicity test has been validated for the detection of all three primary end points, the field of genetic toxicology has developed batteries of tests to ensure that each end point is adequately assessed. A brief description of the genetic toxicology tests that are commonly included in the test batteries is presented here.

6.3.1 The Ames Assay

Mutagenicity evaluations in bacteria are a required component of the genetic toxicology test batteries specified by regulatory guidelines [i.e., ICH S2(R1)], as well as the key test discussed in the GTI guidances to determine if an impurity is DNA reactive. Results from bacterial mutagenicity studies are used to assess the ability of the test article to cause point mutations in DNA (see Table 6.1). These tests are not designed to detect large deletions, SCAs, or NCAs. Despite the phylogenetic distance between bacteria and humans, bacterial mutagenicity assays provide the most heavily weighed of the genetic toxicity data obtained from in vitro test systems. This may be due, at least in part, to the fact that positive results in bacterial mutagenicity tests have a higher specificity for predicting a carcinogenic outcome in rodent 2-year studies than is the case with other in vitro mutagenicity tests [8]. For a brief overview of the historical development of bacterial mutagenicity assays, see the works of Hartman [9] and MacPhee [10].

6.3.1.1 Bacterial Tester Strains and Their Characteristics

The most commonly used test systems for bacterial mutagenicity studies are strains of *Salmonella typhimurium* and *Escherichia coli* (see Table 6.2). The *Salmonella* strains contain preexisting mutations in various genes of the histidine operon that render these bacteria incapable of synthesizing the essential amino acid histidine. Treatment of the *Salmonella* tester strains with a mutagen can cause *reverse mutations* whereby the preexisting mutations are reversed back to a wild-type DNA sequence. In such a case, the phenotype of the *S. typhimurium* bacteria is reverted from His^- (unable to grow in the absence of exogenously supplied histidine) to His^+ (able to grow without exogenous histidine). After reversion from His^- (histidine auxotrophy) (i.e., not able to synthesize) to His^+ (histidine prototrophy) (i.e., able to synthesize), the revertants can be readily detected as colonies that form after plating the treated bacteria on agar that is deficient in histidine [11–14] after a period of incubation.

The *E. coli* bacteria commonly used in mutagenicity testing are derived from the WP2 strain. This strain contains a preexisting nonsense ochre mutation [15] such that the bacteria are unable to synthesize the amino acid tryptophan [16]. Treatment of the *E. coli* tester bacteria with mutagens can cause the bacteria to revert from tryptophan auxotrophy (Trp^-) to tryptophan prototrophy (Trp^+). In the case of the *E. coli* tester strain, this phenotypic reversion can occur as the result of either a mutation that directly reverses the preexisting nonsense ochre mutation or an extragenic mutation elsewhere in the bacterial genome that suppresses the preexisting nonsense mutation. In either case, the mutations can be detected as colonies that form when the treated bacteria are plated on agar that is deficient in tryptophan.

TABLE 6.2
Bacterial Tester Strains Commonly Used in Mutagenicity Testing

			S.typhimurium					
			Additional Mutations and Genotypic Features			Spontaneous Mutation Incidence[a]		
Strain	**Mutation**	**Mutagenic Sensitivity**	**Cell Wall[b]**	**DNA Repair[c]**	**R-Factor Plasmid[d]**	**Range**	**Mean ± SD**	**Decision Threshold[e]**
TA98	*hisD3052*	Frameshift mutation	*rfa*	*ΔuvrB*	+	4–57	18 ± 7	2×
TA100	*hisG46*	Base-pair substitution	*rfa*	*ΔuvrB*	+	51–225	132 ± 32	2×
TA1535	*hisG46*	Base-pair substitution	*rfa*	*ΔuvrB*	—	3–58	17 ± 7	3×
TA1537	*hisC3076*	Frameshift mutation	*rfa*	*ΔuvrB*	—	1–28	8 ± 4	3×
				E. coli				
WP2	*trpE*	Base-pair substitution	—	*ΔuvrA*	—	6–72	28 ± 11	2×

Note: SD, standard deviation.

[a] The spontaneous data (ranges and means) shown are from an approved cathode-ray oscilloscope. Although the ranges and means will vary somewhat between laboratories and within a laboratory over time, the relative values among the strains will be similar to the values shown.

[b] The *rfa* mutation causes a partial loss of the lipopolysaccharide barrier that coats the surface of the bacteria. As a result, permeability to large molecules is increased. The presence of the *rfa* mutation is confirmed by assessing the sensitivity of the bacteria to crystal violet.

[c] Deletions of the *uvrA* or *uvrB* genes eliminate the bacterial DNA excision repair system needed to remove lesions from chemically damaged DNA. The presence of these deletions can be confirmed by assessing the sensitivity of the bacteria to ultraviolet light.

[d] The R-factor plasmid (pKM101) confers increased sensitivity to a variety of mutagens. The plasmid also carries an ampicillin resistance gene; therefore, ampicillin resistance is used to confirm that the strains retain the plasmid.

[e] The decision threshold represents the numerical fold increase in revertant number on plates with treated bacteria versus untreated bacterial control plates.

It is important to appreciate that the preexisting mutations are different among the various tester strains (see Table 6.2). As a result, the bacterial strains exhibit differential sensitivity to various types of mutational events (e.g., base-pair substitutions vs. frameshift mutations). Consequently, several bacterial strains are typically used in mutagenicity testing in an effort to increase the overall sensitivity of the assay. A commonly used panel of bacterial strains includes the *S. typhimurium* strains TA100, TA98, TA1535, and TA1537 and the *E. coli* strain WP2*uvrA*.

6.3.1.2 Conduct of Bacterial Mutagenicity Assays

The typical conduct of a bacterial mutagenicity assay begins with the overnight growth of appropriate tester strains so that cultures in late log phase (i.e., with cells still actively dividing) are obtained. The mutagenic potential of the test article can then be assessed in these bacteria, with the most common test methods being either the plate incorporation method or the preincubation method.

In the plate incorporation assay, an aliquot portion (e.g., 100 μL) of the bacterial cell suspension is treated with 100 μL of a solution or suspension containing a range of concentrations of the test article. This is mixed briefly, combined with 2 mL of molten (~42°C) top agar, and poured onto the surface of a 100 mm petri plate containing 25 mL of base agar. The plates are then incubated for 48–72 hours in darkness to allow histidine prototrophic bacteria to grow and form discrete colonies. The mutagenic activity of the test article is demonstrated by an increase in the number of revertant colonies in the treated cultures compared to their concurrent vehicle controls.

The preincubation method is similar to the plate incorporation approach except that the mixture of bacteria and test article is incubated (typically 30–60 minutes at 37°C) prior to being combined with the top agar and poured onto the base agar in the petri plate.

It is common practice to include two independent trials in a bacterial mutagenicity assay (each in the absence and presence of the exogenous S9 metabolic activation system), with one trial using the plate incorporation method and the other using the preincubation approach. For pharmaceutical API testing, the ICH S2(R1) guideline requires only one trial. For impurity testing, the EMA Q&A document specifies a test conducted to regulatory acceptable standards; it is expected that ICH M7 will provide further clarification on the requirements for an acceptable test for GTI assessment.

The logistical points and other study design features to consider when conducting bacterial mutagenicity assays are summarized in Table 6.3. A more thorough description of the technical aspects of bacterial mutagenicity assays may be found in the works of Maron and Ames [17] and the Organisation for Economic Co-operation and Development (OECD) guideline 471 (1997).

6.3.1.3 Interpretation of Results

A positive result indicative of mutagenicity is typically defined as a treatment-related increase in revertant numbers of either 2× (for the *S. typhimurium* strains TA98 and TA100 and the *E. coli* strain WP2*uvrA*) or 3× (for the *S. typhimurium* strains TA1535 and TA1537) compared to control revertant numbers. A lack of mutagenic activity is indicated when the numbers of revertants in treated cultures do not meet the threshold required to define a positive response.

Because of the large number of bacterial mutagenicity assays that have been completed, some commonly encountered problems have been identified (Table 6.4). It is important to consider these factors when interpreting results from mutagenicity assays. Measures to assist in the recognition and resolution of these problems are discussed in Table 6.4.

TABLE 6.3
Points to Consider in the Conduct of Bacterial Mutagenicity Assays

Protocol Item	Comments
Test system	• Unless a specific mechanistic question is being addressed, a bacterial mutagenicity assay will include several tester strains. Typically, these will be the *S. typhimurium* strains TA98, TA100, TA1535, and TA1537 and the *E. coli* strain WP2*uvrA*. • The assay includes testing cultures plated in triplicate with and without S9 activation. • A valid assay generally includes two independent trials. One trial usually uses the plate incorporation method, whereas the other is a preincubation test. The ICH S2(R1) guideline permits a single trial if a clear result (positive or negative) is obtained in the initial trial.
Dose selection	• The maximum dose should be 5000 μg/plate unless solubility or toxicity limitations necessitate lower doses. • Dose intervals should be approximately half log or closer to avoid "window" effects, which can mask a positive response. This may occur when there is an absence of mutagenicity at low doses, a positive response with mid-doses, and an absence of colony formation due to excessive toxicity at higher doses.
Solvent controls and positive indicators	• Concurrent vehicle controls must be included in each trial for each tester strain. • Compatible solvents are DMSO, ethanol, acetone, and glycerol. • Cultures treated with positive indicators (i.e., known mutagens) are required in each trial for each tester strain. Their purpose is to demonstrate the sensitivity of the bacteria to mutagens and the functional activity of S9.
Marker checks	• Each trial should include untreated bacterial cultures that are evaluated for their characteristic phenotypic markers (e.g., the *rfa* mutation, the uvrB deletion, and the presence of the pKM101 plasmid [see Table 3]).
Test article requirements	• A standard five-strain assay conducted in two independent trials typically requires 1 g of test material.
Time requirements	• The time interval between plating and scoring an individual trial is approximately 72 hours. Plating and scoring for two independent trials can be completed within 1 week.

DMSO, dimethyl sulfoxide.

For GTIs, according to the FDA and EMA guidances, a positive result in the Ames assay indicates the need to control the impurity to TTC levels. It is important to understand that positive bacterial mutagenicity results are not a definitive determination of carcinogenicity. Kirkland and others [18] analyzed the correlation between carcinogenicity and bacterial mutagenicity results. The bacterial mutagenicity results for 176 chemicals that were negative in the 2-year rodent bioassay were analyzed;

TABLE 6.4
Points to Consider When Interpreting the Results of Bacterial Mutagenicity Assays

Interpretation of Results	Comments
Decision criteria	• Typically, a positive response requires an increase in mean revertant number in treated cultures of twofold (*S. typhimurium* strains TA98 and TA100 and *E. coli* strain WP2*uvrA*) or threefold (*S. typhimurium* strains TA1535 and 1537) compared to concurrent controls. These may vary slightly from one laboratory to another.
Background lawn	• Thinning, clumping, or absence of the background lawn is an indication that the test article is cytotoxic to the bacteria. The background lawn may be visualized microscopically as a fine dispersion of bacteria embedded on the top of the agar on the culture plate. Changes in the appearance of the background lawn that are indicative of toxicity may be used to justify doses less than the limit dose of 5000 μg/plate for affected tester strains.
Weakly positive dose responses	• Weakly positive mutagenic responses (i.e., increases in mean revertant number that only slightly exceed the thresholds defined by the decision criteria and that occur only at the very upper end of the dosing range) may be due to mutagenic impurities rather than the test article itself. Follow-up testing with more highly purified samples of the test article should be considered.
Nutritional survivors	• Cytotoxicity that causes a thinning or an absence of the background lawn may result in a slight increase in the concentration of histidine in the agar medium. This increase occurs as a result of lysed bacteria releasing their intracellular contents (including amino acids) into the surrounding agar. On rare occasions, this slight increase in histidine can support colony formation by nonmutated bacteria and may result in a false positive response. This situation can be recognized by the formation of colonies (usually smaller in size than normal) on a plate devoid of a background lawn. The character of the bacterial colonies (auxotrophic nutritional survivors vs. prototrophic mutants) can be determined by replica plating the cultures on agar devoid of histidine. • Test articles (e.g., protein biologics) that can release histidine into the agar growth medium may also cause colony formation by nutritional survivors [19]. • Colony formation by nutritional survivors does not seem to be as problematic with *E. coli* as it is with *S. typhimurium* strains (Kirkland et al. 2007).

of the 176 compounds that were not carcinogenic, 46 were still positive in the Ames assay. Although limited, these data are important to consider when determining the path forward if a positive bacterial mutagenicity result for an impurity is encountered. Positive results in bacteria do not always signify positive results in rodent carcinogenicity studies.

6.4 CHROMOSOMAL ABERRATION

SCA tests are intended to identify agents that have the potential to cause abnormalities in chromosomes that are visible microscopically (Table 6.1) [see OECD guidelines 473 and 475 and ICH guideline S2(R1)]. Although there are many different types of aberrant structural changes that may occur in chromosomes, they are collectively referred to as *clastogenic damage*. It is important to realize that SCA assays are not reliable for detecting point mutations or changes in chromosome numbers.

6.4.1 Test Systems for Chromosomal Aberration Studies

Assays for SCAs can be conducted using either in vitro or (less commonly) in vivo test systems:

- The in vitro studies are carried out using mammalian cell cultures. The more frequently used cell types include Chinese hamster ovary cells, Chinese hamster lung cells, and cultured human lymphocytes, although other cell types may also be considered. Advantages associated with the commonly used in vitro systems include the presence of a relatively stable genome with a large number of chromosomes for analysis, a lengthy record of use in genetic toxicity testing, and well-established historical ranges for spontaneous chromosomal aberrations in control (untreated) cells. They also avoid the use of animals for toxicity testing. However, they are deficient in physiological processes (e.g., drug-metabolizing capability) present in intact mammals, which may affect clastogenicity; for this reason, endogenous S9 metabolic activation is typically a feature of these assays.
- The in vivo test systems for chromosomal aberration assays are almost always limited to rats and mice. Although bone marrow cells are typically the target cells for evaluation, cells from other potential target organs can be considered. Although options for other mammalian species may be weighed, they are very rarely used and should be considered only if there is some compelling rationale to justify their use.

6.4.2 Conduct of Structural Chromosomal Aberration Tests In Vitro

An in vitro assay for SCAs typically includes an initial dose rangefinder to assess the cytotoxicity of the test article as well as to evaluate its effects on the ability of the cell to traverse the cell cycle and complete mitotic division. The rangefinder is followed by the clastogenicity assessment, which most commonly includes three parts (or "arms") that are conducted simultaneously. In the rangefinder, subconfluent cell cultures are treated with various concentrations of the test article. Some of the treated cultures are then incubated for 3 to 4 hours, whereas others are incubated for a time interval that is equivalent to approximately 1.5

cell cycles. Because the commonly used cell types lack many of the enzymes needed to metabolize some promutagenic test articles to their mutagenic form, the rangefinder is conducted in both the absence and the presence of an exogenous S9 metabolic activation system. The cultures are then assessed for cytolethality (as defined either by a visual inspection or by cell counts) and compared to the solvent (vehicle) control. The results of the rangefinder are used to guide dose selection for the chromosomal aberration test, as follows:

- The highest concentration for the chromosomal aberration test should not exceed a concentration that causes either a greater than or equal to 50% decrease in mitotic index or a greater than or equal to 50% level of toxicity. Although it is not necessary to evaluate doses that exceed these toxicity limits, it is important that a high dose selected on the basis of cytotoxicity should approach the 50% threshold.
- In the absence of dose-limiting effects on mitotic indices or cytotoxicity, the maximum concentration to be tested in the chromosomal aberration test should be defined by test article solubility limitations (as determined by the presence of visually evident precipitation) or by a limit concentration (either 0.5 mg/mL or 1 mM, whichever is the lowest).

It is particularly important to be aware that some intertrial variation may occur in the degree of cytotoxicity caused by a test article. Consequently, results of the initial rangefinder may provide only a general indication of the range of concentrations that should be evaluated for SCAs. To address this uncertainty, the concentrations used in an SCA assessment may extend above the cytotoxic levels identified in the rangefinder. The actual concentrations to be evaluated for SCAs should be selected based on a cytotoxicity assessment that is conducted concurrently with the same cell cultures that were plated for the SCA evaluation.

Following completion of the rangefinder, the potential of the test article to induce SCAs is typically evaluated in subconfluent cell cultures using a three-armed testing protocol. These arms include (1) a 3- to 4-hour treatment of the test system in the absence of S9 metabolic activation, (2) a 3- to 4-hour treatment in the presence of S9, and (3) an 18-hour treatment in the absence of S9. Following their respective treatment intervals, the cell cultures are washed, fresh medium with a metaphase-arresting agent (e.g., colchicine) is added, and the cultures are incubated for an additional 1–3 hours. At this point, a high proportion of the cells would have progressed through the cell cycle until being arrested in metaphase where their chromosomes can be visualized. The cells are then harvested, treated with a hypotonic solution to cause swelling, fixed with a solution of methanol and acetic acid, and dropped onto glass slides. Upon impact, the swollen cells burst and the chromosomes adhere to the glass. After staining with Giemsa stain, the chromosomes are examined for SCAs by light microscopy in a blinded fashion. Typical protocols require cultures to be plated in triplicate, with one culture being used for cytotoxicity confirmation and two cultures being used for SCA analysis. Typically, 100 cells per culture are evaluated.

6.4.3 Chromosomal Aberration Studies In Vivo

The test systems for in vivo SCA assays are usually rodents (e.g., rats, mice, or hamsters), although other nonclinical species can be considered if there is a specific need. Caution is recommended if a *nonstandard* mammalian species is used, since a robust historical control database would likely be lacking for rarely used test systems. Cells for chromosomal analysis are most commonly collected from bone marrow, although other tissues also may be considered. The dosing route should be selected to mimic the intended route for humans or to ensure maximal exposure to the target cell population. The top dose should be defined by toxicity. For relatively nontoxic test articles, the ICH S2(R1) guideline defined 2000 mg/kg as an acceptable limit dose.

A mitotic index should be determined in at least 1000 cells per animal for all treated animals as a measure of cytotoxicity. If an indication of toxicity in the target cells is detected in the absence of SCAs, the information can be used to demonstrate that the test article indeed reached the target cells and to strengthen the argument that the test article is truly devoid of clastogenic activity.

A valid test should include concurrent negative and positive controls. Note that the recently revised ICH S2(R1) guideline indicates that positive controls are not needed for every study, particularly after a laboratory has established competence in the conduct of the assay. Results from negative control cultures should fall within the historical control range for the test facility. Results from positive control cultures should demonstrate the sensitivity and responsiveness of the test system to genotoxic insult. Important points to consider when conducting chromosomal aberration assays and interpreting their results are summarized in Table 6.5.

6.4.4 Interpretation of Results

A positive finding requires a dose-responsive, statistically significant increase in the number of cells with chromosomal aberrations.

There are several additional points to be aware of when evaluating the results of SCA studies:

- Although statistical methods are used, statistical significance should not be the only determining factor for a positive response. The experience and judgment of the study director are also important for determinations of biological significance.
- With a cytotoxicity greater than or equal to 50%, increased SCAs may occur secondarily to nonspecific changes associated with cell death. In such cases, a false positive result may be reported.
- SCAs include chromatid aberrations and chromosome aberrations. A chromatid is either of the two daughter strands of a replicated chromosome that are joined by a centromere and separate in cell division to become individual chromosomes. Both chromosome and chromatid aberrations should be considered collectively when assessing SCA data.
- Gaps (defined as nonstaining sites that are less than the width of a chromatid) are recorded in SCA evaluations, but they are not otherwise considered when determining the clastogenic potential of the test article.

TABLE 6.5
Points to Consider in the Conduct and Interpretation of Chromosomal Aberration Assays

Conduct of the Assay	Comments
In vitro	• A variety of cell lines or primary cell cultures, including human cells, may be used (e.g., Chinese hamster fibroblast, human or other mammalian peripheral blood lymphocytes). • Cultures are generally seeded in duplicate. • The assay includes testing cultures with and without S9 activation. • A valid assay generally includes a dose rangefinding experiment followed by an assessment for chromosomal aberrations. • The maximum dose should be 500 μg/mL. Solubility or toxicity limitations may necessitate lower doses. For in vitro studies, the high dose should have a toxicity near 50%. Dose intervals should be approximately half log or closer to avoid window effects, which can mask a positive response. • Cultures should be treated for 3–6 hours with and without metabolic activation and 1.5× the normal cell cycle time without metabolic activation.
In vivo	• Rats, mice, and Chinese hamsters are commonly used, although any mammalian species may be considered. Bone marrow is the most common target tissue. • Each treated and control group must include at least five animals per sex. • A valid assay generally includes a dose rangefinding experiment followed by an assessment for chromosomal aberrations. • The maximum dose should be 2000 mg/kg. Toxicity limitations may necessitate lower doses. For a high dose of X, dose intervals should be ½X and ¼X. If no substantial differences in toxicity between sexes are seen, then a single sex (usually males) is sufficient. • Animals should be exposed for 1.5× the normal cell cycle time (rodents 12–18 hours) and 24 hours after the first sampling time.
Solvent/vehicle controls and positive indicators	• Concurrent vehicle controls must be included in each trial. • Positive indicators (i.e., known mutagens) are required in each trial. Their purpose is to demonstrate the sensitivity of the cells to mutagens.
Analysis	• The cell is the unit of measure, and results are expressed as numbers of aberrant cells. • Generally, 100 cells (metaphases) are scored per replicate (in vitro) or animal (in vivo).
Interpretation of Results	
Decision criteria	• There are several criteria for determining a positive result. These include a concentration-related increase or a reproducible increase in the number of cells with chromosome aberrations. Statistical methods may be used as an aid in evaluating the results. Biological relevance of the results may also be considered.
Reasons for nongenotoxic positive responses	• False positive results may be associated with the generation of reactive oxygen species; changes in pH and osmolality; a cytotoxicity greater than 60%; aneugens; or an inhibition of DNA polymerases, topoisomerases (I and II), kinases, and DNA synthesis.

6.5 RODENT IN VIVO MICRONUCLEUS TESTS

The rodent bone marrow micronucleus (MN) test is used to detect the ability of a test article to cause either SCAs (clastogenic damage) or NCAs (see Table 6.1). The mutational damage associated with these genetic toxicity end points may be caused by an interaction of the test article with either the chromosomes themselves or the mitotic spindle apparatus in the treated cells [see OECD guideline 474 and ICH guideline S2(R1)].

The micronucleus assay most commonly uses an in vivo test system (usually rats or mice). Although the target cells for mutagenic insult are bone marrow erythroblasts, the genotoxic potential of the test article is assessed in erythrocytes collected either from the marrow or from the peripheral circulation. Ordinarily, the full complement of chromosomes in a cell is contained within the nuclear membrane. Following treatment with a test article, chromosomal fragments or whole chromosomes may become separated from the nucleus as the result of clastogenic or aneugenic action, respectively. In such cases, the chromosomal material (either a fragment or an entire chromosome) may then become encased in its own nuclear membrane to form a micronucleus. As the erythroblast matures to form a polychromatic erythrocyte (PCE), the main cell nucleus and any chromosomes contained therein are normally extruded. The micronuclei that remain behind in the cytoplasm of the enucleated cell can be readily visualized and counted microscopically or quantitated by flow cytometry.

6.5.1 Test Systems for Micronucleus Studies

Both rats and mice are considered appropriate for use in the bone marrow micronucleus test. In principle, newly formed erythrocytes can be evaluated from other mammalian species that have shown an adequate sensitivity to detect clastogens and aneugens in bone marrow or peripheral blood. However, care must be taken to ensure that an adequate historical control database is available to assist in the interpretation of results.

6.5.2 Conduct of Micronucleus Tests

A micronucleus assay typically includes an initial dose rangefinder in both male and female rodents to guide dose selection for the main study.

The high dose for the main study may be determined either by the maximum tolerated dose of the test article or (in the absence of toxicity) by the limit dose of 2000 mg/kg. The limit of solubility may also be considered as a potential factor in dose selection, but a range of formulations should be evaluated before basing dose selection on solubility limitations. Once the top dose for the main study is determined, the remaining dose levels are generally ½X and ¼X the high dose. The main study should include a concurrent vehicle control group, three treated groups, and a positive control group with five animals per sex per group. If the toxicity profile from the rangefinder is the same for males and females, then males only may be used for the main study.

The test article treatment schedule can be defined in either of two ways:

- The animals may be treated with the test article for 2 consecutive days (~24 hours apart) with a harvest of bone marrow cells 24 hours after the last dose or collection of peripheral blood 36–48 hours after the last dose.
- The animals may receive a single treatment. Samples of bone marrow would then be collected at 24 and 48 hours after treatment, or peripheral blood samples could be collected 36 and 72 hours after treatment.

A single treatment of the positive control group 24 hours prior to the harvest of bone marrow cells (or 36 hours for peripheral blood) is sufficient. Note that the recently revised ICH S2(R1) guideline indicates that positive controls are not needed for every study, particularly after a laboratory has established its competence in the conduct of the study. Bone marrow cells are usually obtained from the femurs immediately following sacrifice, and peripheral blood is usually collected from the tail vein. Slides are prepared and then stained. The use of a DNA-specific stain (e.g., acridine orange) can eliminate some artifacts associated with using a non-DNA-specific stain (e.g., Giemsa stain). Normally, 2000 PCEs are counted per animal and the number of PCEs that contain a micronucleus is tallied. A spontaneous micronucleated polychromatic erythrocyte (MN PCE) frequency range is generally 0–4 MN PCE/2000 PCE. Important points to consider when conducting chromosomal aberration assays and interpreting their results are summarized in Table 6.6.

6.5.3 Interpretation of Results

The two parameters that are important to consider when assessing data from a micronucleus test include the incidence of MN PCEs and the ratio of PCEs to normochromatic erythrocytes (NCEs):

- The incidence of MN PCEs in the treated animals should be compared to that of the concurrent vehicle control. A statistically significant dose-related increase in MN PCEs is most often the evidence for genotoxic activity.
- The relative proportion of PCEs to the more mature NCEs should be evaluated, and the PCE to NCE ratio should be determined. Typically, the PCE to NCE ratio will be in the range of 0.5–1.0. A decrease in the PCE to NCE ratio is evidence that the test article reached the bone marrow and elicited a biological effect on the target cells. Although a depressed PCE to NCE ratio is not required for a valid assay, it nevertheless can be used to strengthen conclusions for negative (i.e., nongenotoxic) results.

It is important to note that micronuclei data alone will not distinguish between clastogenic and aneugenic activity. However, such a distinction can be made through the use of antikinetochore antibodies (AKAbs). This approach is based on the assumption that intact chromosomes retain their kinetochores, whereas chromosomal fragments are far more likely to be acentric and lacking kinetochores. Consequently, micronuclei that are predominantly AKAb positive are most probably the result of

TABLE 6.6
Points to Consider in the Conduct and Interpretation of Micronucleus Assays

Conduct of the Assay	Comments
Model	• Rats, mice, and (less commonly) Chinese hamsters are the usual test systems, although any mammalian species may be used. Bone marrow or peripheral blood is the target. • Each treated and control group must include at least five animals per sex. • A valid assay generally includes a dose rangefinder study followed by a test for micronucleus induction.
Dose selection	• The maximum dose should be 2000 mg/kg unless toxicity necessitates lower doses. • Dose intervals should be X, ½X, and ¼X. If no substantial differences in toxicity between sexes are seen, then a single sex (usually males) is sufficient.
Solvent/vehicle controls and positive indicators	• Concurrent vehicle controls must be included in each trial. • A positive indicator (i.e., known mutagen) is required for the definitive trial.
Treatment schedule	• Option 1 (multiple daily doses): animals are dosed on 2 consecutive days, and bone marrow samples are collected once 18–24 hours after the last dose. Peripheral blood should be collected between 36 and 48 hours after the last dose. • Option 2 (single dose administration): animals are dosed a single time. Bone marrow samples are collected 24 and 48 hours post treatment. Peripheral blood should be collected 36 and 72 hours post treatment.
Analysis	• PCE is the unit of measure. Generally, 2000 cells are scored per animal. Results are expressed as numbers of MN PCEs per 1000 PCEs examined. • Toxicity is measured as a decreased PCE to NCE ratio. Animals with a PCE to NCE ratio < 0.3 should not be used for the analysis.
Interpretation of Results	
Decision criteria	• There are several criteria for determining a positive result. These include a concentration-related increase or a reproducible increase in the incidence of MN PCEs. Statistical methods may be used as an aid in evaluating the results. Historical control data and biological relevance of the results may also be considered.
Reasons for nongenotoxic positive responses	• False positive results may be associated with compounds that cause hyperthermia or hypothermia, or that stimulate increases in cell division of bone marrow erythrocytes [20].

aneugenic action, whereas micronuclei that are AKAb negative may be attributed to clastogenic events [21]. This mechanistic difference is important for risk assessment and risk management, since the dose response for aneugenicity is generally regarded as having a threshold [22,23].

With respect to clastogenicity as assessed by the rodent micronucleus assay, this can also be a threshold effect, although demonstration of a threshold effect is more challenging than with aneugens [24]. The rodent micronucleus assay has similar specificity with respect to correlation with rodent carcinogenicity as the Ames assay. Benigni and others [25] analyzed the in vivo micronucleus results for 40 chemicals that were negative in the 2-year rodent bioassay; 10 out of 40 were positive in the micronucleus assay but negative in the 2-year rodent bioassay giving a specificity (40/[10 + 40]) of 80%, which is the probability that the micronucleus test will be negative if a chemical is negative in the 2-year rodent bioassay.

6.6 COMET TEST

The comet test (also known as the single cell gel electrophoresis assay) is a method for detecting and measuring DNA strand breaks in individual mammalian cells [26]. Currently, there are no established OECD or other regulatory guidelines that specify study design features for the comet test. Nevertheless, the comet assay is included in ICH guideline S2(R1) as an option for an in vivo test to measure DNA damage that can result in DNA strand breakage.

DNA strand breakage measured with the comet methodology is not a primary mutagenicity end point (see Table 6.1). Nevertheless, the comet test is particularly appealing because it can be used to evaluate in vivo DNA damage (i.e., DNA reactivity) on a tissue-specific basis. As a point for awareness, the standard comet methodology is not effective in detecting agents with interstrand DNA cross-links or DNA–protein cross-links.

The conceptual basis of the comet assay stems from the observation that, under the influence of an appropriately applied electric current, negatively charged molecules such as DNA will migrate in the direction of the anode. During electrophoresis through a matrix such as agarose, the DNA strands move by biased reptation with electrophoretic mobilities that are inversely related to their molecular weights [27]. As a result, shorter pieces of DNA will move through greater distances than longer pieces when subjected to electrophoresis. Consequently, DNA damage that causes an increase in either single- or double-stranded breaks within the genomic DNA of a cell can be detected as an increase in the length or intensity of the tail, which can be visualized following electrophoresis. This DNA has the appearance of a comet, from which the assay derives its name (see Figure 6.1).

6.6.1 Test Systems for Comet Assays

Rats or mice are the preferred test system, although other mammalian species may also be used when justified. The comet assay can be applied to any tissue or cell type of the experimental animal. The recommended tissues are as follows:

- The tissue or cells of the test system that are at the point of contact with the test article at the time of dosing: for example, this would be the stomach for oral dosing, skin for dermal applications, or the respiratory tract for inhaled materials.
- Liver—this is a major organ for the metabolism of absorbed compounds and typically receives a high internal exposure to the test article following oral administration.
- Peripheral blood lymphocytes—these cells can be readily collected and provide a convenient cell type for assessing mutagenic activity systemically.
- Known or suspected target organs for toxicity or pharmacodynamic activity—these will vary with different test articles.

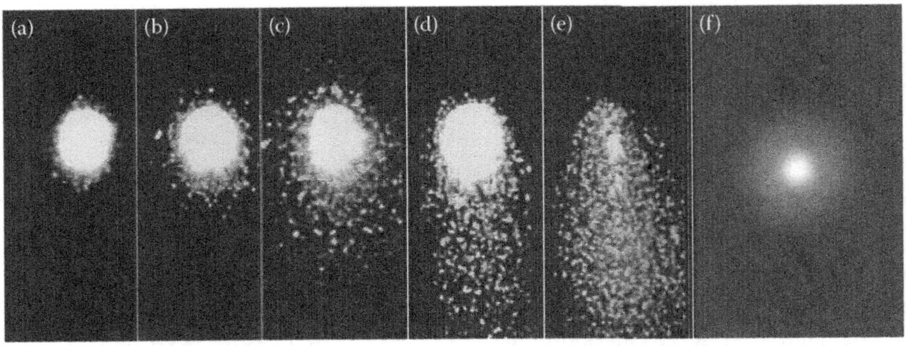

FIGURE 6.1 Representative examples of images from comet tests. The images depict results from individual cells that were collected from treated animals, suspended in agarose on slides, subjected to denaturing conditions (pH ≥ 13), and electrophoresed [(a) through (e)] or maintained at a pH ≥ 13 without electrophoresis (f). The slides were then treated with ethidium bromide to stain the DNA. (a) Cell with no evidence of DNA damage. Note the well-defined comet head and absence of a tail. (b) through (d) Cells with increasing levels of DNA damage, as indicated by an increase in tail length and tail intensity. (e) Cloud cell (sometimes called a hedgehog) indicative of excessive toxicity. Note the absence of a head, with the stain intensity being located entirely in the tail. (f) Diffusion cell (also called a halo). This is also an indication of a cell that was severely damaged by excessive toxicity. The halo resulted from the passive diffusion of highly fragmented DNA strands into the surrounding agarose matrix. (Images (a) through (e) adapted from Avishai N, Rabinowitz C, Rinkevich B. Use of the comet assay for studying environmental genotoxicity: Comparisons between visual and image analyses. *Environ Mol Mutagen.* 2003;42(3):155–65 [28]; image (f) provided by Covance, Harrogate, United Kingdom.)

Other tissues may be collected and evaluated on an as needed basis to address other specific concerns that may arise.

6.6.2 Conduct of Comet Tests

There are no universally accepted methods defined by regulatory agencies for the conduct of comet assays. Despite this gap, comet tests are generally performed in either of two ways (see Table 6.7):

1. Animals may be treated once with the test article, with tissue/organ samples being obtained at 2–6 and 16–26 hours after dosing. The shorter sampling time should be sufficient to detect rapidly absorbed materials as well as unstable or direct-acting compounds. The later sampling time is intended to detect compounds that require more time to be absorbed, distributed, and metabolized.
2. Animals may be treated multiple times at 24-hour intervals, with tissue/organ samples being obtained once approximately 2–6 hours after the last administration of the test article.

Dose selection is based on the same considerations as those used for the in vivo bone marrow micronucleus test. Indeed, combination studies may be conducted in which comet assessments and micronucleus evaluations are done using the same

TABLE 6.7
Points to Consider in the Conduct and Interpretation of Comet Assays

Conduct of the Assay	Comments
Model	• Rats and mice are commonly used, although any mammalian species may be considered. Tissues/organs for analysis include tissue from the point of contact during dosing, liver, peripheral blood, and known or suspected target tissues/organs. • Each treated and control group should include at least five animals per sex. Some protocols include additional animals to be used as replacements. • A valid assay generally includes a dose rangefinder study followed by a test for comet formation.
Dose selection	• The maximum dose should be 2000 mg/kg unless toxicity limitations necessitate lower doses. • Dose intervals should be X, ½X, and ¼X. If no substantial differences in toxicity between sexes are seen, then a single sex (usually males) is sufficient.
Solvent/vehicle controls and positive indicators	• Concurrent vehicle controls must be included in each trial. • A positive indicator (i.e., known mutagen) is required for the comet test.
Treatment schedule	• Option 1: animals may be treated once with the test article, with tissue/organ samples being obtained at 2–6 and 16–26 hours after dosing. • Option 2: animals may be treated multiple times at 24-hour intervals, with tissue/organ samples being obtained once approximately 2–6 hours after the last administration of the test article.
Analysis	• Single cells are measured for tail intensity, tail length, and tail moment. • Generally, 100 cells per animal are scored for each treatment group.
Interpretation of Results	
Decision criteria	• There are several criteria for determining a positive result. These include a concentration-related increase or a reproducible increase in the tail intensity and/or tail moment. Statistical methods may be used as an aid in evaluating the results. Historical control data and biological relevance of the results may also be considered.
Reasons for nongenotoxic positive responses	• Excessive cytotoxicity (as indicated by clouds and diffusion cells) may confound comet data interpretation. Animals with > 30% clouds and/or > 30% diffused cells should be excluded from the comet analysis.

animals. In a combined comet + micronucleus test, the animals would typically receive three treatments that would occur 48, 24, and 2–6 hours prior to sample collection.

Once the tissues are collected, they are processed to isolate single cells. These cells are then suspended in a thin layer of agarose and spread onto microscope slides. The slides are treated under alkaline conditions (pH ≥ 13) to cause the cells to lyse and to permit the unwinding and strand separation of double-stranded DNA. One set of slides is then subjected to electrophoresis for comet evaluations, and another set is incubated under alkaline conditions without electrophoresis to permit the formation of *diffusion cells* (described in Section 6.6.3). During electrophoresis, the negatively charged DNA is drawn away from the nucleus and toward the anode of the electrophoresis apparatus. Shorter DNA fragments migrate further than longer undamaged DNA. Following electrophoresis, the slides are stained with ethidium bromide to allow the visualization of the comet tails formed by the DNA. The degree of DNA migration into the comet tail is a measure of the extent of DNA damage incurred by the cells [29].

To quantitate how much DNA damage has occurred, comets are measured for tail intensity, tail length, and tail moment (see Figure 6.1). Tail intensity is the measure of the amount of DNA in the tail based on the incorporation of a DNA-specific stain (e.g., ethidium bromide). A specialized camera, microscope, and software are used to acquire the tail intensity value. The tail moment is calculated as the product of the length of the migrated DNA (comet tail) and the intensity.

6.6.3 Interpretation of Results

The tail intensity and tail moment in the treated animals should be compared to those of the concurrent vehicle control. A statistically significant dose-related increase in either tail intensity or tail moment (or both) is evidence for genotoxic activity.

It is important to avoid evaluating dose levels of the test article that may cause excessive cytotoxicity. Indications of cytotoxicity include clouds and diffusion cells (see Figures 6.1e and 6.1f):

- Clouds (also known as *hedgehogs*) result from cells that have been highly damaged. The comet head is very small or entirely missing, and most (or all) of the DNA is found in the tail. Because clouds may be indicative of apoptosis rather than genotoxicity, they are not measured as comets. However, the number of clouds present is determined on the slides that are also scored for comets.
- Diffusion cells (also known as *halos*) are also indicative of severely damaged cells. Diffusion cells are scored on slides containing cells that have been lysed under alkaline conditions, but not subjected to electrophoresis. The halos form with the passive diffusion of very short fragments of DNA away from the site of their former nucleus and into the surrounding agarose matrix. Diffusion cells are included as part of the evaluation for cytotoxicity in comet assays because very small fragments of DNA may be lost during the electrophoresis step and the toxicity may therefore not be detectable as clouds.

Comet protocols often specify that animals with more than 30% clouds and/or more than 30% diffused cells should be excluded from an analysis for genotoxicity.

As noted earlier, the standard comet assay is not an effective method to detect the genotoxicity of cross-linking agents (e.g., cisplatinum). This is because the cross-links (either between complementary DNA strands or between DNA and proteins) act to decrease the electrophoretic mobility of the affected DNA strands. As a result, comets will not form.

As mentioned, although not part of the standard battery of genotoxicity tests, the main feature that makes the comet assay attractive to provide additional context to Ames results is that it is an in vivo measure of DNA reactivity. In an analysis by Sasaki and others [30] of 208 chemicals, there were 11 that were Ames positive but negative in the 2-year rodent bioassay. Of these 11 chemicals, only 2 were also positive in the comet assay. Negative in vivo comet results would be very strong evidence that an Ames positive result is not biologically relevant. In addition to the end point being highly relevant to the Ames assay, the comet assay is also highly correlated with 2-year rodent bioassay results. In the analysis by Sasaki and others, there was a 94% positive correlation with 2-year rodent bioassay data [30,31].

6.7 IN VIVO MUTAGENESIS ASSAYS

In vivo mutagenesis assays can be used to establish an important bridge between Ames assay results and 2-year carcinogenicity tests in rodents. For a DNA-reactive impurity, the key determinant of carcinogenic potential is the ability to induce DNA mutations in an in vivo system. In vivo mutagenesis assays allow one to measure this parameter. Two main systems are used to measure mutagenesis: transgenic mouse models and the *pig-a* assay.

The primary transgenic mouse lines that are used to measure in vivo mutagenesis have multiple copies of either the *lacI* (Big Blue® mouse) or *lacZ* (Muta Mouse) gene integrated into their genome [32,33]. The test article is administered to these mice, and genomic DNA is isolated from the tissue of interest. The phage genes are then excised, packaged, and transfected into *E. coli*, which is then grown in a chromogenic substrate. Bacteria that cannot metabolize the substrate will produce white (colorless) plaques, which indicate a mutation in the phage *lac* gene. Bacteria can also be grown on selective media as a measure of mutagenicity [34].

These assays have not been widely used most likely due to their high cost and relative complexity compared to other genotoxicity tests. However, when in vivo data are needed to characterize the risk posed by Ames positive impurities they are at this point in time the assay of choice; notably, the Muta Mouse assay was used as part of a multipronged approach to demonstrate a threshold to EMS. One of the advantages of these assays is their ability to assess gene mutation in any organ or tissue and gene mutation induced by multiple routes of administration. These assays have been particularly useful in assessing mutagenicity in the skin after dermal exposure [35].

An important emerging in vivo mutagenesis test is the *pig-a* (pig-a stands for phosphatidyl inositol glycan-a) gene mutation assay [36]. The *pig-a* gene codes for a catalytic subunit of the *N*-acetylglucosamine transferase complex that is involved in glycophosphatidyl inositol (GPI) anchor production. GPI anchors express protein markers on the surface of cells; mutations in the *pig-a* gene result in an absence of these surface protein markers in affected cells. The absence of these markers can be quantitated using flow-cytometric techniques, and the decrease in surface marker expression is a surrogate for in vivo gene mutation.

In contrast to the transgenic mouse assays, the *pig-a* assay relies on the detection of mutations in an endogenous gene; hence, any strain of laboratory rodent (or large animal species) can be used for this assay. Because of the ability to use normal laboratory strains of rats and mice, these end points could be integrated as a standard component of repeated dose toxicity studies. Current efforts have focused on assessing mutations in white and red blood cells, which would make the assay amenable to serial sampling [37]. In the future, there is also the potential for the assay to be used to measure in vivo mutations in humans to provide a direct measure of clinical relevance, something that is not possible with any other assay today.

Standard protocols for the *pig-a* assay are currently under development, and international interlaboratory trials are ongoing. The efforts to standardize a protocol and validate the assay were recently the subject of a special issue of *Environmental and Molecular Mutagenesis* [38]. Although a significant amount of assay validation and protocol standardization is needed, the *pig-a* assay represents an important emerging tool to assess DNA reactivity in vivo and may be very useful for determining whether thresholds to DNA-reactive compounds exist in vivo.

6.8 CONCLUSIONS

The assessment of GTIs is likely to continue to be challenging; however, there are important new developments that have the potential to significantly impact the current processes and procedures in this field.

The ICH guidance that is currently in development [ICH M7, *Assessment and Control of DNA Reactive (Mutagenic) Impurities in Pharmaceuticals to Limit Potential Carcinogenic Risk*] will have important implications when it is approved as it would be expected to supplant the current EMA and FDA guidance documents. Regardless, there should be continued emphasis to understand the limitations and assumptions that underlie the TTC and to acknowledge that it is based on a hazard assessment paradigm. The concept of hazard identification for GTIs defines the assays that are recommended (in silico assessment, Ames assay). In most cases, hazard identification is all that is needed for impurity assessment as different synthetic routes can be used and control strategies can be implemented. However, when one needs to characterize the potential mutagenicity risk posed by an impurity, it becomes important to have a thorough understanding of the available genotoxicity assays, their strengths and weaknesses, and how they correlate to the carcinogenicity of DNA-reactive chemicals.

In this chapter, we provide a review of the derivation of the TTC and an overview of the genotoxicity assays that could be used for risk characterization of GTIs. With the continued emergence of the *pig-a* gene mutation assay and the eventual approval of ICH M7, the future of GTI assessment is certain to be dynamic.

REFERENCES

1. Venter JC, Adams MD, Myers EW, Li PW, Mural RJ, Sutton GG, Smith HO, Yandell M, Evans CA, Holt RA, et al. The sequence of the human genome. *Science*. 2001 Feb 16;291(5507):1304–51. Erratum in: *Science* 2001 Jun 5;292(5523):1838.
2. Gold LS, Sawyer CB, Magaw R, Backman GM, de Veciana M, Levinson R, Hooper NK, Havender WR, Bernstein L, Peto R, et al. A carcinogenic potency database of the standardized results of animal bioassays. *Environ Health Perspect*. 1984 Dec;58:9–319.

3. Munro IC, Kennepohl E, Kroes R. A procedure for the safety evaluation of flavouring substances. Joint FAO/WHO Expert Committee on Food Additives. *Food Chem Toxicol.* 1999 Feb–Mar;37(2–3):207–32.
4. Kroes R, Kozianowski G. Threshold of toxicological concern (TTC) in food safety assessment. *Toxicol Lett.* 2002 Feb 28;127(1–3):43–6.
5. Cheeseman MA, Machuga EJ, Bailey AB. A tiered approach to threshold of regulation. *Food Chem Toxicol.* 1999 Apr;37(4):387–412.
6. Kenyon MO, Cheung JR, Dobo KL, Ku WW. An evaluation of the sensitivity of the Ames assay to discern low-level mutagenic impurities. *Regul Toxicol Pharmacol.* 2007 Jun;48(1):75–86.
7. Lutz W. Assessment of human toxicological risk of Viracept patients accidentally exposed to ethyl methanesulfonate (EMS) based on preclinical investigations with EMS and ethylnitrosourea. *Toxicol Lett.* 2009;190:3.
8. Kirkland DJ, Aardema M, Banduhn N, Carmichael P, Fautz R, Meunier JR, Pfuhler S. In vitro approaches to develop weight of evidence (WoE) and mode of action (MoA) discussions with positive in vitro genotoxicity results. *Mutagenesis.* 2007 May;22(3):161–75. Epub 2007 Mar 16.
9. Hartman PE. Early years of the *Salmonella* mutagen tester strains: Lessons from hycanthone. *Environ Mol Mutagen.* 1989;14(16):39–45.
10. MacPhee DG. Development of bacterial mutagenicity tests: A view from afar. *Environ Mol Mutagen.* 1989;16(16):35–8.
11. Ames BN, Sims P, Grover PL. Epoxides of carcinogenic polycyclic hydrocarbons are frameshift mutagens. *Science.* 1972;176:47–9.
12. Ames BN, Gurney EG, Miller JA, Bartsch H. Carcinogens as frameshift mutagens: Metabolites of 2-acetylaminofluorene and other aromatic amine carcinogens. *Proc Natl Acad Sci USA.* 1972;68:3128–32.
13. Ames BN, Lee FD, Durston WE. An improved bacterial test system for the detection and classification of mutagens and carcinogens. *Proc Natl Acad Sci USA.* 1973;70:782–6.
14. Ames BN, Durston WE, Yamasaki E, Lee FD. Carcinogens are mutagens: A simple test system combining liver homogenates for activation and bacteria for detection. *Proc Natl Acad Sci USA.* 1973;70:2281–5.
15. Mortelmans K, Ricco ES. The bacterial tryptophan reverse mutation assay with *Escherichia coli* WP2. *Mutat Res.* 2000;455:61–9.
16. Green MHL, Muriel WJ. Mutagen testing using Trp^+ reversion in *Escherichia coli. Mutat Res.* 1976;38:3–32.
17. Maron DM, Ames BN. Revised methods for the *Salmonella* mutagenicity test. *Mutat Res.* 1983;113:173–215.
18. Kirkland D, Aardema M, Henderson L, Müller L. Evaluation of the ability of a battery of three in vitro genotoxicity tests to discriminate rodent carcinogens and non-carcinogens. I. Sensitivity, specificity and relative predictivity. *Mutat Res.* 2005 Jul 4;584(1–2):1–256. Erratum in: *Mutat Res.* 2005 Dec 7;588(1):70.
19. Kirkland D, Pfuhler S, Tweats D, Aardema M, Corvi R, Darroudi F, Elhajouji A, Glatt H, Hastwell P, Hayashi M, et al. How to reduce false positive results when undertaking in vitro genotoxicity testing and thus avoid unnecessary follow-up animal tests: Report of an ECVAM Workshop. *Mutat Res.* 2007 Mar 30;628(1):31–55.
20. Tweats DJ, Blakey D, Heflich RH, Jacobs A, Jacobsen SD, Morita T, Nohmi T, O'Donovan MR, Sasaki YF, Sofuni T, et al. IWGT Working Group. Report of the IWGT working group on strategies and interpretation of regulatory in vivo tests I. Increases in micronucleated bone marrow cells in rodents that do not indicate genotoxic hazards. *Mutat Res.* 2007 Feb 3;627(1):78–91.

21. Sarrif AM, Bentley KS, Fu LJ, O'Neil RM, Reynolds VL, Stahl RG. Evaluation of benomyl and carbendazim in the in vivo aneuploidy/micronucleus assay in BDF1 mouse bone marrow. *Mutat Res*. 1994;310:143–9.
22. Kirkland DJ, Muller L. Interpretation of the biological relevance of genotoxicity test results: The importance of thresholds. *Mutat Res*. 2000;464:137–47.
23. Muller L, Kasper P. Human biological relevance and the use of threshold-arguments in regulatory genotoxicity assessment: Experience with pharmaceuticals. *Mutat Res*. 2000;464:19–34.
24. Elhajouji A, Lukamowicz M, Cammerer Z, Kirsch-Volders M. Potential thresholds for genotoxic effects by micronucleus scoring. *Mutagenesis*. 2011 Jan;26(1):199–204.
25. Benigni R, Bossa C, Worth A. Structural analysis and predictive value of the rodent in vivo micronucleus assay results. *Mutagenesis*. 2010 Jul;25(4):335–41. Epub 2010 Mar 1.
26. Singh NP, McCoy MT, Tice RR, Schneider EL. A simple technique for quantitation of low levels of DNA damage in individual cells. *Exp Cell Res*. 1988;175:184–91.
27. Sealy PG, Southern EM. 1982. Gel electrophoresis of nucleic acids. In: Rickwood D, Hames BD, editors. *Gel Electrophoresis of Nucleic Acids: A Practical Approach*. Washington, DC: IRL Press Ltd. pp. 39–76.
28. Avishai N, Rabinowitz C, Rinkevich B. Use of the comet assay for studying environmental genotoxicity: Comparisons between visual and image analyses. *Environ Mol Mutagen*. 2003;42(3):155–65.
29. Olive PL, Banath JP, Durand RE. Heterogeneity in radiation-induced DNA damage and repair in tumor and normal cells using the 'comet' assay. *Radiation Res*. 1990;122:86–94.
30. Sasaki YF, Sekihashi K, Izumiyama F, Nishidate E, Saga A, Ishida K, Tsuda S. The comet assay with multiple mouse organs: Comparison of comet assay results and carcinogenicity with 208 chemicals selected from the IARC monographs and U.S. NTP Carcinogenicity Database. *Crit Rev Toxicol*. 2000 Nov;30(6):629–799.
31. Brambilla G, Mattioli F, Robbiano L, Martelli A. Genotoxicity and carcinogenicity testing of pharmaceuticals: Correlations between induction of DNA lesions and carcinogenic activity. *Mutat Res*. 2010 Jul–Sep;705(1):20–39. Epub 2010 Mar 3.
32. Gossen JA, de Leeuw WJ, Tan CH, Zwarthoff EC, Berends F, Lohman PH, Knook DL, Vijg J. Efficient rescue of integrated shuttle vectors from transgenic mice: A model for studying mutations in vivo. *Proc Natl Acad Sci USA*. 1989 Oct;86(20):7971–5.
33. Kohler SW, Provost GS, Fieck A, Kretz PL, Bullock WO, Sorge JA, Putman DL, Short JM. Spectra of spontaneous and mutagen-induced mutations in the *lacI* gene in transgenic mice. *Proc Natl Acad Sci USA*. 1991 Sep 15;88(18):7958–62.
34. Gossen J, Vijg J. Transgenic mice as model systems for studying gene mutations in vivo. *Trends Genet*. 1993 Jan;9(1):27–31.
35. Dean SW, Brooks TM, Burlinson B, Mirsalis J, Myhr B, Recio L, Thybaud V. Transgenic mouse mutation assay systems can play an important role in regulatory mutagenicity testing in vivo for the detection of site-of-contact mutagens. *Mutagenesis*. 1999 Jan;14(1):141–51.
36. Schuler M, Gollapudi BB, Thybaud V, Kim JH. Need and potential value of the *pig-a* in vivo mutation assay—A HESI perspective. *Environ Mol Mutagen*. 2011 Dec;52(9):685–9.
37. Dobrovolsky VN, Miura D, Heflich RH, Dertinger SD. The in vivo *pig-a* gene mutation assay, a potential tool for regulatory safety assessment. *Environ Mol Mutagen*. 2010 Oct–Dec;51(8–9):825–35.
38. Dertinger SD, Heflich RH. In vivo assessment of *pig-a* gene mutation—Recent developments and assay validation. *Environ Mol Mutagen*. 2011 Dec;52(9):681–4.

7 Purification of Lead Compounds for Early Toxicology Profiling such as Ames and Short-Term Non-GLP (Good Laboratory Practice) Toxicology Tests

Zheng Hua, Wesley W. Barnhart, and Kyung H. Gahm

CONTENTS

7.1 INTRODUCTION

In drug discovery programs, lead optimization requires the synthesis and testing of lead candidates for a variety of in vitro and in vivo toxicological studies. These studies examine the toxicological profiles of lead compounds and determine their potential as candidates for further development. The molecular structure of lead compounds may contain one or more chiral centers. At this stage, both the synthetic route and the purification method are not finalized and the lead compounds, though successfully synthesized, may contain various impurities including enantiomers, diastereomers, and achiral impurities. The preparation of lead compounds for early in vitro and in vivo toxicology studies, such as the Ames test and short-term toxicology studies, requires a highly efficient chiral and achiral purification process.

The Ames *Salmonella*/microsome mutagenicity assay (Ames test), a short-term bacterial reverse mutation assay, is a type of genotoxicity assay used worldwide and recommended by the Food and Drug Administration and International Conference on Harmonization for an investigational new drug application [1–5]. In addition to an Ames test conducted according to good laboratory practice (GLP) procedures, an abbreviated or "screening" version of the Ames test is typically used in lead optimization studies to help determine which molecule to bring forward in development. Currently, no industry-wide guidelines specifically define the purity requirement of compounds for this assay. High-purity (both chiral and achiral) materials are recommended in lead optimization studies by many pharmaceutical companies to eliminate false positive results. The challenge in purification lies in removing not only the chiral impurities but also an array of achiral impurities, which are not well characterized at this point.

Since the introduction of a commercially available instrument in the early 1990s, supercritical fluid chromatography (SFC) has been gradually accepted as an efficient tool for purification. SFC is a type of normal phase chromatography utilizing compressed carbon dioxide in conjunction with organic solvent as the mobile phase. High diffusivity and low viscosity of the mobile phase allow for rapid flow rates, leading to reduced analysis time without sacrificing efficiency [6–9]. SFC is suitable for purifying materials with stringent purity criteria and aggressive time lines. Advantages include rapid method development and relatively simple scale-up. The reduced amount of solvent waste generated by chromatographic separation and the smaller volume of purified fractions are also outstanding features of this technique. Additionally, volatile organic solvents such as methanol (MeOH), ethanol (EtOH), or isopropanol (IPOH) utilized as organic additives to carbon dioxide are easier to remove compared to an aqueous mobile phase [10–12]. The use of volatile base additives, such as diethylamine (DEA) or isopropylamine, in the SFC mobile phase can dramatically enhance the peak shape of basic compounds, resulting in improved resolution.

Purity of lead compounds is better assessed using at least two orthogonal systems such as reversed phase liquid chromatography–mass spectrometry (LC/MS) and supercritical fluid chromatography–mass spectrometry (SFC/MS) systems. In our laboratory, SFC/MS is the primary means for the assessment of chiral purity as well as the tool to provide additional information on achiral purity. Achiral purity is generally assessed by reversed phase LC/MS systems.

In this chapter, the focus is on describing and illustrating the procedure on how to obtain high-purity lead compounds for either Ames or short-term toxicology tests using SFC purification. A systematic approach to develop a successful SFC purification method is discussed. Three different case studies are presented; examples include single- and multistep purifications of typical mixtures, as well as a multistep purification of a highly complex, impure mixture. The advantage of coupling two columns (identical or different) in SFC to enhance separation is also illustrated.

7.2 GENERAL EXPERIMENTAL PROCEDURES

A general purification procedure is shown in Scheme 7.1. The initial steps in Scheme 7.1 are a standard SFC/MS screen with four chiral columns and a standard

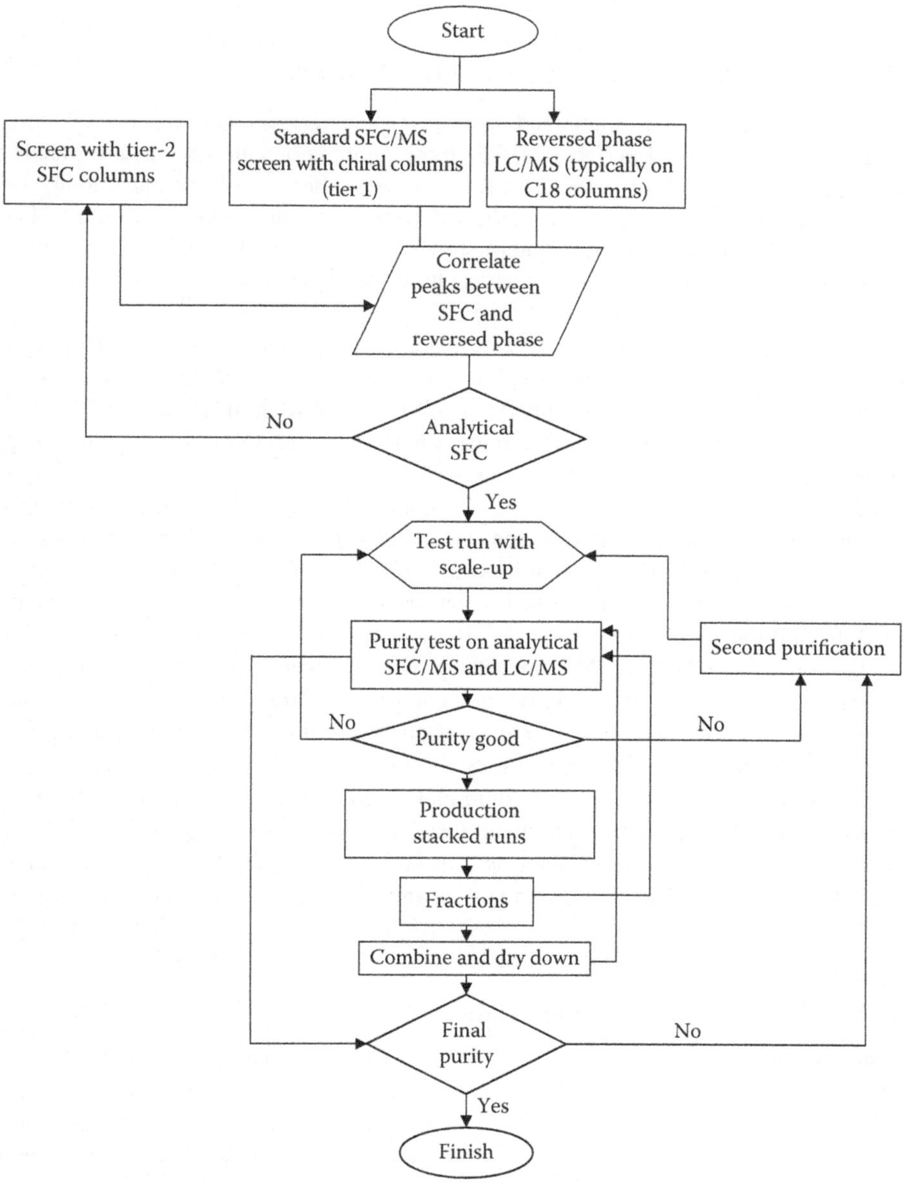

SCHEME 7.1 General purification flow chart.

reversed phase LC/MS analysis. The strategy is to identify and distinguish achiral and chiral impurities in the original crude sample by utilizing two orthogonal modes of separation (SFC and reversed phase chromatography). Tier-1 screening refers to the initial SFC screening using four chiral columns. If a separation is not available with the initial tier-1 screening, 10 additional columns are employed for further method development (tier-2). Once a successful separation method is developed and optimized, scale-up and purification are performed. Details of the process are further discussed in Sections 7.2.1 through 7.2.4 and 7.3.

7.2.1 Sample Characterization and Tier-1 Screening

Tier-1 screening refers to the initial SFC screening using the four columns (4.6 × 150 mm, 5 μm particle size) listed in Table 7.1. The most common organic solvents are MeOH, EtOH, and IPOH with or without 0.2% DEA, depending on compound stability or basicity. A generic 6- or 12-minute gradient (5%–50% organic) applies to all tier-1 SFC analytical columns. The flow rate is normally 3–5 mL/min, and the oven temperature is typically set at 35°C–40°C. The oven temperature and outlet pressure are kept consistent for all gradient analyses.

A commonly used C18 column is Phenomenex's (Phenomenex Inc, Torrance, California) Gemini® (4.6 × 150 mm, 5 μm particle size). The mobile phase consists of A (0.1% trifluoroacetic acid [TFA] or 20 mM ammonium acetate [pH 5.5] in water) and B (0.1% TFA or 5 mM ammonium acetate in MeCN). A typical reversed phase high-performance liquid chromatography (HPLC) generic gradient starts at 10% B and increases to 90% B over 20 minutes. The flow rate is 1.0 mL/min, and the column is at ambient temperatures. The method is usually used to identify achiral impurities greater than 0.1% at both 215 and 254 nm. Occasionally, a different reversed phase method is developed to ensure complete resolution of the product peak from impurities.

The reversed phase LC/MS analysis provides information on the number of achiral impurities and their relative percentage peak areas. When analyzing the result from SFC/MS screening, the masses observed in LC/MS are used to extract selected ion chromatograms from the SFC total ion current chromatogram. Since relatively short gradient times and columns (15 cm compared to 25 cm in purification scale-up) are used in the analytical SFC gradient screening, a partial separation is considered a "hit" when no other options are available. If a successful method is identified at this stage, method optimization and scale-up are performed. Otherwise,

TABLE 7.1
Tier-1 Columns (4.6 × 150 mm, 5 μm)

Column Name	Manufacturer
Chiralpak® AD-H	Chiral Technologies, Inc., West Chester, Pennsylvania
Chiralpak® AS-H	Chiral Technologies, Inc., West Chester, Pennsylvania
Chiralcel® OD-H	Chiral Technologies, Inc., West Chester, Pennsylvania
Chiralcel® OJ-H	Chiral Technologies, Inc., West Chester, Pennsylvania

if desired separation is not available with the initial tier-1 screening, more columns are employed for further method development.

7.2.2 Tier-2 Screening with 10 Additional Columns

The majority of samples can be purified by methods obtained through tier-1 screening; however, there are cases that require more extensive method development. Tier-2 may utilize additional columns (see Table 7.2) and is conveniently conducted overnight. The same three solvents (MeOH, EtOH, and IPOH) used in tier 1 may also be utilized in tier 2. Column switching may be performed through an external 12-port valve (VICI International AG, Schenkon, Switzerland), which is connected to the flow path and computer controlled.

Polysaccharide-based chiral stationary phases (CSPs), coated (Sepapak®, CelluCoat™, and AmyCoat™) and immobilized (Chiralpak® IA, IB, and IC), are included in the tier-2 screen. In addition to the polysaccharide-based CSPs, the synthetic polymer–based P-CAP™ column is also part of the screen. A pyridine column is used to obtain the achiral profile of a sample mixture. In situations where only achiral impurities are present, achiral columns such as cyano, diol, and unmodified silica (Princeton Chromatography, Inc., Cranbury, New Jersey) are used in place of selected chiral columns in tier 2. The user can choose the columns suitable to their needs as well, and the selection of tier-2 columns continues to evolve as needed.

7.2.3 Further Method Development

Ideally, a purification method should provide the best separation with the shortest total purification time. The best case is when all chiral and achiral impurities are separated from the peak of interest in a single run with a reasonable cycle time. In other cases, a number of strategies such as column coupling; multistep purification; and the evaluation of other organic solvents, such as MeCN, *n*-propanol, and *n*-butanol, must be utilized to achieve the desired separation.

TABLE 7.2
Tier-2 Columns (4.6 mm × 150 mm, 5 μm)

Column Name	Manufacturer
Pyridine	Princeton Chromatography
Chiralpak IA	Chiral Technologies, Inc.
Chiralpak IB	Chiral Technologies, Inc.
Chiralpak IC	Chiral Technologies, Inc.
Sepapak®-2-HR	Enantiosep GmbH, Münster, Germany
Sepapak®-3-HR	Enantiosep GmbH
Sepapak®-4-HR	Enantiosep GmbH
(S,S) or (R,R) P-CAP	Sigma Aldrich (St. Louise, Missouri)
5-AmyCoat	Kromasil (AkzoNobel, Bohus, Sweden)
5-CelluCoat	Kromasil (AkzoNobel, Bohus, Sweden)

Through tier-1 and tier-2 screening, partial but inadequate separations may be observed. If a single column length is not able to provide the desired resolution, columns can be coupled to enhance separations. A low column pressure drop in SFC allows column coupling as a common practice. The coupled column chemistries can be identical or different and used in combinations such as achiral–chiral or chiral–chiral [13]. Coupling of identical columns is a simple way of enhancing resolution; however, coupling of different types of columns may require a significant amount of time in method development to ensure its reproducibility and robustness. In such cases, it is generally considered a better strategy to perform multistep purification to avoid potential complications.

Multistep purification is necessary for the following two situations: (1) a single method is available but requires a very long purification run time and (2) no single column can provide a robust separation due to the complexity of the sample. For Ames samples, complexity often arises from many low-level impurities of diverse retention times and, thus, multistep purification is required. As an example of the first situation, it may be necessary to first eliminate the polar impurities to reduce overall purification time when the polarities of the components are quite varied. This strategy allows the second step chiral separation to utilize a lower percentage of organic solvent to enhance chiral resolution with a reasonable cycle time. Another situation exists when a large number of impurities are present in the sample and multiple columns with varied selectivities are required. It is often observed that a good chiral separation is achieved but an achiral impurity coelutes with the desired enantiomer/diastereomer, thus making multistep purification necessary.

In addition to the column coupling and multistep purification strategies, other solvents such as MeCN, alcohol–MeCN mixture, *n*-propanol, and *n*-butanol have been used in our laboratory to explore different selectivities from those achieved with the first-pass solvents in the standard tier-1 and tier-2 screening conditions [13,14]. It is often necessary to adjust the ratios of protic alcohols versus aprotic MeCN to achieve final purification methods.

7.2.4 Optimization, Scale-Up, and Final Analysis

It is convenient to finalize a method on an analytical system before scaling up the preparative SFC. Since the tier-1 and tier-2 screenings are conducted using a generic gradient, some preliminary work must be done to obtain an isocratic method, which will later be scaled up. During this step, the elution order of many components including achiral/chiral impurities may be altered. Isocratic methods are primarily used for purification because of the ability to stack injections.

For any analytical SFC method developed, test scale-up is necessary on the preparative columns and instruments. Analytical SFC methods are utilized as guidelines for preparative methods in terms of flow rate and chromatographic characteristics such as resolution and selectivity. Throughput in preparative chromatography is optimized after considering three major factors: percentage of organic modifier, loading, and cycle time. The solubilities of a sample in organic solvents, as well as in supercritical fluids, are also important factors in the scale-up activity, because the typical ratio of the sample to both mobile and stationary phases is much higher in preparative chromatography [14].

It is typical to complete a test run with three to five stacked injections once the percentage of the organic solvent and the sample load amount per injection are determined. Any peak disturbances from the stacked injection must be carefully examined, since it may affect ultraviolet (UV)-based collection. During the initial scale-up experiment, it is important to observe signs of sample precipitation in the preparative SFC system, which can be possibly due to the difference of solubility between the organic solvent and the SFC mobile phase.

After the fraction is collected from the test scale-up run, the chiral purity and achiral purity of the fraction are assessed in analytical SFC and reversed phase HPLC systems, respectively. If the tested fractions do not meet the desired purity, then modification of the method is required. Once the test scale-up run is successful, the instrument can be run under optimal conditions over an extended period of time. When the sample size is particularly large, and the run is expected to go overnight, the instrument is observed closely during daytime operation to ensure that there is no subtle increase in system pressure normally due to sample precipitation within the system. Sample loading should be reduced to avoid precipitation in the purification system; sometimes, the percentage of organic solvent can be increased to maximize the solubility under SFC conditions.

Achiral and chiral purities are spot-checked during the purification. Fractions are then dried down and subsequently subjected to a vacuum oven at 40°C overnight. Final chemical purity is measured with the reversed phase HPLC system using the dried sample. Chiral purity is checked in the analytical SFC system using the dried sample to confirm that no racemization occurred during the drying process.

7.3 EXAMPLES OF SUPERCRITICAL FLUID CHROMATOGRAPHY PURIFICATION TO GENERATE HIGH-PURITY LEAD COMPOUNDS

In this section, three examples of successful SFC purifications, which were performed to generate high-purity lead compounds for screening, non-GLP Ames, or short-term toxicology tests, are presented. These examples include single- and multistep purifications of typical mixtures, as well as the multistep purification of a highly complex, impure mixture.

7.3.1 Example of a Single-Step Purification

A relatively impure sample was submitted for purification to achieve the appropriate purity for the Ames test. Reversed phase HPLC analysis showed five closely eluting small impurities. Purity of the crude mixture was approximately 80% at UV 215 nm (see Figure 7.1), and the masses of the impurities were identified using mass spectrometry under electrospray ionization conditions. Molecular weight information for the impurities was used to facilitate the development of an analytical SFC method.

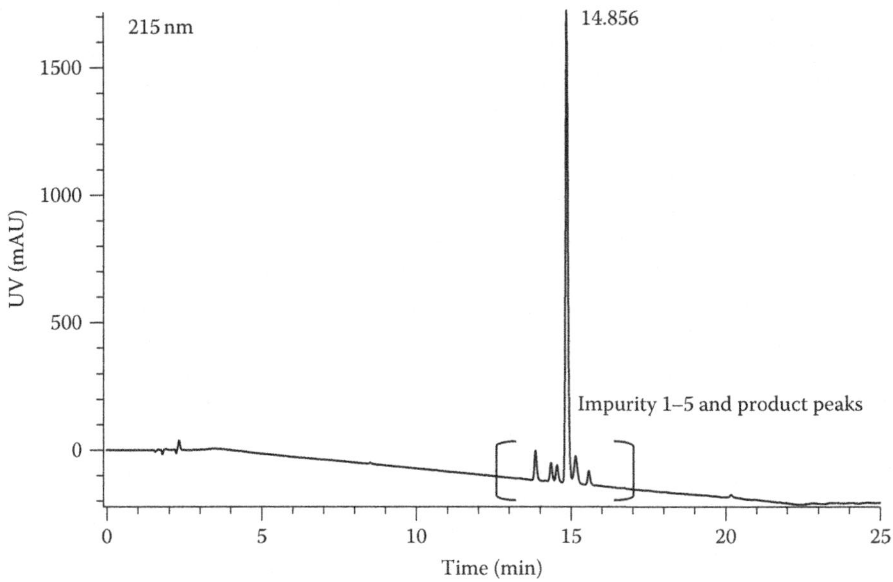

FIGURE 7.1 Reversed phase C18 high-performance liquid chromatography analysis for the crude mixture at 215 nm using a Gemini C18 (4.6 × 150 mm, 5 μm). Mobile phase A: water with 20 mM ammonium acetate (pH 5.5); B: MeCN with 5 mM ammonium acetate (pH 5.5). The gradient was 10% B to 90% B in 20 minutes and held at 90% B for 5 minutes, returned to 5% B in 1 minute, and reequilibrated for 4 minutes. Flow rate was 1.0 mL/min.

After the initial SFC/MS screening, an SFC method was identified to separate out all impurities, including the opposite enantiomer. As illustrated in Figure 7.2, the masses of the impurities were extracted to show that the impurities were resolved from the desired peak. This was an ideal situation in which all achiral impurities and the chiral impurity could be resolved in a single purification. Although the mass ($M + 1 = 633$) was observed in LC/MS, this peak was not observed in SFC/MS possibly due to sensitivity or because this peak did not elute during the analysis time.

It is clear from Figure 7.2 that the achiral impurities observed in reversed phase LC/MS were resolved in an isocratic SFC method with the AD-H column except the impurity of $M + 1 = 633$. The peak ($M + 1 = 496$) at 2.1 minutes is not a fragment from the target molecule since it has a different retention time in LC/MS. The peak ($M + 1 = 496$) at 3.1 minutes might be a fragment ion from the target molecule. The time interval of 2.5–3.5 minutes represents the "window of opportunity" where only the lead compound eluted with no overlap from its enantiomer or other impurities. This method was then scaled up on the semi-prep instrument with a flow rate of 60 mL/min with 30% methanol, which provided adequate resolution (see Figure 7.3) for the 30 mg injection. Figure 7.3 clearly demonstrates that the separation was maintained between the analytical and preparative runs.

The collected main product peak fraction was analyzed by reversed phase analytical HPLC and SFC on completion of the purification. It was found that chemical purity was greater than 99.5% (see Figure 7.4) and chiral purity was greater than 99.5%.

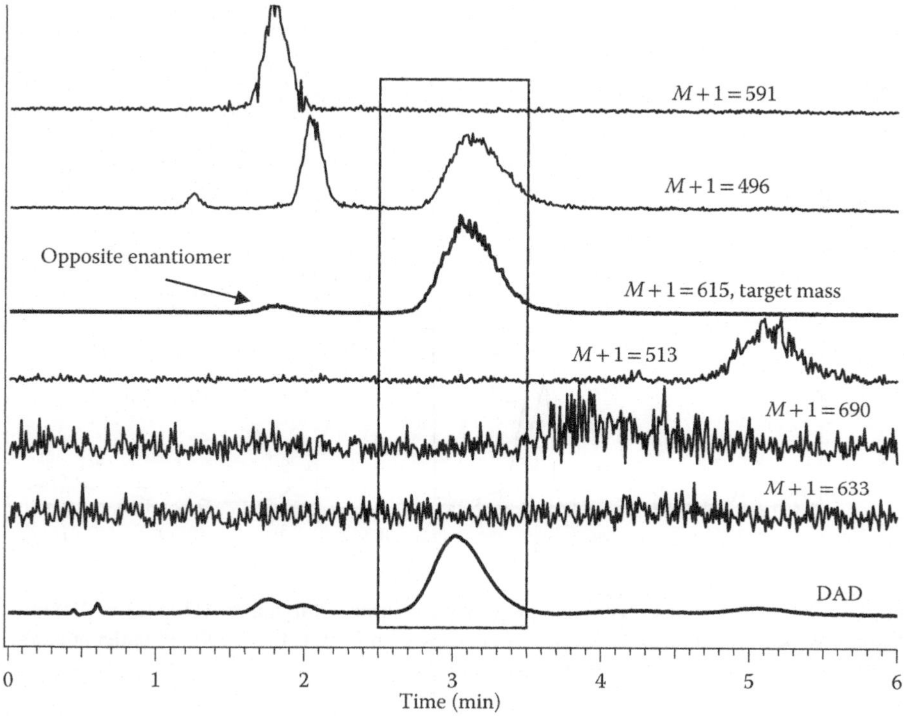

FIGURE 7.2 Analytical supercritical fluid chromatography chromatograms of the crude sample showing the separation of all five impurities and the opposite enantiomer from the lead compound. The bottom trace is the averaged diode array detector (DAD) chromatogram of the crude sample. Other traces represent the extracted ion chromatograms of the impurities. AD-H (4.6 × 150 mm, 5 μm) was used with 20% methanol in CO_2. The flow rate was 4.0 mL/min. Outlet pressure was set to 100 bar, and oven temperature was 35°C.

This case study is one of the many examples where single-step SFC purification was sufficient to obtain a high-purity sample when chiral and many achiral impurities were originally present. Although a lead compound may be chiral, a chiral column can often separate closely related achiral impurities, as was clearly demonstrated in this example.

7.3.2 Example of a Multistep Purification for an Ames Study

In this example, 2.7 g of a crude chiral sample was purified to high purity for the Ames test. Initial reversed phase C18 HPLC analysis revealed multiple achiral impurities (see Figure 7.5). After tier-1 screening, a chiral separation method, using an AD-H column with methanol, showed the presence of a small amount of the opposite enantiomer. An isocratic chiral method was then developed to resolve the opposite enantiomer and achiral impurities in the same step. During scale-up, however, tailing from the major peak was observed when no DEA was used. The typical preparative chromatogram that demonstrates the separation of chiral and several achiral impurities is shown in Figure 7.6. Analytical chiral supercritical fluid chromatograms are shown in Figure 7.7.

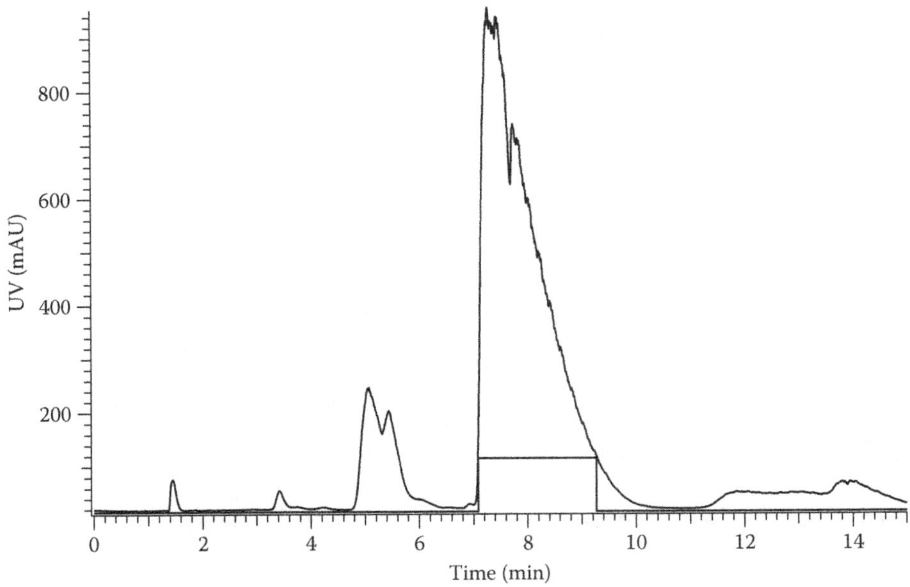

FIGURE 7.3 Test scale-up run on a preparative column. Chiralpak® AD-H (21 × 250 mm, 5 μm) was used. Conditions: 30% methanol in CO_2, 60 mL/min, 40°C, 100-bar outlet pressure. The injection amount was 30 mg. A 220 nm wavelength was used.

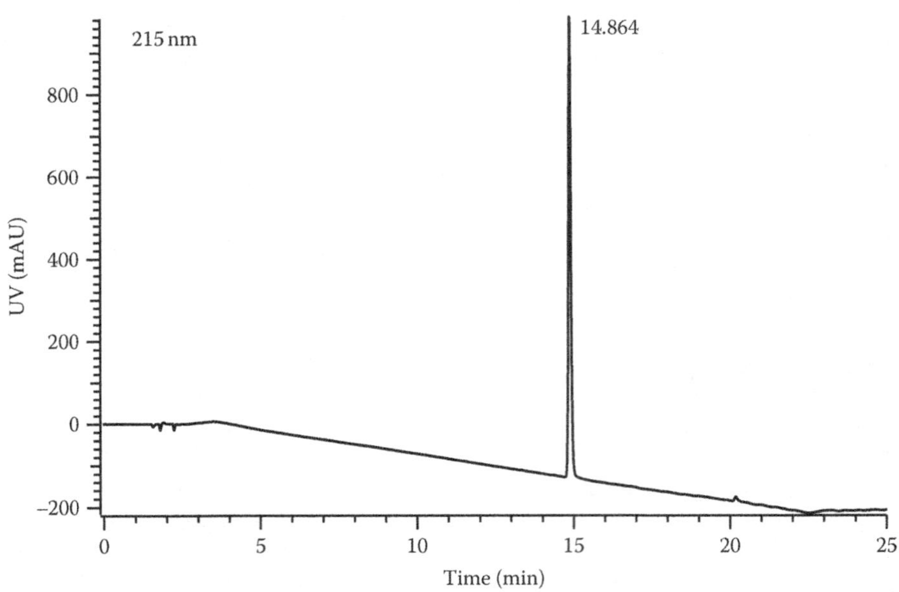

FIGURE 7.4 Postpurification achiral analysis of the dried sample. The chromatographic conditions were the same as in Figure 7.1.

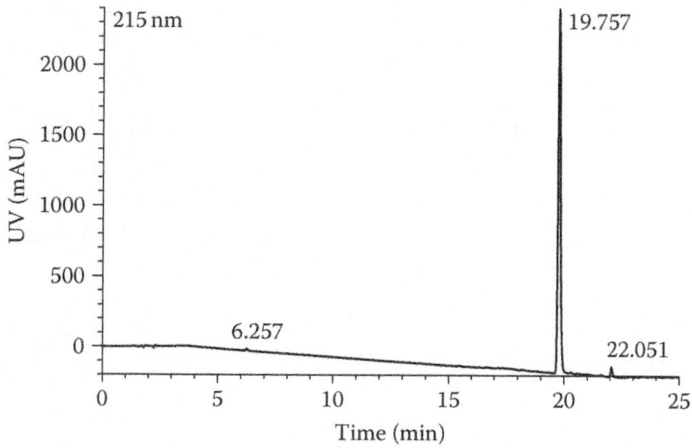

FIGURE 7.5 Reversed phase C18 high-performance liquid chromatography chromatogram of the crude sample containing several small achiral impurities. The chromatographic conditions were the same as in Figure 7.1.

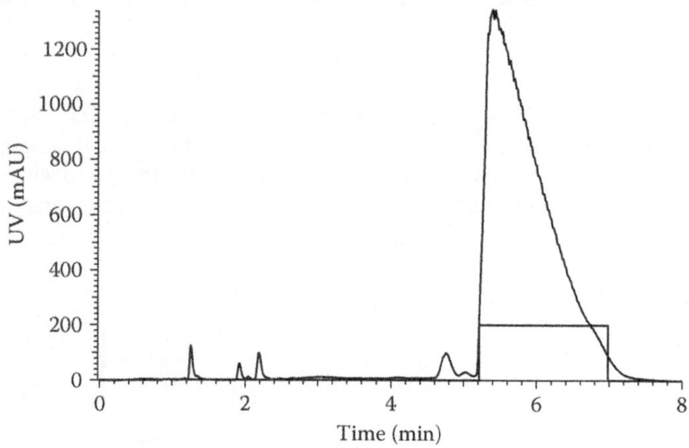

FIGURE 7.6 Chromatogram showing supercritical fluid chromatography scale-up purification using an AD-H (21 × 250 mm, 5 μm) column with 40% methanol in CO_2. Flow rate was 60 mL/min. Column oven temperature was set to 40°C. Approximately 15 mg was loaded per injection.

A portion of the achiral impurity at 22 minutes (see Figure 7.5), however, remained in the collected fraction after the first SFC purification (see Figure 7.8), requiring additional purification. Tier-2 SFC screening was conducted, and the (R,R) P-CAP™ column [15] was found to effectively remove the impurity from the main peak (see Figure 7.9).

The narrow peak width in addition to the large injection amount (80 mg) with a short cycle time (approximately 2 minutes) allowed the second purification to be completed in a very short period of time. Reversed phase analysis (see Figure 7.10) demonstrated that the impurity observed in the initial purification was completely removed.

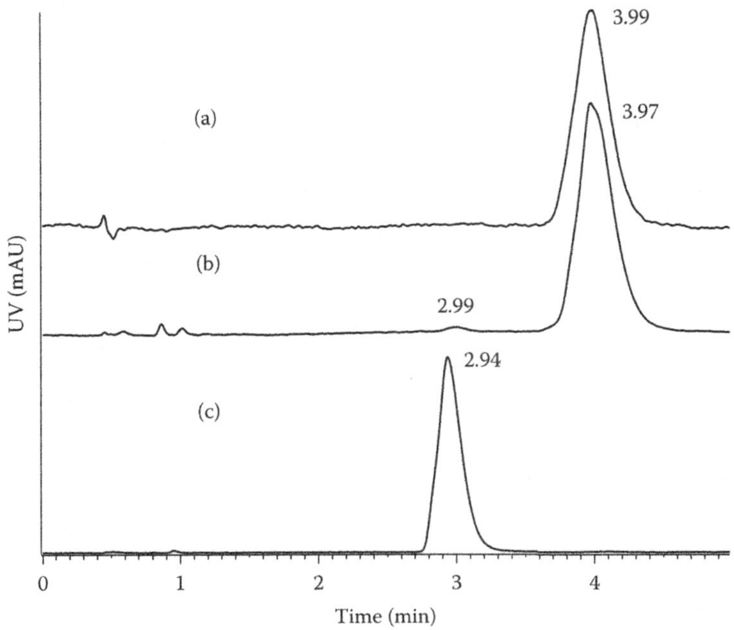

FIGURE 7.7 Analytical chiral supercritical fluid chromatography (SFC) analysis of the dried sample after the first purification: (a) purified sample, (b) the original sample, and (c) the opposite enantiomer standard. SFC conditions: AD-H (4.6 × 150 mm, 5 μm) with 30% methanol in CO_2, 4.0 mL/min, outlet pressure was 100 bar, and column oven temperature was 35°C.

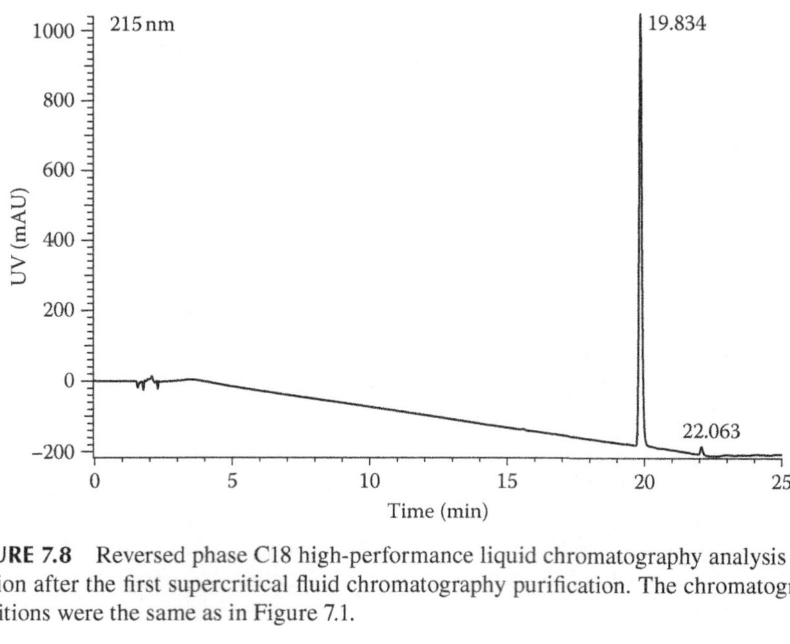

FIGURE 7.8 Reversed phase C18 high-performance liquid chromatography analysis of the fraction after the first supercritical fluid chromatography purification. The chromatographic conditions were the same as in Figure 7.1.

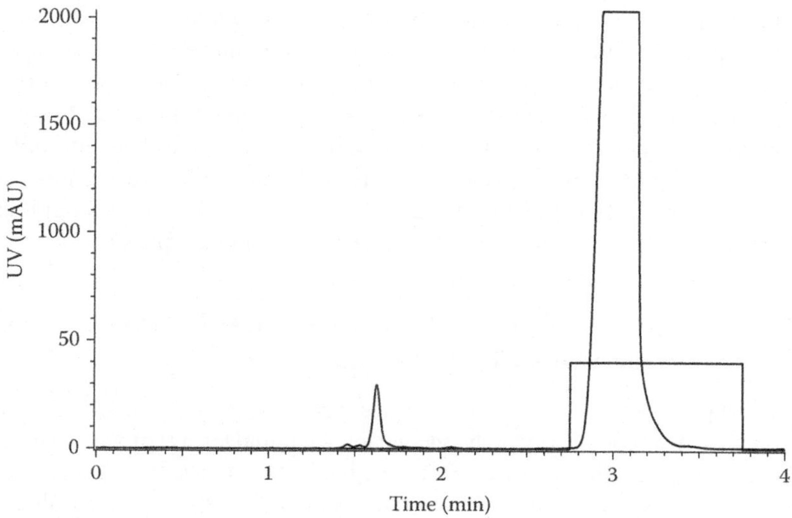

FIGURE 7.9 Preparative chromatogram of the second supercritical fluid chromatography purification using the (R,R) P-CAP (21 × 250 mm, 5 μm) with 35% MeOH in CO_2. Flow rate was 60 mL/min, and injection amount was 80 mg. Column oven temperature was set to 40°C.

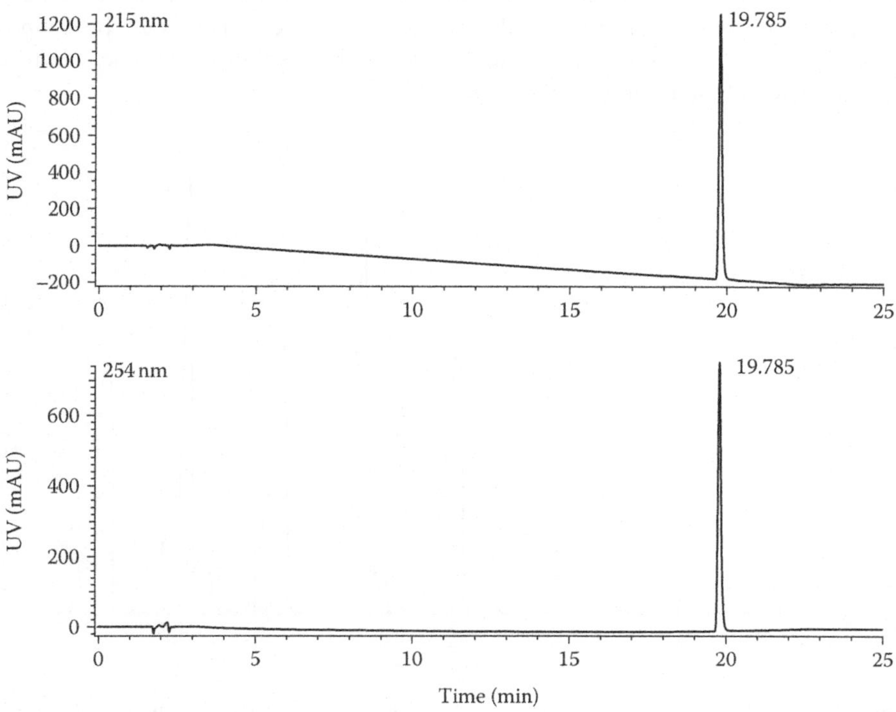

FIGURE 7.10 Reversed phase C18 high-performance liquid chromatography analysis of the second supercritical fluid chromatography purification detected at 215 and 254 nm. The chromatographic conditions were the same as in Figure 7.1.

In this example, a two-step purification was performed to remove both chiral and achiral impurities. Initial purification was capable of removing the chiral and most of the achiral impurities. The second purification method was rapidly developed due to the reduced sample complexity and the diversity of columns in tier-2 screening. Although the achiral impurity in the second purification could also be removed using reversed phase HPLC, SFC provided a shorter run time without using any acid or base additives. In addition, the time required to evaporate methanol was significantly less than that would be required to remove the aqueous mobile phase.

7.3.3 Example of a Multistep Purification of a Very Complex Mixture for a Short-Term Toxicology Study

In this example, two different lots of samples were purified using different methods due to the increased complexity of the second lot. An initial lot (1.0 g) was purified with two OJ-H columns coupled in series (total length was 50 cm) in single-step purification (see Figure 7.11). A second lot, 40 g of crude sample, was prepared for a short-term toxicological study. Unlike the first lot, the second lot was a more complex mixture that contained the opposite enantiomer, diastereomers, and achiral impurities. A chromatogram of the reversed phase C18 HPLC analysis is shown in Figure 7.12.

The purity of the main peak based on reversed phase HPLC analysis was 45% at 215 nm (see Figure 7.12). Because the main peak was a racemic mixture, the content of the desired enantiomer in the crude sample would be less than 23%. The sample was analyzed with the tier-1 screen, and two additional pyridine and cyano columns were used to obtain the achiral profile by SFC.

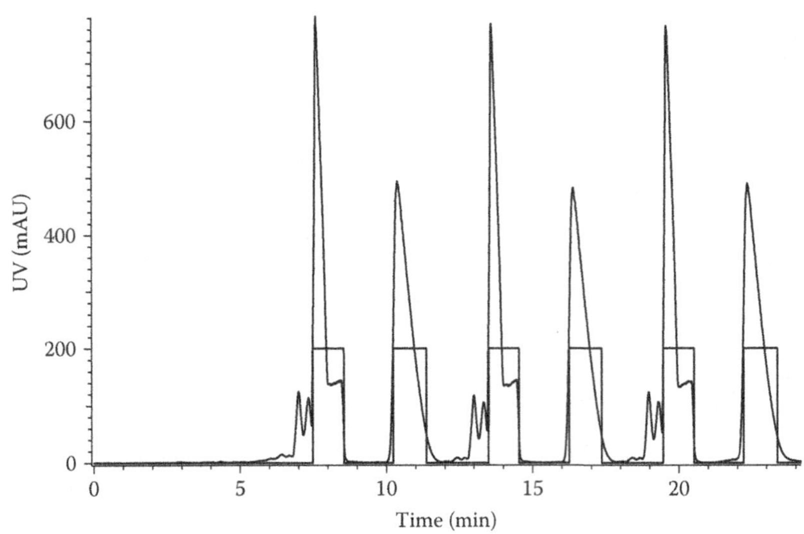

FIGURE 7.11 Preparative supercritical fluid chromatography chromatogram for the purification of lot 1 by coupling two OJ-H (21 × 250 mm, 5 μm) columns in series. Flow rate was 70 mL/min with 10% MeOH in CO_2 using 25 mg per injection. The irregular shape of the first peak was due to solvent effect from the stacked injections.

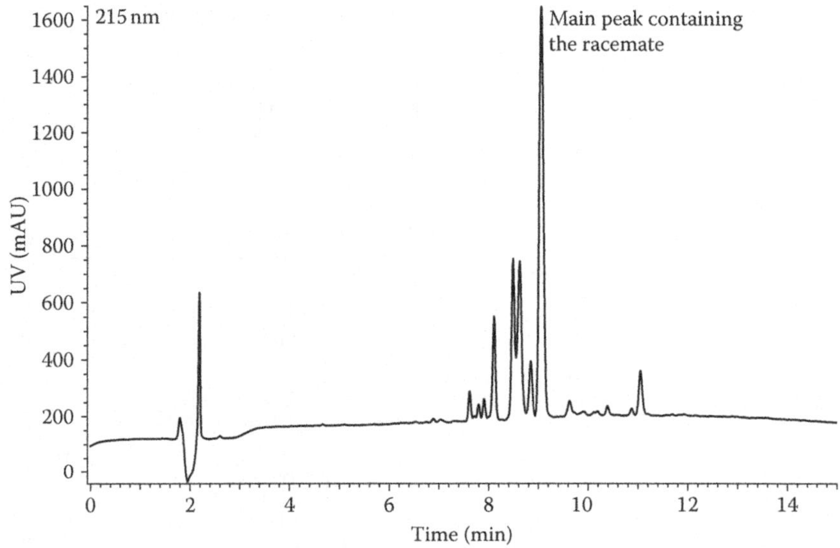

FIGURE 7.12 Reversed phase high-performance liquid chromatography chromatograms of lot 2. The chromatographic conditions were the same as in Figure 7.1, except that a different gradient program was used. The gradient was 10% B to 90% B in 10 minutes and held at 90% B for 5 minutes, returned to 5% B in 1 minute, and reequilibrated for 4 minutes.

A 12-minute-gradient screen was used due to the complexity of the mixture. As shown in Figure 7.13, more components were resolved with OJ-H and AD-H than any other columns studied, and both columns were able to resolve the opposite enantiomer. It was determined that the 4.48-minute peak on OJ-H and the 6.08-minute peak on AD-H corresponded to the desired enantiomer as confirmed by the injection of the authentic standard. The elution order of enantiomers was reversed between AD-H and OJ-H. A test run using AD-H/IPOH was chosen as the first step in purification, since the peak at 6.08 minutes in the AD-H chromatogram (shown in the bottom chromatogram in Figure 7.13) was most resolved from any other sample components.

The preparative SFC chromatogram obtained by a stacked injection of 120 mg of sample is shown in Figure 7.14. An aliquot of the purified fraction was reinjected onto the preparative SFC system. It was found that approximately 10% achiral impurities were still present in the fraction, thus requiring another step of purification.

A robust and fast second purification method was found by using an achiral column with a cyano modified functional group. An example preparative SFC chromatogram for the second purification is shown in Figure 7.15. The main peak in the second purification was found to be pure (>99.5%) after checking the fraction via the reversed phase HPLC method (see Figure 7.16) as well as the chiral analytical SFC method (see Figure 7.17).

This example demonstrates that different purification strategies were necessary to purify different lots of the same lead compound. Coupling two identical columns enabled a single-step purification of the relatively simple lead candidate mixture

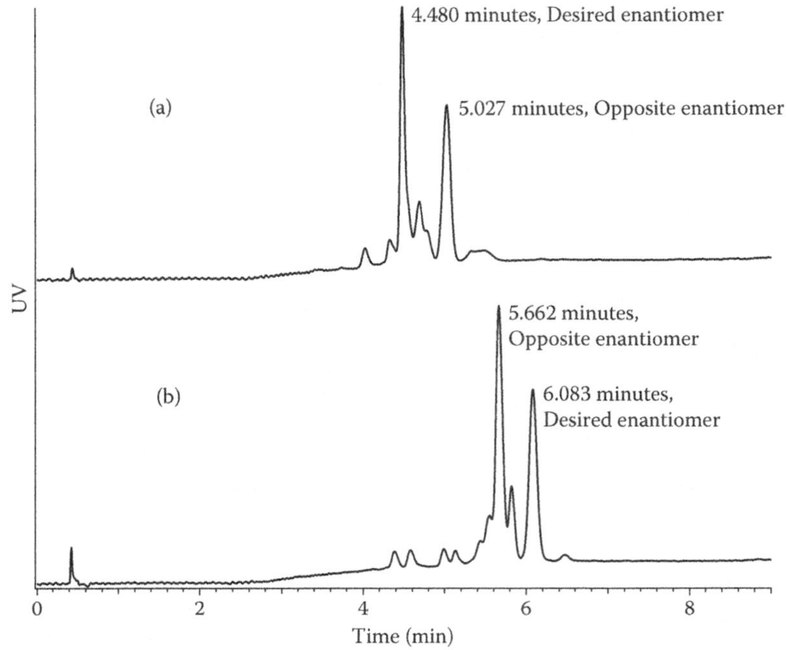

FIGURE 7.13 Representative chromatograms of initial supercritical fluid chromatography screening for lot 2. (a) Using OJ-H column with IPOH in CO_2. (b) Using AD-H column with IPOH in CO_2.

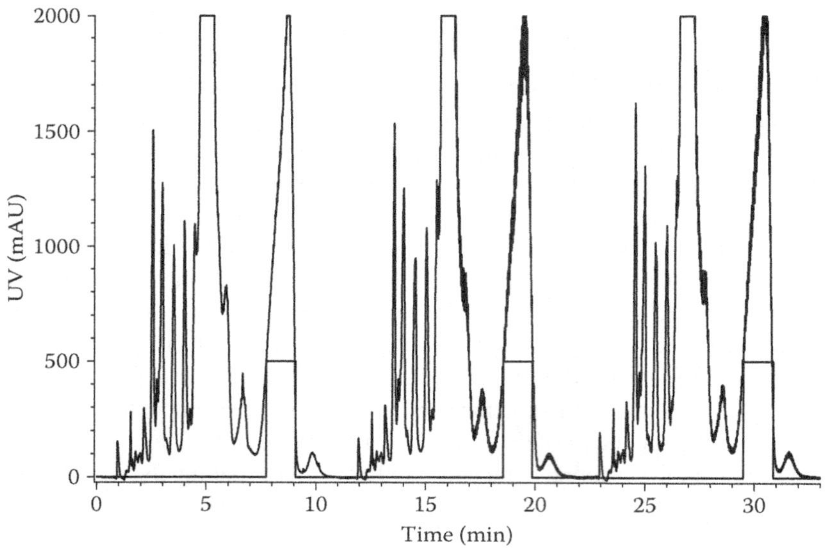

FIGURE 7.14 Example preparative supercritical fluid chromatography chromatogram showing the stacked injections of lot 2. Conditions: AD-H (21 × 250 mm, 5 μm), 22% IPOH, 65 mL/min, 120 mg per injection, and cycle time was 11 minutes.

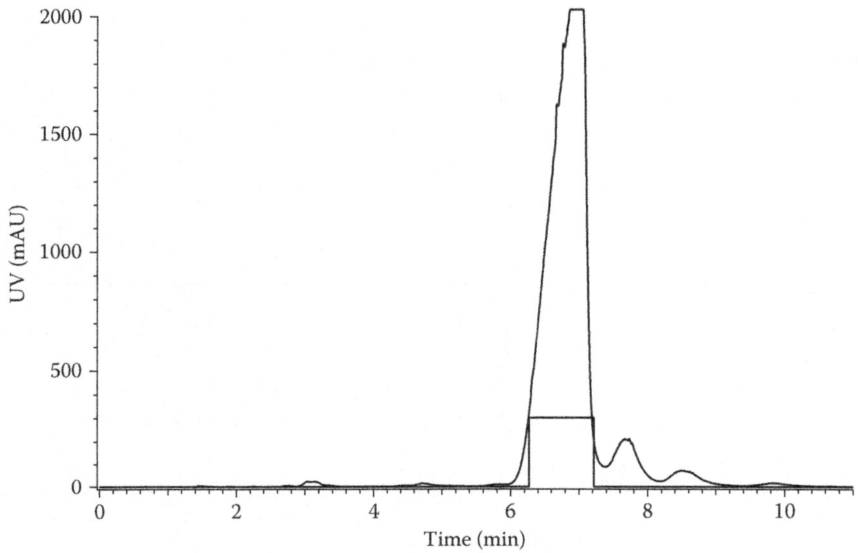

FIGURE 7.15 Second-step supercritical fluid chromatography purification. Conditions: cyano column (21.2 × 250 mm, 5 μm), 18% IPOH, 65 mL/min, and 30 mg per injection.

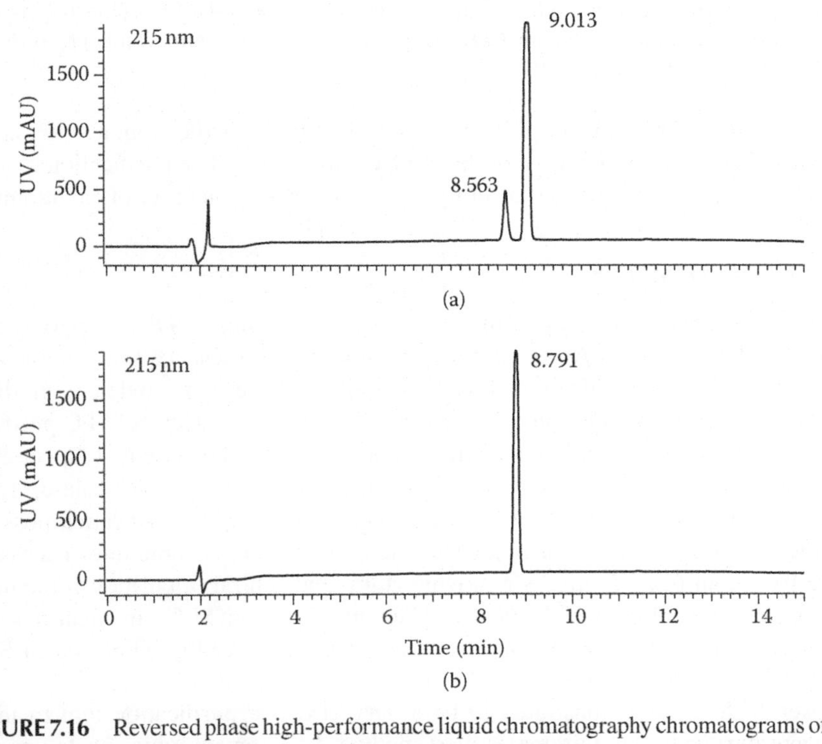

FIGURE 7.16 Reversed phase high-performance liquid chromatography chromatograms of the final product (a) before and (b) after the second supercritical fluid chromatography purification.

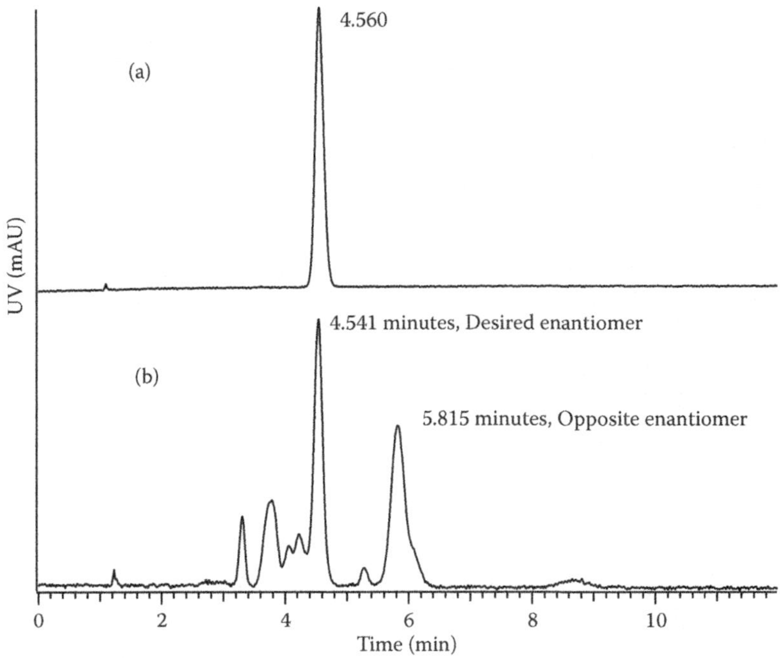

FIGURE 7.17 Chiral supercritical fluid chromatography analysis for (a) purified enantiomer after second purification and (b) the original sample. Chiralcel OJ-H (4.6 × [150 + 250] mm, 5 μm). Conditions: 12% methanol in CO_2 and flow rate was 4.0 mL/min. Outlet pressure was 100 bar and oven temperature 40°C.

(lot 1). With the known peak of interest from the initial purification, a better strategy was devised for purification of the more complex lot 2. The purification of lot 2 was greatly facilitated due to the ability to reverse the elution order of enantiomers.

7.4 CONCLUSIONS

It is demonstrated in this chapter that high-purity, for example, 97% to up to 99.5%, lead candidates for both Ames and short-term toxicology tests can be obtained by SFC purification. Streamlined method development, such as tier-1 and tier-2 gradient column screening, provides an efficient way to establish a successful SFC purification method. Diverse chiral and achiral columns are key components to meet even the most challenging purification requirements. A relatively simple scale-up from analytical methods and stacked injections makes efficient SFC purification possible within a reasonable period of time. Multistep purification is sometimes necessary to purify a complex mixture by removing chiral and achiral impurities in the most efficient way. Coupling two identical or different columns in SFC can enhance separation and provides an additional advantage; it is more readily employed in SFC compared to reversed phase HPLC due to low pressure drop.

Overall, SFC is demonstrated to be a very effective purification tool to meet stringent purification requirements. Fast method development, quick scale-up, and column coupling are all attributes critical to the successful use of SFC purification to separate mixtures of various complexities for toxicological studies. Additional advantages including reduced organic solvent consumption, together with easy evaporation, make SFC an attractive tool to generate high-purity lead compounds for early toxicological tests.

REFERENCES

1. ICH. *Genotoxicity: A Standard Battery for Genotoxicity Testing of Pharmaceuticals S2B*, Harmonized tripartite guideline, International Conference on Harmonization of Technical Requirement for Registration of Pharmaceuticals for Human Use, Brussels, Belgium, **1997**, 1–8.
2. OECD. *OECD Guidelines for the Testing of Chemicals,* 471 Bacterial Reverse Mutation Test, Paris, France: OECD Publishing, **1997**, 1–11.
3. Department of Health and Human Services, *Guidance for Industry, Investigators, and Reviewers*, Exploratory IND Studies, Rockville, Maryland: Food and Drug Administration Center for Drug Evaluation and Research, **2006**.
4. Mortelmans, K.; Zeiger E. The Ames *Salmonella*/microsome mutagenicity assay. *Mutat. Res.* **2000**, 455, 29–60.
5. Bogen, K. Application of alternative models of revertant variance to Ames-test data for 121 mutagenic carcinogens. *Mutat. Res.* **1994**, 455, 29–60.
6. Erickson, B. SFC in flux. *Anal. Chem.* **1997**, 69, 683A–686A.
7. White, C. Integration of supercritical fluid chromatography into drug discovery as a routine support tool. Part I. Fast chiral screening and purification. *J. Chromatogr. A* **2005**, 1074, 163–173.
8. White, C.; Burnett, J. Integration of supercritical fluid chromatography into drug discovery as a routine support tool. Part II. Investigation and evaluation of supercritical fluid chromatography for achiral batch purification. *J. Chromatogr. A* **2005**, 1074, 175–185.
9. Phinney, K.W. SFC of drug enantiomers. *Anal. Chem.* **2000**, 72, 204A–211A.
10. Smith, R.M. Supercritical fluids in separation science—the dreams, the reality and the future. *J. Chromatogr. A* **1999**, 856, 83–115.
11. Welch, C.J.; Leonard Jr. W.R.; DaSilva, J.O.; Biba, M.; Albaneze-Walker, J.; Henderson, D.W.; Liang, B.; Mathre, D.J. Preparative chiral SFC as a green technology for rapid access to enantiopurity in pharmaceutical process research. *LC-GC, North Am.* **2005**, 23, 16–29.
12. Jusforgues, P.; Shaimi, M. Preparative supercritical fluid chromatography. *Analusis Mag.* **1998**, 26, M55–M60.
13. Barnhart, W.; Gahm, K.; Thomas, S.; Notari, S.; Semin, D.; Cheetham, J. Supercritical fluid chromatography tandem-column method development in pharmaceutical sciences for a mixture of four stereoisomers. *J. Sep. Sci.* **2005**, 28, 619–626.
14. Gahm, K.H.; Tan, H.; Liu, L.; Barnhart, W.; Eschelbach, J.; Notari, S.; Thomas, S.; Semin, D.; Cheetham, J. Purification method development for chiral separation in supercritical fluid chromatography with the solubilities in supercritical fluid chromatographic mobile phases. *J. Pharmaceut. Biomed. Anal.* **2008**, 46, 831–838.
15. Han, X.; Berthod, A.; Wang, C.; Huang, K.; Armstrong, D.W. Super/subcritical fluid chromatography separations with four synthetic polymeric chiral stationary phases. *Chromatographia*, **2007**, 65, 381–400.

8 Analysis of Genotoxic Impurities in Pharmaceuticals by Gas Chromatography–Mass Spectrometry

Jie Zheng and Adam P. Schellinger

CONTENTS

8.1 INTRODUCTION

The manufacturing process of pharmaceutical products, especially active pharmaceutical ingredients (APIs), often involves the use of reactive materials (e.g., starting materials, intermediates, catalysts, and reagents). These reactive materials along with certain reactive by-products resulting from the manufacturing process or introduced from the starting materials, intermediates, catalysts, and reagents could remain at trace levels in the final pharmaceutical products. Based on their structures and reactivity, some of these compounds have been classified as genotoxic impurities (GTIs).

Due to the concern for patient safety, the issue of GTIs in pharmaceutical products has attracted increasing attention from the industry[1–5] as well as regulatory agencies.[6–11] The European Medicines Evaluation Agency[9] and Food and Drug Administration[11] guidelines have led to an expectation that a control strategy is developed that demonstrates or justifies control of these undesired GTIs to an acceptable level (often low parts per million) in the final pharmaceutical products as appropriate for the stage of development. Analysis of GTIs in pharmaceutical products often poses significant challenges.[12–15] First, the control limits associated with GTIs (e.g., low parts per million) are significantly lower than anything previously measured (e.g., 0.05%) by up to three orders of magnitude. Even though the implementation of staged threshold of toxicological concern (TTC) guidelines provides some degree of relief for GTI control limits for pharmaceutical products dosed less than 1 year in clinical settings, extremely high sensitivity is still required for GTI analysis.[16] Second, complex matrix effects may arise from in-process samples, APIs, or excipients based on the typically high concentrations at which they need to be prepared. These matrix effects hinder the ability to observe and quantify genotoxic analytes and need to be overcome. Third, many GTIs are highly reactive and are thus prone to degradation during the analysis, which makes quantification very complicated. Fourth, the structure and physicochemical properties of certain GTIs are similar to those of the intermediate or API. Thus, highly specific analysis is often desired. Finally, these selective and sensitive methods must be amendable for testing of pharmaceutical products at all stages of development, including commercial manufacturing. The need to conduct these trace-level analyses after the drug is on the market further challenges the analytical chemist. These methods need to be rugged and robust so that they can be successfully transferred and executed at commercial manufacturing sites around the world.[17] Beyond method development, another challenge is successful implementation of the method at the testing site by ensuring that there are appropriate instrument resources (e.g., mass spectrometers) and experienced analysts capable of troubleshooting, instrument maintenance, and conducting routine testing.

To overcome these challenges for GTI analysis, many pharmaceutical companies have set up highly skilled analytical groups and developed various methodologies and practices based on gas chromatography–flame ionization detection (GC-FID),[18,19] gas chromatography–electron capture detection (GC-ECD),[20] gas chromatography–mass spectrometry (GC-MS),[21,22] high-performance liquid chromatography (HPLC)-UV,[23–25] HPLC-fluorescence,[26] HPLC-MS,[27–34] nuclear magnetic resonance,[35] and ion chromatography.[36] Initially, many of these methods were developed for individual GTIs and products or even specific manufacturing stages or steps.[37–39] As these GTI analytical methodologies and practices evolved, more and more generic analytical methodologies have emerged for the analysis of individual classes of GTIs based on their functional groups linked to genotoxicity.[22,28,34,40] Furthermore, very recently strategic approaches for GTI analysis have been established to accommodate fast-growing pipelines and compressed development time lines.[14,16,41]

Among all of the analytical methodologies that have been reported for GTI analysis, GC-MS is the most widely adopted one because many of the GTIs are volatile to semivolatile and low-molecular weight compounds.[14,41] GC-MS is well suited for GTI analysis because this technique provides chromatographic selectivity, sensitivity, and specificity for many different classes of GTIs. A typical capillary gas chromatography (GC) column can yield more than 100,000 theoretical plates,[42] and a wide range of stationary phases are available to obtain appropriate chromatographic selectivity and thus resolution of GTIs from other components. By operating under the selective ion monitoring (SIM) mode, electron ionization provides superior specificity and detection sensitivity in GC-MS.[17] For example, mass spectrometry (MS) detection could offer two to three orders of magnitude more sensitivity than flame ionization detection, the most commonly used GC detector.[4] Although electron capture detection (ECD) and MS detection provide comparable sensitivity for analytes with halogen or strong electron-withdrawing groups, the specificity of ECD is clearly limited to this subset of GTIs. Furthermore, the availability of robust and low-cost GC-MS systems from several major instrument suppliers has helped to reduce operational costs and simplify the process of method transfer to commercial testing sites or third party manufacturing laboratories.

This chapter aims to provide an overall review of GC-MS applications for the analysis of GTIs in pharmaceutical products. Also, this chapter provides a comprehensive review of various sample preparation approaches for the analysis of GTIs by GC-MS.

8.2 OVERVIEW FOR GAS CHROMATOGRAPHY–MASS SPECTROMETRY APPLICATIONS FOR GENOTOXIC IMPURITY ANALYSIS

GC-MS methods for determining GTIs in pharmaceutical products need to demonstrate sufficient specificity, linearity, sensitivity, precision, and accuracy for their intended use. Table 8.1 summarizes the GC-MS applications for GTI analysis reported in the literature. Generally, the success of these GC-MS applications depended on (1) sample preparation and introduction, (2) chromatography, and (3) detection and quantitation.[14]

Compared to HPLC-based applications (such as liquid chromatography–mass spectrometry [LC-MS]), sample preparation is often more critical for GC-MS as the analytes of interest need to be properly vaporized and introduced into the GC column prior to separation. Thus, it is important to choose appropriate sample preparation approaches based on the physicochemical properties of target GTI analytes and sample matrixes. The direct injection approach (or the so-called dilute and shoot approach) is widely used since it requires the least effort for sample handling and manipulation. However, the major drawback of direct injection is the potential for matrix interference. Since most direct injectors only deliver microliter quantities of solutions, the sample needs to be diluted and injected at a high concentration (e.g., 5–50 mg/mL) to achieve the desired sensitivity. As most pharmaceutical samples (intermediates, APIs, or drug products) are nonvolatile, direct injection of these materials often leads to a buildup of residue that can undergo thermal degradation, contaminate the inlet liner, and lead to interferences in the chromatography. The application of direct injection is limited for isolated intermediates and APIs. Direct injection is not applicable for analyzing drug product samples (e.g., formulated solution, tablet, or suspension) because excipients in the drug product may lead to severe interferences when directly injected. In addition, direct injection is typically not applicable when the sample cannot achieve sufficient solubility based on the required GTI control limit. Instead, one could consider headspace injection and other approaches such as liquid–liquid extraction (LLE), solid-phase extraction (SPE), and solid-phase microextraction (SPME) because these methodologies are very effective in reducing matrix interferences from the sample and enriching the target GTIs. When highly reactive GTIs are being analyzed, selective derivatization might improve the stability and volatility of the GTIs.

TABLE 8.1
GC-MS Applications for GTI Analysis

Author	GTI	Sample Preparation	GC Phase	Quantitation for GC-MS
		Dilute and Shoot		
Ramjit et al.[21]	Methyl mesylate and ethyl mesylate	100 mg of free base or bismesylate salt of DPI 201-106 was dissolved in a mixture of 1 mL of acetonitrile and 0.2 mL DMF	DB-WAX	SIM mode using *m/z* 79 for MMS, EMS, and IS PMS
Raman et al.[43]	CDP and DMS	PPS sample was dissolved in H_2O/methanol 3:1 (v/v) and directly injected into GC	DB-624	SIM mode using *m/z* 187 for CDP and *m/z* 126 for DMS
Sarat et al.[45]	MMS and EMS	API samples were dissolved in methanol/chloroform at 80:20 (v/v) at 600 mg/mL	DB-624	SIM mode using *m/z* 79 for MMS and EMS
Li and Sluggett[46]	CCMTHP	25 mg of β-lactam API sample was dissolved in 5 mL acetonitrile	DB-5ms	SIM mode *m/z* 49
Ramakrishna et al.[47]	MMS and EMS	100 mg of pure INM solid (or tablet powder equivalent to 100 mg of INM) was dissolved in 10 mL with *n*-hexane	DB-1	Scan mode with *m/z* range of 35–130 amu
David et al.[48]	(1) Cinnamonitrile and 3-ethoxy-2-cyclohexenone (2) 2-bromoethanol, 2-iodoethanol, 4-chloro-1-butanol, 2-(2-chloroethoxy)ethanol and 11-bromo-1-undecanol)	(1) Dissolve carbamazepine in CH_2Cl_2 at 50 mg/mL and directly inject into GC (2) Mix 100 μL of 50 mg/mL carbamazepine in pyridine with 100 μL BSTFA. Heat at 70°C for 0 minute. Add 500 μL CH_2Cl_2 and 500 μL water. Inject CH_2Cl_2 layer into GC	HP-5ms for first-dimension separation; DB-17ms for second-dimension separation	SIM mode
		Extraction		
Zheng et al.[49]	DMS	Dissolve 750 mg of a water-soluble API intermediate with 30 mL of 0.1 M NaCl. Add 3 mL of MTBE spiked with d_6-DMS as IS for extraction	DB-624	SIM mode using *m/z* 95 for DMS and *m/z* 100 for IS d_6-DMS

Nassar et al.[50]	EMS	BMS-214662 mesylate API and its parenteral formulation samples were first spiked with IS IMS and then extracted with chloroform	DB-WAX	SIM mode using *m/z* 79 for EMS and *m/z* 123 for IS IMS
Wollein and Schramek[51]	MMS, EMS, IMS, MBS, and EBS	Powder or solid dosage forms of the finished drug product containing 25 mg of API were extracted with *n*-hexane spiked with MTS as an IS	Rxi 5-Sil MS	SIM mode with MTS as IS
Colon and Richoll[52]	Methyl and ethyl esters of methanesulfonic, benzenesulfonic, and *p*-toluenesulfonic acids	Mesylate, besylate, or tosylate salts of API samples were dissolved in 10 mL of 20 mM phosphate buffer and extracted by SPME	DB-1701	SIM mode
Garcia et al.[24]	2-CE	Dissolve cloperastine fendizoate sample in methanol (pH 9 by NH_3) and load to SAX-SPE cartridge. Analyze eluent by GC-MS	VF-23ms	SIM mode using *m/z* 80
Martano et al.[54]	β-IO, CC, DHA, and TMT	Extract cell culture media samples with a Strata Phenyl SPE column and then elute with 10:90 (v/v) tetrahydrofuran/*n*-hexane. Analyze SPE eluent by GC-MS	DB-20 (WAX)	SIM mode using linalool and methylisoeugenol as ISs
		Headspace		
Strohmeyer and Sluggett[55]	TEMPO	Dissolve filibuvir sample in acetonitrile/water 1:1 (v/v) at 100 mg/mL; analyze SPE eluent by headspace GC-MS	VF-624ms	SIM mode using *m/z* 156 ($M^{+\bullet}$) and *m/z* 142 (M^{+}–CH_2•)
Teasdale et al.[56]	MMS	Headspace vial containing 10 µL of MSA and 100 µL of methanol was heated at 78°C for 2 hours and equilibrated at 105°C for 15 minutes	DB-VRX	Scan mode (scan range: 10–300 *m/z*).

(Continued)

TABLE 8.1 (*Continued*)
GC-MS Applications for GTI Analysis

Author	GTI	Sample Preparation	GC Phase	Quantitation for GC-MS
		Derivatization		
Lee et al.[40]	Methyl, ethyl, and isopropyl methanesulfonate esters and dimethyl sulfonate	Dissolve API sample as methanesulfonate salt in H_2O at 100 mg/mL; add sodium thiocyanate immediately and derivatize at 85°C	SGE 50QC3/BPX5	SIM mode for both alkylthiocyanates and alkylisothiocyanates
Alzaga et al.[22]	Alkyl mesylates, besylates, tosylates, and sulfates	Dissolve APIs or intermediate samples in DMSO/water (1:1, v/v). Add deuterated ISs and pentafluorothiophenol; then conduct derivatization at 105°C for 15 minutes	RH-624	SIM mode using molecular ions of derivatives from pentafluorothiophenol; use deuterated ISs
Jacq et al.[57]	EMS	EMS formed from ethanol and MSA is analyzed after derivatization with pentafluorothiophenol using headspace GC-MS	RH-624	SIM mode using *m/z* 228 for EMS derivative and *m/z* 233 for IS deuterated EMS derivative
Sun et al.[59]	Hydrazine	Dissolve 0.5 g of benzoic acid and 0.5 mL of acetone or acetone-d_6 in 10 mL of NMP as derivatization reagent. Mix 10 μg of API sample with 100 μL of derivatization reagent and 10 μL of 0.1% EDTA in a 10 mL headspace vial. Heat at 100°C for 10 minutes for derivatization	DB-624	SIM mode (*m/z* 112 for acetone derivative, and *m/z* 124 and *m/z* 106 for acetone-d_6 derivative)

Chromatography is used to resolve GTIs from diluent and other matrix peaks. Typically, low-bleeding, midpolarity to low-polarity phase capillary columns, such as DB-5 and DB-624, are utilized for most of the GC-MS applications. These phases have high inertness, which is beneficial for analyzing reactive GTIs. The high efficiency of these columns also improves detection sensitivity.

GC-MS detection is generally performed on a single-quad mass spectrometer with an electron impact source acquiring signal in the SIM mode. It is essential to investigate the GTI MS fragment profile and carefully select appropriate diagnostic ions. The selection of these ions is critical for achieving the desired specificity and sensitivity. For less complicated sample preparation (e.g., direct injection), the external standard method typically provides successful quantitation. However, when extensive sample preparations are involved (e.g., extraction or derivatization), it is recommended to introduce an internal standard (IS). Isotopic analogs of the desired GTI are generally preferred for ISs as they provide similar recovery and chromatographic behavior.

Sections 8.2.1 through 8.2.4 further discuss the individual GC-MS applications in Table 8.1 for the analysis of GTIs. The sections are organized according to the sample preparation approach chosen.

8.2.1 Dilute and Shoot

The direct injection approach is the most convenient injection mode for GTI analysis since it requires minimal sample handling. This approach requires GTI analytes to have sufficient volatility and stability on injection and vaporization at a hot inlet. As a large amount of the generally less volatile API is introduced into the GC inlet, the GC inlet may require frequent maintenance to address potential contamination issues.

Ramjit and others[21] developed a GC-MS method to determine methyl mesylate (MMS) and ethyl mesylate (EMS) in the free base and the bismesylate salt of DPI 201-106, a positive inotropic agent used for treating heart failure. This report is one of the pioneering studies investigating residual levels of alkyl mesylates in intermediates and APIs. A 100 mg sample was dissolved in 1 mL of acetonitrile with 0.2 wmL *N*,*N*-dimethylformamide (DMF) to improve solubility. A DB-WAX column (30 m × 0.25 mm internal diameter [i.d.], 0.25 μm film thickness) was used for separation. Analyses were performed using a splitless injection mode and an injection port temperature of 200°C. Quantitative GC-MS was performed in SIM mode using *m*/*z* 79 as the common fragment ion for both MMS and EMS, whereas *n*-propyl mesylate (PMS) was introduced as the IS. The authors found 0.51 and 1.31 ppm for MMS and EMS, respectively, in the bismesylate salt and did not detect either of these alkylating agents in the free base sample.

Raman and others[43] reported a GC-MS method for determining two process impurities, 2-(chloromethyl)-3,4-dimethoxypyridine hydrochloride (CDP) and dimethyl sulfate (DMS), in pantoprazole sodium (PPS). The control limit for CDP and DMS was calculated at 37.5 ppm based on the TTC and 40 mg daily dose of PPS. The PPS sample was dissolved in H_2O/methanol 3:1 (v/v) and was directly injected onto a DB-624 column (60 m × 0.32 mm, 1.8 μm film thickness). Quantitation was conducted under the

SIM mode, which is mass selective and thus highly selective. The practical quantitation limit (PQL) for the method was 3 ppm for each analyte, and the method gave a linear response for CDP and DMS concentrations from 3 to 45 ppm with acceptable recovery. The same group also reported another GC-MS method for determining two GTIs, methyl camphorsulfonate and ethyl camphorsulfonate, in esomeprazole magnesium (EOM).[44] The EOM sample was prepared at a concentration of 50 mg/mL in 1,3-dimethyl imidazolidin-2-one. The separation was conducted on a DB-5 column (30 m long × 0.32 mm i.d., 1.0 μm film thickness). The quantitation was carried out under SIM mode. The limit of detection (LOD) of method was found as 3 ppm.

Sarat and others[45] developed and validated a GC-MS method for the determination of methyl methanesulfonate (MMS) and ethyl methanesulfonate (EMS) at parts-per-million levels in undisclosed APIs. The API samples were dissolved in methanol/chloroform at 80:20 (v/v) at 600 mg/mL and directly injected onto a DB-624 capillary column (30 m × 0.53 mm, 3 μm film thickness) under splitless mode. The SIM mode was used to quantify both the methanesulfonate esters with the common *m/z* 79 fragment ion. The PQL for the method was 0.52 ppm for methyl methanesulfonate and 0.54 ppm for ethyl methanesulfonate. A linear response from the PQL value to 1.5 ppm was observed along with good recovery for the spiked API samples. The authors discussed the criteria for choosing appropriate organic solvents, including purity, sufficient solubility for the samples (to achieve the desired sample concentration), compatibility with GTIs, and no interferences in chromatography. As many pharmaceuticals exist in salt forms, mixed solvents provide better solubility than pure organic solvents. Also, the authors demonstrated acceptable precision without the use of an IS.

Li and Sluggett[46] reported a direct injection GC-MS method for determining the trace level of carbonic acid chloromethyl tetrahydro-pyran-4-yl ester (CCMTHP) in a β-lactam API. The authors evaluated several analytical techniques including LC-MS, GC-FID, GC-ECD, and GC-MS during their method development and found that GC-MS provided the best detection sensitivity. About 25 mg of the API sample was dissolved in 5 mL of acetonitrile. The separation was conducted on a DB-5ms column (30 m long × 0.25 mm i.d., with 1.0 μm film thickness). The authors evaluated the impact of splitless versus split injection, API matrix effects, liner type, and injection temperature. The β-lactam API undergoes thermal degradation, and one of the resulting degradants, tetrahydropyranol (THP), interferes the analysis for CCMTHP. Li applied a 10:1 split ratio to limit the amount of nonvolatile species introduced to the GC column but maintain acceptable sensitivity. To reduce background noise, quantitation was conducted in the SIM mode using *m/z* 49, which is not observed from the THP degradant (Figure 8.1). To minimize carryover and contamination, Li set the inlet temperature to 250°C. The method showed a linear response from 10 to 1000 ppm with a limit of quantitation (LOQ) at 10 ppm. The spiked recovery at 10 ppm was 82%.

Ramakrishna and others[47] developed a GC-MS method for the identification and determination of two GTIs (MMS and EMS) in imatinib mesylate (INM). A total of 100 mg of pure INM solid (or tablet powder equivalent to 100 mg of INM) was weighed in a 10 mL volumetric flask and dissolved with *n*-hexane. The separation of MMS and EMS was obtained on a DB-1 capillary column (30 m × 0.25 mm i.d., 0.25 μm film thickness). An injection volume of 2 μL was found to provide acceptable sensitivity and peak shape. Data were collected under the scan mode with an *m/z*

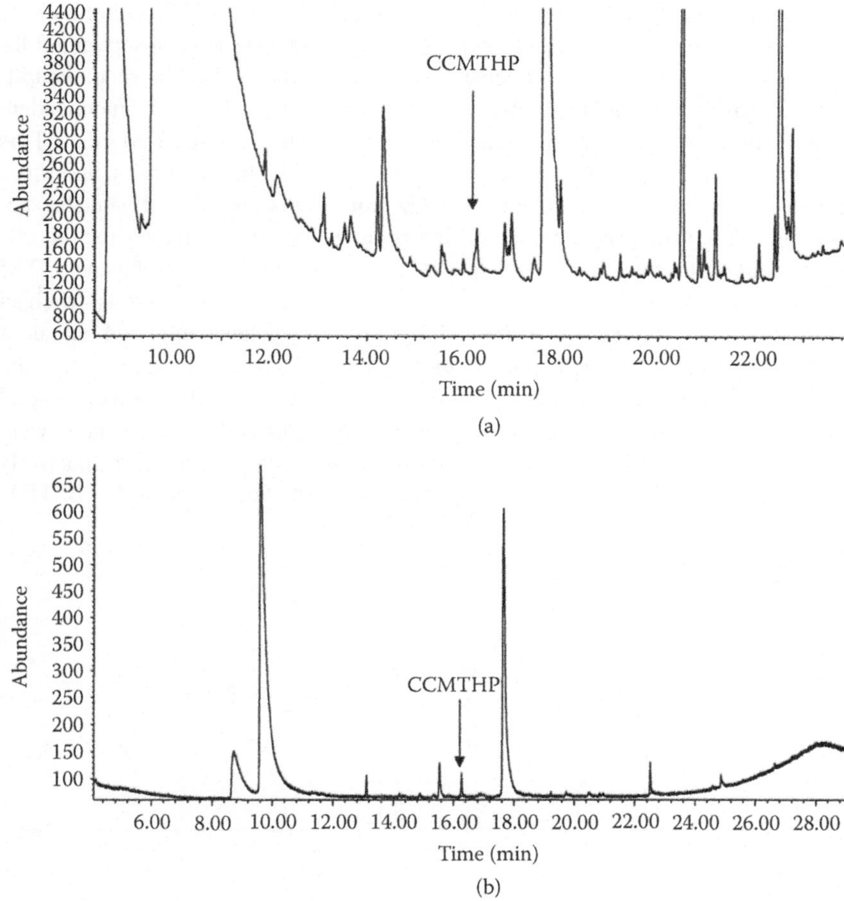

FIGURE 8.1 (a) Gas chromatography–mass spectrometry (GC-MS) detection of active pharmaceutical ingredient (API) sample with single ion monitoring (SIM) at $m/z = 49, 55, 69$, and 84; (b) GC-MS detection of API sample with SIM at $m/z = 49$. (From Li and Sluggett, *J. Pharm. Biomed. Anal.*, 39, 486–494, 2005.)

range of 35–130 amu. The authors validated the method per International Conference on Harmonization of Technical Requirements for Registration of Pharmaceuticals for Human Use guidelines. For both the mesylate esters, the LOD and LOQ values were found to be 0.3 and 1.0 μg/mL, respectively. Further, the response was linear within the range of 1–15 μg/mL for both compounds. The method specificity was tested by performing recovery studies of MMS and EMS at 1, 10, and 15 μg/mL with three different batches of INM tablets. The spiked recovery was between 98% and 100% with no observed interference from the excipients.

The direct injection technique requires the introduction of concentrated sample solutions to the GC inlet, which often leads to contamination. David and others[48] reported a two-dimensional capillary GC method using Deans switching to address the contamination issue. The authors chose an HP-5ms column (30 m × 0.25 mm i.d., with 0.25 μm film thickness) for the first-dimension separation followed by a DB-17ms column (30 m × 0.25 mm i.d., with 0.25 mm film thickness) as the second-dimension separation for heart-cut fractions from the first dimension that contain the GTIs. The second capillary column was installed in a low-thermal-mass oven, which allowed for independent temperature-programmed analysis. Quantitation was conducted under SIM mode. This two dimensional GC (2D-GC) setup significantly reduced the amount of undesired materials (such as APIs, solvents, and derivatization agents) introduced onto the second column or into the MS detector. In this fashion, the contamination for the GC column/system was reduced, which helped to improve the method's robustness. The authors used the 2D-GC-MS system to analyze two classes of GTIs, that is, Michael reactive acceptor GTIs (cinnamonitrile and 3-ethoxy-2-cyclohexenone) and haloalcohols [2-bromoethanol, 2-iodoethanol, 4-chloro-1-butanol, 2-(2-chloroethoxy)ethanol, and 11-bromo-1-undecanol]. Using the direct injection 2D-GC-MS approach, good resolution and high sensitivity for cinnamonitrile and 3-ethoxy-2-cyclohexenone were achieved with a LOD of 0.01 and 0.05 ppm (microgram per gram API), respectively (Figure 8.2). Derivatization with N,O-bis(trimethylsilyl)trifluoroacetamide (BSTFA)

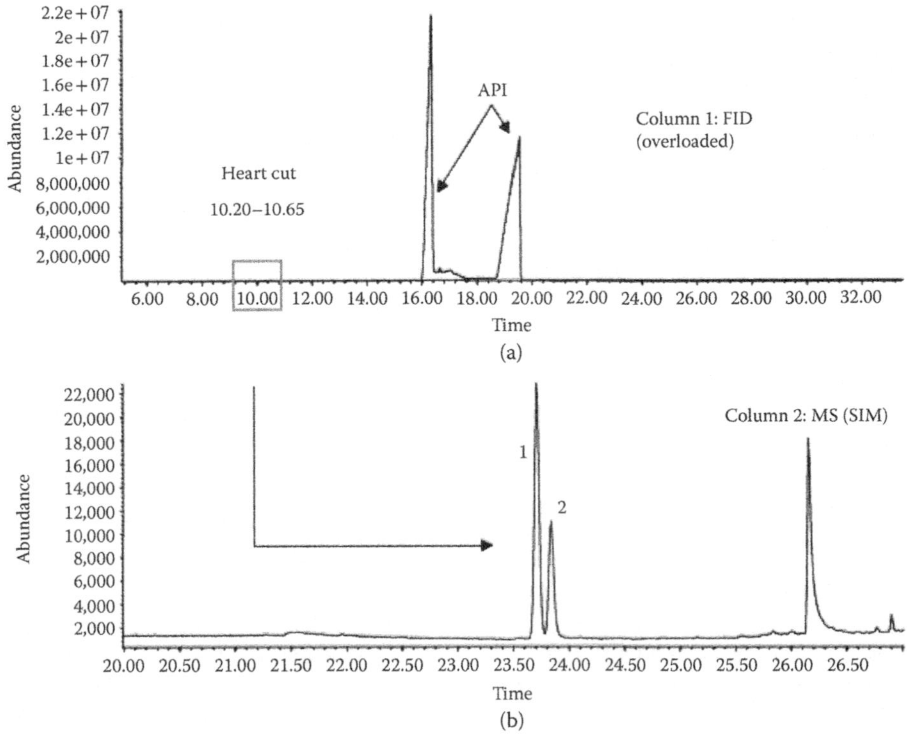

FIGURE 8.2 Analysis of Michael reactive acceptors in carbamazepine at 1 ppm level. (a) Monitor flame ionization detection signal of the first-dimension separation. (b) Separation of the heart-cut fraction on the second-dimension column (DB-17) using mass spectrometry (single ion monitoring) detection and independent oven temperatures. Peak identification: (1) cinnamonitrile and (2) 3-ethoxy-2-cyclohexenone. (From David et al., *Anal. Bioanal. Chem.*, 396, 1291–1300, 2010.)

helped to overcome the low volatility of the haloalcohols. Good linearity and sensitivity were observed with LODs all lower than 0.5 ppm.

8.2.2 Extraction

Although direct injection is very convenient, various extraction techniques, including LLE, SPE, SPME, or liquid-phase microextraction (LPME), are often considered, especially when sample solubility is limited or when there are severe interferences from the sample matrix. Although certain extraction techniques are labor intensive and often raise concern for recovery, they can also be effective in enriching GTI analytes and reducing the matrix effects.

DMS is an alkylating reagent commonly used in organic syntheses and pharmaceutical manufacturing processes. However, in-process testing for DMS is challenging because of its reactivity and complex matrix effects. Zheng and others[49] reported a GC-MS method for determining DMS in a water-soluble API intermediate, which was a methylsulfate salt. DMS and an IS, d_6-DMS, were extracted from the matrix with methyl *tert*-butyl ether (MTBE). GC separation was conducted on a DB-624 column (30 m long × 0.32 mm i.d., 1.8 μm film thickness). MS detection was performed on a single-quad Agilent mass spectrophotometric detector (MSD) equipped with an electron impact source while the MSD signal was acquired in SIM mode. This GC-MS method showed a linear response for DMS from 1.0 to 60 ppm. The practical quantitation limit for DMS was 1.0 ppm, and the practical detection limit was 0.3 ppm. The relative standard derivation for the analyte response was 0.1% for six injections of a working standard equivalent to 18.6 ppm of DMS. The spike recovery ranged from 102.1% to 108.5% for a sample of API intermediate spiked with 8 ppm of DMS. The DMS level was 5.9 ppm, which was much lower than the targeted control limit in the isolated intermediate. Also, it suggests that the quenching process and subsequent work-up are efficient for removing the unreacted DMS.

During method development, Zheng and others[49] screened several solvents for LLE. MTBE was selected because it provided a clean GC-MS background and a higher recovery for DMS. To compensate for the loss of DMS during the LLE procedure and injection, the authors investigated the use of three structurally similar analogs including two isotopic analogs, $[^{13}C_2]$-DMS and d_6-DMS, as ISs (Figure 8.3). The two isotopic analogs, $[^{13}C_2]$-DMS and d_6-DMS, showed similar electron ionization–mass spectrometry (EI-MS) spectra, recovery, and chromatographic retention as DMS, which made them more suitable for use as the IS. However, since $[^{13}C_2]$-DMS and d_6-DMS coeluted with DMS, the SIM ions for quantitation must be chosen carefully to avoid those overlapping ions in the EI-MS spectra of the analyte and IS. The quantitation limit for DMS when using $[d_6]$-DMS as an IS is five times better than using $[^{13}C_2]$-DMS as the IS. This is because of the relative abundance of the quantitative ions chosen for each IS. In the DMS EI-MS spectrum, the *m/z* 95 ion (SIM ion for d_6-DMS) is much more abundant than *m/z* 125 (SIM ion for $[^{13}C_2]$-DMS) (Figure 8.3a).

Nassar and others[50] reported a GC-MS method to investigate the formation and rate of hydrolysis of EMS in the BMS-214662 mesylate drug substance and its parenteral formulation. The BMS-214662 mesylate drug substance was manufactured, and one step involved crystallization in ethanol. Also, the parenteral formulation contains

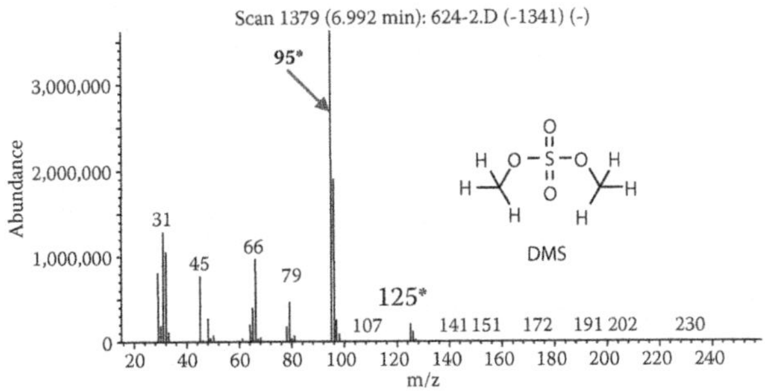

m/z (Rel. abundance)	Rationalization
126 (3%)	$M^{+\cdot}$
125 (6%)	$[M-H]^+$
96 (51%)	$[M-CH_2O]^+$
95 (100%)	$[M-CH_3O]^+$
79 (12%)	$[CH_3SO_2]^+$
66 (25%)	$[H_2SO_2]^+$
45 (21%)	$[CH_3OCH_2]^+$

(a)

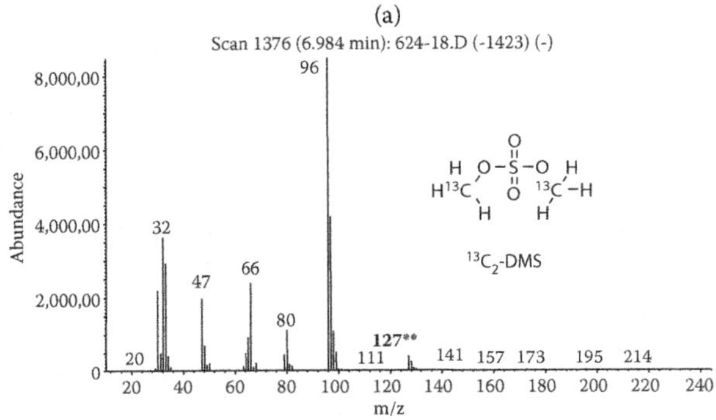

m/z (Rel. abundance)	Rationalization
128 (3%)	M^{+}.
127 (6%)	$[M-H]^+$
97 (52%)	$[M-{}^{13}CH_2O]^+$
96 (100%)	$[M-{}^{13}CH_3O]^+$
80 (14%)	$[{}^{13}CH_3SO_2]^+$
66 (27%)	$[H_2SO_2]^+$
47 (24%)	$[{}^{13}CH_3O^{13}CH_2]^+$

(b)

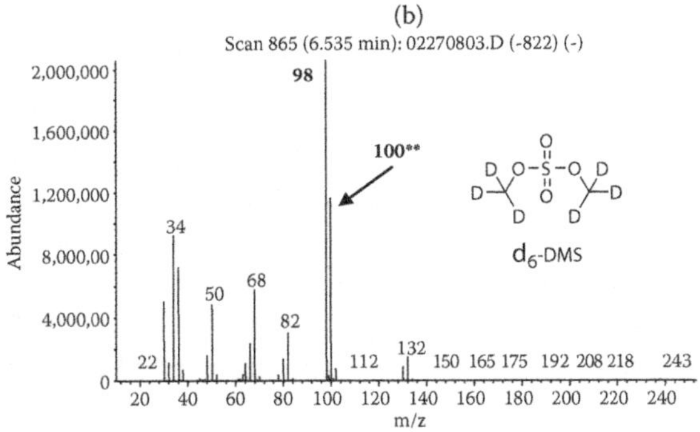

m/z (Rel. abundance)	Rationalization
132 (7%)	$M^{+\cdot}$
130 (4%)	$[M-D]^+$
100 (57%)	$[M-CD_2O]^+$
98 (100%)	$[M-CD_3O]^+$
82 (15%)	$[CD_3SO_2]^+$
68 (28%)	$[D_2SO_2]^+$
50 (23%)	$[CD_3OCD_2]^+$

(c)

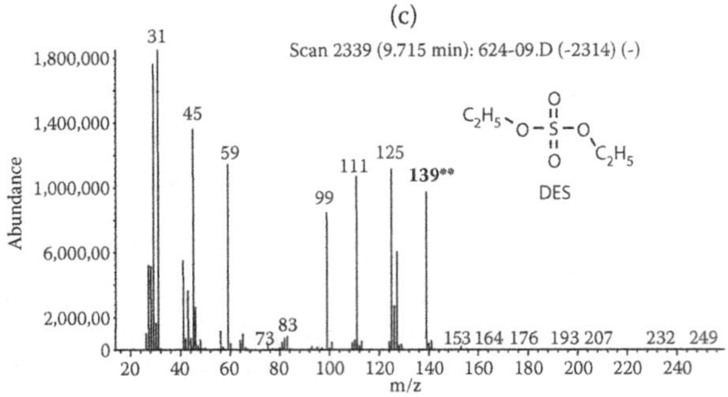

m/z (Rel. abundance)	Rationalization
153 (1%)	$[M-H]^+$
139 (51%)	$[M-CH_3\cdot]^+$
127 (31%)	$[M-H-C_2H_2]^+$
125 (60%)	$[M-H-C_2H_4]^+$
111 (56%)	$[M-H-C_3H_6]^+$
99 (45%)	$[M-H-C_4H_6]^+$
59 (60%)	$[C_3H_7O]^+$
45 (72%)	$[C_2H_5O]^+$
31(100%)	$[CH_3O]^+$

(d)

FIGURE 8.3 Electron ionization–mass spectrometry spectra for dimethyl sulfate (DMS), [$^{13}C_2$]-DMS, d_6-DMS, and DES. Some of the major ions, relative abundance, and rationalization are tabulated as insets. (*Diagnostic ion for DMS. **Diagnostic ion for internal standard. From Zheng et al., *J. Pharm. Biomed. Anal.*, 50, 1054–1059, 2009.)

300 mg/mL of ethanol in a pH 4 buffer. The concern is that a reaction between residual methane sulfonic acid (MSA) and ethanol would form EMS. In this study, the formulation samples were first spiked with isopropyl mesylate (IMS) as an IS and were then extracted with chloroform. After centrifugation, the chloroform layer was injected into a DB-WAX column (30 m long × 0.25 mm i.d., with 0.25 μm film thickness) for analysis. Quantitation was conducted under SIM mode using *m/z* 79 for EMS and *m/z* 123 for IMS (IS). The LOQ was 50 ppb (vs. BMS-214662 mesylate API). The authors detected EMS in five batches of BMS-214662 mesylate API ranging from 0.2 to 0.8 ppm. EMS levels in the parenteral formulation showed no significant increase after storage at 25°C for 18 weeks or at 60°C for 6 weeks. The authors concluded that EMS formation between ethanol and MSA may not occur in the BMS-214662 formulation under the storage conditions.

Wollein and Schramek[51] reported a direct injection GC-MS method for the simultaneous determination of MMS, EMS, IMS, methyl besylate (MBS), and ethyl besylate (EBS) in finished drug products. Generally, fine powder or solid dosage forms of the finished drug product containing around 25 mg of API were extracted with *n*-hexane spiked with methyl tosylate (MTS) as an IS. Then, 1 μL of supernatant was immediately analyzed by GC-MS with a Restek Rxi 5-Sil MS (30 m × 0.32 mm, 0.5 μm film thickness) capillary column. The SIM parameters were set at *m/z* 65, 80, and 95 for MMS; *m/z* 79, 97, and 109 for EMS; *m/z* 43, 79, and 123 for IMS; *m/z* 77, 141, and 172 for MBS; *m/z* 51, 77, and 141 for EBS; and *m/z* 91, 155, and 186 for MTS, although only one ion was used for quantitation. Good sensitivity and linearity (R^2 value ≥ 0.9998) were observed for concentrations between 0.01 and 1.33 μg/mL. Spike recoveries for MMS, EMS, and IMS were greater than 71%. Spike recoveries for MBS and EBS were greater than 94%. MMS, EMS, and IMS were not detected in two lots of bromocriptine mesylate and two lots of doxazosin mesylate tested. The authors suggested the use of *n*-hexane as the LLE solvent due to poor solubility of the API. The choice of *n*-hexane helped to maintain GC-MS detection sensitivity by avoiding matrix interferences. Additionally, *n*-hexane provides better stability for these alkyl sulfonates as solvents with higher polarity or nucleophilic properties were known to degrade the analytes of interest. The disadvantage of using *n*-hexane was that liquid dosage forms could not be analyzed.

Colon and Richoll[52] reported an SPME GC-MS technique to determine methyl and ethyl esters of methanesulfonic, benzenesulfonic, and *p*-toluenesulfonic acids in APIs as a limit test (not more than 5 ppm). Typically, the APIs (which were mesylate, besylate, or tosylate salts) were dissolved in a 4.7 pH phosphate buffer and were then extracted using a PDMS/DVB fiber installed on an SPME device under constant agitation and at room temperature. All samples were analyzed using GC-MS with an SPME injection insert, and the inlet held at 230°C operated in the pulsed splitless mode. The analytes were resolved with a DB-1701 column (30 m × 0.25 mm, 1.0 μm film thickness) and quantified with SIM detection (Figure 8.4). The authors optimized the SPME fiber coating, extraction time, and buffer pH. The validated limit test method provided acceptable reproducibility (relative standard deviation < 6% for five replicate sample preparations), linearity with or without API present (R^2 value > 0.9), and spike recovery (within ±10% spike level). The authors demonstrated successful application of the method in the analysis of four batches of APIs with different structures. Finally, the authors suggested that SPE and LPME could provide viable alternatives to the SPME methodology in cases where the API has limited solubility in aqueous media.

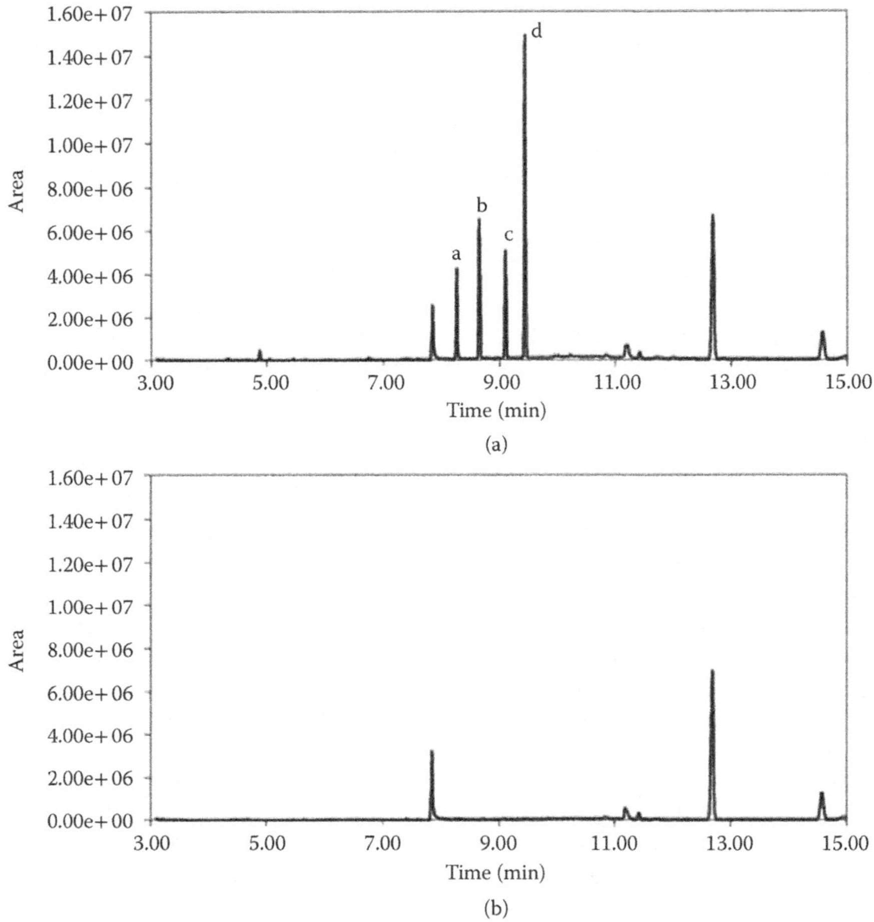

FIGURE 8.4 Sample chromatograms of compound 4 extracted with the optimized solid-phase microextraction procedure. (a) Compound 4 was spiked with 5 ppm (relative to the active pharmaceutical ingredient) of a: methyl besylate, b: ethyl besylate, c: Mp-TS, and d: Ep-TS. (b) Compound 4 without spiking. (Note: Compound 4 is besylate salt of an unspecified active pharmaceutical ingredient, methyl p-toluenesulfonate (Mp-TS) and ethyl p-toluenesulfonate (Ep-TS). From Colon and Richoll, *J. Pharm. Biomed. Anal.*, 39, 477–485, 2005.)

Very recently, Ho and others[53] explored the feasibility of using polymeric ionic liquids (PILs) as a new type of SPME sorbent coatings for the analysis of alkyl halides, aromatics halides, epoxides, and aromatic amines. The separation of all analytes was conducted on an HP-5ms capillary column (30 m × 0.32 mm i.d., 0.25 μm film thickness). A glucaminium-based PIL, *N,N*-didecyl-*N*-methyl-D-glucaminium poly(2-methyl-acrylic acid 2-[1-(3-propylamino)-vinylamino]-ethyl ester), showed good sensitivity for long-chain aliphatic alkyl halides. A poly(1-vinyl-3-propylphenylimidazolium) chloride PIL coating showed good selectivity for larger aliphatic/aromatic analytes. Excellent recovery was observed for both PIL-based coatings with limits of detection ranging from low parts-per-billion to mid parts-per-trillion levels.

Garcia and others[24] reported a GC-MS method for the determination of 2-chloroethanol (2-CE), a well-known alkylating agent, in cloperastine fendizoate. The cloperastine fendizoate sample was dissolved in methanol (basified with ammonia, pH = 9.0) and loaded onto a strong anion exchange (SAX)-SPE cartridge to remove the fendizoate. The eluent was subjected to GC-MS analysis using a VF-23ms capillary column (30 m × 0.25 mm, 0.25 μm film thickness). Quantitation of 2-CE was conducted under SIM mode for *m/z* 80 (the molecular ion of 2-CE). The most abundant fragment ion *m/z* 49 was not selected for quantitation due to matrix interferences. The method gave good specificity and a 1.7 ppm LOD. This GC-MS method was successfully applied for the determination of 2-CE in five different batches of cloperastine fendizoate.

Martano and others[54] developed and validated a GC-MS method for the simultaneous determination of four prominent volatile cleavage products (CPs) of β-carotene, a precursor of vitamin A, in cell culture media. Typically, these CPs, that is, β-ionone (β-IO), cyclocitral (CC), dihydroactinidiolide (DHA), and 1,1,6-trimethyltetraline (TMT), were extracted with a Strata Phenyl SPE column and then eluted with 10:90 (v/v) tetrahydrofuran/*n*-hexane. The SPE eluent was subjected to GC-MS analysis using a DB-20 (WAX) column (30 m × 0.25 mm, 0.50 μm film thickness). The quantitation was conducted under SIM mode using linalool and methylisoeugenol as ISs. The method was linear with a LOD of 53.0 ng/mL. The method also gave reasonable recoveries (71.7%–95.7% at 1.0 μg/mL) and precision (<2.0%, intraday, $n = 5$; <4.8%, interday, $n = 15$).

8.2.3 Headspace

Headspace GC is a common technique in pharmaceutical analysis that has been widely utilized to determine residual organic solvents remaining in pharmaceuticals. As many GTIs are volatile species and the APIs are generally less volatile, headspace GC-MS offers certain advantages over direct injection. Headspace injection often leads to reduced sample contamination as the nonvolatile API sample matrix remains in the vial. Furthermore, a large amount of sample can be placed in the vial, which allows for higher sensitivity without the need for higher sample solubility as in direct injection.

The synthesis of filibuvir, a hepatitis C virus polymerase inhibitor candidate, involves the use of 2,2,6,6-tetramethyl-1-piperidinyloxy (TEMPO) in the second to last step. TEMPO in filibuvir needs to be controlled to 4 or 8 ppm depending on the clinical study duration per the staged TTC. Strohmeyer and Sluggett[55] developed and validated a headspace GC-MS method for detecting low levels of TEMPO in filibuvir. Headspace injection was chosen for analysis because TEMPO is volatile and headspace injection would avoid the contamination of injector and column with the less volatile API. The filibuvir sample was dissolved in acetonitrile/water 1:1 (v/v) at a concentration of 100 mg/mL. The headspace vial was equilibrated at 60°C for 15 minutes. Separation was performed on a VF-624ms capillary column (30 m × 0.32 mm i.d., 1.8 μm film thickness) or a DB-35ms column (15 m × 0.25 mm i.d., 0.25 μm film thickness). Quantitation was conducted in SIM mode using *m/z* 156 ($M^{+}\bullet$) and *m/z* 142 ($M^{+}-CH_2\bullet$). This "fit-for-purpose" method demonstrated good specificity for the detection of TEMPO with filibuvir present and exhibited acceptable linearity over a range of 0.4–6.0 μg/mL (equivalent to 4–60 ppm vs. filibuvir) with LOQ and LOD values at 4 and 2 ppm, respectively. The method also showed appropriate repeatability and spike recovery. A total of 13 batches of filibuvir were subjected to analysis and TEMPO was found to be undetectable (≤2 ppm). By conducting a purge study, the authors demonstrated that the API synthetic process can reject TEMPO well below the desired levels.

Teasdale and others[56] investigated the mechanism for the formation of methyl mesylate from methanol and MSA in an API process. The reaction mixtures were analyzed using a static headspace GC-MS method. A 20 mL headspace vial containing 10 μL of MSA and 100 μL of methanol was heated at 78°C for 2 hours and then equilibrated at 105°C for 15 minutes. A total of 1 mL of headspace gas was injected with 10:1 split ratio. Separations were conducted on a DB-VRX column (60 m × 0.25 mm, 1.4 μm film thickness). Detection was performed in scan mode (scan range: 10–300 *m/z*). The authors were able to gain valuable insight into the reaction profile for these GTIs.

8.2.4 Derivatization

Since many GTIs are highly reactive, they are prone to degradation during analysis. However, these reactive GTIs can be derivatized to provide a stable compound more amendable to analysis. The specificity of GTI analysis may even be enhanced by choosing derivatization reagents that provide a selective reaction with the target GTIs. Furthermore, the resulting derivatized compounds are often less polar, more volatile, and readily analyzed by GC-MS.

Lee and others[40] reported a derivatization headspace GC-MS method to analyze alkylating agents, including methyl, ethyl, and isopropyl methanesulfonate esters and dimethyl sulfonate. In this study, the basic drug substance sample (a methanesulfonate salt crystallized from methanol and ethanol) was dissolved in water at 100 mg/mL and then the derivatization reagent sodium thiocyanate was added immediately prior to sealing the headspace vial. Derivatization was carried out at 85°C, which gave a mixture of alkylthiocyanates and alkylisothiocyanates. These alkylthiocyanate and alkylisothiocyanate derivatives are volatile and readily analyzed by headspace GC (Figure 8.5a). The separation was conducted on a SGE 50QC3/BPX5 0.5 column (50 m long × 0.32 mm i.d., with a film thickness of 0.5 μm). Quantitation of alkylthiocyanates and alkylisothiocyanates was conducted under SIM mode using their molecular ions. The authors found that derivatization with 10% sodium thiocyanate and heating at 85°C for 20 minutes provided the optimum conversion, that is, 100% for methyl methanesulfonate and 74% for ethyl methanesulfonate, respectively. The authors also found that the interference from residual alcohols at less than 1% was almost negligible. The method showed linear responses for concentrations from 0.02 to 0.5 μg/mL and used an IS. Detection limits were found to be 0.02 μg/mL for both methyl and ethyl esters, and 0.05 μg/mL for the isopropyl esters. The spike recoveries for the methyl and ethyl mesylate esters were 108% and 94%, respectively. Certain limitations for this method were also noted. First, the reaction of alkyl mesylates with thiocyanate is complicated due to isomerization (Figure 8.6). Second, the use of sodium thiocyanate for derivatization

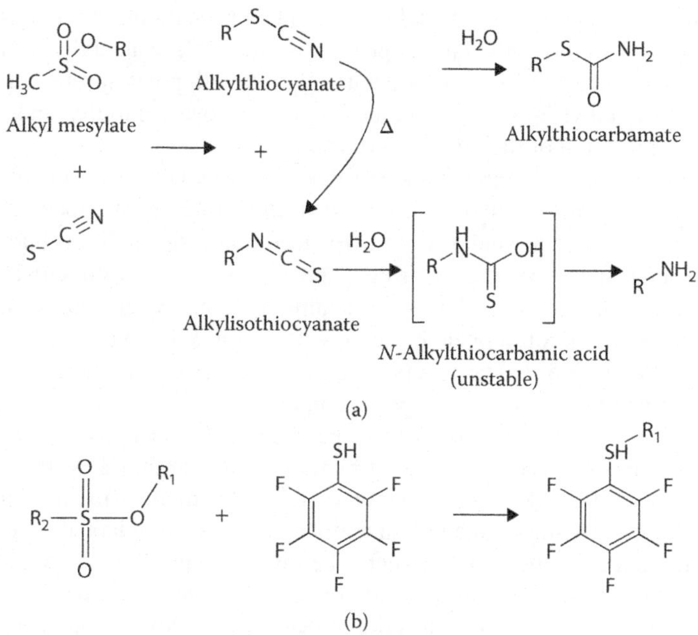

FIGURE 8.5 Example of derivatizations: (a) Reactions of alkyl mesylates with thiocyanate ion in aqueous solution. (b) Derivatization reaction for alkylating agents with pentafluorothiophenol. (From Lee et al., *Analyst*, 128, 857–863, 2003; Alzaga et al., *J. Pharm. Biomed. Anal.*, 45, 472–479, 2007.)

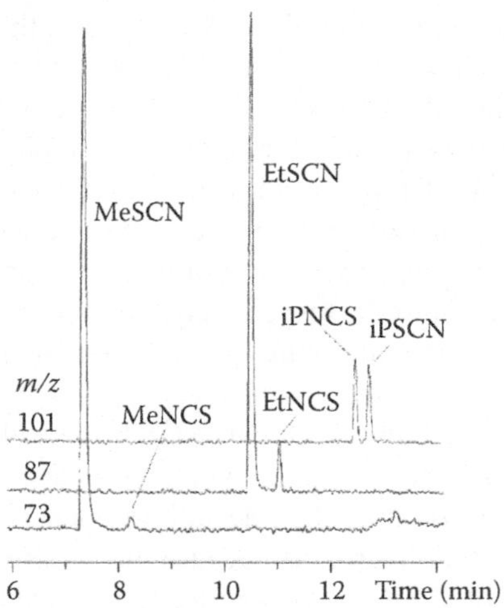

FIGURE 8.6 Selected ion monitoring recording chromatograms of the thiocyanate and isothiocyanate derivatives of methyl, ethyl and isopropyl mesylates (final concentration 0.25 μg/mL). (From Lee et al., *Analyst*, 128, 857–863, 2003.)

requires the samples to be highly soluble in aqueous solutions. As a consequence, partial hydrolysis of these alkylating agents was inevitable, especially for IMS. The authors suggests that replacing the aqueous solution with polar aprotic solvents like DMF or 1,3-dimethyl-2-imidazolidinone would improve solubility and avoid the hydrolytic degradation of the alkyl mesylates.

Alzaga and others[22] developed another derivatization headspace GC method for the simultaneous analysis of several common alkylating agents, including mesylates, besylates, tosylates, and sulfates. To accommodate APIs or intermediates with different solubilities and polarities, the authors used a dimethyl sulfoxide (DMSO)/water mixture (1:1, v/v) as the sample diluent. After adding the deuterated ISs and pentafluorothiophenol, the derivatization was conducted at 105°C for 15 minutes (Figure 8.5b). The reaction of the investigated alkylating agents with pentafluorothiophenol was less complex compared to reactions using sodium thiocyanate.[40] The resulting derivatives from pentafluorothiophenol were analyzed by headspace GC using a RH-624 column (30 m long × 0.32 mm i.d., with 1.8 μm film thickness). The EI-MS signal was monitored under SIM mode using the corresponding molecular ions. A standard addition method with ISs was adapted for quantification to compensate for the matrix effect. The introduction of deuterated analogs as ISs improved the method's accuracy and precision by compensating for variables in the derivatization and extraction/injection processes. In addition, the deuterated ISs reduce the opportunity for reporting false negative results. Compared to an earlier report by Lee and others,[40] this method showed better throughput, selectivity, analyte stability, and detection sensitivity. This procedure was successfully applied to different pharmaceutical matrixes and is particularly suitable for routine analysis with high sample throughput.

As part of industrial collaborative efforts to understand the formation mechanisms and kinetics of sulfonate esters, Jacq and others[57] adopted the derivatization headspace GC-MS approach used by Alzaga and others[22] and developed an automated procedure to monitor the formation of ethyl mesylate from reaction mixtures containing ethanol and MSA. The authors combined a liquid-handling robot with a static headspace module for automatic sample preparation. The EMS formed was analyzed after derivatization with pentafluorothiophenol using headspace GC-MS (Figure 8.7). Linear response was obtained for EMS concentrations between 5 and 500 μg/mL. The limit of detection and limit of quantitation for EMS were 0.5 and 1.0 μg/mL, respectively. Excellent linearity, repeatability, and robustness were obtained, allowing the system to be used in several kinetic studies.[56,58]

Sun and others[59] reported an in situ derivatization headspace GC-MS method for the determination of hydrazine, a known genotoxic compound, in drug substances at low parts-per-million levels. The authors chose acetone or acetone-d_6 as the derivatization reagent for hydrazine because the resulting acetone azine with a moderate boiling point of 133°C is suitable for headspace GC-MS analysis. Note that acetone-d_6 was considered as an alternative reagent when acetone was used as a solvent in the API-manufacturing process. In this study, the derivatization reagent was prepared by dissolving 0.5 g of benzoic acid and 0.5 mL of acetone or

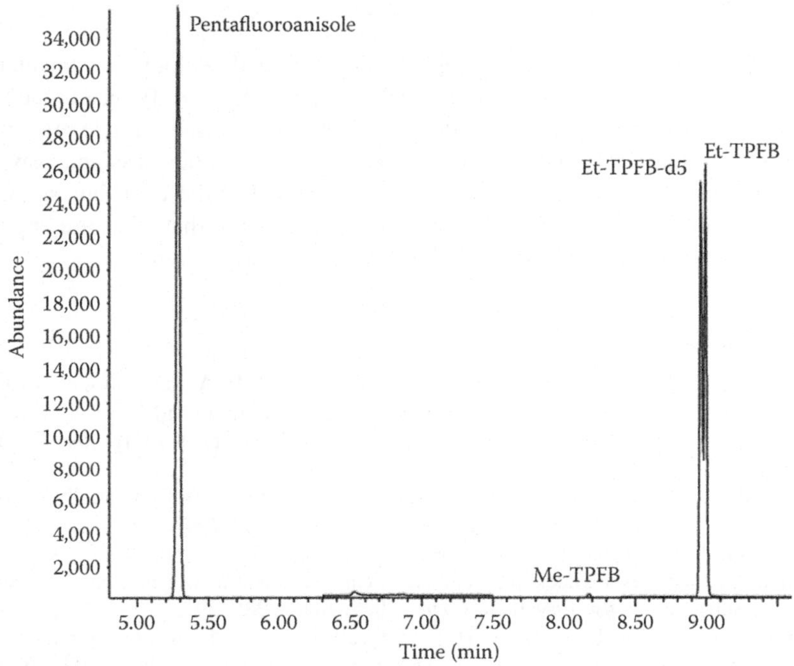

FIGURE 8.7 Chromatogram obtained by static headspace GC-(SIM)MS analysis of methane sulfonic acid/ethanol reaction mixture spiked with ethyl mesylate. Me-TPFB is a trace impurity of derivatized methyl methanesulfonate; ET-TPFB-d5 and Et-TPFB are, respectively, the derivatized internal standard and the derivatized EMS. (From Jacq et al., *J. Pharm. Biomed. Anal.*, 48, 1339–1344, 2008.)

acetone-d_6 in 10 mL of *N*-methyl-2-pyrrolidone (NMP) since NMP was found to be a universal solvent to dissolve the tested APIs. After mixing 10 μg of the API sample with 100 μL of the derivatization reagent and 10 μL of the diluent of 0.1% ethylenediaminetetraacetic acid (EDTA) in a 10 mL headspace vial, derivatization was conducted by incubating the vial at 100°C for 10 minutes. The derivatization reaction was driven to completion during headspace incubation. The resulting acetone azine products were analyzed using an Agilent DB-624 column (25 m long × 0.2 mm i.d., with 1.12 μm film thickness) and were quantified by SIM mode (*m/z* 112 for the acetone derivative and *m/z* 124 and *m/z* 106 for the acetone-d_6 derivative). The method showed a linear response for concentrations ranging from 0.1 to 10 ppm and a LOQ of 0.1 ppm. The authors investigated recoveries of hydrazine from different APIs with various reactive functional groups such as ketones, primary amines, or even Michael acceptors and from APIs with different salt forms, including free bases, HCl salts, and fumarate salts. The spike recovery at the 1 ppm level was found between 79% and 117%. This generic method has been utilized to determine hydrazine in APIs at GlaxoSmithKline, Pennsylvania, during the release and process development.

8.3 CONCLUSION

GC-MS is one of the most widely adopted analytical methodologies for the analysis of GTIs in pharmaceutical products. GC-MS is capable of quantifying a broad range of GTIs in intermediates, APIs, and drug products. Appropriate sample preparation is critical to achieve adequate specificity, sensitivity, and robustness and may even be beneficial by reducing complex matrix effects. Based on these attributes, GC-MS will remain a powerful tool in drug development to ensure that GTIs are appropriately controlled in drug substances and products for years to come.

REFERENCES

1. Muller, L.; Mauthe, R. J.; Riley, C. M.; Andino, M. M.; Antonis, D. D.; Beels, C.; DeGeorge, J.; et al., A rationale for determining, testing, and controlling specific impurities in pharmaceuticals that possess potential for genotoxicity. *Regul. Toxicol. Pharmacol.* **2006,** *44* (3), 198–211.
2. Humfrey, C. D., Recent developments in the risk assessment of potentially genotoxic impurities in pharmaceutical drug substances. *Toxicol. Sci.* **2007,** *100* (1), 24–28.
3. Pierson, D. A.; Olsen, B. A.; Robbins, D. K.; DeVries, K. M.; Varie, D. L., Approaches to assessment, testing decisions, and analytical determination of genotoxic impurities in drug substances. *Org. Process Res. Dev.* **2009,** *13* (2), 285–291.
4. Giordani, A.; Kobel, W.; Gally, H. U., Overall impact of the regulatory requirements for genotoxic impurities on the drug development process. *Eur. J. Pharm. Sci.* **2011,** *43*, 1–15.
5. Robinson, D. I., Control of genotoxic impurities in active pharmaceutical ingredients: a review and perspective. *Org. Process Res. Dev.* **2010,** *14* (4), 946–959.
6. Friscia, O.; Pulci, R.; Fassio, F.; Comelli, R., Chemical reagents as potential impurities of pharmaceutical products: investigations on their genotoxic activity. *J. Environ. Pathol. Toxicol. Oncol.* **1994,** *13* (2), 89–110.
7. Jacobson-Kram, D.; McGovern, T., Toxicological overview of impurities in pharmaceutical products. *Adv. Drug. Deliv. Rev.* **2007,** *59* (1), 38–42.
8. McGovern, T.; Jacobson-Kram, D., Regulation of genotoxic and carcinogenic impurities in drug substances and products. *TrAC Trends Anal. Chem.* **2006,** *25* (8), 790–795.
9. Committee for Medicinal Products for Human Use, *Guideline on the Limits of Genotoxic Impurities*. European Medicines Agency: London, UK, June **2006**; Vol. CPMP/SWP/5199/02, EMEA/CHMP/QWP/251344/2006.
10. Committee for Medicinal Products for Human Use, *Question & Answers on CHMP Guideline on the Limits of Genotoxic Impurities*. European Medicines Agency: London, UK, June **2008**; Vol. EMEA/CHMP/SWP/431994/2007.
11. U.S. Department of Health and Human Services, *Guidance for Industry, Genotoxic and Carcinogenic Impurities in Drug Substances and Products: Recommended Approaches* (draft). Food and Drug Administration, Center for Drug Evaluation and Research: Silver Spring, MD, December **2008**; Vol. http://www.fda.gov/cder/guidance/7834dft.pdf.
12. Raman, N. V. V. S. S.; Prasad, A. V. S. S.; Ratnakar, R. K., Strategies for the identification, control and determination of genotoxic impurities in drug substances: a pharmaceutical industry perspective. *J. Pharm. Biomed. Anal.* **2011,** *55*, 662–667.
13. Parsons, R. L.; Miller, S. A., Analytical considerations for genotoxic and other impurities. *Sep. Sci. Technol.* (San Diego, CA) **2011**, *10*, 171–194. Elsevier, Inc.
14. Baker, A., Development of a strategy for analysis of genotoxic impurities. In: *Genotoxic Impurities*, ed. Teasdale, A., John Wiley & Sons, Inc.: New York, **2010**; pp. 281–303.

15. Callis, C. M.; Bercu, J. P.; DeVries, K. M.; Dow, L. K.; Robbins, D. K.; Varie, D. L., Risk assessment of genotoxic impurities in marketed compounds administered over a short-term duration: applications to oncology products and implications for impurity control limits. *Org. Process Res. Dev.* **2010,** *14*, 986–992.
16. Sun, M.; Liu, D. Q.; Kord, A. S., A systematic method development strategy for determination of pharmaceutical genotoxic impurities. *Org. Process Res. Dev.* **2010,** *14*, 977–985.
17. Elder, D. Analysis of genotoxic impurities: review of approaches. In: *Genotoxic Impurities*, ed. Teasdale, A., John Wiley & Sons, Inc.: New York, **2010**; pp 249–280.
18. Klick, S., Evaluation of different injection techniques in the gas chromatographic determination of thermolabile trace impurities in a drug substance. *J. Chromatogr. A* **1995,** *689* (1), 69–76.
19. Seymour, M., Determination of residual dimethyl sulfate in a lipophilic bulk drug by wide-bore capillary gas chromatography. *J. Chromatogr.* **1989,** *463* (1), 216–21.
20. Skett, P. W., Low-level measurement of potent toxins. In: *Analysis of Drug Impurities*, eds. Smith, R. J., and Webb, M. L., Blackwell Publishing: Oxford, UK, **2007**; pp. 82–123.
21. Ramjit, H. G.; Singh, M. M.; Coddington, A. B., Gas chromatographic/mass spectrometric analysis of methyl methanesulphonate and ethyl methanesulphonate in the bismesylate salt of DPI 201–106, a positive inotropic agent for the treatment of heart failure. *J. Mass Spectrom.* **1996,** *31* (8), 867–872.
22. Alzaga, R.; Ryan, R. W.; Taylor-Worth, K.; Lipczynski, A. M.; Szucs, R.; Sandra, P., A generic approach for the determination of residues of alkylating agents in active pharmaceutical ingredients by in situ derivatization-headspace-gas chromatography-mass spectrometry. *J. Pharm. Biomed. Anal.* **2007,** *45* (3), 472–479.
23. Madireddy, V.; Babu, K. S.; Narayanareddy, P., Stability-indicative, validated, fast HPLC method for quantification of two genotoxic impurities in imatinib mesylate. *Global J. Anal. Chem.* **2011,** *2*, 198–207.
24. Garcia, A.; Ruperez, F. J.; Ceppa, F.; Pellati, F.; Barbas, C., Development of chromatographic methods for the determination of genotoxic impurities in cloperastine fendizoate. *J. Pharm. Biomed. Anal.* **2012,** *61*, 230–236.
25. Trivedi, R. K.; Patel, M. C., Evaluation of pharmaceutical quality of mesalamine delayed release tablets using a new high sensitivity reversed-phase UPLC method for its genotoxic/aniline impurity. *E-J. Chem.* **2011,** *8*, 167–179.
26. Hoogerheide, J. G.; Scott, R. A., Use of 2-mercaptopyridine for the determination of alkylating agents in complex matrices: application to dimethyl sulfate. *Talanta* **2005,** *65* (2), 453–460.
27. Liu, D. Q.; Chen, T. K.; McGuire, M. A.; Kord, A. S., Analytical control of genotoxic impurities in the pazopanib hydrochloride manufacturing process. *J. Pharm. Biomed. Anal.* **2009,** *50*, 144–150.
28. An, J.; Sun, M.; Bai, L.; Chen, T.; Liu, D. Q.; Kord, A., A practical derivatization LC/MS approach for determination of trace level alkyl sulfonates and dialkyl sulfates genotoxic impurities in drug substances. *J. Pharm. Biomed. Anal.* **2008,** *48*, 1006–1010.
29. Bai, L.; Sun, M.; An, J.; Liu, D. Q.; Chen, T. K.; Kord, A. S., Enhancing the detection sensitivity of trace analysis of pharmaceutical genotoxic impurities by chemical derivatization and coordination ion spray-mass spectrometry. *J. Chromatogr. A* **2010,** *1217*, 302–306.
30. Wu, L.; Liu, D. Q.; Vogt, F. G.; Kord, A. S., Gas-phase derivatization via the Meerwein reaction for selective and sensitive LC-MS analysis of epoxides in active pharmaceutical ingredients. *J. Pharm. Biomed. Anal.* **2011,** *56* (Copyright © 2012 American Chemical Society (ACS). All rights reserved.), 1106–1111.

31. Vanhoenacker, G.; Dumont, E.; David, F.; Baker, A.; Sandra, P., Determination of arylamines and aminopyridines in pharmaceutical products using in-situ derivatization and liquid chromatography-mass spectrometry. *J. Chromatogr. A* **2009,** *1216*, 3563–3570.
32. Kakadiya, P. R.; Chandrashekhar, T. G.; Ganguly, S.; Singh, D. K.; Singh, V., Low level determinations of methyl methanesulfonate and ethyl methanesulfonate impurities in emtricitabine active pharmaceutical ingredient by LC/MS/MS using electrospray ionization. *Anal. Chem. Insights* **2011,** *6*, 21–28.
33. Kakadiya, P. R.; Reddy, B. P.; Singh, V.; Ganguly, S.; Chandrashekhar, T. G.; Singh, D. K., Low level determinations of methyl methanesulfonate and ethyl methanesulfonate impurities in lopinavir and ritonavir active pharmaceutical ingredients by LC/MS/MS using electrospray ionization. *J. Pharm. Biomed. Anal.* **2011,** *55*, 379–384.
34. van Wijk, A. M.; Beerman, B.; Niederlaender, H. A. G.; Siebum, A. H. G.; de Jong, G. J., A new approach for generic screening and quantitation of potential genotoxic alkylation compounds by pre-column derivatization and LC-MS/MS analysis. *Anal. Bioanal. Chem.* **2011,** *400*, 1375–1385.
35. Phillips, A., Analysis of genotoxic impurities by nuclear magnetic resonance spectroscopy. In: *Genotoxic Impurities*, ed. Teasdale, A., John Wiley & Sons, Inc.: New York, **2010**; pp. 351–384.
36. Jagota, N. K.; Chetram, A. J.; Nair, J. B., Determination of trace levels of hydrazine in the penultimate intermediate of a novel anti-infective agent. *J. Pharm. Biomed. Anal.* **1998,** *16*, 1083–1087.
37. Elder, D. P.; Lipczynski, A. M.; Teasdale, A., Control and analysis of alkyl and benzyl halides and journal related reactive organohalides as potential genotoxic impurities in active pharmaceutical ingredients (APIs). *J. Pharm. Biomed. Anal.* **2008,** *48* (3), 497–507.
38. Elder, D. P.; Snodin, D.; Teasdale, A., Control and analysis of hydrazine, hydrazides and hydrazones. Genotoxic impurities in active pharmaceutical ingredients (APIs) and drug products. *J. Pharm. Biomed. Anal.* **2011,** *54* (Copyright © 2012 American Chemical Society (ACS). All rights reserved.), 900–910.
39. Elder, D. P.; Teasdale, A.; Lipczynski, A. M., Control and analysis of alkyl esters of alkyl and aryl sulfonic acids in novel active pharmaceutical ingredients (APIs). *J. Pharm. Biomed. Anal.* **2008,** *46* (1), 1–8.
40. Lee, C. R.; Guivarch, F.; Dau, C. N. V.; Tessier, D.; Krstulovic, A. M., Determination of polar alkylating agents as thiocyanate/isothiocyanate derivatives by reaction headspace gas chromatography. *Analyst* **2003,** *128* (7), 857–863.
41. David, F.; Jacq, K.; Vanhoenacker, G.; Sandra, P., Strategic approaches to the chromatographic analysis of genotoxic impurities. In: *Genotoxic Impurities*, ed. Teasdale, A., John Wiley & Sons, Inc.: New York, **2010**; pp. 305–349.
42. Lee, M. L.; Yang, F. J.; Bartle, K. D., *Open Tubular Column Gas Chromatography: Theory and Practice*. John Wiley & Sons: New York, **1984**; p. 445.
43. Raman, N. V. V. S. S.; Reddy, K. R.; Prasad, A. V. S. S.; Ramakrishna, K., Validated chromatographic methods for the determination of process related toxic impurities in pantoprazole sodium. *Chromatographia* **2008,** *68* (5/6), 481–484.
44. Raman, N. V. S. S.; Reddy, K.; Prasad, A. S. S.; Ramakrishna, K., Development and validation of a GC-MS method for the determination of methyl and ethyl camphorsulfonates in esomeprazole magnesium. *Chromatographia* **2008,** *68* (7–8), 675–678.
45. Sarat, M.; Ramakrishna, M.; Suresh, Y.; Harikrishna, S.; Rambabu, C.; Kishore, K.; Reddy, K. N., Low-level determination of residual methyl methane sulfonate and ethyl methane sulfonate in pharmaceuticals by gas chromatography with mass spectrometry. *E-J. Chem.* **2010,** *7* (2), 629–635.
46. Li, H.; Sluggett, G. W., Development and validation of a sensitive GC-MS method for the determination of trace levels of an alkylating reagent in a β-lactam active pharmaceutical ingredient. *J. Pharm. Biomed. Anal.* **2005,** *39* (3–4), 486–494.

47. Ramakrishna, K.; Raman, N. V. V. S. S.; Rao, K. M. V. N.; Prasad, A. V. S. S.; Reddy, K. S., Development and validation of GC-MS method for the determination of methyl methanesulfonate and ethyl methanesulfonate in imatinib mesylate. *J. Pharm. Biomed. Anal.* **2008,** *46* (4), 780–783.
48. David, F.; Jacq, K.; Sandra, P.; Baker, A.; Klee, M. S., Analysis of potential genotoxic impurities in pharmaceuticals by two-dimensional gas chromatography with Deans switching and independent column temperature control using a low-thermal-mass oven module. *Anal. Bioanal. Chem.* **2010,** *396*, 1291–1300.
49. Zheng, J.; Pritts, W. A.; Zhang, S.; Wittenberger, S., Determination of low ppm levels of dimethyl sulfate in an aqueous soluble API intermediate using liquid-liquid extraction and GC-MS. *J. Pharm. Biomed. Anal.* **2009,** *50* (5), 1054–1059.
50. Nassar, M. N.; Cucolo, M.; Miller, S. A., Ethyl methanesulphonate in a parenteral formulation of BMS-214662 mesylate, a selective farnesyltransferase inhibitor: formation and rate of hydrolysis. *Pharm. Dev. Technol.* **2009,** *14* (6), 672–677.
51. Wollein, U.; Schramek, N., Simultaneous determination of alkyl mesylates and alkyl besylates in finished drug products by direct injection GC/MS. *Eur. J. Pharm. Sci.* **2012,** *45*, 201–204.
52. Colon, I.; Richoll, S. M., Determination of methyl and ethyl esters of methanesulfonic, benzenesulfonic, and *p*-toluenesulfonic acids in active pharmaceutical ingredients by solid-phase microextraction (SPME) coupled to GC/SIM-MS. *J. Pharm. Biomed. Anal.* **2005,** *39*, 477–485.
53. Ho, T. D.; Joshi, M. D.; Silver, M. A.; Anderson, J. L., Selective extraction of genotoxic impurities and structurally alerting compounds using polymeric ionic liquid sorbent coatings in solid-phase microextraction: alkyl halides and aromatics. *J. Chromatogr. A* **2012,** *1240* (Copyright © 2012 U.S. National Library of Medicine.), 29–44.
54. Martano, G.; Vogl, C.; Bojaxhi, E.; Bresgen, N.; Eckl, P.; Stutz, H., Solid-phase extraction and GC-MS analysis of potentially genotoxic cleavage products of β-carotene in primary cell cultures. *Anal. Bioanal. Chem.* **2011,** *400* (8), 2415–2426.
55. Strohmeyer, H. E.; Sluggett, G. W., Determination and control of TEMPO, a potentially genotoxic free radical reagent used in the synthesis of filibuvir. *J. Pharm. Biomed. Anal.* **2012,** *62*, 216–219.
56. Teasdale, A.; Eyley, S. C.; Delaney, E.; Jacq, K.; Taylor-Worth, K.; Lipczynski, A.; Reif, V.; et al., Mechanism and processing parameters affecting the formation of methyl methanesulfonate from methanol and methanesulfonic acid: an illustrative example for sulfonate ester impurity formation. *Org. Process. Res. Dev.* **2009,** *13* (3), 429–433.
57. Jacq, K.; Delaney, E.; Teasdale, A.; Eyley, S.; Taylor-Worth, K.; Lipczynski, A.; Reif, V. D.; et al., Development and validation of an automated static headspace gas chromatography–mass spectrometry (SHS-GC-MS) method for monitoring the formation of ethyl methane sulfonate from ethanol and methane sulfonic acid. *J. Pharm. Biomed. Anal.* **2008,** *48* (5), 1339–1344.
58. Teasdale, A.; Delaney, E. J.; Eyley, S. C.; Jacq, K.; Taylor-Worth, K.; Lipczynski, A.; Hoffmann, W.; et al., A detailed study of sulfonate ester formation and solvolysis reaction rates and application toward establishing sulfonate ester control in pharmaceutical manufacturing processes. *Org. Process. Res. Dev.* **2010,** *14* (4), 999–1007.
59. Sun, M.; Bai, L.; Liu, D. Q., A generic approach for the determination of trace hydrazine in drug substances using in situ derivatization-headspace GC-MS. *J. Pharm. Biomed. Anal.* **2009,** *49*, 529–533.

9 Control of Chloroethane in Raw Materials and Drug Substances Using Headspace/Gas Chromatography Analysis

Johanna Ubben and Bing-Shiou Yang

CONTENTS

9.1 INTRODUCTION

Chloroethane is a colorless, flammable gas and is known to induce reverse mutations in Ames tests. These mutations are observed in *Salmonella* TA1535 and TA100.[1]

In 1989, the results of a 2-year study with mice and rats were published as part of the U.S. National Toxicology Program. The results of this study found a high incidence of endometrial uterine carcinomas in female mice. The exposure level for this study was 15,000 ppm of inhaled chloroethane for 6 h/day, 6 days/week for 2 years.[2] Similar findings were also observed in a Chinese hamster in vitro hypoxanthine phosphoribosyltransferase mutagenesis assay at exposure levels of 0.65–2.48 mg/mL.[3] Ebert and others also performed in vivo unscheduled DNA synthesis assays on female mice and in vivo micronucleus assays of male and female mice during this study. The mice were exposed to 25,000 ppm concentrations of chloroethane for 6 h/day over a 3-day period. The results of the in vivo assays did not detect indications of genotoxicity as a result of exposure to chloroethane.

Whereas the genotoxic results are equivocal, chloroethane is considered a potentially hazardous compound and has been observed to produce narcosis effects in humans after a brief exposure to high levels of chloroethane vapor.[4] However, the effects of ingestion are not as well characterized as the inhalation effects.

Chloroethane can form in solutions of HCl and ethanol, used to produce drug substance chloride salts. Due to its physical characteristics, it would not be expected to survive most synthetic processes if an HCL/ethanol solution containing chloroethane as an impurity were used to produce a chloride salt form of the drug substance. However, the toxicity of ingested chloroethane is not well characterized and it is prudent to develop a control test to verify that the compound is not found in a finished active pharmaceutical ingredient (API).

Chloroethane is one of the volatile organic compounds routinely monitored by Environmental Protection Agency methods, such as Method 8021B[5] or Method 8260B.[6] These are methods designed to monitor volatile organic compounds in water, wastewater, and waste solid streams in the environment. These compounds are low-molecular-weight, low-boiling or gaseous compounds at room temperature and are most readily analyzed by gas chromatography (GC) rather than high-pressure liquid chromatography.

Method 8021B uses a Supelco (Bellefonte, Pennsylvania) VOCOL (proprietary stationary phase) as the primary analytical column, with a confirmation column such as a Supelco SPB624 (6% cyanopropylphenyl/94% methyl silicone) stationary phase. Both of these columns have good selectivity with respect to chloroethane. In addition, the 624 stationary phase is the primary analytical column used in USP 467 for the analysis of residual solvents.[7] Therefore, in addition to the selectivity for chloroethane, the selectivity of common residual solvents such as methanol and ethanol is well characterized. In the case of the drug substance being analyzed here, ethanol is an expected residual solvent and methanol is the carrier solvent for the chloroethane standard purchased from Sigma-Aldrich (St. Louis, Missouri). On this stationary phase, chloroethane elutes between methanol and ethanol. It may elute on the tail of a large methanol peak in a sample matrix.

Polyethylene glycol (PEG) columns, commonly referred to as wax columns, may also be used to monitor chloroethane. This is a more polar stationary phase that will elute chloroethane before methanol. This change in elution order between chloroethane and methanol eliminates the problem of the chloroethane eluting on the tail of a large methanol solvent peak. However, many PEG columns have a solid to liquid phase transition at 40°C. Below 40°C, the columns may exhibit loss of capacity and poor retention reproducibility. If the starting GC oven temperature needs to be below 40°C to produce adequate separation of the chloroethane analyte and other impurities from the sample and/or diluent matrix, a PEG column may not be the optimal choice for the analysis.

Sample introduction was also an important consideration during method development for this analysis. The two primary options are liquid injection and headspace injection. Liquid injection techniques can be further subdivided into split, splitless, cool on column (COC), and programmable temperature vaporization (PTV).

In the liquid sampling techniques, the entire sample is transferred from a syringe to the GC inlet, including the high-molecular-weight drug substance. Typically, injection volumes of 1 to 2 μL are used and the injection volume should be small to avoid overloading the column. The most common injection technique used is a split injection in which the majority of the sample is vented to the atmosphere so that only a small portion reaches the analytical column. One issue that can arise when using this technique is discrimination, in which the less volatile compounds transferred to the column are not representative of the original mixture. Discrimination may be avoided by using splitless injection to allow a more representative sample to reach the column, and the splitless injection technique is most suitable for trace analyses. However, for this analysis the analyte of interest is a low-molecular-weight, low-boiling compound and discrimination is not a concern.

In both the split and splitless injection techniques, the GC inlet is heated to temperatures of 200°C or higher to volatilize the sample. Under these conditions, thermal degradation of the drug substance may be a concern, with the potential to introduce unexpected and potentially interfering peaks into an analysis. Even if degradation is not a concern, the drug substance may cause analyte recovery issues as it accumulates in the GC inlet system.

To avoid the problems with thermal degradation or reactivity in a hot inlet, COC or PTV liquid sampling techniques may be considered. COC techniques inject the entire sample directly onto the head of the analytical column or a precolumn attached to the analytical column. In this case, the inlet is set near the GC initial oven temperature, thus avoiding concerns with discrimination or thermal degradation. However, the nonvolatile drug substance can still accumulate at the head of the column and cause recovery issues in an analysis. A PTV injection technique does not inject the sample directly onto the column. It also is typically set to a low inlet temperature at the start of the analysis and is programmed to high temperatures to volatilize and introduce the heavier components of the sample onto the column. If the drug substance accumulates in the liner, the same concerns with recovery due to matrix effects may occur. Both COC and PTV injection techniques are specialized inlets and may not be available on a typical GC system.

Headspace injection introduces only the volatile portion of the sample to the GC inlet. Any compounds, such as the drug substance, that do not volatilize at the chosen vial equilibration temperature will remain in the liquid sample and will not contaminate the GC inlet. Typically, there are fewer issues with analyte recovery using this technique than with liquid injection techniques. Headspace injection is the recommended technique in USP 467 for residual solvents and lends itself well to the analysis of an analyte that exists in the gaseous state at room temperature. The major concern is to choose a vial equilibration temperature that will partition the analyte to the gaseous phase and not introduce the possibility of reaction in the vial. In the case of a drug substance that is an HCL salt and contains ethanol, this reactivity may be a concern.

During the method development of this analysis, the work was based on the existing residual solvent method used to characterize the drug substance. Therefore, a headspace injection technique was chosen to introduce the sample.

Based on these considerations and the fact that the drug substance had been characterized for residual solvents using a Varian VF-624 column, Varian VF-624ms was chosen as the analytical column for this analysis.

A headspace/GC/flame ionization detection (FID) method was developed and validated to perform this analysis. In one case, it was found that chloroethane was trapped within an API at levels as high as 200 ppm. Steps were taken to improve the synthesis to remove as much chloroethane as possible from the API. The developed test procedure was used extensively to provide feedback to the process group as the synthesis was refined to lower the chloroethane and minimize the risks of a potentially hazardous impurity in the API.

9.2 EXPERIMENTAL SECTION

Ampoules of chloroethane in methanol (1000 μg/mL) were purchased from Supelco to use for standard preparation. Dimethylacetamide (DMAc) purchased from Sigma-Aldrich was used as the diluent.

Analyses were performed on an Agilent 6890N equipped with an FID detector and a Markelov HS9000 headspace sampler. The column was a Varian VF-624ms, 30 m × 0.32 mm × 1.8 μm film thickness (df).

9.3 GAS CHROMATOGRAPHY CONDITIONS

Instrument: Agilent 6890N gas chromatograph with an FID detector and Markelov HS9000 Headspace Sampler

Column: Varian VF-624ms, 30 m × 0.32 mm × 1.8 μm df

GC conditions: carrier gas flow rate (He) was 1.2 mL/min; constant flow mode. The oven was programmed from 35°C (5-minute initial hold) to 50°C at 10°C/min, and then from 50°C to 200°C at 50°C/min (3.5-minute hold at 250°C). The injector was set at 240°C. The injection was run at a 35:1 split. A 2 mm internal diameter straight, deactivated liner was used in the injection port. The FID detector was set at 260°C. Hydrogen, air, and helium makeup gas flows were set at 40, 400, and 30 mL/min, respectively.

Headspace conditions: sample oven, needle, transfer lines, and inject solenoid were set at 100°C, 110°C, 120°C, and 150°C, respectively. The GC cycle time was set at 21 minutes. Samples were incubated for 10 minutes with mixing turned on and set at a medium mixing speed. The headspace sampler was used in the one needle on-column mode, with an on-column inject time of 2.4 seconds, a postmix stabilization time of 5 seconds, and a pressure equilibration time of 30 seconds. Helium was used as the headspace carrier gas at 9 psi to pressurize a vial.

Standards: standard was prepared by quantitatively transferring the contents of one or two ampoules of chloroethane into a 20 mL volumetric flask and then diluting it to volume with DMAc. This stock solution was then used to prepare linearity standards spanning a range of approximately 0.08–50 μg/mL, which is equivalent to 3–2000 ppm, relative to a 25 mg sample size of API. For each linearity standard, 1.0 mL of solution was transferred to a 20 mL headspace vial and immediately sealed with a Teflon-lined/silicone septum crimp cap. Two vials at each linearity level were prepared for analysis.

Samples: samples were prepared by accurately weighing approximately 25 mg into a tared 20 mL headspace vial and then pipetting 1.0 mL of DMAc into the vial. The headspace vial was then immediately sealed with a Teflon-lined/silicone septum crimp cap. At room temperature the sample would not completely dissolve, but it was fully dissolved at the headspace oven temperature of 100°C.

For HCl in ethanol samples, approximately 25 mg of the solution was weighed directly into a tared headspace vial containing 1.0 mL of DMAc. These samples were sealed with a Teflon-lined/silicone septum crimp cap prior to final weighing.

9.4 RESULTS AND DISCUSSION

9.4.1 Validation Results

This method was validated for specificity, linearity, recovery, repeatability, and quantitation limit (QL):

Specificity: the method is specific for chloroethane, with a resolution greater than 4 in the QL standard, as shown in Figure 9.1.

Linearity: the method is linear across a range of 0.08 to 50 μg/mL of chloroethane, with a correlation coefficient greater than 0.99. These results are shown in Table 9.1.

Recovery: recovery was assessed at 1600 ppm, the proposed alert level for chloroethane in the API. The recovery results are shown in Table 9.2 and ranged from approximately 102% to 104%, with a relative standard deviation (%RSD) of 1.1. Typical chromatograms of an API control sample and a spiked sample are shown in Figures 9.2 and 9.3, respectively.

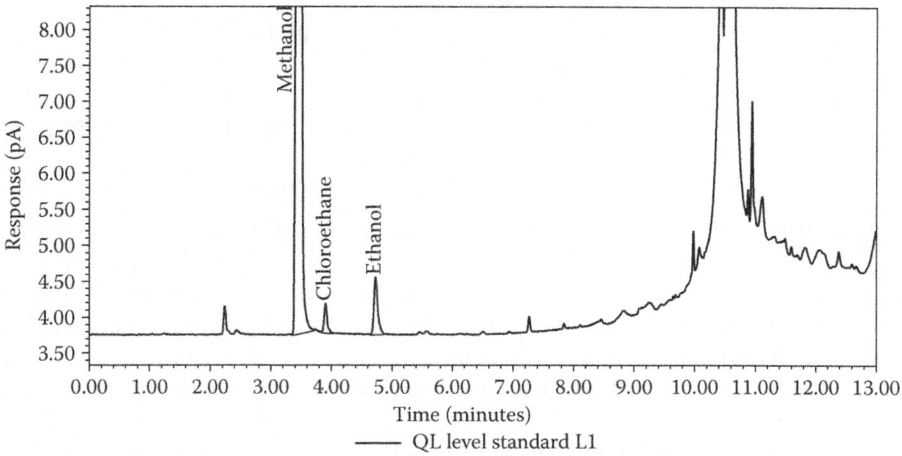

FIGURE 9.1 Resolution for quality limit linearity standard.

TABLE 9.1
Chloroethane Linear Regression Results

Parameter	Result
Correlation coefficient (r)	0.9997
Coefficient of determination (r^2)	0.9994
Slope	5.042294
Intercept	–0.284667
Residual sum of squares	89.710732

TABLE 9.2
Spiked Recovery/Repeatability Results

Sample	Chloroethane (% Recovery)
Spiked sample prep 1	103.95
Spiked sample prep 2	102.91
Spiked sample prep 3	101.74
Average	102.9
%RSD	1.1

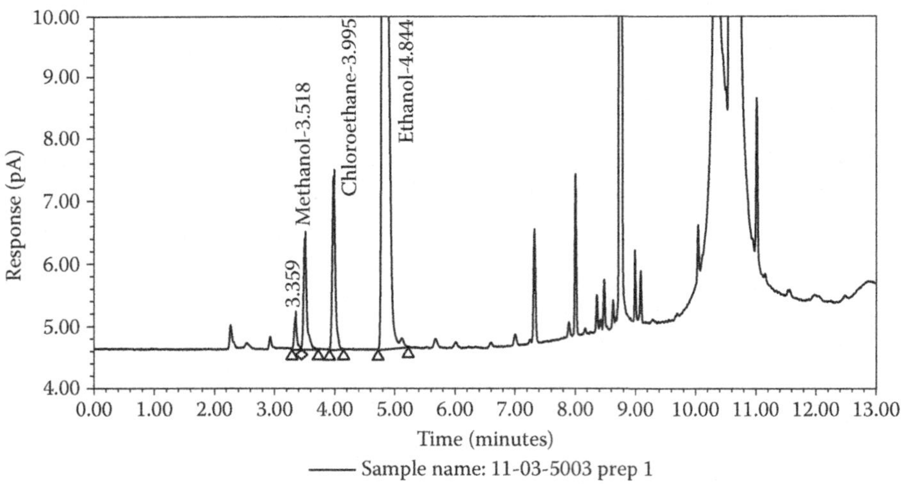

FIGURE 9.2 Chromatogram of a typical active pharmaceutical ingredient (control sample).

QL/detection limit (DL): QL was evaluated by making six injections of the 0.25 μg/mL linearity solution. This linearity standard contained approximately 10 ppm of chloroethane relative to a 25 mg API sample size.

A 10 ppm level standard was detectable with an average peak area of approximately 1.3359 pA × s and %RSD of 1.8. Based on these results, the QL for this compound in this method was set at 10 ppm and the DL was set at 3 ppm. A typical QL chromatogram is shown in Figure 9.4.

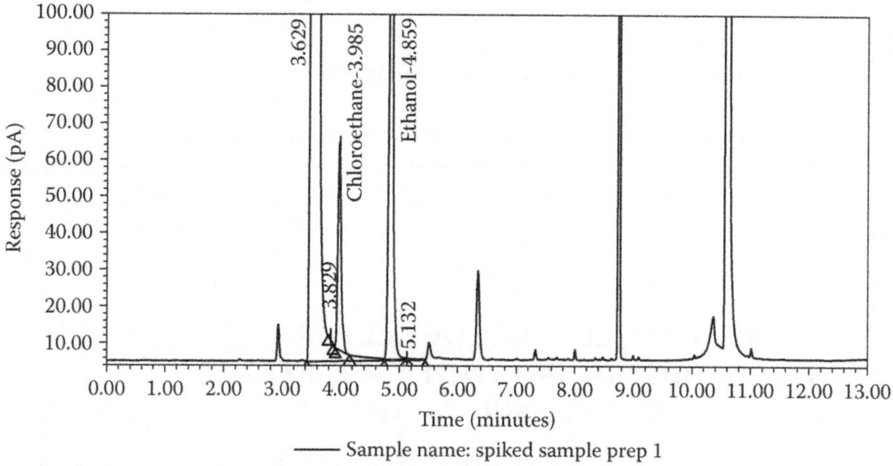

FIGURE 9.3 Chromatogram of a spiked active pharmaceutical ingredient to assess recovery.

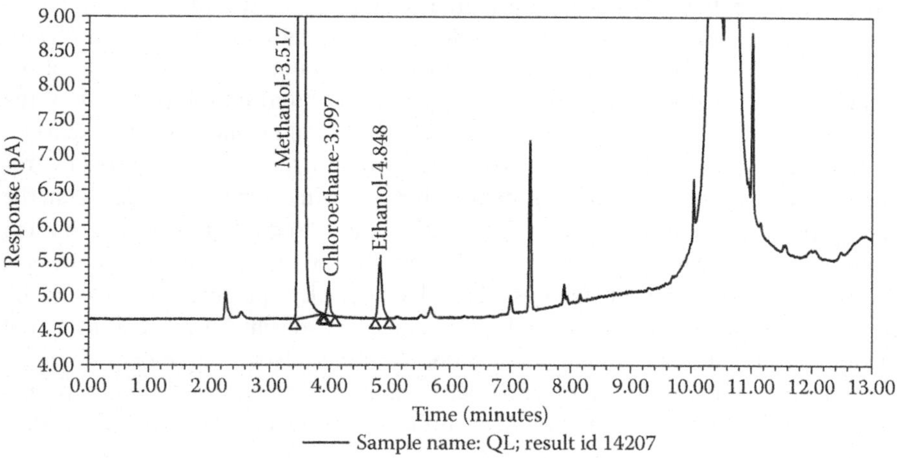

FIGURE 9.4 Chloroethane quality limit solution.

Chloroethane is a potential reaction product of HCL and ethanol, and a kinetic study for this reaction has been performed in the temperature range of 65°C–95°C.[8] The synthetic reaction was carried out at 50°C, and original fate assessments considered that the synthetic conditions did not favor the formation of chloroethane. The headspace equilibration temperature chosen for the analysis was 100°C to maximize analyte sensitivity. It was hypothesized that chloroethane may form in the headspace vial at 100°C in the presence of the drug substance, which is an HCl salt and contains approximately 1% entrapped ethanol. One lot of the drug substance was assessed using a headspace equilibration temperature of 50°C to evaluate the hypothesis. The results of this assessment are shown in Table 9.3.

Although the possibility of a reaction occurring in the headspace cannot be completely ignored, the method as validated is suitable for its intended purpose.

TABLE 9.3
Headspace Equilibration Temperature Assessment

Drug Substance Identification (ID) Number	Headspace Equilibration Temperature (°C)	Chloroethane (ppm)
11-05-4001	100	312.26 ($n = 2$)
11-05-4001	50	312.34 ($n = 1$)

9.5 APPLICATIONS: CONTROL OF EtCL FOR A HYDROCHLORIDE SALT WITH A STRONG TENDENCY TO ENCAPSULATE ETHANOL

Compound A, a development compound in BI, is a hydrochloride salt. A reactive crystallization process in ethanol (EtOH) was developed to produce compound A. The EtOH solvent system was chosen for the crystallization of compound A because it provided (1) better polymorph control (the desired polymorph was difficult to produce in nonalcohol solvents), (2) better tolerance for the residual solvent level (compound A tends to trap solvents, which is likely due to the inherent property of the compound), (3) better solubility (to enable polish filtration) for the free base of the compound prior to reactive crystallization, and (4) good purification and recovery. In early development, HCl in EtOH was used to minimize the amount of H_2O introduced in the process due to its impact on polymorphism control and yield. From the current process, usually less than 1.5% of EtOH was observed in the laboratory and scale-up batches, which was lower than the less than 3% specification used for new drug submissions. A recent study using solid-state nuclear magnetic resonance with compound A made from ^{13}C-labeled EtOH suggested solvent entrapment instead of the formation of a solvate based on the mobility of the solvent. Based on differential scanning calorimetry and thermogravimetric analysis, the entrapped solvent was not released until close to the melting (due to the dissociation of HCl) of the crystals (>150°C).

It has been reported that the genotoxin chloroethane (EtCl) can be generated during the preparation of hydrochloride salt in the presence of ethanol.[9] EtCl was detected at parts-per-million levels for compound A produced from the current process. To control the EtCl level for future scale-up batches, a series of studies were conducted, in collaboration with chemical development and analytical development colleagues, to investigate the potential sources of EtCl. The potential sources of EtCl are (1) formation and storage of HCl in EtOH solution, (2) salt formation/crystallization, and (3) storage of compound A. In this study, a variety of samples have been analyzed using the GC method described in Section 9.4, including HCL/ethanol solutions, API mother liquors (MLs), and finished API. The chloroethane results from testing these solutions were used to optimize the final salt formation and recrystallization of the finished API.

The HCl in EtOH solution was prepared by bubbling HCl gas through EtOH solution at a low temperature. In terms of potential generation of EtCl from this process and the stability of the solution, several variables were investigated, including forming HCl solution with lower concentrations (from 6 to ~2 N), with agitation (to minimize the chance of having local "hot spots"); maintaining lower temperatures (with cooling and slowing the gas flow rate); or adding H_2O to study the effect of H_2O on the inhibition of EtCl formation.[10] The HCl in ethanol solutions were prepared and analyzed on the same day of manufacture under various conditions. Aliquots of these solutions were submitted for analysis on the day of manufacture. Selected solutions were then reanalyzed over a 10-day period. The results of this study are shown in Table 9.4.

It was found that formation of HCl in EtOH solution by bubbling HCl gas at low temperatures still formed trace amounts of EtCl (~50–70 ppm). The EtCl level increased during storage even when stored at low temperatures (e.g., 58–131 ppm after 2 days and then to 222 ppm after 10 days when stored at ~0°C–5°C). A sample left at ambient condition for greater than 3 months had 13.15 wt% EtCl. No clear dependence on the effect of concentration of the EtCl level during formation (in the range of 1.62–2.1 N) was observed, but the solution stability was worse with higher concentrations. The formation of HCl in EtOH solution with approximately 5 wt% H_2O did not inhibit the formation of EtCl, but it did improve the stability during storage. These results indicated that HCl in ethanol solutions needed to be freshly prepared to minimize the formation of chloroethane in the solution, prior to use in forming the salt of the API.

To further investigate the formation of EtCl during the salt formation/crystallization process and in the solid phase, several API mother liquors and finished API samples were tested and the results are summarized in Tables 9.5 and 9.6. Several aspects of the salt formation process were studied, including (1) the effect of crystallization on the purging of EtCl, (2) whether EtCl was formed during salt formation, and (3) whether lower temperatures (e.g., 20°C vs. 50°C) can reduce the amount of EtCl formed. The possibility of using an alternative HCl source (e.g., aqueous HCl) was also studied. It was found that the current process (1 equiv.; 50°C) using "fresh" HCl in EtOH solution produced compound A with 40–80 ppm EtCl. The higher level (312 ppm) of EtCl observed in lot A resulted from the atypical HCl in EtOH solution used that contained approximately 13 wt% of EtCl.

TABLE 9.4
Chloroethane Found in HCl/Ethanol Solutions

Sample	Lot	Storage	EtCl by GC
~6 N HCl in ethanol (stored ~4 months)	A	Ambient	13.15 wt%
1.62 N HCl in ethanol (time 0 [t_0])	B	Ambient	74 ppm
1.62 N HCl in ethanol (t_0)	C	5°C	58 ppm
1.62 N HCl in ethanol (t_0 + 2 days)	C	5°C	131 ppm
1.62 N HCl in ethanol (t_0 + 10 days)	C	5°C	222 ppm
2.1 N HCl in ethanol (t_0)	D	5°C	52 ppm
2.1 N HCl in ethanol (t_0 + 8 days)	D	5°C	853 ppm
1.79 N HCL 95/5 ethanol/water (t_0)	E	5°C	72 ppm
1.79 N HCL 95/5 ethanol/water (t_0 + 7 days)	E	5°C	92 ppm

TABLE 9.5
Chloroethane Found in API Mother Liquors

Mother Liquor Lot	HCl/Ethanol Lot	Salt Formation Process	Chloroethane (ppm)
17	Aqueous 12 N HCl	H_2O/EtOH	111
18	D	EtOH, 1.1 equiv. HCL	192
19	E	H_2O/EtOH, 1.1 equiv.	197
20	D, aged 8 days	EtOH, room temperature (RT), 1.1 equiv. (0.3 equiv.)	219
21	D, aged 8 days	EtOH (first portion, 0.4 equiv.)	427
22	D, aged 8 days	EtOH, RT(first portion, 0.3 equiv.)	124

TABLE 9.6
Chloroethane Found in Finished API

API Lot	HCl/Ethanol Lot[a]	Salt Formation Process	Chloroethane (ppm)
1	A	EtOH	312
2	B	EtOH	44
3	Aqueous 12 N HCl	H_2O/EtOH	40
4	D	EtOH, 1.1 equiv. HCL	252
5	E	H_2O/EtOH, 1.1 equiv.	228
6	D, aged 8 days	EtOH, RT, 1.1 equiv. (0.3 equiv.)	150
7	D, aged 8 days	EtOH (first portion, 0.5 equiv.)	148
8	D, aged 8 days	EtOH (first portion, 0.4 equiv.)	109
9	D, aged 8 days	EtOH, RT (first portion, 0.3 equiv.)	79
10	N/A	Slurry API lot 3 in EtOH, 50°C, 3 hours	36
11	N/A	Recrystallize lot 1 in H_2O/EtOH	14

[a] See Table 9.4 for a description of HCL/ethanol lots.

Formation of compound A using aqueous HCl (12 N) was investigated. With a specially designed seeding strategy, polymorph control was achieved but a lower yield was observed. Solid and ML samples from this study contained 40 and 111 ppm of EtCl, respectively. This indicated potential EtCl formation during the salt formation as no EtCl was expected in the aqueous HCl. Formation of compound A using approximately 10% overcharge of HCl in EtOH or H_2O/EtOH produced a higher level of EtCl. At a lower temperature (20°C vs. 50°C), the EtCl level was reduced. Formation of compound A at a lower supersaturation before seeding (0.4 vs. 0.5 equiv.) reduced EtCl formation. Most importantly, it was found that recrystallization of compound A from H_2O/EtOH can reduce the EtCl from approximately 312 to 14 ppm. Minimal or no EtCl formation during the recrystallization was observed based on the levels in the solid and ML samples. The limitation of EtCl removal via recrystallization from a higher EtCl content seemed to be approximately 14 ppm and may indicate that the compound has some tendency to trap EtCl.

To investigate whether EtCl is formed in the solid sample of compound A, as the compound is an HCl salt and has the tendency to trap a certain amount of EtOH, a stress test was conducted by slurrying compound A with approximately 44 ppm EtCl in EtOH at 50°C for 3 hours. This experiment resulted in only approximately 33 ppm of EtCl (no increase). The EtCl level did not increase for a solid sample of compound A stored at ambient condition for more than 3 weeks. These studies indicate no EtCl formation for compound A in the solid phase when stored at ambient condition.

The original expectation was that because EtCl is normally a gas, even if it is formed during the contact of HCl and EtOH, the level in the final crystallized product should be low. The fact that compound A still tends to contain a certain level of EtCl (e.g., 14–427 ppm) even after recrystallization reveals that EtCl has a strong affinity to the compound. It is also possible that the residual EtCl was attributed to the EtCl dissolved in EtOH that was entrapped in the crystal lattice.

The level of EtCl in compound A produced from the previous scale-up campaigns was below the original staged threshold of toxicological concern (194 ppm for single dose). However, to further control the level of EtCl, the following strategies were established and implemented in the most recent campaign: (1) the HCl in EtOH solution was prepared on the same day before crystallization with a target concentration of approximately 2 N (instead of the 6–7 N used in the previous campaign) to minimize the formation of EtCl during storage and (2) the crystallization temperature was reduced to 20°C–25°C from 50°C with no overcharge of HCL. These strategies worked as expected in a recent 4 kg batch and successfully controlled the residual EtCl level to 37 ppm, lower than the 78 ppm from the previous batch and much lower than the later defined regulatory limit of 1606 ppm.

For future campaigns, new strategies were investigated to further control the EtCl level during scale-up, including (1) the isolation of crude compound A (instead of crude free base) followed by the recrystallization of compound A from the H_2O/EtOH system and (2) the application of aqueous HCl instead of HCl in EtOH. Both pathways showed promising results in terms of controlling polymorph, residual solvent level, and EtCl level (≤40 ppm).

9.5.1 Stability Study

Because compound A is observed to have measurable chloroethane levels, chloroethane was monitored during a stability study of the API to determine if the level increased, decreased, or remained stable. These data are shown in Table 9.7. Although it is not clear why there is more variability in the 25°C/60% relative humidity (RH) results, generally it appears that the chloroethane level is slowly decreasing for the API stored at 25°C/60% RH and more rapidly decreasing at the 40°C/75% RH storage conditions.

TABLE 9.7
Stability Results for Compound A

Time Point	Chloroethane (ppm)	
	25°C/60% RH	40°C/75% RH
Initial testing	40	40
3 months	42	20
6 months	28	12
9 months	36	Not tested
12 months	22	Not tested

9.6 CONCLUSION

A method was developed and validated to control chloroethane in APIs. This method was successfully used to optimize the synthetic process for the salt formation and final crystallization of an API. The levels of chloroethane were reduced by an order of magnitude during this optimization. The proposed specification for the phase development of this API was 1600 ppm. The chloroethane impurity was controlled two orders of magnitude below the proposed specification for the impurity in the API.

REFERENCES

1. Zeiger E.; Anderson B.; Haworth S.; Lawlor T.; Mortelmans K. *Salmonella* Mutagenicity Tests: V Results from the Testing of 311 Chemicals. *Environmental and Molecular Mutagenesis*. 1992, 19(Suppl. 21), 2–141.
2. National Toxicology Program, *Toxicology and Carcinogenesis Studies of Chloroethane (Ethyl Chloride) (CAS No 75-003) in F344/N Rats and B6C3F1 Mice (Inhalation Studies)*, U.S. Department of Health and Human Services, Research Triangle Park, North Carolina, Technical Report Series No. 346, 1989.
3. Ebert R.; Fedtke N.; Certa H.; Wiegand H.J.; Regnier J.F.; Marshall R.; Dean S.W. Genotoxicity Studies with Chloroethane. *Mutation Research*. 1994, 322, 33–44.
4. Agency for Toxic Substances and Disease Registry, *Toxicological Profile for Chloroethane*, U.S. Department of Health and Human Services, Public Health Service, Atlanta, Georgia, December 1998.
5. USEPA SW-846, *Test Methods for Evaluating Solid Waste, Method 8021B Aromatic and Halogenated Volatiles by Gas Chromatography Using Photoionization and/or Electrolytic Conductivity Detectors*, US EPA, Washington, DC, December 1996.
6. USEPA SW-846, *Test Methods for Evaluating Solid Waste, Method 8260B Volatile Organic Compounds by Gas Chromatography/Mass Spectrometry (GC/MS)*, US EPA, Washington, DC, December 1996.
7. United States Pharmacopeia and National Foundry, *General Chapter 467, Residual Solvents, USP-35/NF-30, Rockville, MD, USP, 2011.*
8. Zvetkova, V.I.; Firson, A.P.; Chirkov, H.M. Determination of the Real Rate Constant during the Decomposition of Aliphatic Alcohols. *Doklady Akademia Nauk SSSR*, 1959, 124, 139.

9. Yang Q.; Haney B.P.; Vaux A.; Riley D.A.; Heidrich L.; He P.; Mason P.; et al. Controlling the Genotoxins Ethyl Chloride and Methyl Chloride Formed during the Preparation of Amine Hydrochloride Salts from Solutions of Ethanol and Methanol. *Organic Process Research & Development*, 2009, 13, 786–791.
10. Teasdale A.; Delaney E.J.; Eyley S.C.; Jacq K.; Taylow-Worth K.; Lipczynski A.; Hoffmann W.; et al. A Detailed Study of Sulfonate Ester Formation and Solvolysis Reaction Rates and Application Toward Establishing Sulfonate Ester Control in Pharmaceutical Manufacturing Processes. *Organic Process Research & Development* 2010, 14, 999–1007.

10 Quantification of Genotoxic Impurities in Active Pharmaceutical Ingredients

Roy Helmy, Sophie Strickfuss, Mohammad Al-Sayah, Simon Hamilton, Xiaodong Bu, Claire Lee, Tiebang Wang, and Christopher Welch

CONTENTS

The presence of genotoxic impurities (GIs) within pharmaceutical products has been a subject of considerable recent interest following the issuance of guidelines from European [1,2] and U.S. [3] regulatory agencies. These guidelines typically mandate routine control of both documented GIs and potential genotoxic impurities (PGIs) at the parts-per-million level until suitable scientific data are acquired to dismiss the concern, significantly adding to the number of compounds impacted by these guidelines [4–6]. This chapter describes the challenges surrounding the trace analysis of GIs; however, the majority of the principles presented are similarly applicable to the analysis of PGIs.

Responding to this new GI trace analysis mandate has required considerable innovation in the creation of new methods for pharmaceutical synthesis, purification, and analysis. In particular, the accurate and reproducible trace analysis of GIs has been especially challenging, as many of the traditional methods used for routine pharmaceutical analysis are simply not suitable for GI analysis at the parts-per-million level. This problem is further exacerbated by the fact that many GIs owe their potential genotoxicity to an ability to chemically react with DNA. This propensity toward reactivity renders these GIs extremely difficult to analyze, as samples can easily degrade under standard sample preparation or analysis protocols. Nevertheless, analytical chemists within the pharmaceutical industry and academia have risen to the challenge, creating new methods and approaches for accurately measuring GIs—a vital first step to inform the development of a suitable control strategy. Excellent reviews of current GI methods based on alerting functional groups in the chemical structure of analytes have been published [7–12]. In this chapter, we survey the state of the art of this rapidly evolving analysis field, pointing out a number of current challenges, existing solutions, and still unmet needs.

10.1 A BRIEF INTRODUCTION TO GENOTOXIC IMPURITIES

All active pharmaceutical intermediates (APIs) will always contain impurities as a result of the imperfect nature of chemical reactivity. Typically, these API impurities originate from incomplete consumption of starting materials and intermediates in chemical reactions, formation of by-products, or decomposition of reagents and intermediates. Among this host of potential impurities, a few known as GIs are of special concern owing to their ability to react with DNA and cause genetic mutations. Additionally, genotoxins may also produce their effects by interfering with DNA replication, potentially leading to tumor development or other major health problems. When this damage can be transferred from cell to cell, or from generation to generation, the compounds are said to be mutagenic [5]. At very low levels of exposure, protective mechanisms can sometimes "correct" DNA damage, resulting in threshold effects [13]. However, demonstration of such a threshold of activity for genotoxic compounds is very difficult; hence, for most compounds a linear dose–reactivity relationship is assumed.

To ensure the safety of patients and healthy volunteers in clinical trials for developmental new drugs, the levels of impurities present in APIs must be controlled within appropriate limits. International Conference on Harmonisation of Technical Requirements for Registration of Pharmaceuticals for Human Use (ICH) guidelines Q3A and Q3B were collaboratively created by industry and regulators to address control of impurities in APIs and drug products, respectively [14,15]. These guidelines define the reporting, identification, and qualification thresholds of impurities in APIs and final drug products based on the projected clinical dose, route of administration, and additional mitigating factors. These guidelines recognize that the ICH-defined impurity thresholds are not appropriate guidance for impurities that are "unusually toxic" such as GIs.

To address this shortcoming, the Committee for Proprietary Medicinal Products of the European Medicines Evaluation Agency released a guidance document in 2007, which defined a new framework to address GIs in APIs [1]. The EMEA guidelines recommended that the structures of impurities detected in APIs, as well as those highly likely to form in APIs via chemistry arguments, are assessed in silico for PGI structural alerts with respect to an established public access database. The genotoxicity of these identified PGIs can be confirmed using a suitable bacterial reverse mutation test like the Ames test, but long-term in vivo studies in animals are required to determine whether a compound is a carcinogen [5,16,17].

For compounds with sufficient evidence for a threshold-related mechanism, limits can be calculated following the "permitted daily exposure" (PDE) approach as per ICH guideline Q3C for residual solvents [18]. This approach calculates a PDE derived from the "no observed effect level" or the "lowest observed effect level."

For compounds without sufficient evidence for a threshold-related mechanism, the EMEA guideline recommends that all due diligence is performed to avoid the presence of genotoxic compounds in an API, either by eliminating the particular reagent of concern from the synthesis or by using alternative synthetic processes to avoid the formation of the GI. This approach can often be difficult to implement, as the highly reactive nature of many synthetic intermediates and reagents (e.g., benzyl bromides and epoxides) is essential to their function in the chemical synthesis, but is often also the underlying cause of their genotoxicity. During early development of a new API, where understanding of chemical processes and impurity control are still relatively limited, such efforts are further complicated and become impractical. In cases where the presence of the GI cannot be avoided, levels should be reduced using reaction workup steps and/or purification technologies [5,19].

To establish acceptable levels of GIs in APIs, the EMEA guidance based its approach on a previous approach used by the Food and Drug Administration (FDA) to establish acceptable levels of contaminants leaching from food packaging [20], renaming it the "threshold of toxicological concern" (TTC). The TTC approach was established so as to set acceptable limits for GIs in APIs, which will not expose patients to a significantly increased risk of developing cancer during a lifetime of taking a given medication. Based on the carcinogenic potency in rodents of over 700 carcinogens, exposures of less than 0.15 μg/day were estimated to be a "virtually safe dose," unlikely to increase a lifetime cancer risk by more than 1 in 10^6, for all but the most potent carcinogens. Using this rationale, the EMEA recommended a limit of 1.5 μg/day for chronic exposure to a GI/PGI, representing an excess lifetime cancer risk of 1 in 10^5. This small level of added patient risk was justified by the significant positive health benefits received by the patient taking the medicine. The acceptable concentration of the GI in the API is calculated based on the expected dose as follows:

$$\text{Concentration limit (ppm)} = \frac{\text{TTC}\ (\mu\text{g/day})}{\text{Dose}\ (\text{g/day})} \tag{10.1}$$

The EMEA guidance acknowledges that lower TTC values should be used for compounds of high potency and that higher TTC values can be justified for short-term exposure. The EMEA guidance does not provide specific guidance for the limits of PGIs in investigational APIs during shorter duration clinical trials, leaving its applicability to development somewhat open to interpretation.

To address this issue, the Pharmaceutical Research and Manufacturers of America (PhRMA) proposed a staged TTC approach, which extrapolated the allowable daily intake (ADI) for GIs from lifetime levels [21]. This approach targeted a 1 in 10^6 level of added patient risk for early clinical trials less than 1 year in duration, since healthy volunteers do not receive any benefits from exposure to the API. For longer exposures, the risk is kept as 1 in 10^5 as only patients receiving a benefit from exposure to the API are likely to be used in such long-term clinical studies. The ADIs were therefore set as shown in Table 10.1.

The staged TTC approach was adopted by the EMEA in 2010 [2]. The daily allowable levels were calculated in a way similar to what was proposed by the PhRMA group, but a factor of 2 was introduced to account for deviations from the linear model. A similar TTC approach was also considered acceptable by the FDA in a draft guidance document in 2008 [3].

10.2 QUANTITATIVE ANALYSIS OF GENOTOXIC IMPURITIES

10.2.1 Method Development

Trace level (parts-per-million) quantitation of GIs presents the pharmaceutical analytical chemist with many technical challenges. First, an analytical technique appropriate for the properties of a GI (volatility, thermal stability, presence of a chromophore, hydrophobicity, etc.) must be selected as the basis for the analysis method to be developed. Second, the reactive nature and stability of the GI must be adequately addressed during method development to ensure that requisite reproducibility and accuracy are achieved. Third, the clinical dose and duration of the study must be understood well to guide the development of an adequately sensitive analysis method meeting project needs. The targeted allowable GI limit will directly impact the selection of critical parameters of the analysis method to be developed, including choice of detection technique (ultraviolet [UV], light scattering, electrochemical detection, mass spectrometry, etc.). Additionally, certain components of the sample matrix may also present substantial method development challenges due to analytical interference from the API itself, process impurities, or degradation products. The elimination of these interfering matrix components is typically achieved by (1) isolation of the analyte of interest by sample preparation, (2) chromatographic resolution, or (3) using a more selective detector.

A wide range of analytical techniques can be used to analyze GIs, depending on the properties of the analyte. Due to the high structural diversity of GIs, and the complexity of the sample matrix, no single approach is applicable to address all problems. However, analytical laboratories within the pharmaceutical industry and academia have developed systematic strategies to guide GI method development, which have proved to be quite useful [5,6,11,22–25]. The majority of these strategies

TABLE 10.1

Daily Allowable Limits for GIs as Specified by PhRMA, EMEA, and FDA

	Single Dose (µg/day)	<14 Days (µg/day)	14 Days to 1 Month (µg/day)	1–3 Months (µg/day)	3–6 Months (µg/day)	6–12 Months (µg/day)	>12 Months (µg/day)
PhRMA recommendations [21]	120	120	120	40	20	10	1.5
EMEA limits [2]	120	60	60	20	10	5	1.5
FDA limits [3]	120	120	120	20	10	5	1.5

Sources: Reprinted from Mueller et al., *Regul. Toxicol. Pharmacol.*, 44, 198–211, 2006 [21]; EMA Committee for Medicinal Products for Human Use, (EMA/CHMP/SWP/431994/2007 Rev. 3), 23 September 2010 [2]; U.S. Department of Health and Human Services, Food and Drug Administration, Center for Drug Evaluation and Research, December 2008 [3]. With permission.

consist of the following two steps: (1) evaluation of the volatility of the analyte, which typically informs the choice of chromatographic technique to be used, and (2) evaluation of the detection technique based on the properties of the analyte (e.g., presence of a chromophore and presence of a halogen atom within the molecule).

Traditionally, API impurity analysis in pharmaceutical laboratories has been carried out using high-performance liquid chromatography–ultraviolet (HPLC-UV) for nonvolatile compounds and gas chromatography–flame ionization detection (GC-FID) for volatile compounds; therefore, method development for GIs also typically begins from a chromatographic analysis dictated in large part by analyte volatility. For low parts-per-million detection of GIs, the use of nonspecific detectors such as flame ionization detection (FID) or UV may not be feasible owing to insufficient detector sensitivity and artifacts produced by minor interferences from the sample matrix. Hyphenated mass spectrometry techniques such as gas chromatography–mass spectrometry (GC-MS) and liquid chromatography–mass spectrometry (LC-MS) have gained popularity in GI analysis due to their superior sensitivity and selectivity [25]. For GC-MS, both electron ionization (EI) and chemical ionization ionization modes have been employed, with EI being the more popular approach [24,26]. Ionization modes for LC-MS include electrospray ionization (ESI) and atmospheric pressure chemical ionization (APCI). For some nonpolar and low-polarity compounds that are not efficiently ionized by either ESI or APCI, atmospheric pressure photoionization (APPI) may be used. Selectivity is usually improved by conducting selected ion monitoring (SIM) on a single quadrupole mass spectrometry (MS) or multiple reaction monitoring (MRM) on an MS-MS instrument. LC-MS-MS methods are typically used in the analysis of very complex samples where masses are likely to overlap despite the high resolution of the MS instrument. MRM mode provides a better signal to noise (*S/N*) ratio and reduced baseline offset as compared to SIM mode, resulting in better sensitivity and selectivity [24].

10.2.2 Sample Preparation and Derivatization

For situations where the chromatographic resolution of the analyte from the sample matrix is not sufficient for the development of a suitably sensitive and selective method, sample preparation is critical. Several extraction/preconcentration techniques to isolate and/or concentrate the analyte of interest from the sample matrix have been reported. In particular, techniques relying on the partitioning of the analyte of interest between two phases to enrich one of the phases in the analyte have been investigated. The simplest of these techniques, liquid–liquid extraction (LLE), where the analyte is portioned between two immiscible liquids, often requires an additional concentration step before analysis. Liquid-phase microextraction (LPME) uses a capillary hollow membrane filled with microliters of an extracting solvent, which is introduced in a liquid sample prepared in an immiscible solvent. The analyte is concentrated in the extracting solvent, which typically can be directly analyzed via chromatography. Additionally, solid-phase extraction (SPE) can be employed to partition the analyte of interest and other sample components between a solid stationary phase and a liquid mobile phase. Various stationary phases are commercially available to support SPE, including a large variety of reversed phases, normal phases, and ion exchange phases. An alternative to SPE, solid-phase microextraction (SPME) is a solvent-free extraction technique in which a polymer-coated fused silica fiber is exposed to a gas (headspace) or liquid sample (immersion). The analyte is partitioned between the sample matrix and the silica fiber stationary phase coating and subsequently desorbed at a high temperature in a gas chromatography (GC) injection port or redissolved in a solvent prior to analysis by GC or LC. These techniques can be applied successfully to the analysis of GIs but typically require method optimization and validation to address matrix effects before routine application to samples of interest, often making these techniques difficult to transfer between samples [27].

Derivatization may be used to improve analyte properties by enhancing volatility or stability, reducing polarity, or improving the extraction profile. However, development of derivatization procedures requires optimization of the reaction parameters (temperature, solvent, time, etc.) to maximize recovery, stability of the derivatized species, and selectivity of the derivatization reaction [28,29]. A generic derivatization procedure that produces the same product for a given class of GIs can be used, as in the method reported by Alzaga and others [28], where the same product is obtained from different alkylating agents. With this procedure, the derivatization reaction, shown in Figure 10.1, gives the same product for the methyl ester of methane sulfonic acid or of benzene sulfonic acid.

In this case, a good understanding of the impurities that may be present in the sample is required to ensure that the results obtained from the analysis can be correctly used to calculate the level of GIs present in the sample. Several in situ derivatization procedures have been designed to occur during the sample heating required for GC headspace sampling, thus simplifying the sample preparation procedure. Examples include the in situ derivatization of hydrazine [30] or of alkylating reagents [28,29]. Optimization of the parameters can be very complex as they must be suitable for both the derivatization and the headspace sampling [28].

10.2.3 Gas Chromatography

GC coupled with FID is commonly used in the analysis of volatile GIs [7,9,10,31]. For example, several mesylate esters were successfully analyzed in a drug substance using a simple GC-FID method with a limit of quantitation (LOQ) of 5 ppm

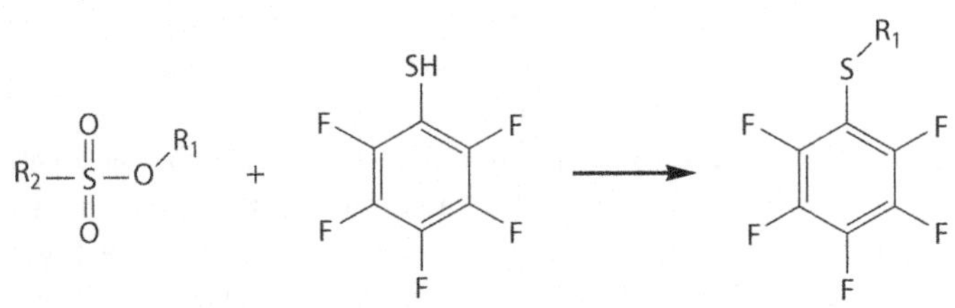

FIGURE 10.1 Derivatization reaction used by Alzaga and others for the derivatization of alkylating agents. (Reprinted from Alzaga et al., *J. Pharm. Biomed. Anal.*, 45, 472–479, 2007. With permission.)

in the drug substance [31]. If the GI has sufficient vapor pressure, headspace GC or dynamic headspace is usually preferred to minimize the contamination of the injection port or column head by nonvolatiles such as APIs [25]. As an example, a recently published report described the analysis of the oxidation catalyst 2,2,6,6-tetramethyl-1-piperidinyloxy in 13 batches of filibuvir drug substance using headspace-GC-MS [32]. Development of a headspace GC method requires optimization of the headspace parameters (temperature, solvent, and duration of heating); but the resulting method can often be applied to a range of similar compounds, such as structurally related alkylating agents [28,29]. In cases where the GI is not volatile enough for headspace GC analysis, direct injection techniques may be used [33]. In such instances, frequent cleaning of the injector is recommended, as the buildup of API impurities and associated thermal degradation products can interfere with GI quantitation [26,31]. In addition, the thermal stability of GIs must be assessed and appropriately accounted for to allow proper quantitation [33]. For GIs bearing a halogen such as alkyl and benzyl halides, electron capture detectors (ECDs) have been investigated due to the unique halogen selectivity of these detectors [7]. This technique is particularly useful due to its specificity if no other component of the sample matrix contains halogen atoms, but the overall sensitivity of the approach typically rivals that of FID for monohalogenated compounds [26]. The sensitivity of ECD detection can vary significantly depending on the nature and number of halogen atoms present on the molecule; hence, most published methods opt for MS detection [7,24]. The use of a nitrogen–phosphorus specific detector, also known as a thermionic specific detector, has also been reported for the analysis of trace formaldehyde in antibiotics after derivatization to the oxime with hydroxylamine [34].

GC-MS applications have been widely used in the low-level quantitation of volatile GIs when other nonselective detectors do not offer the required sensitivity. Sulfonic acids (methane sulfonic acid, benzene sulfonic acid, and *p*-toluene sulfonic acid) are commonly used in the synthesis of active pharmaceutical ingredients as acid catalysts, or as counterions to form salts to upgrade purity or control pharmaceutical properties. Crystallization in solvent systems containing alcohols can potentially lead to the formation of corresponding sulfonate esters, which are well-publicized GIs [21]. Similarly, the presence of other strong inorganic acids (e.g., HCl) in solvent systems containing alcohols can lead to the formation of alkyl halides, which are also considered GIs [21]. These volatile sulfonate esters and small polar alkyl halides have been analyzed using very sensitive GC-MS methods such as the one shown in Figure 10.2 [35]. This method utilizes an Rtx®-200 column and if used in the SIM mode can achieve detection limits as low as 1 ppm. Excellent linearity, precision, and spike and recovery values have all been reported.

GC-MS has also been used for the quantitation of hydrazine at levels as low as 0.1 ppm. In one approach developed by Sun and others [30], hydrazine was first converted to an acetone azine by in situ derivatization with acetone, which was then analyzed by headspace GC-MS. Residual parts-per-million levels of alkylating agents, epoxides, halogenated sulfides, and halogenated amines have all been successfully analyzed by GC-MS as well [7,24,25,28].

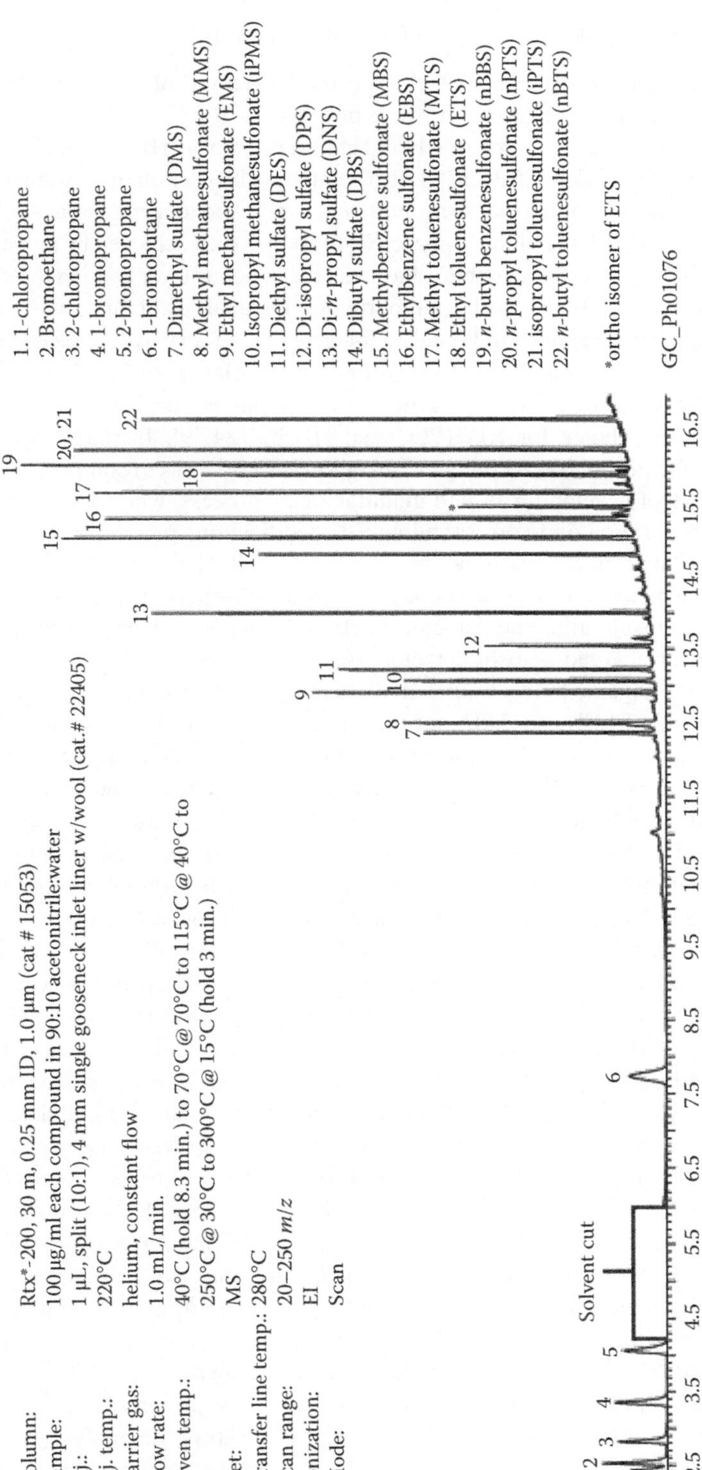

FIGURE 10.2 Small, polar alkyl halides are well resolved on the fluorinated Rtx®-200 column, as are less volatile sulfonate esters. (Reprinted from Restek Corporation, Bellefonte, Pennsylvania. With permission.)

10.2.4 High-Performance Liquid Chromatography

HPLC is widely used as a separation technique for the analysis of nonvolatile GIs. The most widely used separation mode is reverse-phase (RP) HPLC [7,9,10]. For extremely polar analytes, hydrophilic interaction liquid chromatography (HILIC) can be used to achieve sufficient retention [25,36,37]. HILIC employs eluents rich in acetonitrile with low aqueous content, a mobile phase mixture that is believed to form a water-rich layer at the surface of the polar stationary phase. Separation is then achieved by partitioning the polar analytes between the mobile phase and the aqueous layer coating the stationary phase. The use of HILIC has been reported for the analysis of quaternary ammonium ions, which were either compounds of interest (quaternary hydrazine derivatives [37]) or derivatization products (of alkyl sulfonates and dialkyl sulfates [38]).

For the detection of analytes bearing a chromophore, the UV detector is the conventional first choice for RP-HPLC and HILIC [24,39]. In particular, liquid chromatography–ultraviolet techniques can be very useful tools to control GIs during the early stages of developing an API manufacturing process, where clinical studies are shorter in duration and corresponding GI specifications may be looser and not require quantitation at low parts-per-million limits [39]. However, UV detection is often neither sensitive nor selective enough to achieve low parts-per-million levels of detection. As a result, alternate detectors such as evaporative light scattering detectors (ELSDs) or charged aerosol detectors (CADs) may be used, as applied to the analysis of carcinogenic pyrrolizidine alkaloids in plant products [40]. However, the examples of using ELSDs or CADs for the analysis of GIs in APIs are quite limited owing to the lack of robustness and/or selectivity needed for GI analysis. One major disadvantage of ELSDs is that the response can be highly dependent on analyte volatility and eluent composition. CAD response is also dependent on analyte volatility [41]. CAD is becoming more popular due to its more universal response, wide dynamic range, precision, and simplicity. The use of a chemiluminescent nitrogen detector has also been reported for the analysis of two nitrogen-bearing GIs, hydrazine and 1-1-dimethylhydrazine, as shown in Figure 10.3 [36].

LC-MS and LC-MS-MS applications have been widely used in low-level quantitations of nonvolatile GIs when UV detectors, ELSDs, or CADs do not afford the required sensitivity or specificity. Analytes containing basic nitrogens can often be easily detected by atmospheric pressure ionization LC-MS in the positive ion mode due to their high proton affinity. Similarly, acidic analytes can often be detected in the negative ion mode via deprotonation [39]. Analytes with weak acidic groups are better ionized by negative mode APCI than ESI [24]. For neutral compounds, coordination ion spray–MS can be used to detect adducts (Li^+, Na^+, K^+, and NH_4^+) in the positive ion mode [24]. APPI can also be efficiently used to ionize neutral and low-polarity analytes. For very reactive intermediates, the compound of interest is often converted to a stable, easily ionized derivative prior to analysis. Direct analysis of GIs such as aziridines, arylamines, and aminopyridines by LC ESI-MS in SIM mode has been reported [25]. Aldehydes and ketones can be derivatized with 2,4-dinitrophenyl hydrazine, and the corresponding derivatives can be analyzed by LC-MS using ESI in the negative ion mode [25]. Figure 10.4 shows the analysis of 1 ppm solutions of derivatized formaldehyde, propionaldehyde, hexanal, and nonanal [25].

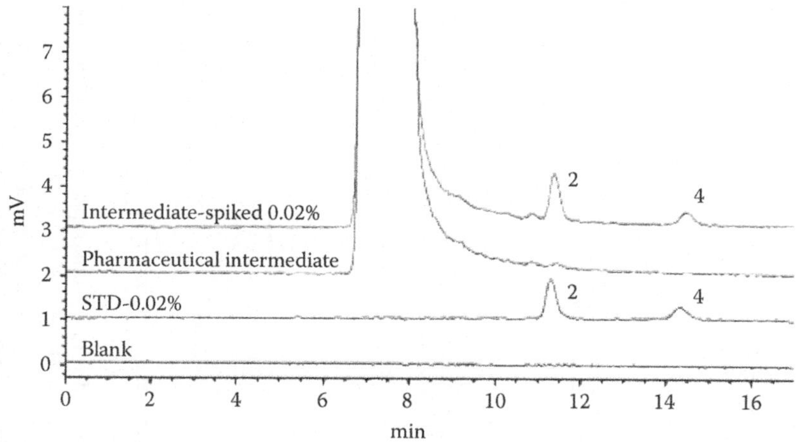

FIGURE 10.3 Simultaneous analysis of 1,1-dimethylhydrazine and hydrazine in an active pharmaceutical ingredient on a hydrophilic interaction liquid chromatography column. The drug sample contained 10 mg/mL of the intermediate in dimethyl sulfoxide (DMSO)/ethyl alcohol (30/70, v/v). The spiked intermediate sample consisted of 10 mg/mL intermediate, 2 μg/mL 1,1-dimethylhydrazine, and 2 μg/mL hydrazine in DMSO/ethyl alcohol (30/70, v/v). The standard (STD) solution had 2 μg/mL each of 1,1-dimethylhydrazine and hydrazine in water/ethyl alcohol (30/70, v/v). 2: 1,1-Dimethylhydrazine and 4: hydrazine. Column temperature was 30°C with a flow rate of 0.4 mL/min using a splitter and a chemiluminescent nitrogen detector (CLND). The CLND system was set at 1050°C combustion furnace, 50 mL/min argon, 280 mL/min oxygen, 75 mL/min makeup (argon), 30 mL/min ozone, 5°C cooler, gain ×1, and 750 V on Photomultiplier (PMT). Other conditions were as in the study by Liu and others. (Reprinted from Liu et al., *J. Chromatogr. A.*, 1216, 2362–2370, 2009. With permission.)

GC-FID and GC-MS are highly sensitive techniques for the analysis of sulfonic acid esters, but many of the aromatic sulfonates have very high boiling points and poor GC sensitivity. LC-MS has been demonstrated as a highly sensitive analytical technique, with limits of detection (LODs) less than 0.1 ppm being easily achievable [42].

Figure 10.5 shows the methyl, ethyl, and isopropyl esters of toluenesulfonic acid at 1 ppm concentration with respect to the API. The MS utilized SIM in the positive ion electrospray mode, monitoring *m/z* 204, 218, and 232 (the ammonium adducts of each compound). The instrument's scan switches between SIM ions at appropriate times in the chromatographic run to capture the relevant peaks. Although all three ions could be monitored simultaneously, selecting a specific ion by time increases the duty cycle of the instrument and thus improves the sensitivity. Good linearity and recovery are observed [42].

10.2.4.1 Multiple Reaction Monitoring

Once the mainstay of low-level metabolite and peptide quantitation, triple quadrupole mass spectrometers are now finding uses in the determination of GIs. MRM experiments allow the differentiation of isobaric compounds based on their unique fragment ions generated after a collisionally induced dissociation, known as a transition. This highly selective detection technique can eliminate background

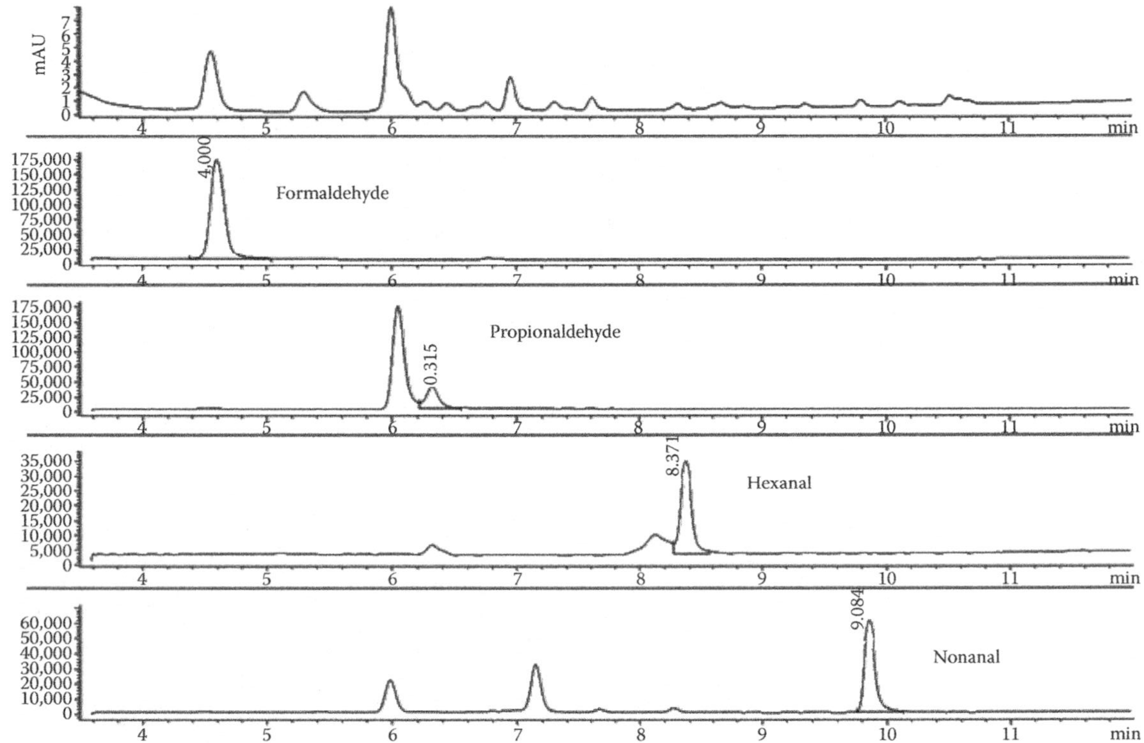

FIGURE 10.4 Liquid chromatography–mass spectrometry analysis of 1 ppm solutions of aldehydes derivatized with 2,4-dinitrophenyl hydrazine. The chromatogram is shown on top as well as the extracted ion chromatograms of formaldehyde (ion 209.1), propionaldehyde (ion 237.1), hexanal (ion 279.1), and nonanal (ion 321.1). Conditions: column: Zorbax StableBond Phenyl column (15 cm × 3 mm inside diameter × 3.5 μm, Agilent, Santa Clara, California), using a 10 mM ammonium acetate in water/acetonitrile gradient from 36% acetonitrile to 100% acetonitrile at 12.5 minutes with a 0.65 mL/min flow rate. Other conditions were as listed by David and others. (Reprinted from David et al., *LC-GC Eur.*, 22, 552, 554–556, 558–561, 2009. With permission.)

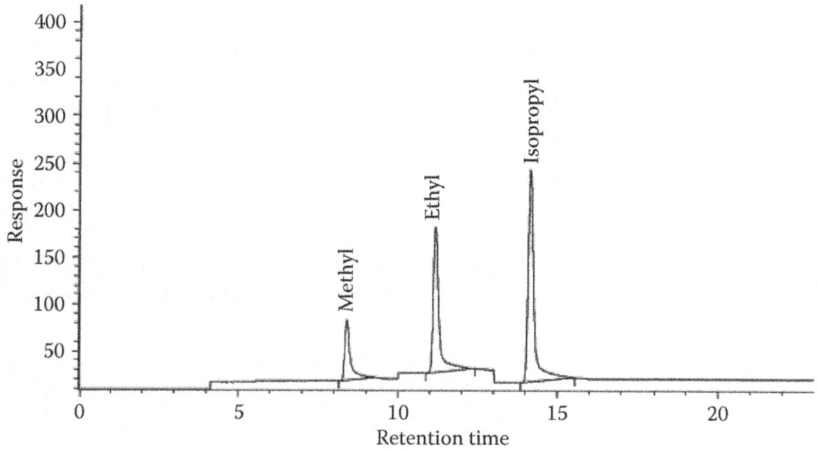

FIGURE 10.5 High-performance liquid chromatography–mass spectrometry chromatogram of an active pharmaceutical ingredient spiked with 1 ppm tosyl esters using a 50 μL injection volume and a 50 mg/mL sample solution. Conditions: Zorbax RXC8 25 cm × 0.46 cm column (5 μm particles); flow rate: 1.5 mL/min, split to 0.3 mL/min into the MS source. Eluent A: 0.1 w/v% ammonium acetate in 25 v/v% methanol in water; eluent B: 0.1 w/v% ammonium acetate in 95 v/v% methanol in water. The gradient was linear from 23% to 50% B in 16 minutes and 50% to 100% B in 1 minute, which was followed by 100% B hold for 6 minutes and reequilibration for 4 minutes. The column temperature was 45°C. Electrospray in positive mode was used with single ion monitoring SIM for $[M^{+}NH4]^{+}$ ions. The fragmentor voltage (*V*), gain (EMV), and dwell time (millisecond) were 70, 1.0, and 580, respectively. The drying gas temperature (degree Celsius), drying gas flow (liter per minute), nebulization pressure (pounds per square inch gauge [psig]), and capillary voltage (*V*) were 350, 10, 40, and 4000, respectively. Other conditions were as per the study by Taylor and others. (Reprinted from Taylor et al., *J. Chromatogr. A.*, 1119, 231–237. 2006. With permission.)

interference from chromatography systems and significantly improve signal to noise, and therefore improve detection limits. Triple quadrupole mass analyzers can be coupled to both HPLC and GC systems, and in both cases they can enable the analysis of low-level GIs in complex matrices. In addition, the fragment ions and neutral loses observed can be used as indications of similar molecular functionality.

In our laboratories, we recently demonstrated the selectivity and sensitivity of a triple quadrupole mass spectrometer utilizing an MRM scan. A number of sulfonate esters were detected to sub–parts per million levels with a very fast chromatographic separation. In each case, the monitored transition was that associated with the loss of the alkylsulfonate substituent; an example of such a transition is shown in Figure 10.6.

A further application of the MRM experiment includes the determination of a substituted hydrozinopyridine. The available HPLC-UV method was only capable of a GI detection limit of 100 ppm, with a target API concentration of 2 mg/mL (Figure 10.7). Monitoring the transition, using an MRM scan, from *m/z* 144 to *m/z* 117 (Figure 10.8) using a Sciex API3000 triple quadruple mass spectrometer enabled

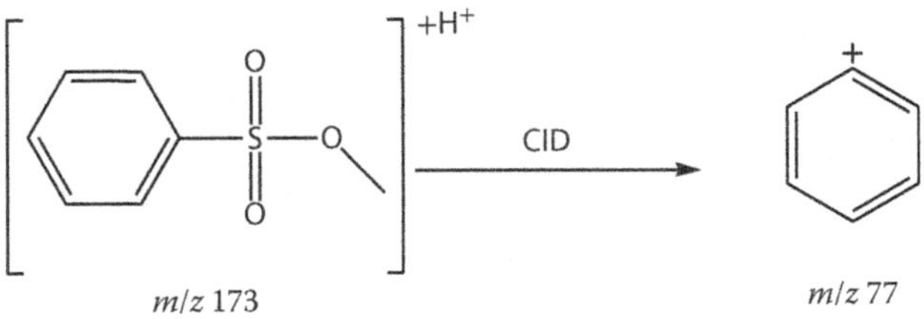

FIGURE 10.6 Schematic of the fragmentation pathway of benzenesulfonic acid methylester. Abundant transition from *m/z* 173 to *m/z* 77 by collisionally induced dissociation enables parts-per-million detection limits.

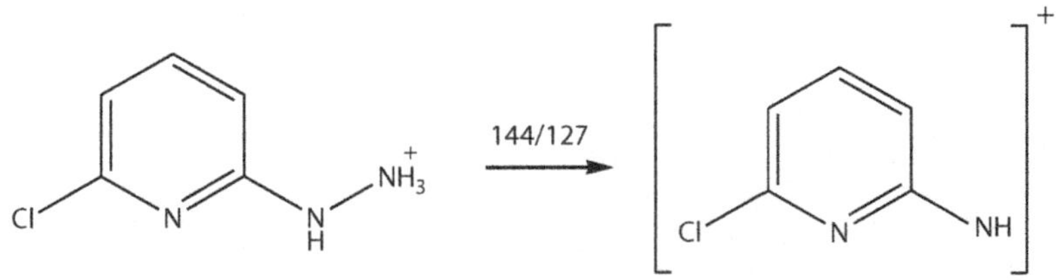

FIGURE 10.7 Ultraviolet chromatogram of 100 ppm chlorohydrazinopyridine. Peak elutes approximately 7 minutes. Waters Xterra RP8, 15 cm × 4.6 mm, 3.5 μm. Mobile phase A: phosphate buffer at pH 5.8. B: acetonitrile. Gradient method in 30 minutes, flow = 1.0 mL/min, λ = 230 nm, and injection = 10 μL.

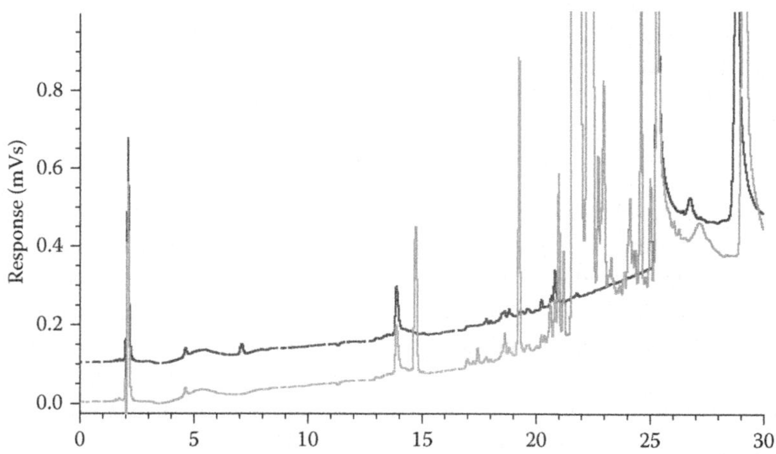

FIGURE 10.8 Schematic of the fragmentation pathway of substituted hydrozinopyrdine. Abundant transition from *m/z* 144 to *m/z* 127.

the detection limit to be reduced to 1 ppm with very little method development (Figure 10.9). The huge improvement in selectivity enabled significant reductions in chromatographic run time, provided there was no ion suppression associated with coeluting peaks. In this example, a recovery greater than 80% was achieved in a 6-minute isocratic separation.

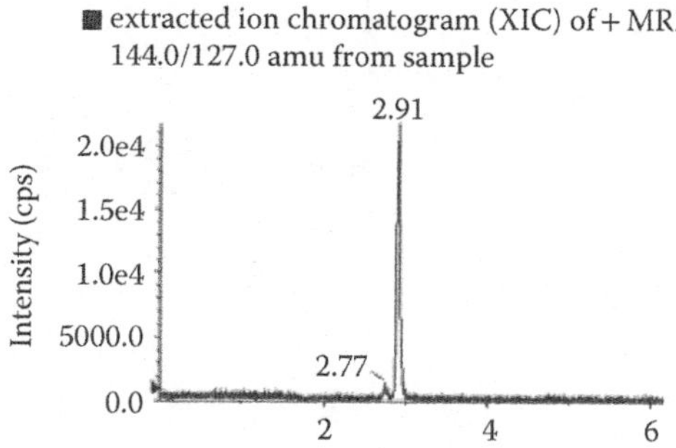

FIGURE 10.9 Total ion chromatogram of 1 ppm (with respect to active pharmaceutical ingredient concentration). Chlorohydrazinopyridine. Conditions: Waters Xterra RP8, 15 cm × 4.6 mm, 3.5 μm. Mobile phase A: 0.05% (v/v) formic acid in water; B: 0.05% (v/v) formic acid in acetonitrile. Isocratic 20:80 (v/v) A/B for 5 minutes. Flow: 0.5 mL/min; injection: 10 μL; mass spectrometry: electrospray ionization positive, multiple reaction monitoring (144/127).

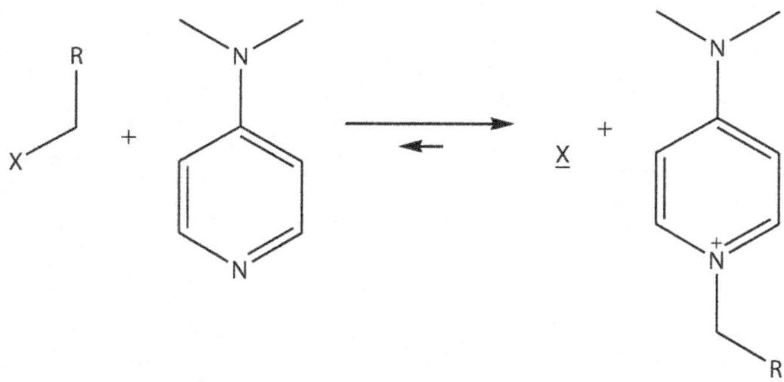

FIGURE 10.10 Reaction scheme of 4-dimethylaminopyridine derivatization of alykyl-halides to afford quaternary nitrogen species.

Generic screening methods for alkyl halides have been demonstrated [43], using 4-dimethylaminopyridine (4-DMAP) as a derivatizing agent. The reaction scheme is shown in Figure 10.10.

The nature of the derivatizing agent is such that any potential reactive alkylating agents will be derivatized. This facet of the sample preparation ensures that all potential alkylating agents in the sample matrix will react and therefore be detected, not just those species initially targeted from a theoretical scrutiny of the synthetic route. The resulting quaternary nitrogen species provided excellent sensitivity in ESI+MS and can be easily resolved from the sample matrix using HILIC. The derivatives of a number of GIs are shown in Table 10.2 along with the fragment ions and DMAP fragments.

Initial screening can be conducted using a precursor ion scan, or a neutral loss scan of *m*/*z* 122, which is indicative of the 4-DMAP fragment. Peaks displaying the fragments of *m*/*z* 121 and 123 or a neutral loss of *m*/*z* 122 can be assumed to have reacted with the derivatizing agent and thus are a potential alkylating species. The main drawback to this approach is that alkyl halides with the same alkyl group molecular weight, but different halogens, cannot be differentiated since the resulting derivatives are isobaric. However, knowledge of the synthetic process would enable a structure to be proposed.

TABLE 10.2
Overview of the Compounds Amenable to Derivatization and Subsequent Mass Information Obtained by Orbitrap Detection

Component		Fragment (m/z)			Loss of 122 Da in MS^2
Code	Name	Derivative	Specific Fragment	DMAP-Related Fragment	(Yes/No)
S1	4-(Dibromomethyl)-benzoic acid ethyl ester	363[a]	n.a.	n.a.	n.a.
S2	Ethyl 4-(bromomethyl)benzoate	285	257, 135	123	Yes
S3	Methyl 4-(bromomethyl)benzoate	271	149	121	Yes
S4	4-(Bromomethyl)benzoic acid	257	135	123	Yes
S5	Component A	419	391, 253	n.d.	No[b]
S6	Ethyl 4-(chloromethyl)benzoate	285	163	123	Yes
S7	Methyl 4-(chloromethyl)benzoate	271	149	121	Yes
S8	4-(Chloromethyl)benzoic acid	257	135	123	Yes
S9	Component B	419	391, 253	n.d.	No[b]
S10	9-Chloromethyl-anthracene	313	191	n.d.	Yes
S11	9,10-bis-Chloromethyl-anthracene	361	239	123	Yes
S15	Methyl iodide	137	n.d.	121	No
S16	3-(Bromomethyl)-l,1′-biphenyl	289	167	n.d.	Yes
S17	Clindamycin B	n.d.	n.a.	n.a.	n.a.
S18	2-Chloro-*N,N*-diethylamine	222	100	n.d.	Yes
S19	1-(3-Chloropheny1)-4-(3-chloropropy1)piperazine	359	237	n.d.	Yes
S20	2,6-Dichlorobenzyl bromide	281	159	123	Yes
S21	Chloroacetyl chloride	199[a]	n.d.	123	No
S22	2-Chloroethyl methyl ether	181	149	123	No
S24	Methyl *p*-toluene sulfonate	137	n.d.	121	No
S26	1-(4-Azidophenyl)-2 bromoethanonone	282	254	n.d.[c]	No
S28	Tertiary butyl bromoacetate	237	181	123	No

Sources: Reprinted from van Wijk et al., *Anal. Bioanal. Chem.*, 400, 1375–1385, 2011. With permission.

Note:

[a] Only under water-free conditions

[b] Fragment m/z 269, i.e., neutral loss of 122 in MS3

[c] Fragment m/z 123 in MS3 n.a., not applicable; n.d., not detected.

Additional examples of the use of MRM MS have been reported for the analysis in APIs of quaternary hydrazine derivatives [37] and methyl methanesulfonate (MMS) and ethyl methanesulfonate [44]. For both these examples, improved sensitivity was reported thanks to the removal of interferences present in the sample matrix.

10.2.4.2 High-Resolution Mass Spectrometry

The development of highly sensitive, high-resolution mass spectrometers such as quadrupole time of flight (QToF) and Orbitrap instrumentation has led to a paradigm shift in metabolite identification and quantitation. Since metabolite analysis combines the analysis of small molecules in complex matrices, it stands to reason that these techniques will also lend themselves to the analysis of GIs in drug substances and products. The main benefit of QToF and Orbitrap mass analyzers is that all ions are detected, rather than scanning a specific *m/z*, as is the case for quadrupole instruments. Thus, the need to know the expected GIs prior to running the experiment becomes less important. The data can also be interrogated retrospectively if additional GIs are expected to be present. In addition, the very high resolution of these instruments can reduce the effect of isobaric background interference, by selecting an extracted ion chromatogram over a very narrow mass window, for example, *m/z* 0.01.

10.2.5 Ion Chromatography

Ionic GIs can be particularly challenging to analyze due to their high polarity and are often derivatized to allow analysis by GC or HPLC [9,36]. The use of ion chromatography (IC) as an alternative technique has been reported for the direct analysis of a variety of GIs or GI precursors [45–48], and IC methods have been shown to be sensitive enough for the determination of GIs, as, for example, shown in Table 10.3 for the analysis of monoethylsulfate in Indinavir sulfate drug substance [48].

TABLE 10.3
Comparison of CE and IC Method Validation Data for the Analysis of Monoethylsulfate in Indinavir Sulfate Drug Substance

Parameter	CE	IC
System precision (%RSD of injection)	<5	<2
Linearity range (μg/mL)	Between 4 and 15	Between 0.075 and 1.2
Recovery (%)	89.2	97.1
LOD (ng/mL)	880[a]	24
LOQ (ng/mL)	2422	74
Run time	8 minutes	25 minutes

Sources: Reprinted from Prasanna et al., *J. Pharm. Biomed. Anal.*, 50, 1065–1069, 2009. With permission.

Note: LOD and LOQ values were calculated based on residual standard deviation (RSD) method from linearity data from sensitivity experiment.

[a] Precision was not established.

An advantage of IC is that the two principal detectors typically used, conductivity and amperometry, allow for direct detection without the need for derivatization. However, applications to GI analysis in APIs remain limited to a relatively small subset of ionic compounds.

10.2.6 Element-Specific Analysis by High-Resolution Inductively Coupled Plasma–Mass Spectrometry

Much of the focus on the assessment of GI alerts has been on specific classes of alkyl esters of sulfonates and phosphonates, haloalkenes, and primary halides, as shown in Figure 10.11 [21]. Each of these GIs contains elements that are not typically encountered in APIs, allowing the possibility of element-specific detection approaches.

These nonspecific approaches have been used to quantitate sulfur, phosphorus, or halogen-containing GIs using high-resolution inductively coupled plasma–mass spectrometry (HR-ICP-MS). The total amount of any of the aforementioned elements can be measured and compared to TTC limits, assuming that the residual levels of these elements detected are due solely to the GIs. Compared to traditional approaches, direct determination of total sulfur, phosphorous, and halogens provides a rapid and straightforward method for analysis of GIs that requires only minimal method development. In addition, the approach is equally applicable to volatile and nonvolatile GIs. However, if total residual sulfur, phosphorous, and so on limits exceed TTC limits, more specific analyses are required to deconvolve the non-compound-specific result.

10.2.7 Sulfur, Phosphorus, Chlorine, Bromine, and Iodine Detection

The new-generation sector field HR-ICP mass spectrometers are capable of a resolution as high as 14,000 ($m/\Delta m$), which is crucial to achieving a robust, interference-free, and accurate determination of these target elements at subparts per million levels [49–55]. Table 10.4 lists the major isobaric interferences and the resolution required to resolve them for all the isotopes of sulfur, phosphorous, Cl, and Br [56–58]. For example, accurate ICP-MS determination of $^{37}Cl^+$ is impossible with quadrupole ICP-MS because of the interference from $^{36}Ar^1H^+$. Figure 10.12 illustrates the contrast of the $^{37}Cl^+$ spectrum at medium and high resolutions. Clearly, at medium resolution $^{37}Cl^+$ and $^{36}Ar^1H^+$ interferences cannot be totally baseline resolved, whereas analyses at higher resolutions are adequate to avoid the overlap.

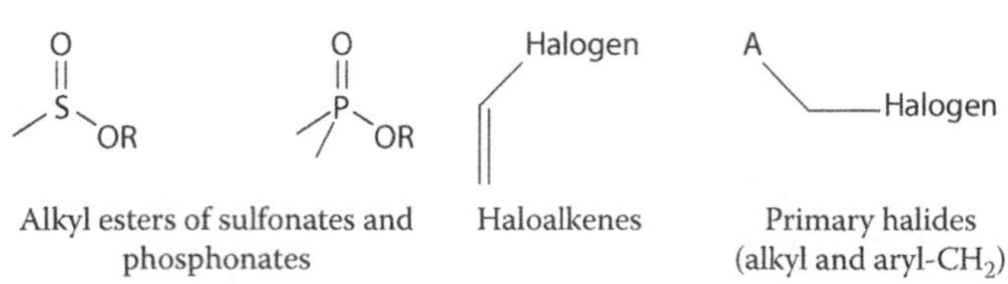

FIGURE 10.11 Some examples of sulfur-, phosphorus-, and halogen-containing genotoxic impurities. (Reprinted from Mueller et al., *Regul. Toxicol. Pharmacol.*, 44, 198–211, 2006. With permission.)

TABLE 10.4
Resolution Required to Separate Analyte Ions from Interfering Ions

Isotope	Interference	Resolution Required ($m/\Delta m$)
^{32}S	$^{16}O^{16}O$	1801
	$^{64}Zn^{2+}$	4562
	$^{15}N^{16}O^{1}H$	1040
	$^{14}N^{18}O$	1508
^{34}S	$^{18}O^{16}O$	1297
	$^{17}O^{16}O^{1}H$	1000
	$^{33}S^{1}H$	2977
	$^{68}Zn^{2+}$	6238
^{31}P	$^{14}N^{16}O^{1}H$	970
^{35}Cl	$^{18}O^{16}O^{1}H$	1058
^{37}Cl	$^{1}H^{36}Ar$	3899
^{79}Br	$^{40}Ar^{38}Ar^{1}H$	5405
^{81}Br	$^{40}Ar^{40}Ar^{1}H$	4964

Similarly, a high resolution is required to fully resolve $^{81}Br^{+}$ from the interference of $^{40}Ar^{40}Ar^{1}H^{+}$, as shown in Figure 10.12.

10.2.8 Instrumental Limit of Detection

The instrumental LODs of all five elements at medium- and high-resolution modes by HR-ICP-MS are listed in Table 10.5 [59]. The results clearly demonstrate that subparts per million LODs can be achieved with HR-ICP-MS. It is also interesting to note that although an increase in the mass resolution did lead to lower signal intensities for all four elements, this did not necessarily translate into higher LODs. This is because a higher resolution not only reduces ion transmission but also reduces the background transmission.

10.2.8.1 Comparison of GC-FID, GC-MS, and HR-ICP-MS

MMS, a known GI, is used here to demonstrate that the use of HR-ICP-MS provides results that are comparable with those provided by the more commonly used LC- and GC-based techniques. As shown in Table 10.6, although not as sensitive as GC-MS, HR-ICP-MS is more than adequate for the required TTC limits of up to a 12-month duration study.

10.2.9 Acceptable Elemental Limits Calculation for Genotoxic Impurity Alerts

The information needed for the calculation of total elemental limits of GI alerts includes the following: molecular weight of the GI, projected human dose (milligrams per day), and staged TTC limits (micrograms per day) based on duration of

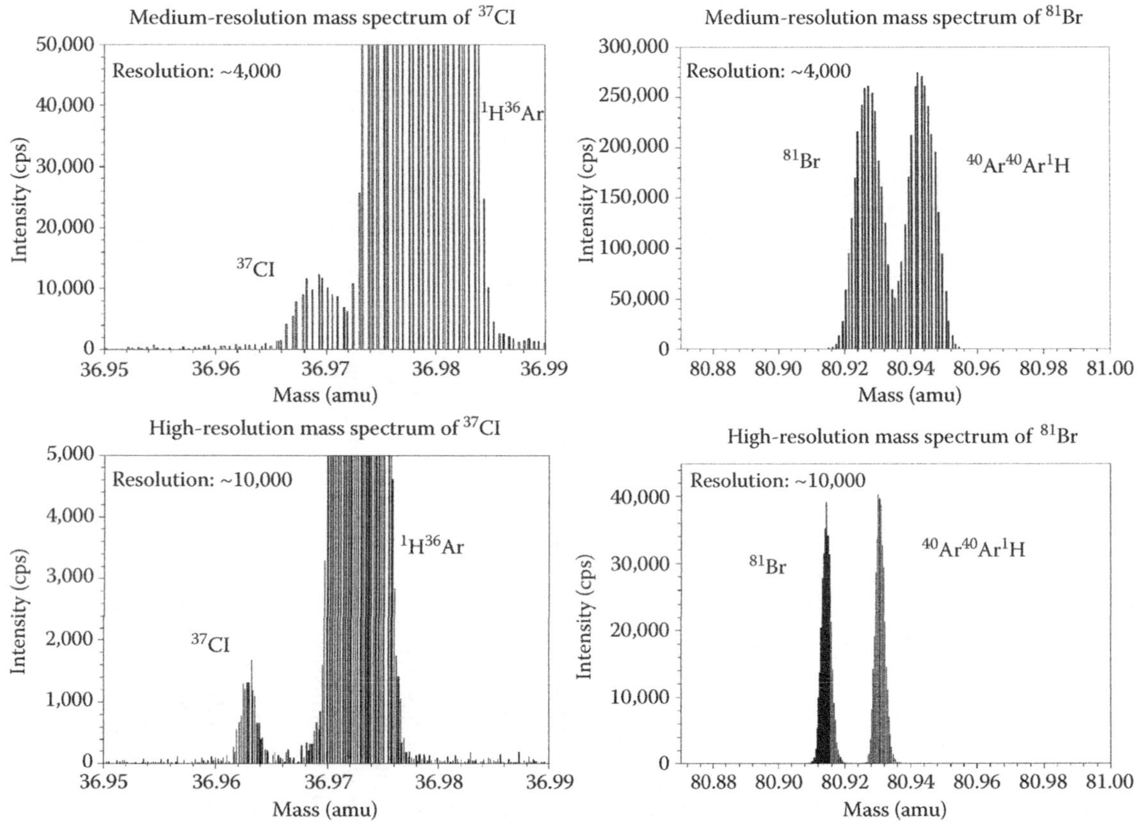

FIGURE 10.12 High-resolution inductively coupled plasma–mass spectrometry spectral comparison of ^{37}Cl and ^{81}Br at medium- and high-resolution modes. (Reprinted from Bu et al., *J. Anal. At. Spectrom.*, 18, 1443–1451, 2003. With permission.)

TABLE 10.5
Detection Limits in Medium Resolution and High Resolution

Element	MR Limits of Detection		HR Limits of Detection	
	LOD (µg/L)	LOD in solid[a] (µg/g)	LOD (µg/L)	LOD in solid[a] (µg/g)
^{35}Cl	7.01	1.40	3.25	0.65
^{37}Cl	N/A	N/A	4.18	0.84
^{79}Br	0.23	0.05	0.08	0.02
^{81}Br	N/A	N/A	0.10	0.02
^{127}I	0.03	0.01	0.05	0.01
^{32}S	1.41	0.28	1.40	0.28
^{34}S	2.39	0.48	1.94	0.39
^{31}P	2.15	0.43	1.98	0.40

Note: HR, high resolution; MR, medium resolution; N/A, not applicable.
[a] LODs for the solid samples are based on 5 mg/mL target sample concentration.

TABLE 10.6
Comparison of GC-FID, GC-MS, and HR-ICP-MS

	GC-FID Method	GC-MS Single-Quad Method	HR-ICP-MS Total S Method
Run time (minute)	18	8	3
Target conc. (mg/mL)	100	10	5
LOD (ppm)	13	1	4
Detection	FID 220°C	Mass Spec Aux 280°C SIM 80 *m/z*, 109 *m/z*, 203 *m/z*	Mass Spec S32 elemental, medium resolution

exposure [21]. To calculate acceptable elemental limits for GI alerts, use the following equation:

$$\text{Total Limit}\left(\mu g/g\right)=\frac{\text{AW of the Element}}{\text{MW of the GI}}\times\frac{\text{TTC Limit}\left(\mu g/\text{day}\right)}{\text{Projected Dosage}\left(g/\text{day}\right)}$$

where MW stands for molecular weight.

For example, the total elemental S limit for MMS alerts of a Merck API based on a 1-month study and a 2.7 g/day clinical dose is given as follows:

$$\text{Acceptable total S}=\frac{32}{114}\times\frac{60\,\mu g/\text{day}}{2.7\,g/\text{day}}=6.2\,\mu g/g=6.2\text{ ppm}$$

For GIs containing specific elements not otherwise present in the API or sample matrix, HR-ICP-MS offers a powerful addition to the GI analysis toolbox, in particular for GIs for which standards are not available.

10.2.10 Validation

Depending on the stage of development of the program, or whether the analysis is carried out on the final API or on an early process intermediate, GI methods validation does not necessarily need to be carried out to the same extent. The simplest approach is the limit test approach, which is a pass/fail test using the toxicology limit as a one point calibration. Four samples can demonstrate the suitability of this approach: a blank, the API sample, the one point calibration standard, and the API sample spiked with the standard. The blank injection can demonstrate the lack of interference, and the spiked sample assesses the percentage recovery to ensure method suitability. The result would be reported as less than one point calibration standard, 40 ppm in this case as shown in Figure 10.13. This approach only enables the assessment of selectivity of the method, as well as the recovery from the sample matrix. It does not provide any information on method linearity or precision and does not allow tracking of the GI during process development [22].

On the other hand, especially if used on commercialized APIs, GI methods might need to be fully validated to meet the regulatory expectations of marketed products. Validation might include specificity, sensitivity, linearity, precision, accuracy, and stability. Regulatory and ICH guidelines provide guidance on the level of validation depending on whether a limit test or a quantitative method is used [60]. Additional validation may be deemed necessary, in particular to assess the robustness of the method, especially if the method needs to be transferred between sites, or to a manufacturing environment.

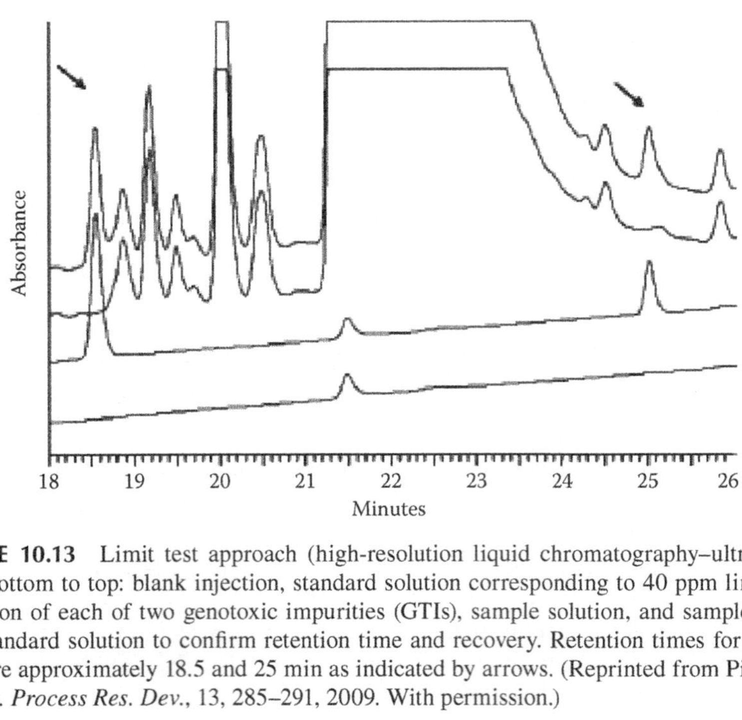

FIGURE 10.13 Limit test approach (high-resolution liquid chromatography–ultraviolet). From bottom to top: blank injection, standard solution corresponding to 40 ppm limit concentration of each of two genotoxic impurities (GTIs), sample solution, and sample spiked with standard solution to confirm retention time and recovery. Retention times for the two GTIs are approximately 18.5 and 25 min as indicated by arrows. (Reprinted from Pierson et al., *Org. Process Res. Dev.*, 13, 285–291, 2009. With permission.)

10.3 FUTURE OUTLOOK AND CONCLUSIONS

The analysis of GIs in pharmaceutical ingredients and products has been a topic of intense focus since the introduction of the EMEA regulations in 2006. Since that time, analytical chemists have been rapidly developing improved tools for dealing with the GI analysis issue, with hyphenated mass spectrometric methods rapidly becoming the preferred analytical methods. While pharmaceutical analytical chemists strive to develop generic or systematic approaches to method development to support broad drug candidate portfolios, GI analysis still remains a very labor intensive process requiring methods optimization to address sample-specific properties. At the present time, the field of GI analysis is evolving, with a steady stream of better and more sensitive analytical approaches being developed for addressing this problem. Clearly, accurately predicting the future in such circumstances is difficult, but it is fairly certain that the development of systematic approaches toward GI analyses is here to stay.

REFERENCES

1. Committee for Medicinal Products for Human Use, Committee for Medicinal Products, European Medicines Agency, (CPMP/SWP/5199/02, EMEA/CHMP/QWP/251344/2006). "Guideline on the limits of genotoxic impurities," 28 June 2006.
2. EMA Committee for Medicinal Products for Human Use, (EMA/CHMP/SWP/431994/2007 Rev. 3). "Questions and answers on the "Guideline on the limits of genotoxic impurities,"" 23 September 2010.
3. U.S. Department of Health and Human Services, Food and Drug Administration, Center for Drug Evaluation and Research (CDER). "Guidance for industry. Genotoxic and carcinogenic impurities in drug substances and products: Recommended approaches. Draft Guidance," December 2008.
4. Snodin, D.J. "Genotoxic impurities: From structural alerts to qualification," *Org. Process Res. Dev.* 2010, 14, 960–976.
5. Robinson, D.I. "Control of genotoxic impurities in active pharmaceutical ingredients: A review and perspective," *Org. Process Res. Dev.* 2010, 14, 946–959.
6. Sun, M.; Liu, D.Q.; Kord, A.S. "A systematic method development strategy for determination of pharmaceutical genotoxic impurities," *Org. Process Res. Dev.* 2010, 14, 977–985.
7. Elder, D.P.; Lipczynski, A.M.; Teasdale, A. "Control and analysis of alkyl and benzyl halides and other related reactive organohalides as potential genotoxic impurities in active pharmaceutical ingredients," *J. Pharm. Biomed. Anal.* 2008, 48, 497–507.
8. Elder, D.P.; Snodin, D.; Teasdale, A. "Analytical approaches for the detection of epoxides and hydroperoxides in active pharmaceutical ingredients, drug products, and herbals," *J. Pharm. Biomed. Anal.* 2010, 51, 1015–1023.
9. Elder, D.P.; Snodin, D.; Teasdale, A. "Control and analysis of hydrazine, hydrazides and hydrazones. Genotoxic impurities in active pharmaceutical ingredients (APIs) and drug products," *J. Pharm. Biomed. Anal.* 2011, 54, 900–910.
10. Elder, D.P.; Teasdale, A.; Lipczynski, A.M. "Control and analysis of alkyl esters of alkyl and aryl sulfonic acids in novel active pharmaceutical ingredients," *J. Pharm. Biomed. Anal.* 2008, 46, 1–8.
11. Elder, D.P.; Snodin, D.; Teasdale, A. "Analytical approaches for the detection of epoxides and hydroperoxides in active pharmaceutical ingredients, drug products, and herbals," *J. Pharm. Biomed. Anal.* 2010, 51, 1015–1023.

12. Teasdale, A., Ed. *Genotoxic Impurities—Strategies for Identification and Control*; John Wiley & Sons: Hoboken, NJ, "Impurities in new drug substances," 2010.
13. Delaney, E.J. "An impact analysis of the application of the threshold of toxicological concern concept to pharmaceuticals," *Regul. Toxicol. Pharmacol.* 2007, 49, 107–124.
14. International Conference on Harmonisation of Technical Requirements for Registration of Pharmaceuticals for Human Use, "Impurities in new drug products," 2006. ICH Q3A.
15. International Conference on Harmonisation of Technical Requirements for Registration of Pharmaceuticals for Human Use, 2006. ICH Q3B.
16. Kirkland, D.; Aardema, M.; Henderson, L.; Mueller, L. "Evaluation of the ability of a battery of three in vitro genotoxicity tests to discriminate rodent cardinogens and non-carcinogens. I. Sensitivity, specificity and relative predictivity," *Mutat. Res.* 2005, 584, 1–256.
17. U.S. Department of Health and Human Services, Food and Drug Administration, Center for Drug Evaluation and Research. "Guidance for industry and review staff. Recommended approaches to integration of genetic toxicology study results," January 2006.
18. International Conference on Harmonisation of Technical Requirements for Registration of Pharmaceuticals for Human Use, "Impurities: Guideline for residual solvents Q3C (R5)," 2011.
19. Maddula, S.R.; Kharkar, M.; Manudhane, K.; Kale, S.; Bhori, A.; Lali, A.; Dubey, P.K.; Sarma, K.R.J.; Bhattacharya, A.; Bandichhor, R. "Preparative chromatography technique in the removal of isostructural genotoxic impurity in Rizatriptan: Use of physicochemical descriptors of solute and adsorbent," *Org. Process Res. Dev.* 2009, 13, 683–689.
20. US Food and Drug Administration (FDA). "Food additives: Threshold of regulation for substances used in food-contact articles," 21 CFR, 170.39, revised April 1, 2013.
21. Mueller, L.; Mauthe, R.J.; Riley, C.M.; Andino, M.M.; De Antonis, D.; Beels, C.; De George, J., et al. "A rationale for determining, testing, and controlling specific impurities in pharmaceuticals that possess potential for genotoxicity," *Regul. Toxicol. Pharmacol.* 2006, 44, 198–211.
22. Pierson, D.A.; Olsen, B.A.; Robbins, D.K.; DeVries, K.M.; Varie, D.L. "Approaches to assessment, testing decisions, and analytical determination of genotoxic impurities in drug substances," *Org. Process Res. Dev.* 2009, 13, 285–291.
23. Raman, N.V.V.S.S.; Prasad, A.V.S.S.; Ratnakar, R.K. "Strategies for the identification, control and determination of genotoxic impurities in drug substances: A pharmaceutical industry perspective," *J. Pharm. Biomed. Anal.* 2011, 55, 662–667.
24. Liu, D.Q.; Sun, M.; Kord, A.S. "Recent advances in trace analysis of pharmaceutical genotoxic impurities," *J. Pharm. Biomed. Anal.* 2010, 51, 999–1014.
25. David, F.; Jacq, K.; Vanhoenacker, G.; Sandra, P.; Baker, A. "Method selection for trace analysis of genotoxic impurities in pharmaceuticals," *LC-GC Eur.* 2009, 22, 552, 554–556, 558–561.
26. Li, H.; Sluggett, G.W. "Development and validation of a sensitive GC-MS method for the determination of trace levels of an alkylating reagent in a beta-lactam active pharmaceutical ingredient," *J. Pharm. Biomed. Anal.* 2005, 39, 486–494.
27. Colon, I.; Richoll, S.M. "Determination of methyl and ethyl esters of methanesulfonic, benzenesulfonic, and *p*-toluenesulfonic acids in active pharmaceutical ingredients by solid-phase microextraction (SPME) coupled to GC/SIM-MS," *J. Pharm. Biomed. Anal.* 2005, 39, 477–485.
28. Alzaga, R.; Ryan, R.W.; Taylor-Worth, K.; Lipczynski, A.M.; Szucs, R.; Sandra, P. "A generic approach for the determination of residues of alkylating agents in active pharmaceutical ingredients by in situ derivatization-headspace-gas chromatography-mass spectrometry," *J. Pharm. Biomed. Anal.* 2007, 45, 472–479.

29. Lee, C.R.; Guivarch, F.; Nguyen, V.D.C.; Tessier, D.; Krstulovic, A.M. "Determination of polar alkylating agents as thiocyanate/isothiocyanate derivatives by reaction headspace gas chromatography," *Analyst* 2003, 128, 857–863.
30. Sun, M.; Bai, L.; Liu, D.Q. "A generic approach for the determination of trace hydrazine in drug substances using in situ derivatization-headspace GC-MS," *J. Pharm. Biomed. Anal.* 2009, 49, 529–533.
31. Li, W. "Trace analysis of residual methyl methansulfonate, ethyl methanesulfonate and isopropyl methanesulfonate in pharmaceuticals by capillary gas chromatography with flame ionization detection," *J. Chromatogr. A.* 2004, 1046, 297–301.
32. Strohmeyer, H.E.; Sluggett, G.W. "Determination and control of TEMPO, a potentially genotoxic free radical reagent used in the synthesis of filibuvir," *J. Pharm. Biomed. Anal.* 2012, 62, 216–219.
33. Klick, S. "Evaluation of different injection techniques in the gas chromatographic determination of thermolabile trace impurities in a drug substance," *J. Chromatogr. A.* 1995, 689, 69–76.
34. Manius, G.J.; Wen, L.F.L.; Palling, D. "Three approaches to the analysis of trace formaldehyde in bulk and dosage form pharmaceuticals," *Pharm. Res.* 1993, 10, 449–453.
35. Rigdon, A.; Lake, R.; Lee, C.; Helmy, R.; Strulson, C.; Figus, M. "Two options for analyzing potential genotoxic impurities in active pharmaceutical ingredients," *Restek Advantage Chromatography Newsletter.* 2008, 3, 20–21.
36. Liu, M.; Ostovic, J.; Chen, E.X.; Cauchon, N. "Hydrophilic interaction liquid chromatography with alcohol as a weak eluent," *J. Chromatogr. A.* 2009, 1216, 2362–2370.
37. Hmelnickis, J.; Pugovics, O.; Kazoka, H.; Viksna, A.; Susinskis, I.; Kokums, K. "Application of hydrophilic interaction chromatography for simultaneous separation of six impurities of mildronate substance," *J. Pharm. Biomed. Anal.* 2008, 48, 649–656.
38. An, J.; Sun, M.; Bai, L.; Chen, T.; Liu, D.Q.; Kord, A. "A practical derivatization LC/MS approach for determination of trace level alkyl sulfonates and dialkyl sulfates genotoxic impurities in drug substances," *J. Pharm. Biomed. Anal.* 2008, 48, 1006–1010.
39. Liu, D.Q.; Chen, T.K.; McGuire, M.A.; Kord, A.S. "Analytical control of genotoxic impurities in the pazopanib hydrochloride manufacturing process," *J. Pharm. Biomed. Anal.* 2009, 50, 144–150.
40. Schaneberg, B.T.; Molyneux, R.J.; Khan, I.A. "Evaporative light scattering detection of pyrrolizidine alkaloids," *Phytochem. Anal.* 2004, 15, 36–39.
41. Yuabova, Z.Y.; Holschlag, D.R.; Rodriguez, S.A.; Qin, C.; Papov, V.V.; Qiu, F.; McCaffrey, J.F.; Norwood, D.L. "Genotoxic impurities: A quantitative approach," *J. Liq. Chromatogr. Relat. Technol.* 2008, 31, 2318–2330.
42. Taylor, G.E.; Gosling, M.; Pearce, A. "Low level determination of *p*-toluenesulfonate and benzenesulfonate esters in drug substance by high performance liquid chromatography/mass spectrometry," *J. Chromatogr. A.* 2006, 1119, 231–237.
43. van Wijk, A.M.; Beerman, B.; Niederlaender, H.A.G.; Siebum, A.H.G.; de Jong, G. J. "A new approach for generic screening and quantitation of potential genotoxic alkylation compounds by pre-column derivatization and LC-MS/MS analysis," *Anal. Bioanal. Chem.* 2011, 400, 1375–1385.
44. Kakadiya, P.R.; Reddy, B.P.; Singh, V.; Ganguly, S.; Chandrashekhar, T.G.; Singh, D.K. "Low level determination of methyl methanesulfonate and ethyl methanesulfonate impurities in lopinavir and ritonavir active pharmaceutical ingredients by LC/MS/MS using electrospray ionization," *J. Pharm. Biomed. Anal.* 2011, 55, 379–384.
45. Fernando, P.N.; Egwu, I.N.; Hussain M.S. "Ion chromatographic determination of trace hydroxylamine in waste streams generated by a pharmaceutical reaction process," *J. Chromatogr. A.* 2002, 956, 261–270.

46. Jagota, N.K.; Chetram, A.J.; Nair, J.B. "Determination of trace levels of hydrazine in the penultimate intermediate of a novel anti-infective agent," *J. Pharm. Biomed. Anal.* 1998, 16, 1083–1087.
47. Jagota, N.K.; Nair, J.B.; Kurtulik, P.T. "Ion chromatography of methanesulfonic acid in pharmaceuticals," *J. Pharm. Biomed. Anal.* 1995, 13, 1291–1295.
48. Prasanna, S.J.; Sharma, H.K.; Mukkanti, K.; Sivakumaran, M.; Kumar, K.S.R.P.; Kumar, V.J. "Validation of a sensitive ion chromatography method for determination of monoethylsulfate in Indinavir sulfate drug substance," *J. Pharm. Biomed. Anal.* 2009, 50, 1065–1069.
49. Aggarwal, S.K. *Advanced Trace Analysis,* Edited by Susanta Lahiri; Alpha Science International: Oxford, UK, 2010.
50. Lewen, N. "The use of atomic spectroscopy in the pharmaceutical industry for the determination of trace elements in pharmaceuticals," *J. Pharm. Biomed. Anal.* 2011, 55, 653–661.
51. Moldovan, M.; Krupp, E.M.; Holliday, A.E.; Donard, O.F.X. "High resolution sector field ICP-MS and multicollector ICP-MS as tools for trace metal speciation in environmental studies: A review," *J. Anal. At. Spectrom.* 2004, 19, 815–822.
52. Evans, P.; Wolff-Briche, C.; Fairman, B. "High accuracy analysis of low level sulfur in diesel fuel by isotope dilution high resolution ICP-MS, using silicon for mass bias correction of natural isotope ratios," *J. Anal. At. Spectrom.* 2001, 16, 964–969.
53. Golik, V.M.; Golik, S.V.; Trepachev, S.A.; Kuz'mina, N.V. "Determination of B, Si, P, S, Cl, and Br in uranium materials by inductively coupled plasma mass spectrometry," *J. Anal. Chem.* 2010, 65, 1511–1516.
54. Marchante-Gayon, J.M.; Muniz, C.S.; Alonso, J.I.G.; Sanz-Medel, A. "Analysis of biological materials by double focusing-inductively coupled plasma-mass spectrometry (DF-ICP-MS)," *Adv. At. Spectrosc.* 2002, 7, 117–177.
55. Shah, M.; Caruso, J.A. "Inductively coupled plasma mass spectrometry in separation techniques: Recent trends in phosphorus speciation," *J. Sep. Sci.* 2005, 28, 1969–1984.
56. Carr, J.E.; Dill, A.E.; Kwok, K.; Carnahan, J.W.; Webster, G.K. "LC-ICP-MS for nonmetal selective detection of pharmaceuticals," *Curr. Pharm. Anal.* 2008, 4, 206–214.
57. Vanhaecke, F. "ICP-MS alternative means for the elimination of interferences," *Anal. Bioanal. Chem.* 2002, 372, 20–21.
58. Vanhaecke, F.; Moens, L. "Recent trends in trace element determination and speciation using inductively coupled plasma mass spectrometry," *Fresenius' J. Anal. Chem.* 1999, 364, 440–451.
59. Bu, X.; Wang, T.; Hall, G. "Determination of halogens in organic compounds by high resolution inductively coupled plasma mass spectrometry (HR-ICP-MS)," *J. Anal. At. Spectrom.* 2003, 18, 1443–1451.
60. International Conference on Harmonisation of Technical Requirements for Registration of Pharmaceuticals for Human Use, "Validation of analytical procedures: Text and methodology Q2 (R1)," 2005.

11 Analytical Testing and Control for Genotoxic Impurities in Drug Substances

Jane Li and Larry Wigman

CONTENTS

11.1 INTRODUCTION

The heightened awareness and need for control of genotoxic impurities (GTIs) present increasing challenges and concerns during drug development. Their toxicity at trace levels requires accurate analysis of these impurities at the parts-per-million level or lower, posing significant challenges to analytical methodologies. For both the pharmaceutical industry and the health authorities alike, the characterization and control of GTIs to levels considered acceptable for patient safety is an evolving and complicated process. Drug companies strive to find ways to avoid the introduction of GTIs in drug substances. However, the intrinsic nature of synthetic chemistry renders this avoidance an insurmountable task due to the necessary use of electrophilic agents for carbon–carbon and carbon–nitrogen bond formation and due to formation from side reactions. Therefore, as a common practice, known and potential GTIs are strategically controlled to meet safety requirements by regulatory authorities. One essential part of this strategic effort is the development of sensitive, specific, and robust analytical methodologies to enable the understanding of these impurities, including their origins of generation and fate through the process leading to the final drug substances. Chromatography combined with conventional detection has generally been the method of choice to achieve specificity and robustness with acceptable sensitivity. The advancement in new analytical instrumentation and technologies has pushed sensitivity for GTIs to levels unattainable by conventional methods. This chapter intends to provide an overview of GTIs from regulatory, toxicology, and analytical testing and control perspectives by combining the authors' own experience with approaches shared by the pharmaceutical industry.

11.2 REGULATORY AND TOXICOLOGICAL OVERVIEW

In an ideal world, drugs would contain no impurities since the impurities are generally of no therapeutic value. To minimize impurities, steps should be taken during drug development to avoid, control, or remove potential impurities. Impurities in drugs may originate from many sources, including raw materials, reagents, intermediates, synthetic by-products, leachables, and extractables due to incompatibilities with excipients and packaging, and degradation products. The control and toxicological qualification of impurities has been well understood for many years and harmonized internationally (International Conference on Harmonisation of Technical Requirements for Registration of Pharmaceuticals for Human Use [ICH] Q3 guidelines).[1] In these guidelines, the concept that some impurities "are unusually potent" or "produce toxic or unexpected pharmacological effects" is highlighted; however, the concept of genotoxicity is not specifically addressed.

Genotoxic compounds are generally reactive with or intercalate into DNA, leading to cancer or heritable changes in patients, with carcinogenicity being more easily detected. Genotoxicity includes induced genetic damage and fixation, gene mutation, larger scale chromosomal damage, and recombination and numerical chromosome changes. Genotoxicity is characterized by a standard test battery (ICH S2).[2] Guidelines to specifically address GTIs were first introduced in 2002 by the European Medicines Agency (EMA).[3] This initial guideline acknowledged the gaps in ICH Q3[1] and proposed a default threshold of toxicological concern (TTC) for GTIs to minimize patient risk as follows: "A TTC value of 1.5 μg/day intake of a GTI is considered to be associated with an acceptable risk (excess cancer risk of <1 in 100,000 over a lifetime) for most pharmaceuticals. From this threshold value, a permitted level in the active substance can be calculated based on the expected daily dose. Higher limits may be justified under certain conditions such as short-term exposure periods." The TTC was based on a lifetime exposure; to define a strategy for shorter term exposures, the concept of a "staged TTC" was developed by a Pharmaceutical Research and Manufacturers of America working group and published by Mueller and others.[4] The staged TTC approach has been recognized and incorporated into both Food and Drug Administration (FDA) (draft)[5] and EMA[6] guidances. An internationally harmonized guidance is in preparation.[7] The current guidances were compared and summarized.[8] Although there are some differences between the guidances, the staged TTC concepts are similar, as shown in Table 11.1.

The staged TTC concept allows analytical method development to evolve along with a clinical drug candidate as the drug advances from early phase development to commercialization. Ultimately, the analytical impact of staged TCC means that dose-dependent analysis potentially to low parts-per-million levels is typically needed for the drug substance (e.g., for long-term dosing of a 500 mg/day drug product, limits for GTIs would be 3.0 μg/g). These limits are approximately 1000 times lower than those for a nongenotoxic impurity, as described in ICH Q3.[1] Many GTIs are intrinsically reactive, and process chemists often do not expect them to persist through a multistep synthesis. Therefore, regulatory specifications may not always be needed for these products; however, methods capable of demonstrating the purging power of the chemical process to these low levels are needed to ensure patient safety. Several recent articles have provided overviews of control strategies in active pharmaceutical ingredient (API) development and analytical considerations in developing methods for GTIs.[9–12] ICH guideline M7, *Assessment and Control of DNA Reactive (Mutagenic) Impurities in Pharmaceuticals to Limit Potential Carcinogenic Risk*, is under development as one way to recognize the existence of mechanisms leading to a dose response that is nonlinear or has a practical threshold. This applies not only to compounds that interact with non-DNA targets but also to DNA-reactive compounds, whose effects may be modulated by, for example, rapid detoxification before coming into contact with DNA, or effective repair of induced damage. The

TABLE 11.1
Staged TTC Limits

Duration of Exposure	FDA Limit (μg/day)	EMA Limit (μg/day)
Single dose	120	120
<14 Days	120	60
1–3 Months	20	20
3–6 Months	10	10
6–12 Months	5.0	5.0
>12 Months	1.5	1.5

TABLE 11.2
Classification of GTIs

Class 1	Impurities known to be both genotoxic and carcinogenic	Analytical methods for testing and control of the GTIs are required
Class 2	Impurities known to be genotoxic but with unknown carcinogenic potential	
Class 3	Impurities containing alerting structures, unrelated to the structure of the API, and of unknown genotoxic potential	
Class 4	Impurities containing alerting structures, related to the structure of the API	Impurities are monitored and controlled as nongenotoxic impurities; no GTI methods are required
Class 5	Impurities with no alerting structures, or where sufficient evidence exists that genotoxicity is absent	

regulatory approach to such compounds can be based on the identification of a critical no-observed-effect level and use of uncertainty factors (ICH Q3C) when data are available rather than the more general TTC concept.

GTIs have been widely studied and modeled, and classified into five categories,[13] which are summarized in Table 11.2.

The alerting structures shown in Figure 11.1 are known to be involved in reactions with DNA. The figure does not contain all potential alerting structures, and a more thorough analysis of a particular structure can be performed using various software programs whose results correlate well with the results from ICH S2[2] and are often used by regulatory agencies for predictive purposes.[14–16]

11.3 ANALYTICAL APPROACHES FOR TESTING AND CONTROL OF GENOTOXIC IMPURITIES

To build rational and sufficient control strategies for GTIs, it is crucial to have analytical methods that are sensitive, selective, and robust for analyzing the impurities. The methodology for analyzing GTIs depends on target specifications and expected values for these impurities. In some cases, the level of a GTI is expected to be well below the targeted specification and supported by robust analytical data and the selected method may therefore be described as a limit test to alleviate routine testing burdens.

For most of the potential and confirmed GTIs, analytical methods of superior sensitivity are required to support the development of control strategies because of the low TTC limit of 1.5 μg/day set by the regulatory authorities for commercial products. To achieve the necessary sensitivity for low parts per million–level determinations, techniques with both selective separation and highly sensitive detection are generally required. Thus, hyphenated techniques such as gas chromatography–mass spectrometry (GC-MS), liquid chromatography–mass spectrometry (LC-MS), or LC-MS/MS are often applied to resolve and selectively quantify GTIs.

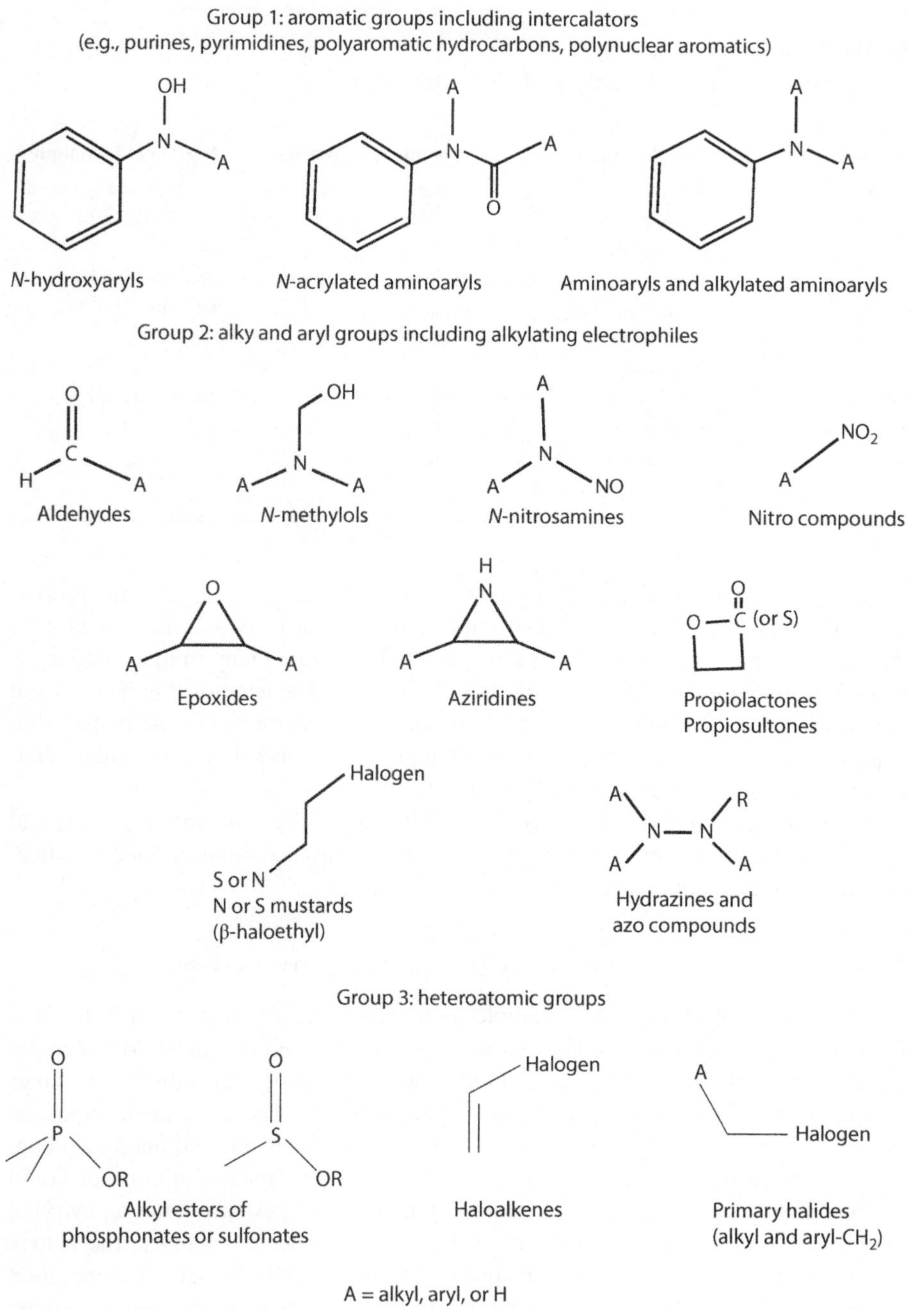

FIGURE 11.1 Alerting structures of genotoxic impurities. (Adapted from Mueller et al., *Regul. Toxicol. Pharmacol.*, 44, 198–211, 2006.)

TABLE 11.3
Properties of GTIs and Analytical Considerations

Genotoxicity Properties	Examples	Chemical Properties	Analytical Techniques
DNA reactive	Alkylating electrophiles, epoxides, azo compounds	Unstable, polar, low molecular weight	Derivatization followed by GC-MS or LC-MS
	Aromatic amines, aromatic nitro compounds	Stable, moderate polarity, variable molecular weight	Direct analysis by GC-MS or LC-MS
DNA intercalation	Polyaromatic hydrocarbons, polynuclear aromatics, cytostatic molecules	Stable, low to moderate polarity, moderate to high molecular weight	Direct analysis by LC-MS

Selection of the appropriate analytical technique is directly related to the properties of the GTI. The great variety of structures, reactivities, responsiveness to selected detection methods, and potential matrix effects has posed many unique challenges to analytical method development for these impurities. The genotoxic properties and the chemical properties of GTIs can be correlated in general terms and paired with appropriate analytical techniques in an attempt to minimize the development challenges. This logic is illustrated in Table 11.3.

Sections 11.3.2 and 11.3.3 are organized following this consideration and logic to describe the development of appropriate analytical methods and provide practical examples for each approach.

11.3.1 Sample Preparations for Genotoxic Impurity Analysis

Direct analysis of GTIs without sample pretreatment has been successful, but it is not without its challenges. In many cases, determination of GTIs down to low parts-per-million levels in the presence of high concentrations of drug substances makes sample preparation very critical in method development. Excessive presence of the matrix can detrimentally affect analytical method development, making it challenging or impossible to achieve the desired sensitivity. Headspace sampling of liquid or solid samples provides a simple sampling method for gas chromatography (GC) analysis of volatile analytes including GTIs. For complicated samples, the matrix needs to be removed by applying an appropriate sample pretreatment or preparation technique. This usually involves extraction of the analytes from the sample matrix. Well-established extraction techniques include liquid–liquid extraction (LLE) and solid-phase extraction or solid-phase microextraction (SPME). These extraction methods are routinely applied for robust and accurate analysis by GC or liquid chromatography (LC). For highly volatile or reactive GTIs that may be labile to GC or LC conditions, sample preparations generally involve derivatization to improve the stability of the GTIs and to render them suitable for accurate analysis. Derivatization has been widely and successfully employed to enhance the method capabilities and detector sensitivities for both high-performance liquid chromatography (HPLC) and GC.

11.3.2 Analysis of Genotoxic Impurities by Gas Chromatography

11.3.2.1 Direct Genotoxic Impurity Analysis by Gas Chromatography

Many methods have been reported for GTI analysis by GC coupled with flame ionization detection (FID), mass spectrometry (MS), or other specific detectors.[17–23] When possible, direct analysis without extensive sample preparation is preferred to streamline the method development and testing requirements. This approach can be achieved by increasing sample concentrations and column loading and has shown to meet the testing needs for different GTIs. Li[18] reported a gas chromatography–flame ionization detection (GC-FID) method for the trace analysis of methyl methanesulfonate (MMS), ethyl methanesulfonate (EMS), and isopropyl methanesulfonate (IMS) in pharmaceutical drug substances. Samples at 40 mg/mL were dissolved in mixtures of acetonitrile and water and directly injected into a gas chromatograph in the splitless mode. The method achieved a detection limit of approximately 1 ppm and a quantitation limit of 5 ppm for the analysis of these esters in pharmaceuticals, although quantitation of these compounds at levels less than 5 ppm would require a more sensitive GC-MS method. In another example for atenolol analysis, residual levels of the GTIs allyl chloride, 1,3-dichloro-2-propanol, and 2,3-dichloro-1-propanol were determined by direct injection GC-SIM-MS (SIM stands for single ion monitoring) analysis of standard and sample solutions.[24] The method was able to achieve a limit of quantitation (LOQ) of 3 ppm in relation to 50 mg/mL atenolol concentration. Both methods suffered the major issue of matrix accumulation and contamination of injector liners, leading to poor method recovery. Therefore, the liners needed replacement or cleaning after every 15–20 injections, making the methods less desirable for routine or high-throughput analysis.

Headspace GC-MS analysis of trace-level benzene, a by-product in the synthesis of a Genentech (South San Francisco, California) API starting material, was performed at sample concentrations of 200 mg/mL. Benzene is a known human carcinogen with an ICH concentration limit of 2 ppm. The analytical separation was performed using an Agilent DB-624 column (30 m × 0.32 mm × 1.8 μm). With selected ion detection at *m/z* 78, benzene levels as low as 0.1 ppm were readily quantitated. Figure 11.2 is a total ion current chromatogram of benzene at a concentration of 0.25 ppm and the corresponding SIM-MS spectrum. The method demonstrated excellent linearity up to 100 ppm ($R^2 = 0.999$) and was successfully applied for the analysis of a number of development samples to ensure sufficient control of benzene when this starting material was carried on to API synthesis.

Novel approaches without extensive extractions or derivatization have been explored and reported in the literature. Sun and others[25] developed a matrix deactivation method for the analysis of some GTIs by GC-MS. This approach was based on the hypothesis that the instability of some analytes at trace levels was caused by

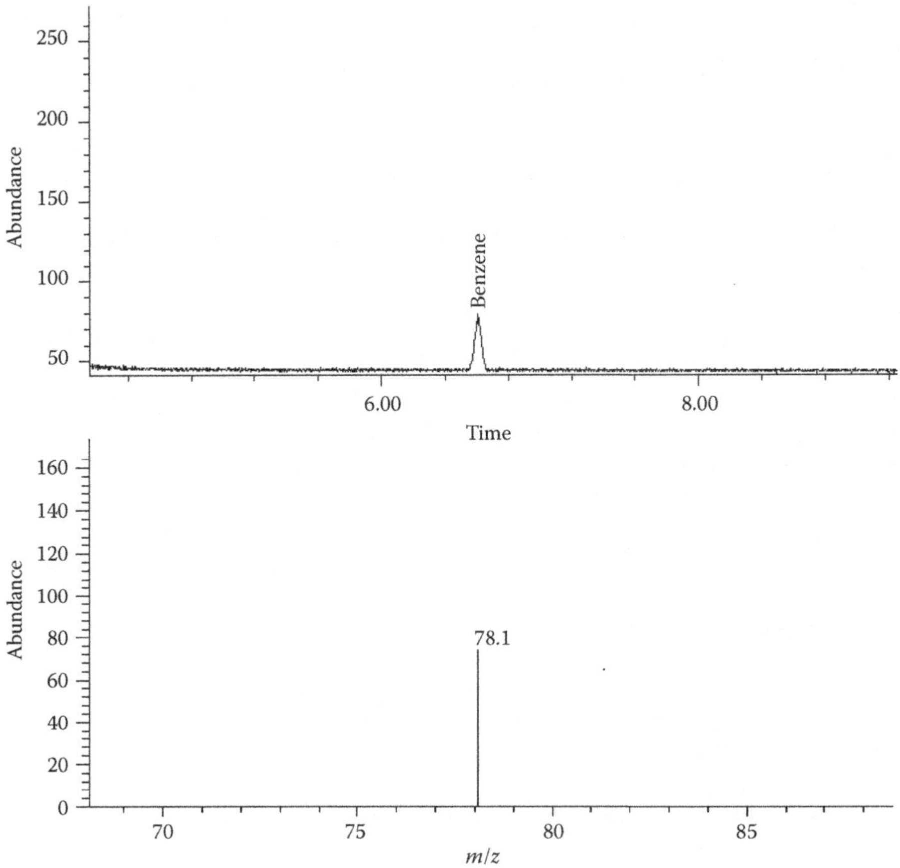

FIGURE 11.2 Gas chromatography chromatogram and mass spectrometry spectrum of benzene.

reactions between the analytes and some low-level reactive species in the sample matrix. Adding chemical scavengers or neutralizing species to the sample matrix would chemically quench these hypothetical reactive species and stabilize the reactive and unstable analytes. This approach was applied for the analysis of trace-level unstable alkylators such as *O*-1-chloroethyl *S*-methyl carbonothioate, resulting in significant improvement of sensitivity to less than 1 ppm.

11.3.2.2 Genotoxic Impurity Analysis by Gas Chromatography with Sample Extraction

GTI analysis by direct injection GC methods suffers significant issues of column contamination and matrix effects that are caused by the injection of high concentrations of API or matrix onto the heated GC column along with low-level analytes. Therefore, samples are often pretreated through various extractions to remove the matrix and/or to enrich the analytes for increased sensitivity. Extraction was almost the default method of choice early on before instruments and detectors such as MS detectors were routinely explored for GTI analysis, and it still holds significance and applicability in enabling many trace analyses.

Numerous methods have been reported for the analysis of low-level alkyl sulfonates in pharmaceutical development due to their prevalence in API synthesis and significance as GTIs.[26–28] The development of one internal drug substance involved the use of methanesulfonic acid (MSA) as a reagent for salt formation and methanol as a solvent, leading to the possibility of MMS formation during the process. A direct injection GC-MS method was developed to analyze the API for residual MMS using external standard for quantitation. The API samples (~100 mg/mL) were prepared in a mixture of water and acetonitrile (95/5, v/v) and extracted with dichloromethane for GC analysis and MS detection in SIM mode. The method was shown to be linear between 0.1 and 2.5 μg/mL (or 1 and 25 ppm relative to sample concentration) with $R^2 \geq 0.99$. The lower limit was verified as LOQ, although the achievable LOQ was lower than 0.1 μg/mL. Acceptable recovery was demonstrated at multiple levels. The method was used to test the API for both development and release purposes. The same method conditions were demonstrated to be suitable for the trace analysis of EMS and IMS that may come from commercial MSA supplies. Figure 11.3 shows the chromatograms for the working standard (1 μg/mL) and the API.

A similar approach was described for the simultaneous determination of alkyl mesylates and alkyl besylates in finished drug products by LLE and direct injection GC-MS.[26] The sulfonates, including MMS, EMS, IMS, methyl besylate, and ethyl besylate, were extracted from finished drug products by *n*-hexane for direct injection into the gas chromatograph for analysis. *N*-hexane as the extraction solvent brought the double benefits of (1) extracting only sparse amounts of API or other related impurities from the matrix and (2) ensuring a longer stability period for the analytes over solvents of higher polarity. Using methyl tosylate as the internal standard for quantitation, the method afforded a quantitation limit of 0.01 μg/mL, corresponding to 0.04 μg/g (or parts per million) in tablet or capsule powders. A direct injection LLE-GC-MS method was also developed for the alkylating agent dimethyl sulfate (DMS) and was applied to study the efficiency of DMS removal from the process.[29] DMS and the internal standard d_6-DMS were extracted together with methyl *tert*-butyl ether to overcome the matrix interference. A sensitive headspace SPME GC-MS method was reported for the determination of alkyl sulfonates at the parts-per-billion level in pharmaceutical APIs and drug products.[27] The standard addition method was employed to ensure uniform sample matrix effect, and the limit of detection (LOD) was achieved at 50 ppb for alkyl mesylates and 5 ppb for besylates and tosylates. In another study, seven alkyl sulfonate esters were determined at the 5 ppm level by GC-MS, and SPME was demonstrated to be superior to liquid-phase microextraction or LLE for the detection of these sulfonate esters in the APIs.[30] A strong anion-exchange (SAX)-SPE GC-MS method was reported[21] for the trace analysis of 2-chloroethanol in cloperastine fendizoate, an antitussive API. After removing fendizoate by SAX-SPE, purified sample solutions were analyzed by GC-MS with selected ion detection at *m/z* 80, corresponding to the molecular ion of 2-chloroethanol. The LOD was achieved at 1.7 ppm for 2-chloroethanol, and the method was used to test multiple batches of the API.

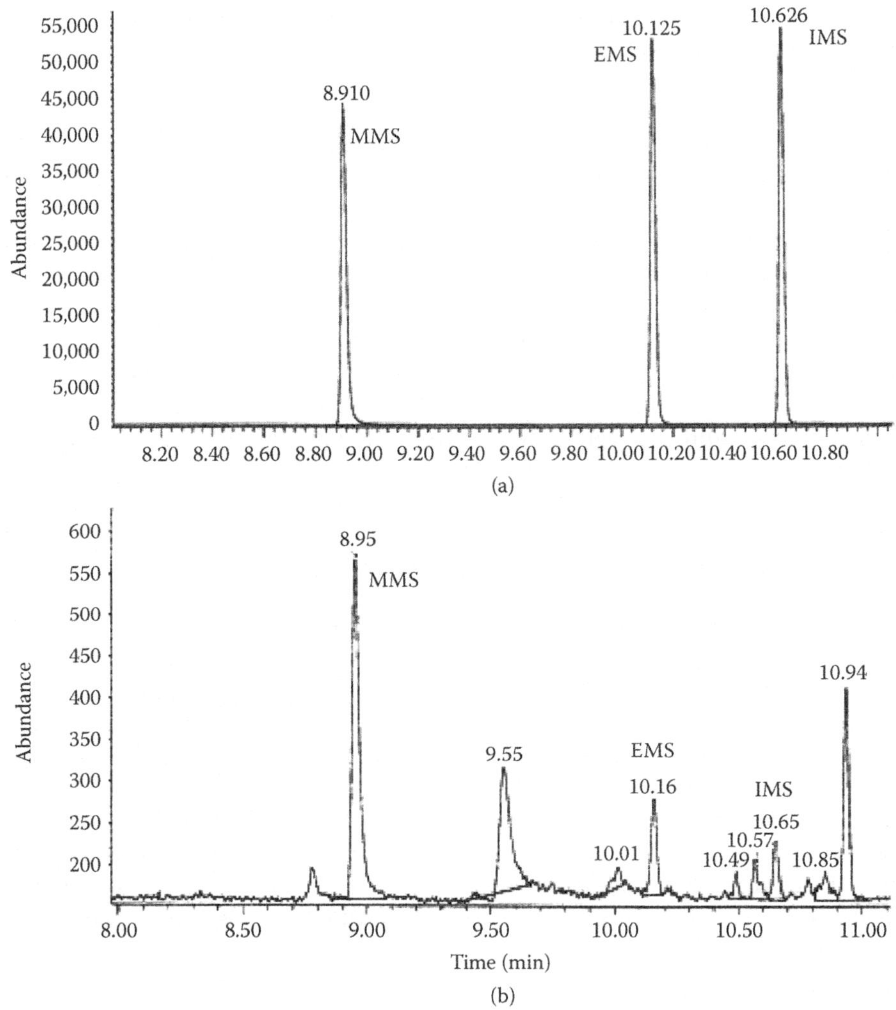

FIGURE 11.3 Chromatograms for standard and active pharmaceutical ingredient (API) solutions. (a) Working standard at 1 μg/mL; (b) API solution at 100 mg/mL. (Courtesy of Dr. Kavita Mistry and Edward Hoff, Genentech, South San Francisco, California.)

11.3.2.3 Genotoxic Impurity Analysis by Gas Chromatography with Derivatization

Various extraction methods have proved to be effective in overcoming the sample matrix issues for GC analysis in a high-temperature environment. However, many GTIs are highly reactive and labile to normal sample preparation or testing conditions. Consequently, low recovery due to analyte decomposition or interaction with the matrix is a significant challenge and concern for the trace-level analysis. Headspace sampling and injection avoids introduction of matrix components (mostly APIs) into the GC system, therefore improving method accuracy and robustness. However, not all GTIs are suitable for the headspace approach, particularly if they have limited volatility and/or thermal stability. Fortunately, chemical derivatization has shown to be a viable alternative in filling the gaps and addressing issues.

Hydrazine is a prime candidate for chemical derivatization for trace-level GC analysis. Its analysis at trace levels has been reviewed and highlighted in a number of articles.[9,10,31–33] Hydrazine is a highly polar and reactive GTI; the reactivity makes it challenging to test hydrazine reliably under normal LC or GC conditions. For GC analysis, hydrazine has very limited volatility (boiling point > 100°C) and does not respond well to FID. Therefore, most hydrazine analyses are performed by derivatization, followed by GC or LC analysis coupled with MS detection. Acetone is known to work well both as a solvent and as a derivatizing agent for the analysis of hydrazines by GC,[34,35] and low parts per billion–level detection has been achieved with a nitrogen-specific detector.[34] Using acetone or acetone-d_6 as a derivatizing agent, a general in situ derivatization-headspace-GC-MS method was developed and validated for the determination of hydrazine in drug substances at low parts-per-million levels.[36] The derivative acetone azine was detected by MS, and acetone-d_6 was used as an alternative reagent in the attempt to differentiate the preexisting acetone azine in drug substances due to the presence of acetone from the manufacturing process. The LOQ was demonstrated to be as low as 0.1 ppm with good recovery. The method was used for hydrazine analysis in a number of API matrices to support API release and process development.

Aldehydes are commonly analyzed by derivatization as well, especially the short-chain aldehydes with no chromophores that are not suitable for direct GC-FID or liquid chromatography–ultraviolet analysis.[37–43] Li and others[42] reported a headspace GC method for the quantification of low-molecular-weight aldehydes in pharmaceutical excipients. C1–C8 aliphatic aldehydes and benzaldehyde were derivatized by *O*-2,3,4,5,6-(pentafluorobenzyl)hydroxylamine hydrochloride (PFBHA), analyzed by static headspace GC for PFBHA aldehyde oximes, and detected by negative chemical ionization mass spectrometry (NCIMS). The method demonstrated detection limits less than or equal to 0.1 ppm (relative to a 10 mg/mL sample concentration) for all aldehydes. Over 30 common excipients were screened for these aldehydes, and formaldehyde and acetaldehyde were found to be the most abundant. This GC-NCIMS method was later expanded to the analysis of aldehydes in excipients used in liquid/semisolid formulations.[43] The method employed the derivatization of aldehydes in an aqueous medium to be compatible with the liquid and surfactant types of excipients followed by direct injection for GC analysis and MS detection.

For the GC analysis of sulfonate and sulfate esters, pentafluorothiophenol (PFTP) is a common derivatizing agent.[44,45] An in situ derivatization headspace GC-MS method was reported[44] for the simultaneous determination of some trace-level alkyl sulfonates and dialkyl sulfates using PFTP. An LOQ less than 1 ppm was achieved for each analyte by this method, and an internal standard was employed for accurate quantitation and robust analysis. The use of deuterated internal standards reduced the interference from matrix, and the method was developed as a potential generic method for the determination of these alkylating agents in different API matrices. The major drawback of the method is the lack of selectivity with respect to the alkyl groups. For the same alkyl group, the same PFTP derivative is formed for mesylates, besylates, tosylates, or sulfates. Therefore, knowledge of the synthetic process is required for targeted determination and control of these GTIs. The same derivatizing agent was used for a method to monitor the formation of EMS from reaction mixtures containing ethanol and MSA.[45] After its formation, EMS was derivatized by PFTP and then analyzed using a static headspace GC-MS method.

Other examples of derivatization for GC analysis include the determination of volatile amines through acid chloride derivatization and detection by nitrogen–phosphorus detection for aziridine (a potential carcinogen) and electron capture detection for 2-chloroethylamine (precursor of aziridine in API synthesis).[46] The bulk of the drug substance was removed by extraction or precipitation, and the derivatization occurred in a water–organic solvent two-phase system. The derivatives were extracted to the organic phase and analyzed by GC. The method achieved detection limits of about 0.2 mg/kg (or parts per million) for aziridine and 0.5 mg/kg (or parts per million) for 2-chloroethylamine; however, the sample preparation procedures were complex due to the multiple steps of extractions that were performed in addition to the derivatization.

11.3.3 Analysis of Genotoxic Impurities by High-Performance Liquid Chromatography

11.3.3.1 Genotoxic Impurity Analysis by High-Performance Liquid Chromatography–Ultraviolet (UV) Spectroscopic Detection

Many pharmaceutical compounds, including API starting materials and intermediates, have alerting structures such as the ones shown in Figure 11.1 and need to be tested as potential GTIs in APIs for process development and control purposes. These compounds are generally nonvolatile and have aromatic or other structural features that render them suitable for analysis by HPLC with ultraviolet (UV) detection. For a particular API, structural similarity between these impurities and the API demands excellent selectivity of the HPLC methods in ensuring accurate quantitation. Reversed-phase HPLC is the method of choice for API impurity analysis and is often optimized to meet sensitivity requirements for genotoxic analysis.[47–50] Severin[48] described an HPLC-UV method for the determination of *N*-nitrosohexamethyleneimine, a potential carcinogen, in tolazamide bulk drug and pharmaceutical dosage forms. *N*-nitrosohexamethyleneimine was extracted with diethyl ether, enriched by evaporation of the solvent, and determined by HPLC-UV at 228 nm. The method afforded a LOD as low as 1 ppb; however, it involved multiple steps of sample treatment. The procedure was labor intensive and prone to mistakes. An HPLC-UV-DAD method was reported for the low-level analysis of 4-amino-2-ethoxy-cinnamic acid and its ethyl ester in a drug substance and in prototype formulations.[47] With sample extraction or derivatization, detection limits of 2–15 ppm in the drug substance were achieved using this method for the potential GTIs 4-amino-2-ethoxy-cinnamic acid, the HCl salt of 4-amino-2-ethoxy-ethyl cinnamate, and 4-bromo-3-ethoxy-nitrobenzene. Conversion of the crystalline free base drug to its partially crystalline potassium salt helped to improve the solubility of the drug substance and enable the preparation of sample solution at 15 mg/mL for increased sensitivity.

For GTIs labile to typical LC conditions or with a low response to UV detection, chemical derivatization is widely applied to enable the analysis by HPLC-UV. As an example of this approach, residual methanesufonyl hydrazide (MSH) in equipment cleaning was determined by HPLC-UV for an API production.[51] MSH was used as the source of a reducing agent in the first step of the API production; therefore, the cleaning samples from the production needed testing for this toxic material. For the analysis, MSH was derivatized with benzaldehyde to form benzalazine, which was then analyzed by HPLC and detected by UV for which LOQ was achieved at 0.3 ppm. An internal method was established for the determination of acid chlorides by HPLC-UV.[52] The analytes were derivatized with methanol to form methyl esters for the analysis and quantitation. For HPLC analysis of aldehydes, the analytes are typically derivatized with 2,4-dinitrophenylhydrazine. The resulting hydrazones are analyzed by HPLC-UV for selectivity and sensitivity.[37,39,40] Nageswari and others[39] applied this derivatization approach for the low-level quantitation of formaldehyde in a drug substance by HPLC-UV. The LOD and LOQ were achieved at 0.5 and 1.5 ppm, respectively, and the method was used to monitor the formaldehyde level in scale-up batches of bulk API.

11.3.3.2 Genotoxic Impurity Analysis by High-Performance Liquid Chromatography–Mass Spectrometry

As is the case for GTI analysis by GC-MS, MS has become the detection of choice for superior sensitivity and selectivity in HPLC analysis of GTIs.[28,53–56] The advancement in both MS instrumentation and methodology for LC applications has routinely pushed GTI method sensitivity to levels unachievable by HPLC-UV.

Both SIM and multiple reaction monitoring (MRM) have demonstrated great utility and success in GTI analysis for pharmaceuticals. Various organohalides were analyzed at trace levels by HPLC via derivatization using 4-dimethylaminopyridine (DMAP)[57,58] or dimethylamine (DMA)[25,59] as the derivatizing agent. The derivatizing agents react with the halides to form quaternary amine salt derivatives. The DMA or DMAP derivatives were resolved from matrix and other components in the samples by hydrophilic interaction liquid chromatography (HILIC) using Waters (Milford, Massachusetts) Atlantis HILIC columns prior to quantitation by either ESI-SIM-MS (ESI stands for electrospray ionization) or tandem MS using MRM. For the analysis of 4-fluorobenzyl chloride (4FBCl),[57] the precursor to product ion transition of the DMAP-derivatized 4FBCl, *m/z* 231 to 109, was used for the selected reaction monitoring for high selectivity and sensitivity. DMAP derivatization was also used as a generic approach for the screening and quantitation of potential genotoxic alkyl halides containing Cl, Br, or iodine in various positions.[58] In addition to monitoring the characteristic precursor to product transitions, the loss of DMAP (observed as a neutral loss of 122 Da) by the derivatives would suggest the presence of any molecule that reacted with DMAP, therefore allowing the screening of unknown potential GTIs at relevant levels. The structure of originating compounds can be further elucidated using other fragmentation information of the same parent molecules, followed by the development of a more specific method for quantitative target analysis when needed. This method could be applied for screening GTIs in different samples including APIs, drug products, chemical process samples, and samples from stability studies, although further studies are needed to demonstrate the applicability. For the DMA derivatizations,[25,59] the derivatives were quantitated by ESI-MS due to their high proton affinity. Sensitivity (signal to noise ratio [*S/N*] > 10) was achieved at less than or equal to 2 ppm for (*S*)-(+)-3-chloro-1,2-propanediol, bis(2-chloroethyl) ether[59] and bis(2-chloroethylamine)[25] with typical sample concentrations of 5 mg/mL. DMA was also used to derivatize (*R*)-2-(5-bromo-2,3-difluoro-phenoxymethyl)-oxirane, for which similar sensitivity was achieved.[59] As the authors[60] pointed out, the spike recovery for all three compounds had to be carefully studied or concurrently determined at the time of sample analysis, and the method *S/N* and analytical results were corrected for recovery of each analyte.[29] A group of commonly encountered alkyl esters of sulfonates or sulfates in drug substances were determined at low parts-per-million levels using a general derivatization LC-MS method.[60] Trialkylamines (methyl and ethyl) were used as derivatizing agents for these esters to form stable quaternary ammonium ions. HILIC was again selected to retain these highly polar derivatives and readily separate them from the API peak. The quantitation was performed by ESI-MS in positive ion mode. The method demonstrated excellent sensitivity for all alkyl esters at a target analyte level of 1 to 2 ppm (API at 5 mg/mL); however, no individual LOD/LOQ was reported. The method was tested for multiple APIs and successfully applied to the determination of various alkyl sulfonates or dialkyl sulfates in APIs and drug products. Wu and others[63] reported the analysis of some epoxides by derivatization with LC-MS. The derivatization of the epoxides was achieved in the gas phase via Meerwein reaction with the ethylnitrilium ion, generated by the protonation of acetonitrile in the mobile phase under typical conditions of atmospheric pressure ionizations. As a result, the gas-phase derivatization allowed direct LC-MS analysis of the epoxides, which would otherwise be challenging due to poor ionization. Commonly used atmospheric pressure ionization techniques including ESI, APCI, and APPI were evaluated for optimal formation of the Meerwein reaction products, and APCI was found to offer the best sensitivity and the most robust detection compared to the other two techniques. Quantitation of the epoxides was performed by both SIM and MRM. Both methods afforded an excellent sensitivity less than or equal to 1 ppm detection limit (relative to 10 mg/mL API); however, the MRM method is preferred for selectivity reason when dealing with complex sample matrices.

Simultaneous analysis of several potential GTIs for one Genentech drug substance was achieved by HPLC-MS-MS in MRM mode. These process impurities were generated at various steps during the API synthesis. Their fate through the synthetic process was thoroughly studied to support the development of late-stage and commercial manufacturing process of the API. Results from spiking studies and analysis of multiple batches of starting materials, intermediates, and the API allowed the establishment of appropriate control strategies for these impurities during API manufacture and supported the TTC limit in the API for confirmed GTIs. For the analysis of these impurities in the API, chromatographic separation was achieved using an ACE 3 C18 column (150 mm × 4.6 mm), mobile phases of acetonitrile and 0.1% acetic acid in water, and a linear gradient of 10%–90% acetonitrile over 25 minutes. The standard addition method was used for the quantitation of each impurity. An internal standard was used to improve the reproducibility of the analysis. Excellent linearity was achieved within the range of 1–50 ppm for each analyte spiked into the 4 mg/mL API solutions, and an R^2 of 0.99 or 1.00 was obtained for all impurities except two, for which R^2 values were shown to be 0.98 and 0.97. The method demonstrated excellent reproducibility of relative standard deviation (RSD) less than 20% based on duplicate standard injections at five concentration levels between 1 and 50 ppm for each analyte. Except one impurity, the average recovery based on duplicate injections of the API solutions at three spiking levels for each analyte was 83% or better with RSD of all individual recovery results below 10%. The "last" impurity had an average recovery of 78% with an RSD of 12% for all the individual recovery results. These results demonstrated the excellent recovery of the method for these impurities. The sensitivity of the method was estimated to be in the low– to mid–parts per billion range for the impurities based on the reproducible responses obtained at the 1 ppm level. The method was used to support the release of the API batches for late-phase development use. Figure 11.4 shows the chromatograms of an unspiked API sample solution at approximately 4 mg/mL and an API solution spiked with 1 ppm of each GTI.

Other examples include analysis of dimethyl sulfate via derivatization with 2-mercaptopyridine and quantification by HPLC-UV or HPLC-MS for improved sensitivity,[53] and determination of arylamines using derivatization with LC-MS.[61]

11.3.4 Novel and Unconventional Analytical Approaches for Genotoxic Impurity Analysis

As shown in Figure 11.1, GTIs carry a wide range of structural features. The unique characteristics of these impurities have led to some creative approaches from sample preparation to detection in advancing GTI analysis and control for drug development. The matrix deactivation method for GC-MS and LC-MS[46] offered a different approach to handle the analyte stability issue caused by the matrix. David and others[62] reported a two-dimensional GC method with Deans switching for the analysis of some potential GTIs in pharmaceuticals. Sample solutions were injected onto an apolar column for first-dimension separation. The GTI fractions were then heart cut and transferred to a polar second-dimension column for further separation prior to MS detection. The second-dimension column was installed in a low-thermal-mass oven module that allowed independent temperature program and control. This setup avoided the introduction of major components in the matrix (APIs, solvents, derivatizing agents, etc.) into the second column or MS detector, greatly reducing contamination or column degradation. The method demonstrated a reproducibility of RSD less than 15% and a sensitivity less than 1 ppm LOD for the testing of some Michael reactive acceptors and haloalcohols, indicating its potential to be an effective tool in GTI testing for pharmaceuticals.

Trace-level analysis of neutral GTIs can be a challenging task due to their poor ionization efficiency in MS. The gas-phase derivatization via Meerwein reaction took advantage of the acetonitrile used for HPLC and resulted in improved sensitivity for some epoxides when detected by MS.[63] Improved sensitivity for some trace-level epoxides was also demonstrated using coordination ion spray–mass spectrometry (CIS-MS). In CIS-MS, neutral molecules are converted to positively or negatively charged complexes in solution or gas phase, resulting in significant improvement in their detection sensitivity.[59,64,65] This ionization method is applicable for both positive

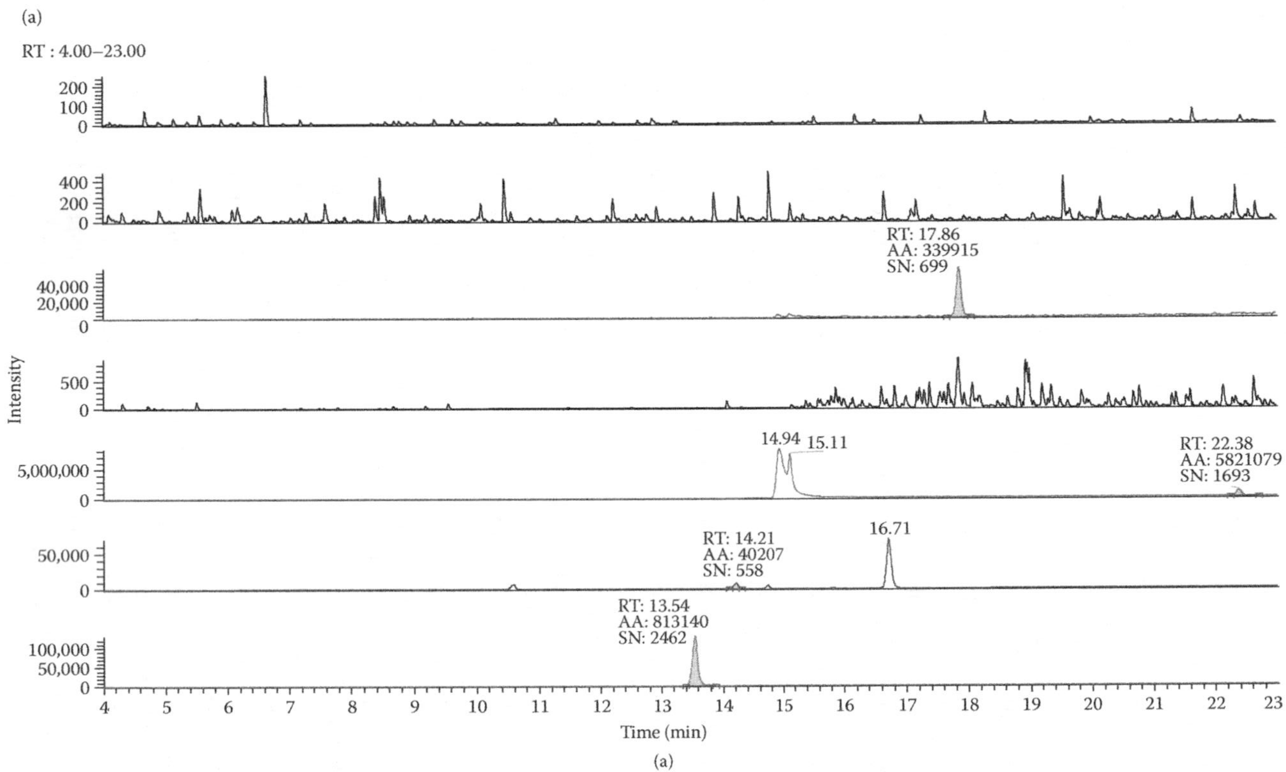

FIGURE 11.4 Chromatograms for analysis of genotoxic impurities (GTIs) by HPLC-MS/MS. (a) Stacked multiple reaction monitoring (MRM) chromatograms for each GTI in the unspiked active pharmaceutical ingredient (API); (b) stacked MRM chromatograms for each GTI spiked at 1 ppm into the API solution. Each shaded line represents one GTI. (Courtesy of Dr. Christine Gu, Genentech, South San Francisco, California.)

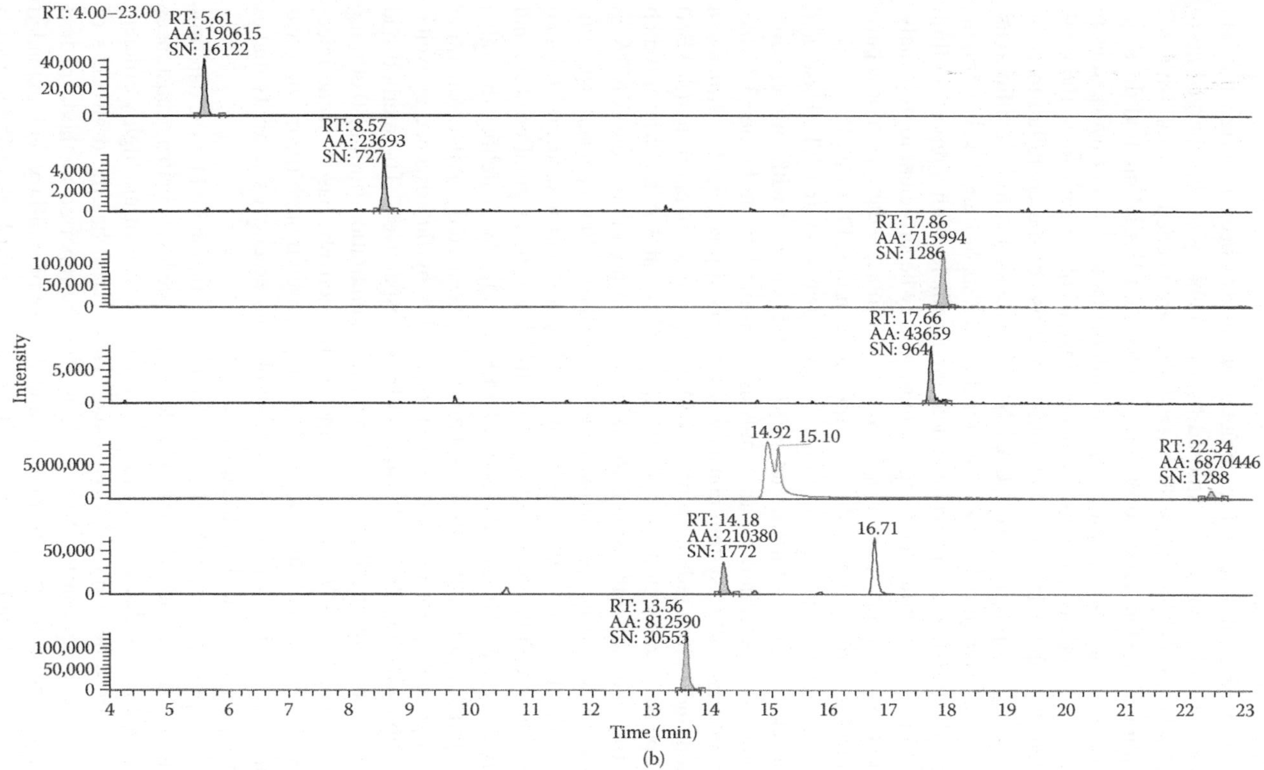

FIGURE 11.4 (*Continued*) Chromatograms for analysis of genotoxic impurities (GTIs) by HPLC-MS/MS. (a) Stacked multiple reaction monitoring (MRM) chromatograms for each GTI in the unspiked active pharmaceutical ingredient (API); (b) stacked MRM chromatograms for each GTI spiked at 1 ppm into the API solution. Each shaded line represents one GTI. (Courtesy of Dr. Christine Gu, Genentech, South San Francisco, California.)

and negative ions. For the analysis of trace-level epoxides,[46] metal ions such as Na^+, K^+, or NH^{4+} were doped into the HPLC mobile phase and formed coordination complexes with the epoxides. The adduct ions were detected by MS for improved sensitivity at a low parts-per-million level.

Applications of special detection techniques have also been reported for GTI analyses. Yuabova and coauthors[50] explored the applicability of the charged aerosol detector (CAD) for GTI analysis. Further studies are needed to establish CAD as a viable option for GTIs. Liu and others[66] described a HILIC method development using alcohol as a weak eluent for the simultaneous analysis of hydrazines in pharmaceutical intermediates. The use of alcohol instead of acetonitrile as the weak eluent provided different selectivity and enabled the use of the chemiluminescent nitrogen detector. The optimized HILIC-CLND (CLND stands for chemiluminescent nitrogen detector) method was applied for the analysis of trace-level hydrazine and 1,1-dimethylhydrazine in a pharmaceutical intermediate, with a reporting limit of 0.02% w/w for both analytes. This reporting limit may be sufficient for the analysis and control of GTIs in intermediates; however, it is short of the typical low parts per million–level sensitivity required for the quantitation of GTIs in APIs.

A fluorescence detector has been used for the analysis of certain GTIs. Hasei and coauthors[67] developed a two-dimensional HPLC system coupled with online reduction and fluorescence detection for the analysis of 3-nitrobenzanthrone (3-NBA). The two-dimensional HPLC system consists of a reversed-phase HPLC dimension for purification and online reduction of 3-NBA to 3-aminobenzanthrone (3-ABA) and a normal-phase dimension for separation and analysis of 3-ABA, with a switch valve connecting the two dimensions. Reduction of the nonfluorescent 3-NBA to fluorescent 3-ABA was achieved in ethanol with a catalyst column packed with alumina coated with platinum. After reduction, 3-ABA was introduced and trapped onto a concentration column by switching the valve from the original position and was then introduced onto the normal-phase HPLC column by switching the valve back to the original position. The normal-phase HPLC separated 3-ABA from other interfering components prior to the detection of 3-ABA by fluorescence at excitation and emission wavelengths of 490 and 560 nm, respectively. The method had a reported detection limit of 0.002 ng and reported quantitation limit of 0.006 ng, which were calculated by *S*/*N* values of 3 and 10, respectively, based on the injections of 3-NBA from the linearity study at 2–2000 pg. Even though the method was developed for the analysis of environmental samples, its excellent sensitivity makes it appealing for pharmaceutical applications also.

A recent paper presented a capillary zone electrophoretic method for the determination of dimethyl sulfate and chloroacetyl chloride at trace levels in drug substances by indirect photometric detection.[68] Capillary electrophoresis offers highly efficient separations and consumes very little reagents.[69] However, it has the drawbacks of poor sensitivity and precision for trace analysis. For the reported method, method optimization for sample injection techniques and the use of capillary with extended light path led to an enhanced sensitivity of LOQ = 10 ppm for both analytes, which was well below the limits required in the APIs based on the TTC approach. This study showed that capillary electrophoresis could be a reasonable complement to the commonly used GC or HPLC method for determination of GTIs.

11.3.5 Development of General Methods for Genotoxic Impurity Analysis

GTI analysis poses many challenges technically and operationally. Method development is often time consuming due to sensitivity and selectivity requirements. When dealing with an extensive development portfolio with many compounds targeting a variety of therapeutic areas, it is not sustainable to develop specific GTI methods for each compound in the portfolio. On the other hand, different pharmaceutical compounds could encounter the same GTIs or impurities of similar structural features. Development of generic or general methods is therefore an effective way of simplifying method development and harmonizing GTI methods across various compounds. Alzaga and others[44] described an in situ derivatization-headspace-GC-MS method as a generic approach for the determination of residual sulfonates and sulfates in APIs. A generic in situ derivatization LC-MS method was developed for the analysis of some arylamines and aminopyridines at the sub–parts per million level relative to APIs.[61] Another laboratory took a standardized approach to develop a "selective ion monitoring of analytes in real time" method for low-level detection of potential GTIs in pharmaceutical compounds.[70] The methods utilized complementary techniques such as GC-MS and LC-MS to enable the analysis of a wide range of GTIs and showed low parts-per-million sensitivity for each technique with sample concentrations less than or equal to 10 mg/mL.

For any particular compound, general methods greatly reduce the effort required for individual method development; however, the applicability of these general methods still needs to be demonstrated or confirmed with respect to specificity, sensitivity, and accuracy.

11.4 SUMMARY

GTIs present unique challenges for the pharmaceutical industry from understanding of the toxicity to accurate analysis and control of the impurities in APIs and drug products.[71–76] Regulatory requirements have come a long way and continue to evolve to safeguard patient well-being.

Analytical method development for GTIs has many factors to consider in order to achieve the required sensitivity. To reliably quantitate the trace-level GTIs in the presence of high matrix concentrations, some sample treatment is required prior to chromatographic analysis. For GTI detection, MS has been established as the detection of choice for superior sensitivity and selectivity that can be achieved by either the SIM or MRM approach.

Over the last few years, quality by design (QbD) has become the new paradigm for pharmaceutical development. The QbD principle has been applied to build the analysis and control strategies for GTIs through API synthetic processes.[53,77,78] Sound strategies demand extensive method development and validation followed by data collection for numerous samples. This could prove overwhelming for many laboratories. How to balance the drug development effort versus the regulatory requirements on GTIs remains a difficult question with few answers.

ACKNOWLEDGMENTS

The authors thank Dr. Nik Chetwyn, Dr. Peter Yehl, and Dr. Christine Gu for their content and technical reviews. Dr. Kavita Mistry, Dr. Christine Gu, and Edward Hoff of Genentech are acknowledged for granting permission to use their data. Kyle Tse of Genentech is acknowledged for contributions to benzene analysis by GC-MS.

REFERENCES

1. International Conference on Harmonisation of Technical Requirements for Registration of Pharmaceuticals for Human Use. Quality Guidelines: Impurity Q3A (R2) (finalized October 2006), Q3B (R2) (finalized June 2006), Q3C (R5) (finalized February 2011). http://www.ich.org/products/guidelines/quality/article/quality-guidelines.html (accessed September 28, 2012).
2. International Conference on Harmonisation of Technical Requirements for Registration of Pharmaceuticals for Human Use. Safety Guidelines: Genotoxicity Studies S2 (R1) (finalized November 2011). http://www.ich.org/products/guidelines/safety/article/safety-guidelines.html (accessed September 28, 2012).
3. European Medicines Agency, Committee for Medicinal Products for Human Use, *Guideline on the Limits of Genotoxic Impurities*. CPMP/SWP/5199/02/draft 2. London, 18 December 2002.
4. Mueller, L.; Mauthe, R. J.; Riley, C. M.; Andino, M. M.; De Antonis, D.; Beel, C.; DeGorge, J., et al. A Rationale for Determining, Testing, and Controlling Specific Impurities in Pharmaceuticals that Possess Potential for Genotoxicity. *Regul. Toxicol. Pharmacol.* 2006, *44,* 198–211.
5. Food and Drug Administration. *Draft Guidance for Industry Genotoxic and Carcinogenic Impurities in Drug Substances and Products: Recommended Approaches* (December 2008). www.fda.gov/downloads/drugs/guidancecomplianceregulatoryinformation/guidances/ucm079235.pdf (accessed March 26, 2014).
6. European Medicines Agency, Committee for Medicinal Products for Human Use, *Guideline on the Limits of Genotoxic Impurities*. EMEA/CHMP/QWP/251344/2006. London, 28 June 2006.
7. International Conference on Harmonisation of Technical Requirements for Registration of Pharmaceuticals for Human Use. Multidisciplinary Guidelines M7 (draft February 2013). http://www.ich.org/products/guidelines/multidisciplinary/article/multidisciplinary-guidelines.html (accessed March 26, 2014).
8. Thayer, A. Genotoxic Impurities. *Chem. Eng. News.* 2010, *88(39)*, 16–26.
9. Parsons, R. L.; Miller, S. A. Analytical Considerations for Genotoxic and Other Impurities. In *Sep. Sci. Technol.*, Volume 10 *Handbook of Modern Pharmaceutical Analysis*; Ahuja, S., Scypinski, S., Eds.; Elsevier: Amsterdam, the Netherlands, 2011; pp. 171–194.
10. Liu, D. Q.; Sun, M.; Kord, A. S. Recent Advances in Trace Analysis of Pharmaceutical Genotoxic Impurities. *J. Pharm. Biomed. Anal.* 2010, *51*, 999–1014.
11. Raman, N. V. V. S. S.; Prasad, A. V. S. S.; Reddy, K. R. Strategies for the Identification, Control and Determination of Genotoxic Impurities in Drug Substances: A Pharmaceutical Industry Perspective. *J. Pharm. Biomed. Anal.* 2011, *55*, 662–667.
12. Pierson, D. A.; Olsen, B. A.; Robbins, D. K.; DeVries, K. M.; Varie, D. L. Approaches to Assessment, Testing Decisions, and Analytical Determination of Genotoxic Impurities in Drug Substances. *Org. Process Res. Dev.* 2009, *13*, 285–291.
13. Robinson, D. I. Control of Genotoxic Impurities in Active Pharmaceutical Ingredients: A Review and Perspective. *Org. Process Res. Dev.* 2010, *14*, 946–959.

14. Dobo, K. L.; Greene, N.; Cyr, M. O.; Caron, S.; Ku, W. Q. The Application of Structure-Based Assessment to Support Safety and Chemistry Diligence to Manage Genotoxic Impurities in Active Pharmaceutical Ingredients during Drug Development. *Regul. Toxicol. Pharmacol.* 2006, *44(3)*, 282–293.
15. Kruhlak, N. L.; Contrera, J. F.; Benz, R. D.; Matthews, E. J. Progress in QSAR Toxicity Screening of Pharmaceutical Impurities and Other FDA Regulated Products. *Adv. Drug Deliver Rev.* 2007, *59*, 43–55.
16. Mayer, J.; Cheeseman, M. A.; Twaroski, M. L. Structure-Activity Relationship Analysis Tools: Validation and Applicability in Predicting Carcinogens. *Regul. Toxicol. Pharmacol.* 2008, *50(1)*, 50–58.
17. Elder, D. P.; Teasdale, A.; Lipczynski, A. M. Control and Analysis of Alkyl Esters of Alkyl and Aryl Sulfonic Acids in Novel Active Pharmaceutical Ingredients (APIs). *J. Pharm. Biomed. Anal.* 2008, *46*, 1–8.
18. Li, W. Trace Analysis of Residual Methyl Methanesulfonate, Ethyl Methanesulfonate and Isopropyl Methanesulfonate in Pharmaceuticals by Capillary Gas Chromatography with Flame Ionization Detection. *J. Chromatogr. A.* 2004, *1046*, 297–301.
19. Ramjit, H. G.; Singh, M. M.; Coddington, A. B. Gas Chromatographic/Mass Spectrometric Analysis of Methyl Methanesulphonate and Ethyl Methanesulphonate in the Bismesylate Salt of DPI 201-206, a Positive Inotropic Agent for the Treatment of Heart Failure. *J. Mass Spectrom.* 1996, *31*, 867–872.
20. Lee, C. R.; Guivarch, F.; Nguyen Van Dau, C.; Tessier, D.; Krstulovic, A. M. Determination of Polar Alkylating Agents as Thiocyanate/Isothiocyanate Derivatives by Reaction Gas Chromatography. *Analyst.* 2003, *128*, 857–863.
21. Garcia, A.; Ruperez, F. J.; Ceppa, F.; Pellati, F.; Barbas, C. Development of Chromatographic Methods for the Determination of Genotoxic Impurities in Cloperastine Fendizoate. *J. Pharm. Biomed. Anal.* 2012, *61*, 230–236.
22. Yang, Q.; Haney, B. P.; Vaux, A.; Riley, D. A.; Heidrich, L.; He, P.; Mason, P., et al. Controlling the Genotoxin Ethyl Chloride and Methyl Chloride Formed during the Preparation of Amine Hydrochloride Salts from Solutions of Ethanol and Methanol. *Org. Process Res. Dev.* 2009, *13*, 786–791.
23. Li, H.; Sluggett, G. W. Development and Validation of a Sensitive GC-MS Method for the Determination of Trace Levels of an Alkylating Reagent in a β-Lactam Active Pharmaceutical Ingredient. *J. Chromatogr. A.* 2005, *39*, 486–494.
24. Mamilla, Y. R; Chetyala, K. R.; Venna, R.; Gajjela, R.; Katragadda, K.; Govindasamy, S.; Mulukutla, S.; Datta, D. A Sensitive and Selective GC-MS Method for Analysis of Process-Related Genotoxic Impurities in Atenolol. *Chromatographia.* 2010, *71*, 733–736.
25. Sun, M.; Bai, L.; Terfloth, G. J.; Liu, D. Q.; Kord, A. S. Matrix Deactivation: A General Approach to Improve Stability of Unstable and Reactive Pharmaceutical Genotoxic Impurities for Trace Analysis. *J. Pharm. Biomed. Anal.* 2010, *52*, 30–36.
26. Wollein, U.; Schramek, M. Simultaneous Determination of Alkyl Mesilates and Alkyl Besilates in Finished Drug Products by Direct Injection GC/MS. *Eur. J. Pharm. Sci.* 2012, *45*, 201–204.
27. Hu, J.; Li, L.; Semin, D.; Cauchon, N.; Cheetham, J. Determination of Alkyl Sulfonates at ppb Level by Headspace SPME/GC-MS Method. In *Abstracts of Papers, 238th ACS Annual Meeting*, Washington, DC, August 16–20, 2009; ANYL-152.
28. Taylor, G. E.; Gosling, M.; Pearce, A. Low Level Determination of *p*-Toluenesulfonate and Benzenesulfonate Esters in Drug Substance by High Performance Liquid Chromatography/Mass Spectrometry. *J. Chromatogr. A.* 2006, *1119*, 231–237.
29. Zheng, J.; Pritts, W. A.; Zhang, S.; Wittenberger, S. Determination of Low ppm Levels of Dimethyl Sulfate in an Aqueous Soluble API Intermediate Using Liquid-Liquid Extraction and GC-MS. *J. Pharm. Biomed. Anal.* 2009, *50*, 1054–1059.

30. Colon, I.; Riholl, S. M. Determination of Methyl and Ethyl Esters of Methanesulfonic, Benzenesulfonic and *p*-Toluenesulfonic Acids in Active Pharmaceutical Ingredients by Solid-Phase Microextraction (SPME) Coupled to GC/SIM-MS. *J. Pharm. Biomed. Anal.* 2005, *39*, 477–485.
31. Elder, D. Analysis of Genotoxic Impurities: Review of Approaches. In *Genotoxic Impurities: Strategies for Identification and Control*; Teasdale, A., Ed.; John Wiley & Sons: Hoboken, New Jersey, 2010; pp. 249–280.
32. Kean, T.; Miller, J. H.; Skellern, G. G.; Snodin, D. Acceptance Criteria for Levels of Hydrazine in Substances for Pharmaceutical Use and Analytical Methods for Its Determination. *Pharm. Sci. Notes.* 2006, *2*, 23–33.
33. Elder, D. P.; Snodin, D.; Teasdale, A. Control and Analysis of Hydrazine, Hydrazides and Hydrazones—Genotoxic Impurities in Active Pharmaceutical Ingredients (APIs) and Drug Products. *J. Pharm. Biomed. Anal.* 2011, *54*, 900–910.
34. Holtzclaw, J. R.; Rose, S. L.; Wyatt, J. R.; Rounbehler, D. P.; Fine, D. H. Simultaneous Determination of Hydrazine, Methylhydrazine, and 1,1-Dimethylhydrazine in Air by Derivatization/Gas Chromatography. *Anal. Chem.* 1984, *56*, 2952–2956.
35. Fortin, D. T.; Chen, R. Developing a Trace Level GC-MS Method for Detecting Methylhydrazine in an Experimental Drug Substance. *J. Chromatogr. Sci.* 2010, *48*, 299–302.
36. Sun, M.; Bai, L.; Liu, D. Q. A Generic Approach for the Determination of Trace Hydrazine in Drug Substances Using In Situ Derivatization—Headspace GC-MS. *J. Pharm. Biomed. Anal.* 2009, *49*, 529–533.
37. Zhong, Q.; Ngim, K.; Goretski, C. Determination of Trace Level Formaldehyde in Methanol. In *Abstract of Papers, 239th ACS National Meeting*, San Francisco, California, March 21–25, 2010; ANYL-124.
38. Manius, G. J.; Wen, L-F. L.; Palling, D. Three Approaches to the Analysis of Trace Formaldehyde in Bulk and Dosage Form Pharmaceuticals. *Pharm. Res.* 1993, *10*, 449–453.
39. Nageswari, A.; Krishna Reddy, K. V. S. R.; Mukkanti, K. Low-Level Quantitation of Formaldehyde in Drug Substance by HPLC-UV. *Chromatographia.* 2012, *75*, 275–280.
40. Soman, A.; Qiu, Y.; Li, Q. C. HPLC-UV Method Development and Validation for the Determination of Low Level Formaldehyde in a Drug Substance. *J. Chromatogr. Sci.* 2008, *46*, 461–465.
41. Wu, P.; Chang, C.; Chou, S. Determination of Formaldehyde in Cosmetics by HPLC Method and Acetylacetone Method. *J. Food Drug Analysis.* 2003, *11*, 8–15.
42. Li, Z.; Jacobus, L. K.; Wuelfing, W. P.; Golden, M.; Martin, G. P.; Reed, R. A. Detection and Quantification of Low-Molecular-Weight Aldehydes in Pharmaceutical Excipients by Headspace Gas Chromatography. *J. Chromatogr. A.* 2006, *1104*, 1–10.
43. Li, Z.; Kozlowski, B. M.; Chang, E. P. Analysis of Aldehydes in Excipients Used in Liquid/Semi-Solid Formulations by Gas Chromatography—Negative Chemical Ionization Mass Chromatography. *J. Chromatogr. A.* 2007, *1160*, 299–305.
44. Alzaga, R.; Ryan, R. W.; Taylor-Worth, K.; Lipczynski, A. M.; Szucs, R.; Sandra, P. A Generic Approach for the Determination of Residues of Alkylating Agents in Active Pharmaceutical Ingredients by In Situ Derivatization-Headspace-Gas Chromatography-Mass Spectrometry. *J. Pharm. Biomed. Anal.* 2007, *45*, 472–479.
45. Jacq, K.; Delaney, E.; Teasdale, A.; Eyley, S.; Taylor-Worth, K.; Lipczynski, A.; Reif, V. D., et al. Development and Validation of an Automated Static Headspace Gas Chromatography-Mass Spectrometry (SHS-GX-MS) Method for Monitoring the Formation of Ethyl Methane Sulfonate from Ethanol and Methane Sulfonic Acid. *J. Pharm. Biomed. Anal.* 2008, *48*, 1339–1344.
46. de Haan, P. E.; de Jong, D.; van den Berg, J. H. M. Determination of Volatile Amines in Drug Substances by Capillary Gas Chromatography. *J. High Resolution. Chromatogr.* 1989, *12*, 604–607.

47. Soman, A.; Jacob, S.; Swanek, F. Low Level Determination of 4-Amino-2-Ethoxyl-Cinnamic Acid and Its Ethyl Ester in a Drug Substance and Its Formulation Prototypes by HPLC-UV-DAD. *J. Chromatogr. Sci.* 2009, *27*, 315–319.
48. Severin, G. Monitoring a Potential Carcinogen in Pharmaceutical Formulations at the Low Parts per Billion Level: High-Performance Liquid Chromatographic Determination of *N*-Nitrosohexamethyleneimine in Tolazamide. *J. Chromatogr.* 1987, *386*, 57–63.
49. Madireddy, V.; Babu, K. S.; Narayanareddy, P. Stability-Indicative, Validated, Fast HPLC Method for Quantitation of Two Genotoxic Impurities in Imatinib Mesylate. *Global J. Anal. Chem.* 2011, *2*, 198–207.
50. Yuabova, Z. Y.; Holschlag, D. R.; Rodriguez, S. A.; Qin, C.; Papov, V. V.; Qiu, F.; McCaffrey, J. F.; Norwood, D. L. Genotoxic Impurities: A Quantitative Approach. *J. Liq. Chromatgr. Relat. Technol.* 2008, *31*, 2318–2330.
51. Li, J.; Yao, J.; He, C. Determination of Residual Level of Methanesulfonyl Hydrazide in Equipment Cleaning for API Production. In *Abstracts of Papers, 240th ACS National Meeting*, Boston, Massachusetts, August 22–26, 2010; ANYL-136.
52. Venkatramani, C. J.; Girotti, J.; Wigman, L.; Chetwyn, N. Quantitative Analysis of Reactive Acid Chloride, A Potential Genotoxic Impurity. In *38th International Symposium on HPLC Separation & Related Techniques,* Anaheim, California, June 16–21, 2012.
53. Liu, D. Q.; Chen, T. K.; MaGuire, M. A.; Kord, A. S. Analytical Control of Genotoxic Impurities in the Pazopanib Hydrochloride Manufacturing Process. *J. Pharm. Biomed. Anal.* 2009, *50*, 144–150.
54. Kakadiya, P. R.; Reddy, B. P.; Singh, V.; Ganguly, S.; Chandrashekhar, T. G.; Singh, D. K. Low Level Determinations of Methyl Methanesulfonate and Ethyl Methanesulfonate Impurities in Lopinavir and Ritonavir Active Pharmaceutical Ingredients by LC/MS/MS Using Electrospray Ionization. *J. Pharm. Biomed. Anal.* 2011, *55*, 379–384.
55. Venugopal, N.; Reddy, A. V. B.; Reddy, K. G.; Madhavi, V.; Madhavi, G. Method Development and Validation Study for Quantitative Determination of 2-Chloromethyl-3,4-Dimethoxy Pyridine Hydrochloride A Genotoxic Impurity in Pantoprazole Active Pharmaceutical Ingredient (API) by LC/MS/MS. *J. Pharm. Biomed. Anal.* 2012, *70*, 592–597.
56. Szekely, Gy.; Henriques, B.; Gil, M.; Ramos, A.; Alvarez, C. Design of Experiments As a Tool for LC-MS/MS Method Development for the Trace Analysis of the Potentially Genotoxic 4-Dimethylaminopyridine Impurity in Glucocorticoids. *J. Pharm. Biomed. Anal.* 2012, *70*, 251–258.
57. Yang, E.; Wang, S.; Bowen, C.; Kratz, J.; Cyronak, M. J.; Dunbar, J. R. Trapping 4-Fluorobenzyl Chloride in Human Plasma with Chemical Derivatization Followed by Quantitative Bioanalysis Using High-Performance Liquid Chromatography/Tandem Mass Spectrometry. *Rapid Commun. Mass Spectrom.* 2005, *19*, 759–766.
58. van Wijk, A. M.; Beerman, B.; Niederländer, H. A. G.; Siebum, A. H. G.; de Jong, G. J. A New Approach for Generic Screening and Quantitation of Potential Genotoxic Alkylation Compounds by Pre-Column Derivatization and LC-MS/MS Analysis. *Anal. Bioanal. Chem.* 2011, *400*, 1375–1385.
59. Bai, L.; Sun, M.; An, J.; Liu, D. Q.; Chen, T. K.; Kord, A. S. Enhancing the Detection Sensitivity of Trace Analysis of Pharmaceutical Genotoxic Impurities by Chemical Derivatization and Coordination Ion Spray Mass Spectrometry. *J. Chromatogr. A.* 2010, *1217*, 302–306.
60. An, J.; Sun, M.; Bai, L.; Chen, T.; Liu, D. Q.; Kord, A. S. A Practical Derivatization LC/MS Approach for Determination of Trace Level Alkyl Sulfonates and Dialkyl Sulfates Genotoxic Impurities in Drug Substances. *J. Pharm. Biomed. Anal.* 2008, *48*, 1006–1010.
61. Vanhoenacker, G.; Dumont, E.; David, F.; Baker, A.; Sandra, P. Determination of Arylamines and Aminopyridines in Pharmaceutical Products Using In-Situ Derivatization and Liquid Chromatography-Mass Spectrometry. *J. Chromatogr. A.* 2009, *1216*, 3563–3570.

62. David, F.; Jacq, K.; Sandra, P.; Baker, A.; Klee, M. S. Analysis of Potential Genotoxic Impurities in Pharmaceuticals by Two-Dimensional Gas Chromatography with Deans Switching and Independent Column Temperature Control Using Low-Thermal-Mass Oven Module. *Anal. Bioanal. Chem.* 2010, *396*, 1291–1300.
63. Wu, L.; Liu, D. Q.; Vogt, F. G.; Kord, A. S. Gas Phase Derivatization via the Meervein Reaction for Selective and Sensitive LC-MS Analysis of Epoxides in Active Pharmaceutical Ingredients. *J. Pharm. Biomed. Anal.* 2011, *56*, 1106–1111.
64. Gao, S.; Zhang, Z. P.; Karnes, H. T. Sensitivity Enhancement in LC/Atmospheric Pressure Ionization MS Using Derivation and Mobile Phase Additives. *J. Chromatogr., B: Anal.Technol. Biomed. Life Sci.* 2005, *825*, 98–110.
65. Bayer, E.; Gfrörer, P.; Rentel, C. Coordination-Ionspray-MS (CIS-MS), a Universal Detection and Characterization Method for Direct Coupling with Separation Techniques. *Angew. Chem., Int. Ed.* 1999, *38*, 992–995.
66. Liu, M.; Ostovic, J.; Chen, E. X.; Cauchon, N. Hydrophilic Interaction Liquid Chromatography with Alcohol as a Weak Solvent. *J. Chromatogr. A.* 2009, *1216*, 2362–2370.
67. Hasei, T.; Nakanishi, H.; Toda, Y.; Watanabe, T. Development of a Two-Dimensional High-Performance Liquid Chromatography System Couples with On-Line Reduction as a New Efficient Analytical Method of 3-Nitrobenzanthrone, a Potential Human Carcinogen. *J. Chromatogr. A.* 2012, *1253*, 52–57.
68. Khan, M.; Jayasree, K.; Krishna Reddy, K. V. S. R.; Dubey, P. K. A Validated CE Method for Determinating Dimethyl Sulfate a Carcinogen and Chloroacetyl Chloride a Potential Genotoxin at Trace Levels in Drug Substances. *J. Pharm. Biomed. Anal.* 2012, *58*, 27–33.
69. Hansen, S. H.; Sheribah, Z. A. Comparison of CZE, MEKC, MEEKC and Non-Aqueous Capillary Electrophoresis for the Determination of Impurities in Bromazepam. *J. Pharm. Biomed. Anal.* 2005, *39*, 322–327.
70. Lee, C.; Figus, M.; Zhang, L.; Marcinko, S.; Helmy, R. Application of "SMART" Method for Low Level Detection of Potential Genotoxins in Pharmaceutical Compounds. In *Abstract of Papers, 238th ACS National Meeting*, Washington, DC, August 16–19, 2009; ANYL-361.
71. Elder, D. P.; Lipczynski, A. M.; Teasdale, A. Control and Analysis of Alkyl and Benzyl Halides and Other Related Reactive Organohalides as Potential Genotoxic Impurities in Active Pharmaceutical Ingredients (APIs). *J. Pharm. Biomed. Anal.* 2008, *48*, 497–507.
72. Elder, D. P.; Snodin, D. J. Drug Substances Presented as Sulfonic Acid Salts: Overview of Utility, Safety and Regulation. *J. Pharm. Pharmacol.* 2009, *61*, 269–278.
73. Giordani, A.; Kobel, W.; Gally, H. U. Overall Impact of the Regulatory Requirements for Genotoxic Impurities on the Drug Development Process. *Eur. J. Pharm. Sci.* 2011, *43*, 1–15.
74. Baker, A. Development of Strategy for Analysis of Genotoxic Impurities. In *Genotoxic Impurities: Strategies for Identification and Control*; Teasdale, A., Ed.; John Wiley & Sons: Hoboken, New Jersey, 2010; pp. 281–303.
75. Snodin, D. J. Genotoxic Impurities: From Structural Alerts to Qualification. *Org. Process Res. Dev.* 2010, *14*, 960–976.
76. Callis, C. M.; Bercu, J. P.; DeVries, K. M.; Dow, L. K.; Robbins, D. K.; Varie, D. L. Risk Assessment of Genotoxic Impurities in Marketed Compounds Administered over a Short-term Duration: Applications to Oncology Products and Implications for Impurity Control Limits. *Org. Proc. Res. Dev.* 2010, *14*, 986–992.
77. Robinson, D. Control of Genotoxic Impurities in Active Pharmaceutical Ingredients: A Review and Perspective. *Org. Process Res. Dev.* 2010, *14*, 946–959.
78. Sun, M.; Liu, D. Q.; Kord, A. S. A Systemic Method Development Strategy for Determination of Pharmaceutical Genotoxic Impurities. *Org. Process Res. Dev.* 2010, *14*, 977–985.

12 Approaches to Assess, Analyze, and Control Genotoxic Impurities in Drug Substance Development

Heewon Lee

CONTENTS

12.1 INTRODUCTION

Impurities are introduced in the form of starting materials, reagents, by-products, intermediates, and degradation products to drug substances. As they do not provide therapeutic benefits to patients, they need to be controlled for the safety and quality of pharmaceutical substances. The International Conference on Harmonisation of Technical Requirements for Registration of Pharmaceuticals for Human Use (ICH) Q3A guideline *Impurities in New Drug Substances* [1] provides guidance on the content and qualification of impurities in new drug substances produced by chemical syntheses for registration applications. The ICH Q3B guideline *Impurities in New Drug Products* [2] provides guidance on the content and qualification of impurities in new drug products produced from chemically synthesized new drug substances for registration applications. Residual solvents are regulated by the ICH Q3C guideline *Impurities: Guideline for Residual Solvents* [3], which recommends acceptable amounts for residual solvents in pharmaceuticals for the safety of the patient. Though these ICH Q3X guidelines are intended for registration applications and not for clinical research stages of development, impurity assessment, analysis, and control during drug development are carried out with these guidelines as guiding principles in the pharmaceutical industry. ICH Q3A lists three thresholds (reporting, identification, and qualification) for impurities. The guideline states that "identification of impurities present at an apparent level of not more than (≤) the identification threshold is generally not considered necessary. However, analytical procedures should be developed for those potential impurities that are expected to be unusually potent, producing toxic or pharmacological effects at a level not more than (≤) the identification threshold." In addition, Footnote 3 of the qualification threshold states that "lower thresholds can be appropriate if the impurity is unusually toxic." It is generally accepted that genotoxic impurities are in the category of unusually potent and unusually toxic impurities.

Genotoxicity is a broad term that refers to any deleterious change in genetic material regardless of the mechanism by which the change is induced [4]. It encompasses effects from mutagenicity through DNA reactivity, DNA damage, and chromosomal damage, both structural chromosome breakage and aneuploidy [5]. Genotoxic impurities with limited toxicological data should be controlled at the level of the (staged) threshold of toxicological concern (TTC) (sTTC stands for staged TTC) according to the European Medicines Agency (EMA) guideline [6,7] and the Food and Drug Administration (FDA) draft guideline [8]. Readers are referred to Chapters 1 and 3 for details on the (s)TTC concept and calculations of acceptable exposures. To distinguish genotoxic impurities from nongenotoxic (nonmutagenic) impurities, nongenotoxic impurities are sometimes referred to as "regular," "typical," or "ordinary" impurities.

Toxicological overviews and regulations of genotoxic impurities are discussed in the literature [9–11]. Some interesting examples of genotoxic impurity assessment and control in pharmaceutical products in the literature are listed in the references section of this chapter [12–31].

This chapter presents the approaches employed to assess genotoxic risks and analyze and control genotoxic impurities in drug substance development with a few case studies discussing the rationale and procedures.

12.2 CLASSIFICATION AND QUALIFICATION OF IMPURITIES

The ultimate risk concern for genotoxic impurities is carcinogenicity; however, carcinogenicity data are not available for most impurities that are encountered. Therefore, available alternative data are used for risk assessment purposes. These alternative data include structure–activity relationships, bacterial reverse mutation (Ames) tests, and a generic determination of virtually safe exposure levels (TTC and sTTC). Mueller and others [5] proposed a classification of impurities with respect to their genotoxic potential. This system is not a general classification of genotoxicity, rather this classification would be used solely for the purpose of deciding whether an impurity possesses a high level of risk and is therefore to be controlled at very low levels of daily intake.

The proposed classification of impurities is as follows:

- Class 1: impurities known to be both genotoxic (mutagenic) and carcinogenic.
  - This group includes known animal carcinogens with reliable data for a genotoxic mechanism and human carcinogens. Published data on their chemical structures exist, demonstrating the genotoxic nature of the impurities.
- Class 2: impurities known to be genotoxic (mutagenic), but with unknown carcinogenic potential.
  - This group includes impurities with demonstrated mutagenicity based on testing of the impurities in conventional genotoxicity tests, but with unknown carcinogenic potential.
- Class 3: alerting structure, unrelated to the structure of the active pharmaceutical ingredient (API) and of unknown genotoxic (mutagenic) potential.
  - This group includes impurities with functional moieties that can be linked to genotoxicity based on structure but have not been tested as isolated compounds. They are identified based on chemistry and using knowledge-based expert systems for structure–activity relationships. The alerting functional moiety is not present in the structure of the parent API. Some generic rule-based alerts may be quite unspecific, and further consideration must be given to chemical structural constraints, chemical environment, or experimental data in the assessment of potential genotoxicity. Due to the uncertain relevance of structural alerts, regulatory action should not be based solely on the presence of a particular functional group; rather, the accuracy for predicted genotoxicity should be evaluated case by case based on the available scientific literature, additional data on the chemical class, and further available (genotoxicity) test results on closely related structures.
- Class 4: alerting structure, related to the API.
  - This group includes impurities that contain an alerting functional moiety that is shared with the parent structure. The genotoxicity of the isolated impurity is unknown, but the genotoxicity of the active principle is characterized through conventional genotoxicity testing. Similar chemical constraints and chemical environment exist for the alerting substructure in the impurity and the API.
- Class 5: no alerting structure or sufficient evidence for absence of genotoxicity.
  - This group is adequately covered by the existing ICH Q3A, Q3B, and Q3C guidelines.

A qualification strategy for each class of impurities proposed [5] is as follows:

- Class 1: carcinogenic impurities should be avoided by modification of formulation options, synthetic route, starting materials, reactants, or purification steps. When it is not practical or realistic to avoid these impurities, specifications should be determined. Compound-specific risk assessment should be conducted if sufficient 2-year rodent data are available to determine the carcinogenic potency of an impurity. The appropriate data could be cancer slope factors, TD50 (the average daily dose estimated to halve the probability of remaining tumor free throughout a 2-year study) values, or maximum tolerated dose information. If insufficient information is available to calculate a compound-specific risk, then general risk assessment based on the (s)TTC concept is performed.
- Class 2: conventional genotoxicity tests are designed to identify genetic hazards, and the data generated are not suitable for quantitative characterization of risk. Impurities for which there is strong evidence of mutagenicity should be assessed on the basis of whether or not there is evidence for a threshold-mediated mechanism.
  - For impurities with sufficient evidence of threshold-based genotoxicity (i.e., compounds that do not react with DNA, such as spindle poisons, topoisomerase inhibitors, and DNA synthesis inhibitors), calculate permitted daily exposures (PDEs) from no-observed-effect levels or lowest-observed-effect levels as described in ICH Q3C for class 2 solvents.
  - For impurities without sufficient evidence for a threshold-related mechanism, including compounds with direct DNA reactivity such as DNA binding or DNA adduct formation, levels should be controlled according to the (s)TTC approach. These compounds are usually bacterial mutagens, unequivocal in vitro clastogens, and frequently positive in in vitro genotoxicity assays.
- Class 3: when an impurity contains a structural alert without experimental data, the impurity may be controlled according to the (s)TTC levels. Alternatively, the impurity may be tested in a genetic toxicology assay (typically a bacterial reverse mutation assay, the Ames test) to gain toxicological understanding of structural predictions. If the assay results are positive (Ames positive), determining the impurity to be a bacterial genotoxin, then evaluate the compound as a class 2 impurity. If the compound is not mutagenic (Ames negative), the experimental data override the structural alert and the impurity is controlled as an ordinary impurity.
- Class 4: when an impurity has a structural alert that is shared with the API, thorough genotoxicity testing of the API is usually sufficient to qualify the structurally similar impurity. When genotoxicity tests on the API are negative, the impurity is controlled as an ordinary impurity.
- Class 5: these impurities are controlled as ordinary impurities.

A summary of the impurity classification and the aforementioned qualification strategy is shown in Table 12.1.

TABLE 12.1
Classification, Qualification, and Risk Assessment of Impurities

Classification	Question	Action
Class 1 impurities: genotoxic carcinogens	Step 1: impurity eliminated?	Yes: no further action
		No: go to step 2
	Step 2: data available for quantitative risk assessment to determine ADI?	Yes: determine ADI
		No: (s)TTC
Class 2 impurities: genotoxic, but carcinogenicity unknown	Threshold mechanism established?	Yes: Determine PDE according to the ICH Q3C procedure for class 2 solvents
		No: (s)TTC
Class 3 impurities: alert—unrelated to API	Impurity genotoxic?	Yes: follow class 2 impurities flow
		No: control as an ordinary impurity
Class 4 impurities: alert—related to API	API genotoxic?	Yes: risk–benefit analysis
		No: control as an ordinary impurity
Class 5 impurities: no alerts		Control as an ordinary impurity

Source: Adapted from Mueller et al., *Regul. Toxicol. Pharm.*, 44, 198–211, 2006.

12.3 FLOW DIAGRAM FOR GENOTOXICITY ASSESSMENT AND CONTROL

Due to the different threshold requirements for "ordinary" ("typical" or "regular," denoting "nongenotoxic" in this context) impurities versus "genotoxic" impurities, determining the genotoxicity (or nongenotoxicity) of an impurity is an important step in establishing control strategies. Figure 12.1 shows a typical flowchart for impurity assessment and control. Prior to the manufacture of the drug substance batch used for human clinical trials, the structures in the synthetic scheme, including starting materials, intermediates, reagents, and known by-products, of the drug substances are submitted for in silico evaluation. If the in silico evaluation of a structure is negative, the impurity is considered nongenotoxic and controlled according to the ICH Q3A/B limits. If the in silico assessment gives a structural alert, then the impurity is assigned to one of the five categories [5]. Depending on which class the impurity is categorized into, different strategies could be applied, as described in Section 12.2. If the compound is not expected to be present above TTC in the drug substance due to the chemical properties of the compound, a justification is written and the compound is controlled "as low as reasonably practicable" (ALARP principle) according to the pharmaceutical assessment described in EMA and FDA guidelines [6,8]. When dealing with an impurity with insufficient toxicological data and the impurity level is expected to be above (s)TTC, one can proceed to control the impurity by

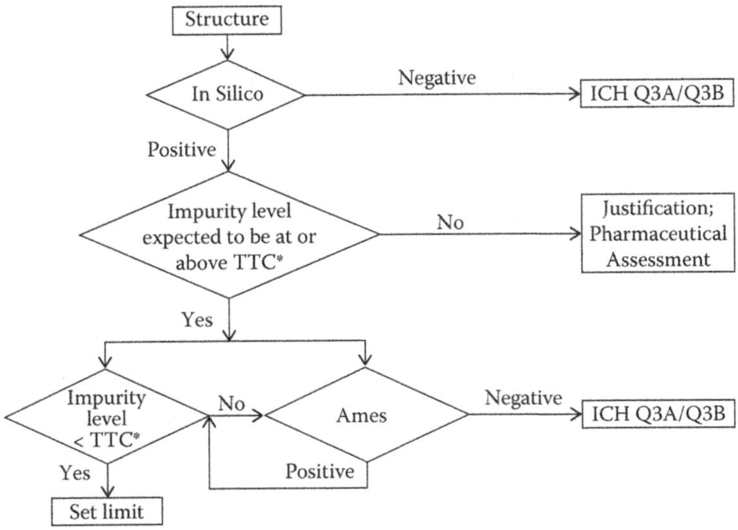

FIGURE 12.1 Flow chart for the assessment of potential and actual impurities. *Staged–threshold of toxicological concern (sTTC) might be applicable depending on the development stages. Compound-specific risk assessment should be performed when adequate toxicity data (long-term studies) are available. See Table 12.1 for classification, qualification, and risk assessment of impurities.

chemical and analytical efforts, or to test the impurity in a genetic toxicology assay (typically the Ames test). If the Ames test result is negative, then the impurity is controlled as an ordinary impurity according to ICH Q3A and Q3B. If the Ames test result is positive, then the impurity is controlled at the level of (s)TTC.

Ames testing and chemical/analytical control could happen sequentially or simultaneously depending on the resource availability and the project time lines. The two possible outcomes at this step are as follows: (1) the impurity is controlled according to (s)TTC regardless of the Ames test result, and (2) the Ames test result is negative and the impurity is controlled as an ordinary impurity according to ICH Q3A/B. Thus, a rational practice at this stage is as follows: if the anticipated Ames test result is positive based on prior knowledge of compounds with similar structures, then one would proceed with chemical/analytical control effort without spending precious resources on Ames testing. If the compound structure is frequently encountered in drug development and it has a reasonable chance to be Ames negative based on prior knowledge, then the compound is submitted for Ames testing. If the Ames test result of the compound is negative, then it saves all future chemical/analytical efforts to control the impurity at (s)TTC levels. If prior chemical/analytical efforts show that controlling the impurity at (s)TTC levels is very challenging, then one would proceed with Ames testing to ensure that the demanding efforts are justified.

The compound may not be expected to be present in the drug substance at the TTC level due to the properties of the compound. For example, a compound with low boiling point is not expected to be present in a drug substance that was manufactured by high-temperature procedures such as distillation or hot filtration. Another example is compounds with high reactivity that can be reasonably expected to be present at a level below TTC in the drug substance. An example of such a compound is acyl chloride. Acyl chloride is a strong alkylating reagent commonly used in chemical synthesis for alkylation purposes. Therefore, acyl chloride could alkylate DNA and RNA and cause mutation and genotoxicity. However, due to its high reactivity, acyl chloride readily reacts with water (or other protic solvents) under process conditions or ambient conditions and becomes carboxylic acid (esters when reacting with alcohols). In this case, acyl chloride is not expected to be present in the drug substance above the TTC level. This is an example of the "process control" described in the option 4 approach for process-related impurity control (see Section 12.4).

12.4 APPROACHES TO CONTROL PROCESS-RELATED IMPURITIES

Control of process-related impurities could be achieved by one of the following four approaches [4]. The central concepts of each approach are set in italics:

- Option 1 includes a test for the impurity in the *drug substance specification* with an acceptance criterion *at or below the acceptable limit* using an appropriate analytical procedure.
- Option 2 includes a test for the impurity in the *specification for a raw material, starting material, or intermediate, or as an in-process control (IPC)*, with an acceptance criterion *at or below the acceptable limit* using an appropriate analytical procedure.
- Option 3 includes a test for the impurity in the *specification for a raw material, starting material, or intermediate, or as an IPC*, with an acceptance criterion *above the acceptable limit* using an appropriate analytical procedure coupled with demonstrated understanding of *fate and purge* and associated process controls that ensure that the level in the drug substance is below the acceptable limit without the need for any additional testing.
- Option 4 involves *understanding of process* parameters and impact on residual impurity levels (including *fate and purge knowledge*) with sufficient confidence that the level of the impurity in the drug substance will be below the acceptable limit such that *no analytical testing* is needed for this impurity. This option relies on *process control* in lieu of analytical testing.

The options and approaches employed to control genotoxic impurities could evolve and change over the course of drug development as we gain process understanding and confidence in the process capability and control.

12.5 FACTORS TO CONSIDER WHEN DEVELOPING ANALYTICAL METHODS TO CONTROL GENOTOXIC IMPURITIES

Frequently used analytical techniques to measure genotoxic impurities are chromatographic methods including high-performance liquid chromatography (HPLC) and gas chromatography (GC) methods. The most typical detector is an ultraviolet

(UV) absorption detector for HPLC and a flame ionization detection (FID) detector for GC. These analytical techniques are well established and provide numerous desirable characteristics including linearity and sensitivity. Furthermore, these instruments are ubiquitous in almost all analytical laboratories, making method transfer straightforward. When a genotoxic impurity is volatile enough for GC analysis, GC offers advantages as most pharmaceutical intermediates and products (and their associated impurities) are not volatile, effectively reducing matrix interferences and possibly improving specificity and sensitivity of the methods.

HPLC-UV and GC-FID methods provide retention time–based specificity; therefore, when overlapping interferences are present in the sample they suffer loss of specificity. This problem is demonstrated in Figure 12.2. Each chromatogram is overlaid with a chromatogram of a hypothetical genotoxic impurity standard at 30 ppm (against a product to be monitored). The genotoxic impurity elutes at 13 minutes of retention time. In Figure 12.2a, the genotoxic impurity elutes at 13 minutes and the product elutes at 24 minutes. There is no interference, and this method is capable of detecting the genotoxic impurity at a 30 ppm level. In Figure 12.2b, the genotoxic impurity elutes at 13 minutes and the product elutes at 27 minutes. Even though the product elutes at a retention time that is well separated from the retention time of the genotoxic impurity, there is interference from a minor component of the product, making this method not suitable. Figure 12.2c shows a product that elutes very close to the genotoxic impurity; there is also interference from a minor component, and the method is not suitable. The interference problem could be a factor in determining where in the synthetic scheme to control the genotoxic impurity. Obviously, it is desirable to avoid the interferences shown in Figure 12.2b and c. It should be noted that even when the product is chromatographically well separated from the genotoxic impurity, as in Figure 12.2b, any minor components (impurities) in the product could present an interference issue. Impurity profile of the product may differ considerably from batch to batch causing an analytical method not reliable, when dealing with trace level (typically parts per million) impurity control. The challenges stemming from varying impurity profiles are severe during the early stages of development, where manufacturing changes are made frequently and process understanding and control is still being established. To manage the variability, analytical methods that provide more specificity than HPLC-UV or GC-FID might be desired. Mass spectral detection provides such specificity [25–27,31]. The EMA guideline [6] states that "detection and/or quantification of these (genotoxic) residues should be done by state-of-the-art analytical techniques," requiring pharmaceutical companies to employ cutting-edge analytical technologies for genotoxic impurity control.

Another important factor to consider is solubility of the product (starting materials, intermediates, or API) in which the genotoxic impurity needs to be controlled. In general, a high solubility of samples is required to detect a very low (typically parts per million) level of genotoxic impurity. Obviously, judicious selection of a suitable dissolving solvent is essential to achieve high solubility.

Consider the following hypothetical synthetic scheme, in which the genotoxic compound A is a starting material of the synthesis:

Genotoxic A (GTA) → intermediate B → intermediate C → intermediate D → API

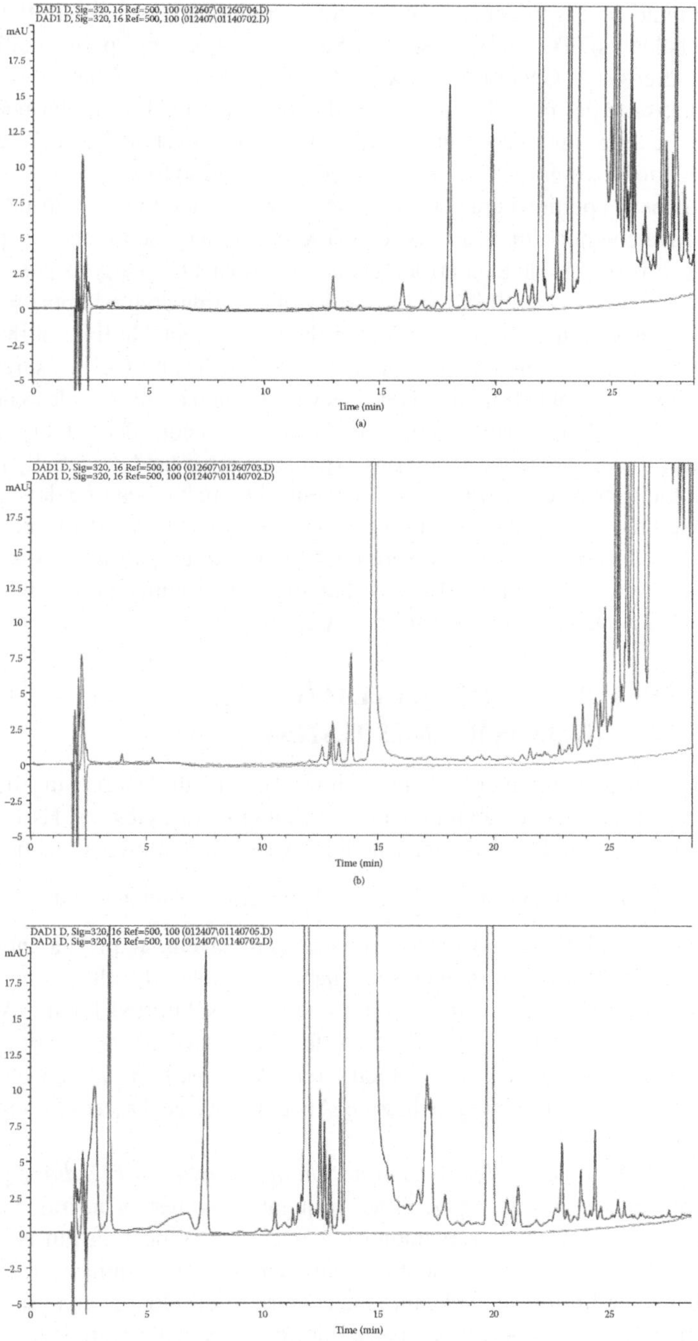

FIGURE 12.2 Generic example of high-performance liquid chromatography/ultraviolet method without and with interferences. Each chromatogram is shown overlaid with a 30 ppm standard (retention time at 13 minutes).

The ultimate goal is to control the genotoxic compound A (GTA) in the API at a level at or below (s)TTC. As discussed in Section 12.4, control points could be any compound after GTA. Generally speaking, the level of GTA becomes lower as more synthetic steps are performed. In other words, the level of GTA in intermediate C is expected to be lower than that in intermediate B, and the level of GTA in intermediate D is expected to be lower than that in intermediate C, and so forth. It would be prudent to choose a control point where GTA is reasonably expected to be controlled, such as after a reactive condition that can destroy GTA or after a crystallization step that can reject GTA. It is preferable to control a genotoxic impurity as early as possible in the synthesis. One reason is to allow the possibility of cleaning up the impurity further in cases where the analytical results are higher than the acceptable limit at the planned control point. If the planned control point is at the API, and the measured level is higher than the acceptable limit, then one has very limited options to clean up the API with an unacceptable impurity profile. Another factor to consider is that a genotoxic impurity may undergo reactive conditions to generate other (and potentially more than one) genotoxic impurities, making impurity control more challenging downstream.

Five case studies of genotoxic impurity control are discussed in Sections 12.6 through 12.10. These case studies were encountered during drug substance development and were selected to demonstrate diverse circumstances and different approaches employed for genotoxic impurity control.

12.6 CASE STUDY #1: INTERMEDIATE PREPARATION FOR AMES TESTING

A synthetic intermediate (intermediate A) showed a structural alert in in silico assessment, and it was decided to submit the intermediate for Ames testing. Hydrazine was one of the reagents used to make intermediate A (see the following reaction):

$$\text{Starting material} + H_2N\text{–}NH_2 \text{ (hydrazine)} \rightarrow \text{intermediate A}$$

Hydrazine is a highly reactive and strong reducing agent, and it is a known mutagen (Ames positive) and carcinogen (see Chapter 3, Table 3.13). It was necessary to control the level of hydrazine in the intermediate A submitted for the Ames test, so that the Ames test results reflected the effects of intermediate A, not hydrazine. A gas chromatography/mass spectrometry (GC-MS) method was developed as a limit test on acetone-derivatized hydrazine. Acetone was used as a dissolving solvent and a derivatizing agent [19].

The ICH Q2A guideline *Text on Validation of Analytical Procedures* [32] states that "testing for impurities can be either a quantitative test or a limit test for the impurity in a sample. Either test is intended to accurately reflect the purity characteristics of the sample. Different validation characteristics are required for a quantitative test than for a limit test." When it is necessary to develop an analytical method rapidly to determine the content of a particular impurity at a certain level, limit tests are often considered. Limit tests require less validation than quantitative tests as the scope of limit tests is narrower. Table 12.2 shows the validation requirements of quantitative and limit tests in the ICH Q2A guideline. A limit test needs to be validated to show specificity and detection limit only. Typically, recovery is measured

TABLE 12.2
Validation Required for Quantitative versus Limit Tests

Type of Analytical Procedure Characteristics	Identification	Testing for Impurities		Assay Dissolution (Measurement Only) Content/Potency
		Quantitative	Limit	
Accuracy	–	+	–	+
Precision				
Repeatability	–	+	–	+
Intermediate precision	–	+[a]	–	+[a]
Specificity[b]	+	+	+	+
Detection limit	–	–[c]	+	–
Quantitation limit	–	+	–	–
Linearity	–	+	–	+
Range	–	+	–	+

Note: "–" signifies that this characteristic is not normally evaluated; "+" signifies that this characteristic is normally evaluated.

[a] In cases where reproducibility has been performed, intermediate precision is not needed.

[b] Lack of specificity of one analytical procedure could be compensated by other supporting analytical procedures.

[c] May be needed in some cases.

to demonstrate that the analytical method is capable of detecting a spiked analyte at the level of interest. When an analyte is reactive, recovery is an especially important characteristic of the analytical method, because appropriate recovery data indicate that the analytical method is able to measure the analyte present in samples.

12.7 CASE STUDY #2: UNEXPECTED IMPURITY IN AN OUTSOURCED INTERMEDIATE

One of the identification methods specified for an outsourced compound was infrared (IR) spectral comparison to a reference standard. This procedure is typically known as "ID by IR." The acquired IR spectrum of the outsourced batch and that of the reference standard are shown in Figure 12.3. The two spectra match well except for an absorbance peak at 1740 cm^{-1}, which is observed only in the IR spectrum of the outsourced batch. Therefore, this batch failed the ID by IR test. An IR absorbance peak at 1740 cm^{-1} suggests a carbonyl functional group. After careful examination of the batch production record from the vendor, diethyl carbamyl chloride (DECC) was proposed as the identity of this impurity. The identity of the impurity was confirmed by GC-MS experiments against a purchased authentic sample of DECC. DECC is a known genotoxin and carcinogen.

A GC-MS method was developed to follow and control the impurity. DECC is a reactive compound similar to acyl chloride. The appropriate selection of sample diluent and preparation is important as DECC can react with the diluent to hinder

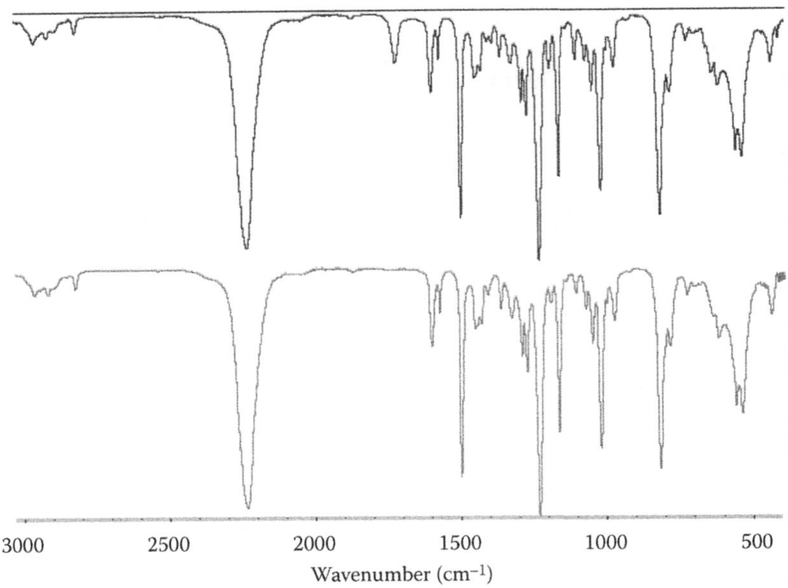

FIGURE 12.3 ID by IR analysis of an outsourced compound. Top trace: an outsourced batch; bottom trace: a reference standard.

adequate analytical measurement. For example, DECC can react with methanol to form methyl carbamate (similar to the formation of ester when acyl chloride reacts with alcohol). The mass spectra from GC-MS of DECC prepared in methanol are shown in Figure 12.4. Intact DECC shows a molecular ion at *m/z* 135 with a chlorine isotope pattern (bottom mass spectrum), whereas methyl carbamate shows a molecular ion at *m/z* 131 without a chlorine pattern (top mass spectrum), illustrating the importance of proper choice of diluent as an important part of analytical method development.

Modification of the batch production procedure was requested to the vendor to avoid the formation of DECC for future batches.

The ICH Q10 guideline *Pharmaceutical Quality System* [33] has a section titled "Management of Outsourced Activities and Purchased Materials" with the following statements:

> The pharmaceutical quality system, including the management responsibilities described in this section, extends to the control and review of any outsourced activities and quality of purchased materials. **The pharmaceutical company is ultimately responsible to ensure processes are in place to assure the control of outsourced activities and quality of purchased materials.** These processes should incorporate quality risk management and include:
>
> (a) Assessing prior to outsourcing operations or selecting material suppliers, the suitability and competence of the other party to carry out the activity or provide the material using a defined supply chain (e.g., audits, material evaluations, qualification);
>
> (b) Defining the responsibilities and communication processes for quality-related activities of the involved parties. For outsourced activities, this

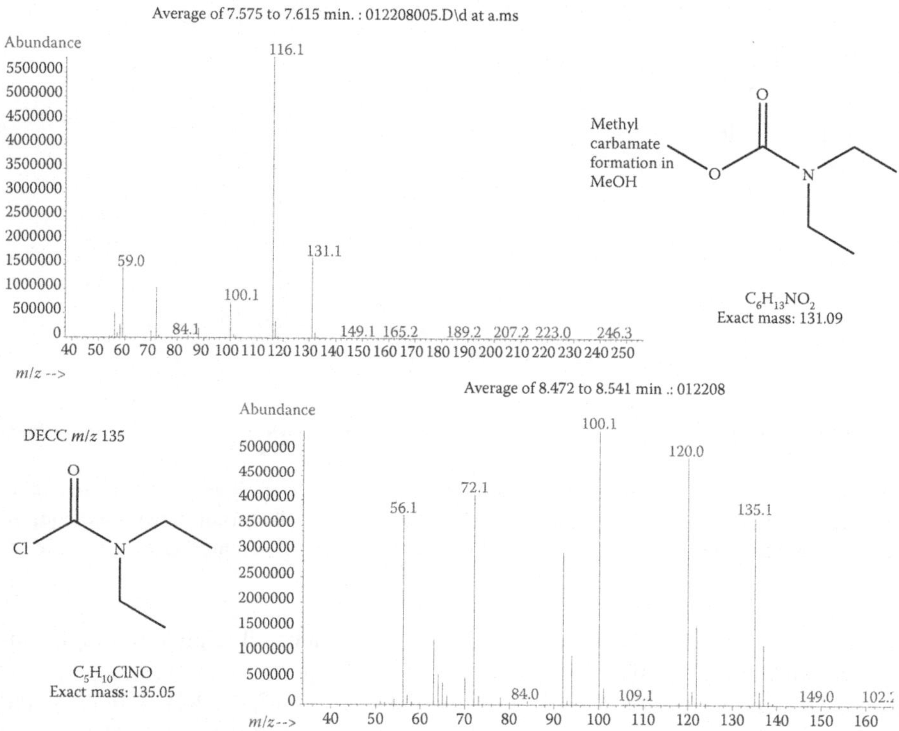

FIGURE 12.4 Gas chromatography/mass spectrometry data of diethyl carbamyl chloride (DECC). MeOH stands for methanol.

should be included in a written agreement between the contract giver and contract acceptor;

(c) Monitoring and review of the performance of the contract acceptor or the quality of the material from the provider, and the identification and implementation of any needed improvements;

(d) Monitoring incoming ingredients and materials to ensure they are from approved sources using the agreed supply chain.

The ultimate responsibility of ensuring quality drug substance lies with the pharmaceutical company even when a genotoxic impurity originates from an outsourced batch. This case study demonstrates that setting well-informed and thoughtful specifications is a critical aspect of proper control of impurities in outsourcing activities, as well as good process understanding and effective communications with vendors.

12.8 CASE STUDY #3: CONTROL OF 2-CHLOROPYRIDINE IMPURITIES BY GAS CHROMATOGRAPHY/MASS SPECTROMETRY

The FDA draft guideline on genotoxic impurities [8] states that "in cases where a class or family of structurally similar impurities is identified and is expected to have similar mechanisms resulting in their genotoxic or carcinogenic potential, the total

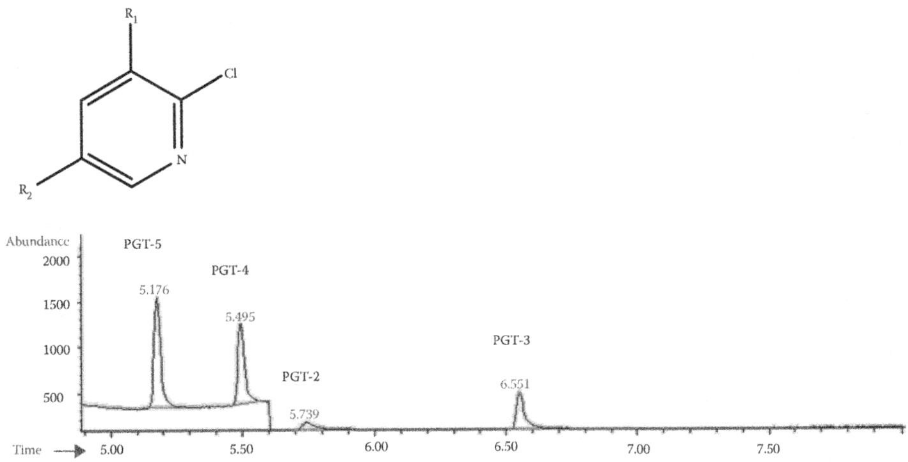

FIGURE 12.5 Gas chromatography/mass spectrometry (GC-MS) of potential genotoxic impurities with the 2-chloropyridine moiety. The GC-MS chromatogram shown was acquired from a standard mixture of 20 ppm each of PGT-2, PGT-3, PGT-4, and PGT-5.

daily exposure to the related compound should be evaluated relative to the recommended threshold exposure."

Five potential genotoxic impurities (named PGT-1, PGT-2, PGT-3, PGT-4, and PGT-5) needed to be controlled in an intermediate. PGT-2 through PGT-5 shared the same 2-chloropyridine functionality as shown in Figure 12.5. The acceptable limit for genotoxic impurity control was 100 ppm. All four potential genotoxic impurities displayed structural alerts in in silico assessment because of the 2-chloropyridine moiety. Due to the structural similarity it was assumed that they would have the same mode of action for genotoxicity, and it was decided to control the four impurities as a sum [16]. PGT-1 did not share the 2-chloropyridine functionality; therefore, the control limit for PGT-1 was set at 100 ppm. A limit test method using GC-MS was developed for the four PGTs that had the 2-chloropyridine moiety. The GC-MS chromatogram acquired in a timed selected ion monitoring (SIM) mode is shown in Figure 12.5. The chromatogram was obtained from a standard mixture made of 20 ppm each of the four potential genotoxic impurities. If all four 2-chloropyridine PGTs showed not more than (NMT) 20 ppm individually, then the sum of the four 2-chloropyridine PGTs is NMT 80 ppm, satisfying the 100 ppm control limit. A limit test HPLC-UV method was developed separately for PGT-1 (data not shown). The summary of three batches is shown in Table 12.3. Lot A and lot B exhibited a level of potential genotoxic impurities below the control limit for all five PGTs. However, lot C showed more than 20 ppm for PGT-3. To meet the NMT 100 ppm control limit with a comfortable safety margin, the 20 ppm limit GC-MS method was revalidated with a 10 ppm limit for PGT-2, PGT-4, and PGT-5. The analytical results with the 10 ppm limit GC-MS method are shown for lot C in Table 12.3. This effort cleared all three lots to be used in the manufacture of the API.

12.9 CASE STUDY #4: CONTROL OF GENOTOXIC INTERMEDIATES BY LC/MS/MS

A nitro aromatic starting material was used to make an API via hydrogenation. The synthetic scheme is shown in Figure 12.6. If the nitro aromatic starting material was not completely consumed, then the residual nitro aromatic starting material would undergo hydrogenation to the corresponding aniline impurity. Both the nitro aromatic starting material and the potential aniline impurity showed structural alerts in in silico assessment as expected (see Chapter 2 for structural alerts).

The 10 ppm limit test LC/MS/MS methods were developed for the nitro aromatic starting material and the corresponding aniline impurity separately. Figure 12.7 shows the specificity advantage of an LC/MS/MS method using a triple quadrupole mass spectrometer [26], compared to UV detection and single quadrupole mass spectrometric detection. All four chromatograms were obtained from a single sample, which was the product dissolved in dimethyl acetamide (used as the diluent) at 100 mg/mL, spiked with the aniline impurity at 10 ppm. The first trace (from the top) shows single reaction monitoring (SRM) of the aniline impurity, clearly showing a good signal to noise (*S/N*) ratio at 10 ppm at a retention time of 5.3 minutes. The second trace is a UV

TABLE 12.3
Control of Potential Genotoxic Impurities with 2-Chloropyridine Functionality (PGT2, PGT3, PGT4, and PGT-5): Analytical Results Summary of Three Batches

Impurity	Lot A (ppm)	Lot B (ppm)	Lot C (ppm)
PGT-1	NMT 50	NMT 50	NMT 50
PGT-2	NMT 20	NMT 20	NMT 10
PGT-3	NMT 20	NMT 20	NMT 40
PGT-4	NMT 20	NMT 20	NMT 10
PGT-5	NMT 20	NMT 20	NMT 10

Note: Required control levels are 100 ppm for PGT-1 and 100 ppm for the sum of PGT-2, PGT-3, PGT-4, and PGT-5.

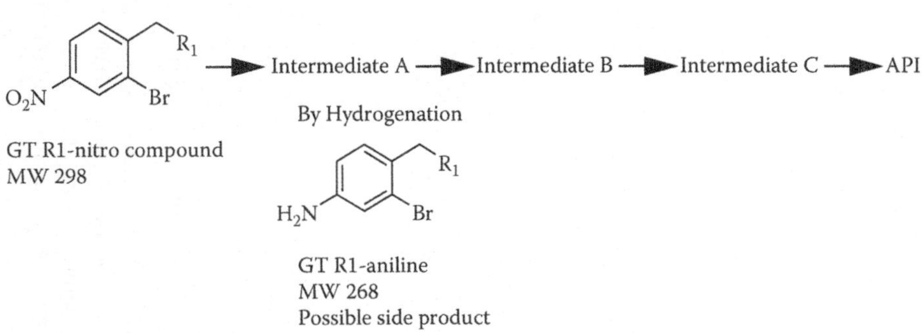

FIGURE 12.6 Potential formation of aniline side product by the hydrogenation of a nitro aromatic starting material.

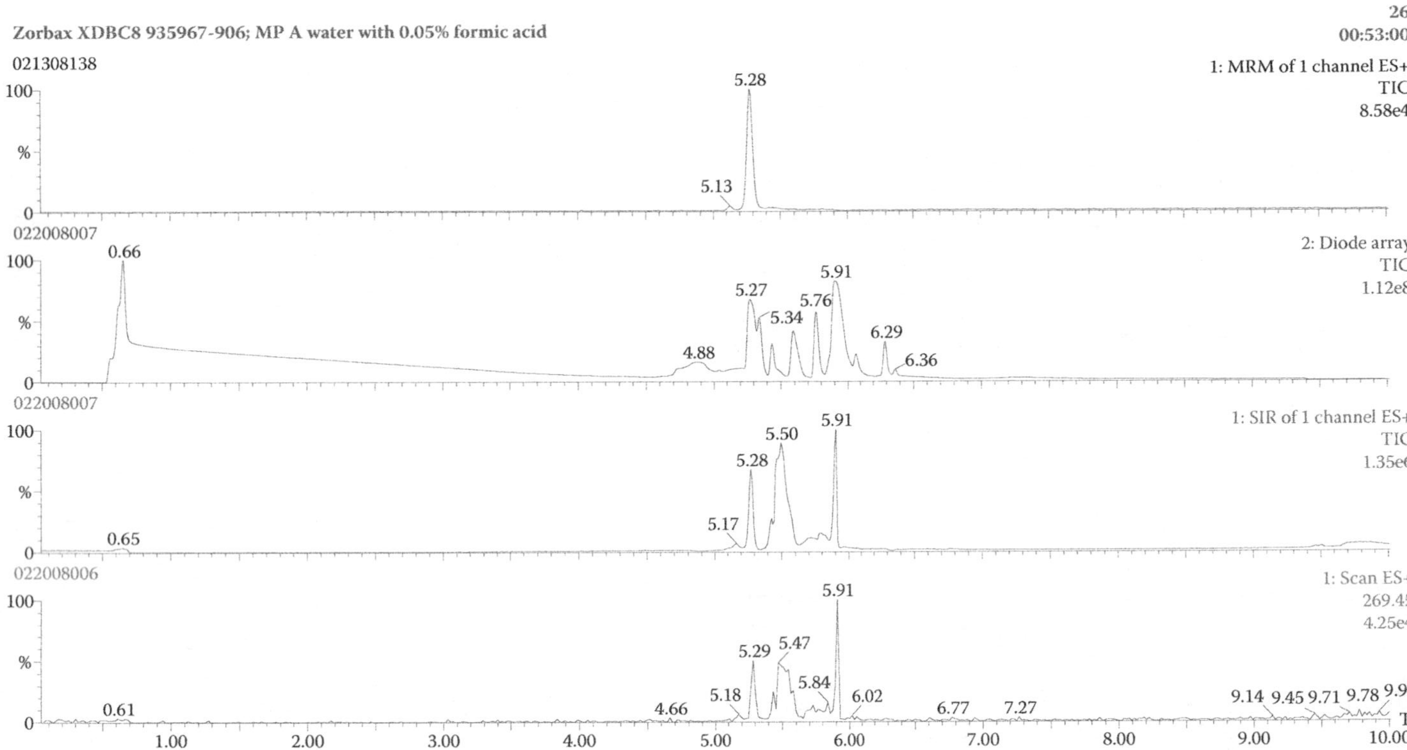

FIGURE 12.7 Specificity advantage of an LC/MS/MS method. The four chromatograms shown were acquired on a single sample of a product at 100 mg/mL in dimethyl acetamide (DMAc) with the aniline impurity spiked at 10 ppm. First trace (from the top): single reaction monitoring chromatogram for the aniline impurity; second trace: ultraviolet chromatogram; third trace: selected ion monitoring chromatogram for the aniline impurity; fourth trace: extracted ion chromatogram from scan mass spectrometry for the aniline impurity. TIC, total ion chromatogram.

chromatogram, and apparently this detection did not have suitable specificity for the aniline impurity. The third trace was acquired in a SIM mode for the aniline impurity. The peak at 5.3 minutes is evident, although not as clear as in the first trace, and other interference also showed up on this channel. The fourth trace was acquired in a full mass spectrometry scan mode, and the molecular ion for the aniline impurity was plotted as an extracted ion chromatogram. The peak at 5.3 minutes is noticeable, but the specificity and *S/N* ratio were not as good as in the SRM or SIM modes. A better specificity, sensitivity, and detection limit could be achieved by using state-of-the-art analytical technologies such as triple quadrupole mass spectrometers.

In this case study, the starting material was a genotoxic compound that needed to be monitored and controlled. The chemical steps used to manufacture the API could potentially convert the genotoxic starting material to different genotoxic side products. This case study illustrates the need to consider and control not just genotoxic intermediates but also potential side products that are genotoxic to ensure the quality and safety of drug substances.

12.10 CASE STUDY #5: CONTROL OF METHANE SULFONIC ACID AND TOSIC ACID IN LIEU OF TESTING CORRESPONDING ALKYL ESTERS

The well-known Viracept case [12–15] in 2007 was caused by ethyl methane sulfonate (EMS) contamination. Viracept is a protease inhibitor used for the treatment of HIV patients, and the contamination was first detected by a bad smell in blister-packed tablets. The final manufacturing step of the Viracept drug substance, nelfinavir mesylate, is the formation of the mesylate salt by adding methane sulfonic acid (MSA) to a suspension of nelfinavir in ethanol. As MSA reacts with nelfinavir, the product is dissolved. Subsequently, nelfinavir mesylate is isolated from the ethanolic solution by spray-drying. However, it was neither MSA nor the manufacturing process itself that caused the EMS contamination found in the drug substance. The root cause of the contamination was that tank drying was not performed after cleaning the MSA hold tank with ethanol. Upon refilling the tank with neat MSA, reaction with residual ethanol over several months led to the formation of significant concentrations of EMS in the MSA that was used to produce the drug substance causing the contamination.

Interactions of strong acids and alcohols during the process of salt formation may produce various alkylating agents such as alkyl halides and esters of alkyl sulfonic acids. The alkylating agents reported in the literature include methyl methane sulfonate (MMS), EMS, besylates, and tosylates [12–16, 20, 22–24, 28]. Most of these alkylating agents are known genotoxins.

One of the intermediates used to make a drug substance was a *p*-toluenesulfonic acid (tosic acid) salt. The process map is shown in Figure 12.8. Crude API in tetrahydrofuran (THF) was synthesized using intermediate A and intermediate B (tosic acid salt). The crude API was precipitated by adding water and isolated. The final crystallization in *n*-propanol and water was performed for purification and form control. Because intermediate B was a tosic acid salt and the final crystallization solvents included *n*-propanol, the formation of *n*-propyl tosylate was a concern as a potential contaminant in the drug substance.

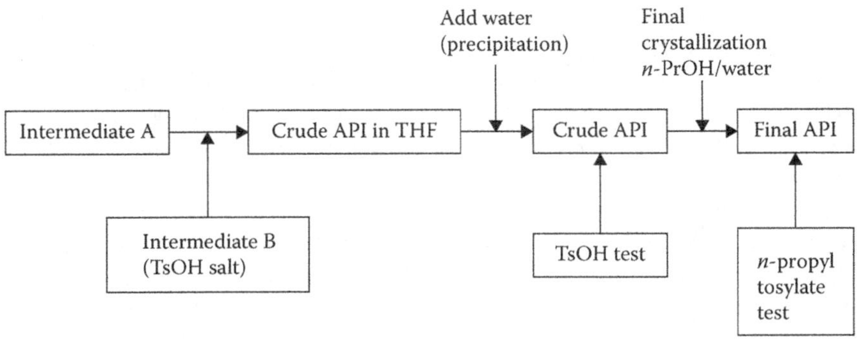

FIGURE 12.8 Control of *n*-propyl tosylate in active pharmaceutical ingredient crystallization.

The *n*-propyl tosylate could be tested in the final API; however, if the detected level was above the acceptable limit then the options to remove *n*-propyl tosylate would have been very limited. Considering the choices for controlling *n*-propyl tosylate in the API it was decided to test tosic acid in the crude API stage, and a limit test HPLC-UV method was developed for this purpose. The limit test was performed as an in-process control (IPC) so that if the level of tosic acid was above the acceptable limit then the process step was repeated until the acceptable limit was reached. This ensured that the level of tosic acid was below the acceptable limit at the crude API stage; thus, the testing of API for *n*-propyl tosylate was not necessary. However, *n*-propyl tosylate was still tested (by GC-MS) in the API in early batches to ensure compliance and to understand the purging capability of the processes. Purge reflects the ability of a process to reduce the level of an impurity, and the purge factor is defined as the level of an impurity at an upstream point in a process divided by the level of the impurity at a downstream point in the process [4]. As more analytical testing data and process understanding accumulated, it was possible to set a target value of tosic acid in the crude API that would provide the API with an acceptable level of *n*-propyl tosylate with confidence.

Another project had an intermediate that was an MSA salt. The subsequent steps involved methanol, ethanol, and *n*-propyl alcohol. Therefore, MMS, EMS, and *n*-propyl methane sulfonate (n-PMS) were potential genotoxic impurities in the drug substance. The strategy of testing MSA (by LC/MS) at an appropriate synthetic step was used successfully to control MMS, EMS, and n-PMS in the drug substance, simplifying and alleviating analytical workload.

12.11 CONCLUSION

The rationale, procedures, and practices employed to assess, analyze, and control genotoxic impurities are discussed with a few case studies. Classification of impurities assesses the level of risk that a genotoxic impurity poses and provides a basis to set priorities for toxicological testing, analysis, and control. The qualification strategy and flowchart for assessment and control present representative procedures and practices. The actual case studies presented demonstrate the different circumstances in which genotoxic impurities are encountered in drug substance development and highlight the various approaches employed to control them properly.

The successful management of genotoxic impurity issues in a pharmaceutical environment involves many different disciplines including toxicology, chemical development, pharmaceutical development, analytical development, drug metabolism and pharmacokinetics, and regulatory affairs. Effective communication and collaboration between disciplines is absolutely essential. Genotoxic impurities present demanding challenges for pharmaceutical development. Proper management of genotoxic impurities is a critical element in achieving the worthwhile goal of ensuring the quality and safety of drug substances and products for patients.

ACKNOWLEDGMENTS

The author acknowledges the following colleagues for their valuable discussions and contributions to the case studies presented in this chapter: Nelu Grinberg, Sherry Shen, Shengli Ma, Earl Spinelli, Xudong Wei, Bing-Shiou Yang, Dan Fandrick, Jason Mulder, Rogelio Frutos, Thomas Tampone, Yibo Xu, Steve Han, Xiao-Jun Wang, Li Zhang, Yongda Zhang, and Maurice Marsini.

REFERENCES

1. International Conference on Harmonisation of Technical Requirements for Registration of Pharmaceuticals for Human Use, *ICH Harmonised Tripartite Guideline, Impurities in New Drug Substances*, ICH Q3A(R2), October 2006.
2. International Conference on Harmonisation of Technical Requirements for Registration of Pharmaceuticals for Human Use, *ICH Harmonised Tripartite Guideline, Impurities in New Drug Products*, ICH Q3B(R2), June 2006.
3. International Conference on Harmonisation of Technical Requirements for Registration of Pharmaceuticals for Human Use, *ICH Harmonised Tripartite Guideline, Impurities: Guideline for Residual Solvents*, ICH Q3C (R5), February 2011.
4. International Conference on Harmonisation of Technical Requirements for Registration of Pharmaceuticals for Human Use, *ICH Harmonised Tripartite Guideline, Assessment and Control of DNA Reactive (Mutagenic) Impurities in Pharmaceuticals to Limit Potential Carcinogenic Risk*, ICH M7 Draft Guideline, February 2013.
5. L. Mueller, R. J. Mauthe, C. M. Riley, M. M. Andino, D. D. Antonis, C. Beels, J. DeGeorge, et al., "A rationale for determining, testing, and controlling specific impurities in pharmaceuticals that possess potential for genotoxicity," *Regul. Toxicol. Pharm.* 44, 198–211 (2006).
6. European Medicines Agency, Committee for Medicinal Products for Human Use, *Guideline on the Limits of Genotoxic Impurities*, January 2007.
7. European Medicines Agency, *Questions and Answers on the 'Guideline on the Limits of Genotoxic Impurities'*, September 2010 (Rev. 3).
8. Food and Drug Administration, Center for Drug Evaluation and Research, *Guidance for Industry, Genotoxic and Carcinogenic Impurities in Drug Substances and Products: Recommended Approaches, Draft Guidance*, December 2008.

9. D. Jacobson-Kram, T. McGovern, "Toxicological overview of impurities in pharmaceutical products," *Adv. Drug Deliver. Rev.* 59, 38–42 (2007).
10. T. McGovern, D. Jacobson-Kram, "Regulation of genotoxic and carcinogenic impurities in drug substances and products," *Trend. in Anal. Chem.* 25(8), 790–795 (2006).
11. J. Grosso, "Zero risk and zero benefit: Some thoughts on the control of genotoxic impurities," *Curr. Opin. Drug Discov. Devel.* 12(6), 730–731 (2009).
12. L. Mueller, T. Singer, "EMS in Viracept—The course of events in 2007 and 2008 from the non-clinical safety point of view," *Toxicol. Lett.* 190, 243–247 (2009).
13. C. Gerber, H. Toelle, "What happened: The chemistry side of the incident with EMS contamination in Viracept tablets," *Toxicol. Lett.* 190, 248–253 (2009).
14. L. Mueller, E. Gocke, T. Lave, T. Pfister, "Ethyl methanesulfonate toxicity in Viracept—A comprehensive human risk assessment based on threshold data for genotoxicity," *Toxicol. Lett.* 190, 317–329 (2009).
15. V. E. Walker, D. A. Casciano, D. J. Tweats, "The Viracept—EMS case: Impact and outlook," *Toxicol. Lett.* 190, 333–339 (2009).
16. D. Elder, J. Harvey, "Is there a real case for cumulative control of structurally related genotoxic impurities?" *Org. Process Res. Dev.* 14, 1037–1045 (2010).
17. G. J. S. Jenkins, S. H. Doak, G. E. Johnson, E. Quick, E. M. Waters, J. M. Parry, "Do dose response thresholds exist for genotoxic alkylating agents?" *Mutagenesis.* 20, 389–398 (2005).
18. S. M. Galloway, J. V. Reddy, K. McGettigan, R. Gealy, J. Bercu, "Potentially mutagenic impurities: Analysis of structural classes and carcinogenic potencies of chemical intermediates in pharmaceutical syntheses supports alternative methods to the default TTC for calculating safe levels of impurities," *Regul. Toxicol. Pharm.* 66, 326–335 (2013).
19. D. T. Fortin, R. Chen, "Developing a trace level GC-MS method for detecting methylhydrazine in an experimental drug substance," *J. Chromatogr. Sci.* 48, 299–302 (2010).
20. Z. Sobol, M. E. Engel, E. Rubitski, W. W. Ku, J. Aubrecht, R. H. Schiestl, "Genotoxicity profiles of common alkyl halides and esters with alkylating activity," *Mut. Res.* 633, 80–94 (2007).
21. G. Eichenbaum, M. Johnson, D. Kirkland, P. O'Neill, S. Stellar, J. Bielawne, R. DeWire, et al., "Assessment of the genotoxic and carcinogenic risks of *p*-nitrophenol when it is present as an impurity in a drug product," *Regul. Toxicol. Pharm.* 55, 33–42 (2009).
22. M. Nassar, M. Cucolo, S. A. Miller, "Ethyl methanesulphonate in a parenteral formulation of BMS-214662 mesylate, a selective farnesyltransferase inhibitor: Formation and rate of hydrolysis," *Pharm. Dev. Technol.* 14(6), 672–677 (2009).
23. Y. Hajikarimian, S. Yeo, R. W. Ryan, P. Levett, C. Stoneley, P. Singh, "Investigation into the formation of the genotoxic impurity ethyl besylate in the final step manufacturing process of UK-369,003-26, a novel PDE5 inhibitor," *Org. Process Res. Dev.* 14, 1027–1031 (2010).
24. A. Teasdale, E. J. Delaney, S. C. Eyley, K. Jacq, K. Taylor-Worth, A. Lipczynski, W. Hoffmann, et al., "A detailed study of sulfonate ester formation and solvolysis reaction rates and application toward establishing sulfonate ester control in pharmaceutical manufacturing processes," *Org. Process Res. Dev.* 14, 999–1007 (2010).
25. H. Lee, S. Shen, N. Grinberg, "Identification and control of impurities for drug substance development using LC/MS and GC/MS," *J. Liq. Chrom. & Related Technol.* 31, 2235–2252 (2008).
26. H. Lee, "Pharmaceutical applications of liquid chromatography coupled with mass spectrometry (LC/MS)," *J. Liq. Chromatogr. R. T.* 28, 1161–1202 (2005).

27. M. Sun, D. Q. Liu, A. S. Kord, "A systematic method development strategy for determination of pharmaceutical genotoxic impurities," *Org. Process Res. Dev.* 14, 977–985 (2010).

28. Q. Yang, B. P. Haney, A. Vaux, D. A. Riley, L. Heidrich, P. He, P. Mason, et al., "Controlling the genotoxins ethyl chloride and methyl chloride formed during the preparation of amine hydrochloride salts from solutions of ethanol and methanol," *Org. Process Res. Dev.* 13, 786–791 (2009).

29. C. Lee, R. Helmy, C. Strulson, J. Plewa, E. Kolodziej, V. Antonucci, B. Mao, C. J. Welch, Z. Ge, M. A. Al-Sayah, "Removal of electrophilic potential genotoxic impurities using nucleophilic reactive resins," *Org. Process Res. Dev.* 14, 1021–1026 (2010).

30. Z. Cimarosti, F. Bravo, P. Stonestreet, F. Tinazzi, O. Vecchi, G. Camurri, "Application of quality by design principles to support development of a control strategy for the control of genotoxic impurities in the manufacturing process of a drug substance," *Org. Process Res. Dev.* 14, 993–998 (2010).

31. J. Zheng, W. A. Pritts, S. Zhang, S. Wittenberger, "Determination of low ppm levels of dimethyl sulfate in an aqueous soluble API intermediate using liquid-liquid extraction and GC-MS," *J. Pharmaceut. Biomed* 50, 1054–1059 (2009).

32. International Conference on Harmonisation of Technical Requirements for Registration of Pharmaceuticals for Human Use, *ICH Harmonised Tripartite Guideline, Validation of Analytical Procedures: Text and Methodology*, ICH Q2(R1), November 2005.

33. International Conference on Harmonisation of Technical Requirements for Registration of Pharmaceuticals for Human Use, *ICH Harmonised Tripartite Guideline, Pharmaceutical Quality System*, ICH Q10, June 2008.

13 Control and Analysis of Genotoxic Impurities in Drug Substance Development

Zhi Chen, Jianbing Zhang, Shan Xiao, Qin Ji, and Frances Liu

CONTENTS

13.1 INTRODUCTION

Impurities arising from synthesis and formulation processes, or from degradation of the active pharmaceutical ingredient (API) and excipients, may be present in finished drug products (DPs). The safety of impurities is a concern for both pharmaceutical companies and regulatory authorities, and the control of impurities is generally addressed by International Conference on Harmonisation of Technical Requirements for Registration of Pharmaceuticals for Human Use (ICH) guidelines from a regulatory point of view. However, these documents do not provide any specific guidance related to the determination of acceptable levels for genotoxic impurities (GTIs).

GTIs are a subset of impurities that may induce genetic mutations, chromosomal breaks, and/or chromosomal rearrangements and have the potential to cause cancer in humans. In 2002, the European Medicines Evaluation Agency (EMEA) published a draft position paper that addressed the issue of GTIs from both toxicological and quality perspectives.[1] The original paper became a draft guideline in 2004, and the concept of threshold of toxicological concern (TTC) was introduced first in this

TABLE 13.1
Staged TTC Limits for GTIs

	Duration of Clinical Trial Exposure					
	Single Dose[a] or ≤14 Days[b]	≤1 Month	≤3 Months	≤6 Months	≤12 Months	>12 Months[b]
Allowable daily intake (μg)	120	60	20	10	5	1.5

[a] EMEA guideline.
[b] FDA draft guideline.

guideline.[2] The TTC concept is part of a risk assessment approach that aims to establish a conservative while virtually acceptable maximum daily intake below which a GTI poses negligible safety concerns to human health. The utilization of TTC establishes a 1.5 μg/day limit based on a lifetime exposure resulting in a cancer risk of 1 in 100,000, a risk level that the EMEA guideline considers justified because of the benefits derived from pharmaceuticals. The EMEA guideline was finalized in 2006[3] and became effective in 2007. Later, durational adjustments to the TTC limit were also accepted by the EMEA for medicinal products in development phases.[4] As shown in Table 13.1, a set of staged TTC limits were published dependent on the duration of exposure in clinical trials. Following the EMEA, the U.S. Food and Drug Administration (FDA) also published its own draft guidance in 2008 providing recommendations on how to evaluate the safety of genotoxic and carcinogenic impurities during clinical development and for marketing applications.[5] The recommended exposure thresholds in this guidance are generally aligned with the staged TTC limits recognized by the EMEA.

13.2 INDUSTRIAL PRACTICE

The pharmaceutical industry recognizes its obligation and strives to make substantial efforts during drug development to control all impurities, including GTIs, at safe levels.

Early in the discovery stage, structural alert assessment is performed to assist lead optimization and final candidate selection. In silico structure–activity relationship (SAR) tools are usually used to predict genotoxicity of the drug candidates. Recently, it has also been recognized that early SAR screening of potential hydrolysis products and known metabolites may facilitate the elimination of high-risk candidates from the pipeline and save time and resources in the development phase.

Prior to the first in-human clinical investigation, synthetic schemes to manufacture the API and key raw materials must be evaluated. Potential impurity structures, including starting materials, isolated intermediates, reagents and solvents, by-products, catalysts, and counterions, are submitted for in silico genotoxicity analysis. For structural alert hits, the impurities that have proved challenging to control are isolated and purified and are investigated by in vitro genotoxicity assays, such as the Ames test. If GTIs are found, the optimum option is to develop an alternative synthetic route that can remove these impurities. If this is not technically feasible, impurity specifications should be set based on maximum administered doses and maximum duration of study using the staged TTC concept. Accordingly, analytical methods with adequate selectivity and sensitivity are commonly required. In cases where testing is not performed, scientific justification must be given, which usually includes the evaluation of compound reactivity, purification steps, point of introduction, and so on. However, analytical data may still be needed later in development to support the indicated rationale. (For some very highly reactive impurities, purge is obvious; so no data would be required. Additionally, for families of GTIs data are often generated for the impurity with the highest probability of presence and the others do not have to be tested.) For formulation development, a similar practice may also apply. While, the impurities of interest are often degradation products which may be introduced in the manufacturing process of the final dosage form or during storage. To identify potential degradation impurities for genotoxicity assessment, stress studies are commonly needed.

The identification and control of GTIs in APIs and formulation is a dynamic approach that needs to be reassessed throughout the drug development process. For example, if the synthesis and/or formulation process has changed the new process needs to be assessed. If the acceptable toxicology limit has changed due to the changes in a clinical program, the capabilities of the manufacturing process and analytical method for control at the new level need to be assessed as well.

13.3 CASE STUDIES

Three industrial case studies are discussed in this section. As mentioned in Section 13.2, when the maximum daily dose and/or the duration of study are changed in a clinical program the limit of GTIs may also have to be updated based on the staged TTC concept. In the first case study, the specification of a GTI in a drug substance was lowered 10-fold when the clinical program went from early phase (phase I to phase IIa) to late phase (phase IIb to phase III). Accordingly, a new API-manufacturing process was developed and risk assessment followed by design of experiments (DOE) was applied to assess the GTI removal capability of the new process. An analytical method with improved sensitivity was also developed to fit its intended use in this case. GTI determination at trace levels requires highly sensitive analytical methods, which may possess tremendous challenges in a quality control (QC) environment. The most conventional analytical instrumentations in pharmaceutical analysis are high-performance liquid chromatography–ultraviolet (HPLC-UV) and gas chromatography (GC) with a flame ionization detector. However, they are often inadequate for accurate analysis of GTIs at low parts-per-million levels, depending on properties of the analytes and sample matrices.[6] In the second and third case studies, attention is focused on specific analytical challenges related to GTI determination.

13.3.1 Case Study 1

Quality by design (QbD) is a systematical approach to understand and optimize product quality attributes to meet specific objectives. The focus of this concept is that quality should be built rather than tested in a product. QbD approaches have been described in a number of regulatory guidelines.[7] For example, ICH Q10, *Pharmaceutical Quality System*, introduced the concept of control strategy, defined as a planned set of controls derived from product and process understanding that ensures process performance and product quality.[8] Hence it is essential to fully understand the manufacturing process to ensure that the quality of both drug substance and DP is appropriate and consistent.[7] As part of pharmaceutical QbD approaches, DOE is efficient in process development for evaluating the effects and possible interactions of several factors. This case study discusses the control strategy to regulate a GTI during the clinical phase transition period. The control strategy included the development, risk assessment, and utilization of the specification settings, purification process, and analytical testing method. DOE was applied in process development to facilitate the full understanding and hence optimization of the purification process for GTI removal.

API #1 was under development to treat certain cancers. By-product A can be formed by the hydrolysis of API #1. This impurity is mutagenic and therefore should be strictly controlled. During early development (phase I to phase IIa), the maximum daily dose of API #1 was 500 mg and the maximum duration of treatment was 6 months. Accordingly, the limit of GTI compound A in the drug substance was set to 20 ppm using the staged TTC limit (see Table 13.1):

$$\frac{10\ \mu\text{g}}{500\ \text{mg}} = 20\ \text{ppm} \tag{13.1}$$

A specific reverse-phase high-performance liquid chromatography (RP-HPLC) method with ultraviolet (UV) detection was used for its analysis. This method has a linear range from 2 to 30 ppm, the limit of quantitation (LOQ) is 2 ppm, and the limit of detection (LOD) is 1 ppm. The chromatograms for LOQ and LOD are shown in Figure 13.1.

When the clinical program entered the late phase (phase IIa and beyond) and was prepared for registration, the drug substance specification of compound A was reassessed in accordance with the EMEA GTI guideline for marketed products. A limit of 2 ppm was proposed based on the new maximum daily dose for API #1 of 800 mg/day:

$$\frac{1.5\mu\text{g}}{800\ \text{mg}} = 1.9\ \text{ppm} \approx 2\ \text{ppm} \tag{13.2}$$

The new specification represented a reduction in the level of compound A from not more than (NMT) 20 ppm to NMT 2 ppm in the drug substance, requiring an update in all regulatory dossiers and the corresponding analytical method for this specification. An RP-HPLC method with mass spectrometry (MS) detection was developed to support the proposed specification. This method has an LOQ of 0.6 ppm, as shown in Figure 13.2.

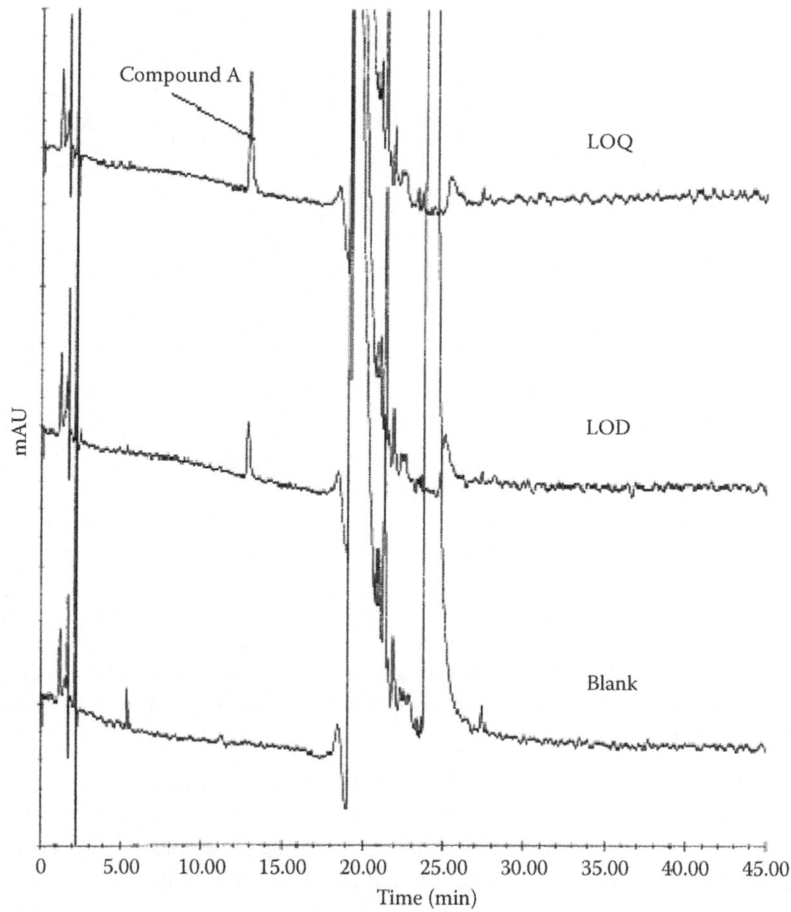

FIGURE 13.1 Chromatograms of limit of quantitation (2 ppm), limit of detection (1 ppm), and blank for genotoxic impurity compound A using a high-performance liquid chromatography–ultraviolet method.

Compound A ranged from less than 2 ppm to 20 ppm in the drug substance in early batches. A new purification process was then developed using charcoal to reduce the level of this compound. A risk analysis of batches that failed specification was conducted to investigate critical parameters for the new process. The following parameters required further investigation based on the risk assessment: starting concentration of compound A, amount of charcoal, aging time, and temperature with charcoal. A DOE was run to test for these parameters, and the results are shown in Table 13.2. The design is a standard half fraction of a 2^4 factorial of resolution 4 with two-way interactions confounded with other two-way interactions. It has three center points added to the design, denoted using "*" in Table 13.2. Figure 13.3 shows the DOE normal plot of the standardized effects, assessing the purification process based on the data in Table 13.2. Basically, compound A starting concentration and charcoal amount had the most significant impact on this processing step. The higher

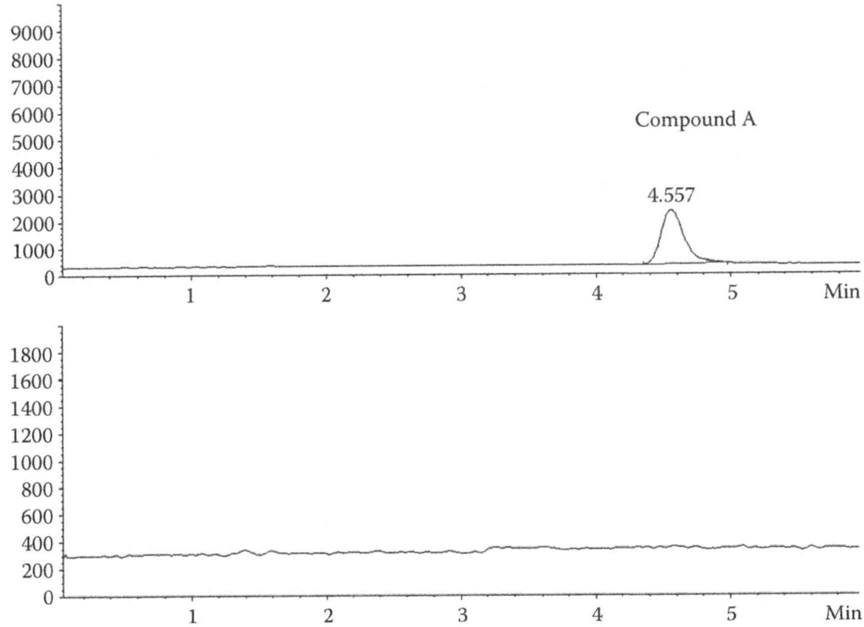

FIGURE 13.2 Chromatograms of limit of quantitation (0.6 ppm, top) and blank (bottom) for genotoxic impurity compound A using a high-performance liquid chromatography–mass spectrometry method.

the compound A starting concentration, the higher the final concentration of compound A in API #1. A larger amount of charcoal was more effective in reducing the compound A level in API #1. Additionally, aging time and the interaction of both compound A starting concentration and charcoal amount had some effect over the process but were not as significant as the other two factors. Based on the DOE results, the following specification was set for the upstream process as part of the control strategy: compound A starting concentration = NMT 55 ppm. Hence, the three replicates at center points with the compound A starting concentration of 55 ppm represented a worst-case scenario in real-life manufacturing.

Using the replicates from the center points, a mean of 1.1 ppm and a standard deviation of 0.3 (0.26) ppm were obtained. The standard deviation represented the combined process and analytical variability. Using three standard deviations (3σ) from the mean to achieve a confidence interval, the process would support a specification of NMT 1.1 + (3 × 0.3) = 2 ppm of compound A after purification. To assess analytical method variability alone, a standard deviation ($\sigma_{analytical}$) of 0.10 ppm was estimated using the analytical method transfer data, where the same three batches were tested using the new high-performance liquid chromatography–mass spectrometry (HPLC-MS) method in different analytical laboratories at different sites (Table 13.3). Using the calculated analytical variability, the standard deviation of the process ($\sigma_{process}$) was determined to be 0.24 ppm using Equation 13.3:

$$\sigma^2_{process} = \sigma^2_{total} - \sigma^2_{analytical} \tag{13.3}$$

TABLE 13.2
DOE Design and Results Assessing the Purification Process to Remove GTI Compound A from the Drug Substance

Batch No.	Compound A Starting Conc. (ppm)	Temp. (°C)	Aging Time (min)	Charcoal Amt. (% rel.)	Compound A Final Conc. (ppm)
1	100	60	210	7	1.1
2*	55	55	120	5	0.8
3	10	60	30	7	0.4
4	100	60	30	3	11.1
5	100	50	210	3	4.9
6	10	60	210	3	0.2
7*	55	55	120	5	1.2
8	10	50	30	3	0.8
9	100	50	30	7	2.3
10*	55	55	120	5	1.3
11	10	50	210	7	0.1

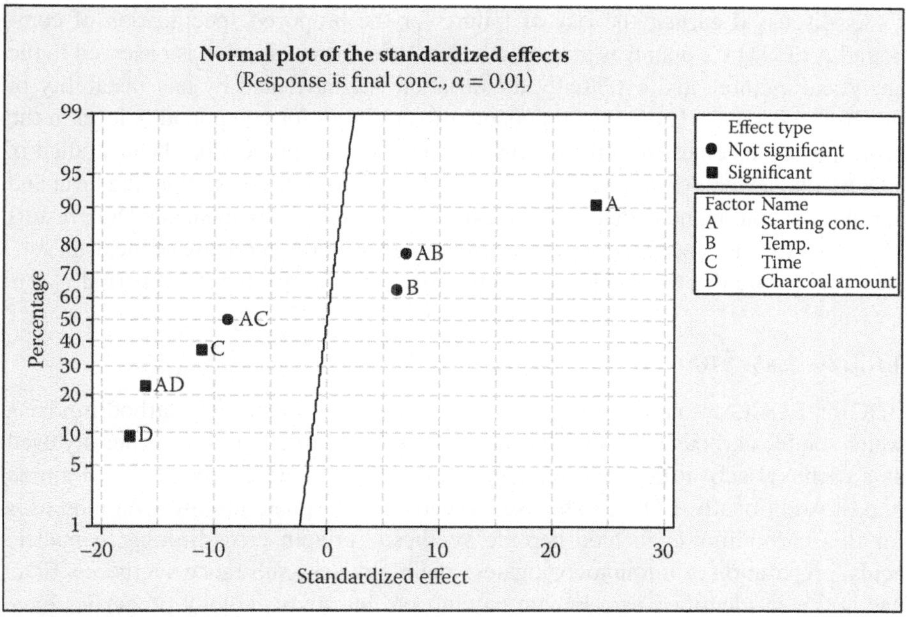

FIGURE 13.3 Design of experiments normal plot of the standardized effects assessing the purification process to remove genotoxic impurity compound A from the active pharmaceutical ingredient. (Effects on the right side of the solid line are positive, whereas those on the left side of the solid line are negative.)

TABLE 13.3
GTI Compound A Levels Determined Using HPLC-MS during Analytical Method Transfer

	Average of Two Replicates (ppm)			
Batch No.	Transferring Laboratory	Receiving Laboratory I	Receiving Laboratory II	Standard Deviation
001	1.11	1.14	1.28	0.09
002	1.49	1.30	1.42	0.10
003	<0.6	<0.6	<0.6	N/A

Note: N/A, not applicable.

Therefore, the main component of variability for compound A level in the drug substance comes from process variability.

To assess the variability in compound A that could be expected throughout the retest period of API #1, three laboratory-scale drug substance batches were placed on supportive stability and three full-scale drug substance batches were placed on commitment stability. In both the stability data sets, no correlation was observed for compound A level with time. Hence, the environment or stability of API #1 with respect to the GTI compound A does not add to the variability for determining the compliance of compound A level to the proposed specification limit.

As discussed earlier, the risk of failure for the proposed specification of compound A of NMT 2 ppm was judged to be low. Very little variation is observed in the analytical method, and essentially no variation is contributed by any instability of the drug substance. The main component of variability for compound A level in the drug substance comes from the process variability. This process has been studied in a DOE, and a worst-case manufacturing condition was used to estimate the level and variation of compound A that could be expected. All risk assessment and DOE results supported the proposed new specification for compound A at release of the drug substance to ensure that the product meets the specification over the retest period.

13.3.2 Case Study 2

EDC or EDCI, acronyms for 1-ethyl-3-(3-dimethylaminopropyl) carbodiimide, a water-soluble carbodiimide usually obtained as the hydrochloride, is generally used as a carboxyl activating agent for the coupling of primary amines to yield amide bonds. Additionally, EDC can be used to activate phosphate groups. Common uses for this carbodiimide include peptide synthesis, protein cross-linking to nucleic acids, preparation of immunoconjugates, and many drug substance syntheses. EDC has not been identified as a known carcinogen, but its toxicology properties have not been fully understood by thorough investigations. Therefore, the limit of EDC in a drug substance needs to be assessed case by case and varied in different study phases. In this case study, a 50 ppm initial limit was posed on EDC in API #2 based on good-laboratory-practices toxicity studies.

The analysis of EDC in the drug substance was a challenge. Conventional HPLC-UV and GC-FID methods are not suitable as EDC lacks strong UV adsorption and volatility. However, the EDC molecule has three nitrogen atoms (one saturated and two unsaturated) and hence can easily adopt H^+ and be ionized and detected by MS. Therefore, a liquid chromatography–mass spectrometry method was developed for the analysis of EDC in API #2. As mentioned earlier, EDC is usually in its hydrochloride form when used in organic syntheses. During method development, it was found that the stability of EDC hydrochloride in the sample solution may become a severe issue. When dissolved in water, EDC hydrochloride can convert to another form following the scheme in Figure 13.4. This alternative form of EDC has decreased H^+ adoptability, and its retention behavior in high-performance liquid chromatography is also altered. Figure 13.5 shows chromatograms of 25 ppm EDC hydrochloride (relative to drug substance) in an acetonitrile/water (80/20, v/v) solution that was repeatedly injected. The intensity of EDC peak in the initial injection of freshly prepared solution was approximately 240,000, and the peak intensity decreased to 220,000 in the second injection. After 24 hours, the EDC peak disappeared. The kinetics of EDC conversion to its alternative form can be significantly accelerated if the drug substance is a hydrochloride salt. In this case study, API #2 is a dihydrochloride salt. Figure 13.6 shows the chromatogram of a freshly prepared drug substance solution spiked with 25 ppm EDC. Apparently, the EDC peak is not observed in the chromatogram.

To solve the solution stability issue, various attempts have been made to maintain the preferable form of EDC in the sample solution, and the most effective approach was to prepare the drug substance containing EDC in an alkaline solvent. In this case, the sample solvent consisted of acetonitrile/water/ammonia at 80/20/0.2 (v/v/v); 20 mM of ammonia formate plus 0.05% of formic acid in water and acetonitrile were used as the mobile phases A and B, respectively, for gradient elution of EDC on a C18 column. EDC was detected in selected ion monitoring (SIM) at +156 *m/z*. Although the initial limit of EDC was set at 50 ppm, the LOQ of the analytical method was validated at 2 ppm, anticipating a decreased limit in later development phases. The chromatograms in Figure 13.7 demonstrated the stability of EDC after 24 hours at 2 ppm level, whereas the chromatograms in Figure 13.8 show the stability of 2 ppm EDC spiked in the drug substance in an alkaline sample solution. In both cases, the difference of EDC peak areas in fresh and aged solutions was less than 10%, which met the predefined acceptance criterion of NMT 30% for solution stability. The accuracy/recovery result could be obtained by comparing the peak areas of EDC in the chromatograms in Figures 13.7a and 13.8a. The recovery value was 95%, which met the predefined acceptance criterion (70%–130%) as well.

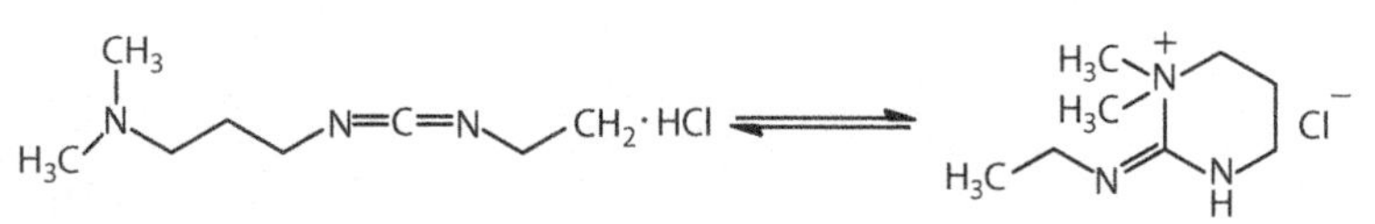

FIGURE 13.4 Form conversion of 1-ethyl-3-(3-dimethylaminopropyl) carbodiimide hydrochloride.

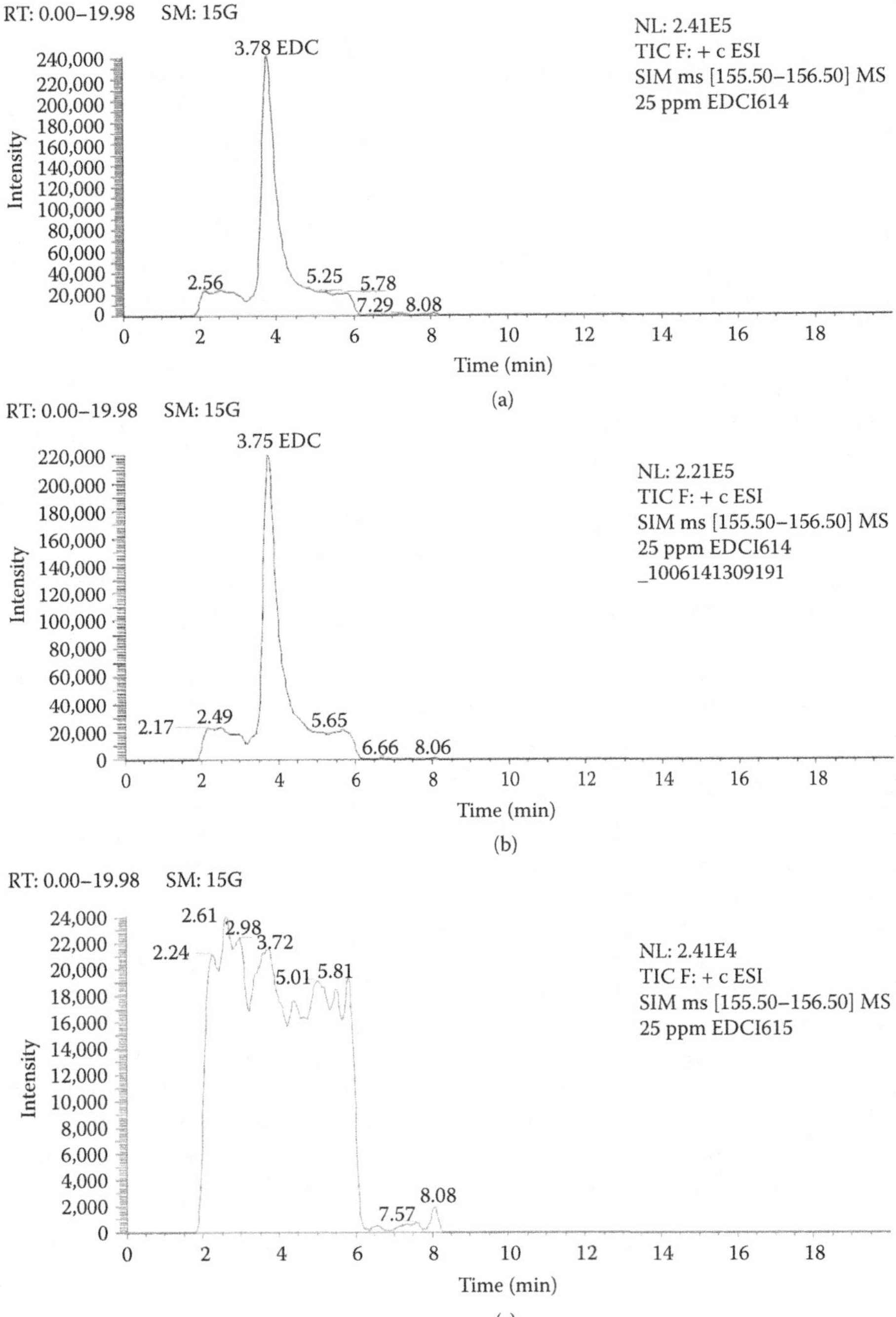

FIGURE 13.5 Chromatograms of 25 ppm 1-ethyl-3-(3-dimethylaminopropyl) carbodiimide (EDC): (a) initial injection, (b) second injection, and (c) injection after 24 hours. EDC is detected by single ion monitoring at +156 *m*/*z*.

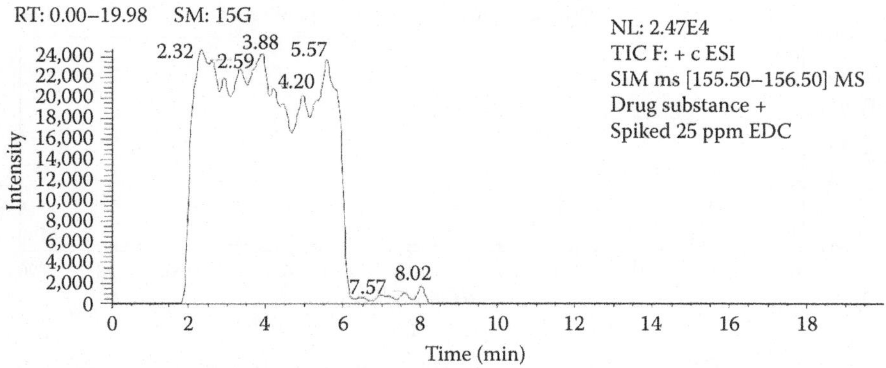

FIGURE 13.6 Chromatogram of 25 ppm 1-ethyl-3-(3-dimethylaminopropyl) carbodiimide (EDC) spiked in API #2 (dihydrochloride salt) at time zero. EDC is detected by single ion monitoring at +156 *m/z*.

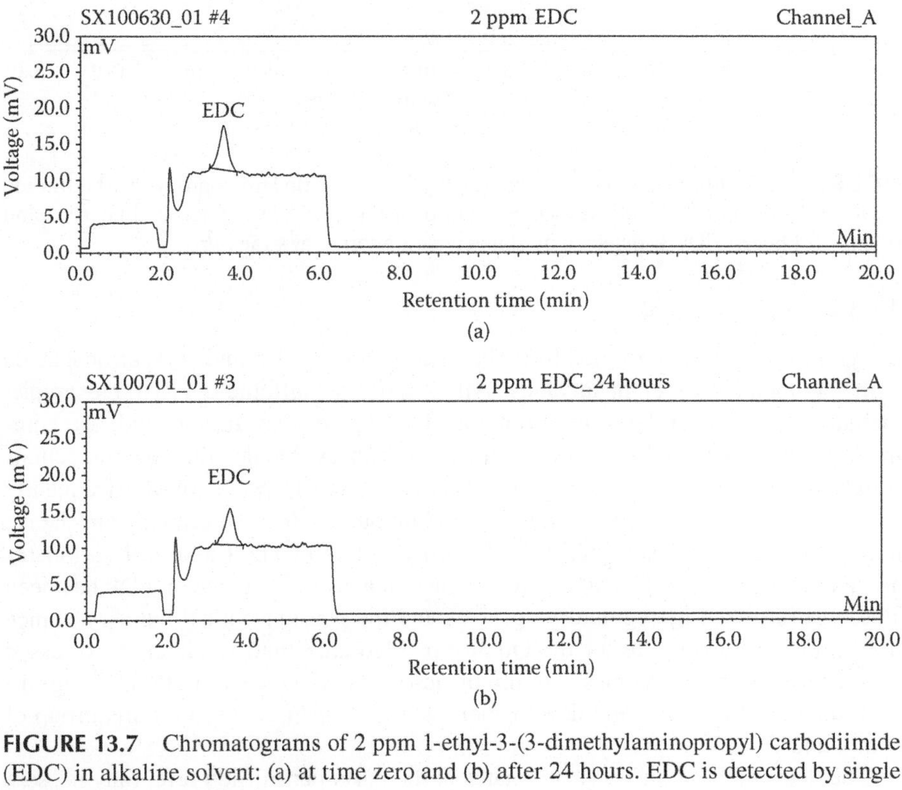

FIGURE 13.7 Chromatograms of 2 ppm 1-ethyl-3-(3-dimethylaminopropyl) carbodiimide (EDC) in alkaline solvent: (a) at time zero and (b) after 24 hours. EDC is detected by single ion monitoring at +156 *m/z*.

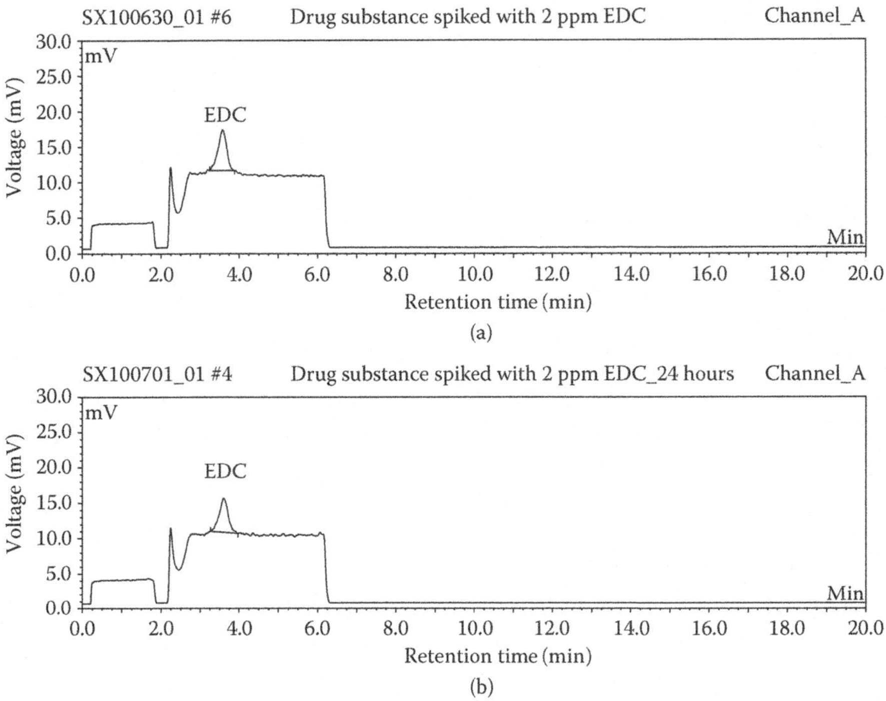

FIGURE 13.8 Chromatograms of 2 ppm 1-ethyl-3-(3-dimethylaminopropyl) carbodiimide (EDC) spiked in API #2 (dihydrochloride salt) in alkaline solvent: (a) at time zero and (b) after 24 hours. EDC is detected by single ion monitoring at +156 *m*/*z*.

13.3.3 Case Study 3

Alkyl sulfonates can be formed from the esterization of alkyl and aryl sulfonic acids with low-molecular-weight alcohols. Alkyl and aryl sulfonic acids, for example, methanesulfonic acid, benzenesulfonic acid, and *p*-toluenesulfonic acid, are common counterions for salt formation in drug substances. Meanwhile, low-molecular-weight alcohols, for example, methanol, ethanol, and isopropanol, are frequently used as organic solvents to synthesize and/or purify drug substances and hence may become residual solvents in APIs. Alkyl esters of alkyl and aryl sulfonates are a known class of GTIs, and their presence in a drug substance or a DP has been closely scrutinized by regulatory agencies and the pharmaceutical industry.[9] Since the European Directorate for the Quality of Medicines and Healthcare expressed its concerns on the alkyl mesylate impurities in mesylate salts in 2000,[10] a significant amount of research and development work has been focused on this group of compounds.[11]

The determination of alkyl sulfonates at the parts-per-million level may possess multiple analytical challenges, which have been reviewed previously in the literature.[11] Briefly, three types of approaches have been developed: (1) direct analysis using GC-FID[12] or GC-MS-SIM[13], (2) direct analysis using HPLC-MS-SIM[14], and (3) indirect analysis with derivatization.[9,15–18] Chemical instability of alkyl sulfonates is the major problem for direct analysis methods, which may raise concerns regarding sensitivity and recovery.[9] Derivatization may stabilize these highly reactive analytes; however, technical issues such as long reaction time,[9] low recovery,[16] and low specificity[15] were still observed.

In this case study, a generic headspace-GC method was developed for the determination of alkyl sulfonate with in situ derivatization. As shown in Figure 13.9, this method used chloride ion as the derivatizing reagent and the derivatization product, alkyl chloride, was analyzed using GC-FID. Factors that may affect the derivatization reaction, for example, choice of derivatizing reagent and its concentration, solvent pH, reaction time, and so on, were evaluated and optimized. This method was successfully validated over the range of 0.5–5 ppm with an LOD of 0.2 ppm. The specificity and sample matrix effect were also studied using a series of marketed APIs as model compounds.

13.3.3.1 Experimental Section

An appropriate amount of a drug substance containing alkyl sulfonate was dissolved in 1 mL of *N*-methyl-2-pyrrolidone (NMP) in a headspace vial. One hundred microliters of the derivatizing reagent was then added, and the vial was sealed immediately. After in situ incubation at 80°C for 15 minutes, the headspace sample was analyzed on a DB-624 column (Agilent J&W, Santa Clara, California) with the oven temperature held at 35°C for 10 minutes. Two analytes, methyl methanesulfonate (MMS) and ethyl methanesulfonate (EMS), were studied in this case. Their derivatization products, methyl chloride and ethyl chloride, elute at about 2.5 and 5.6 minutes, respectively.

13.3.3.2 Results and Discussion

13.3.3.2.1 Using Hydrochloric Acid as the Derivatizing Reagent

The capability of hydrochloric acid (HCl) as the derivatizing reagent was evaluated first. HCl at various concentrations (1, 2, and 3 N, respectively) was added into headspace vials containing MMS and EMS, respectively, and analyte recovery values were obtained. As shown in Figure 13.10, the recovery of MMS was approximately 100% at all three HCl concentration levels, whereas for EMS the recovery increased slightly from 80% to 90% when the HCl concentration varied from 1 to 3 N. Therefore, HCl is capable of derivatizing both MMS and EMS with satisfactory recoveries.

Alcohols are frequently used as organic solvents to synthesize and/or purify drug substances, and hence they may become residual solvents in APIs. As shown in Figure 13.11, alcohols may also react with chloride ion to form alkyl chloride in the

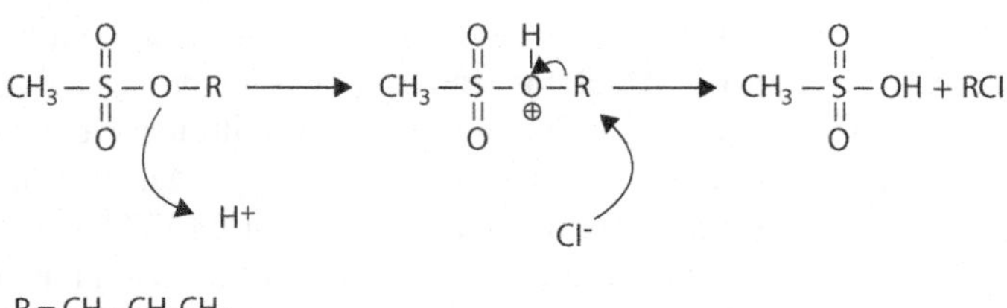

FIGURE 13.9 Derivatization reaction of alkyl sulfonate with chloride ion.

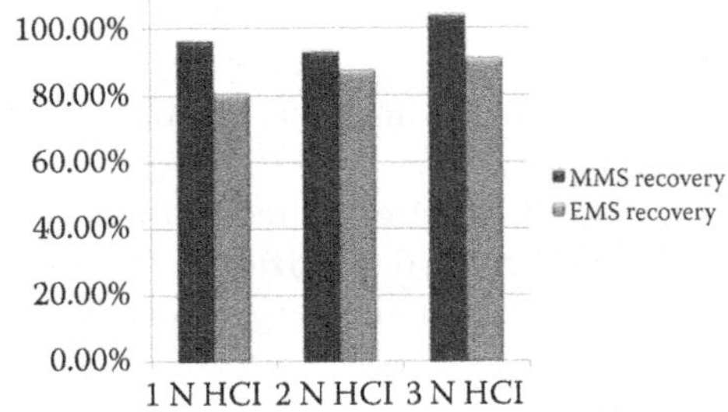

FIGURE 13.10 Effect of HCl concentration on derivatization.

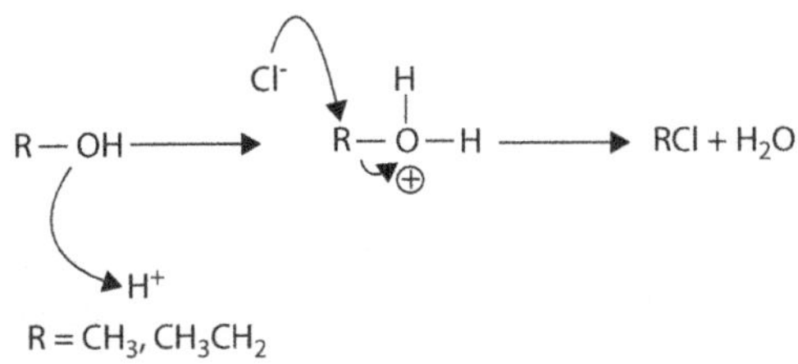

FIGURE 13.11 Reaction of alcohol with chloride ion in acidic condition.

acidic condition and hence interfere with the accurate determination of alkyl sulfonate by giving positively biased results.

To evaluate if residual alcohols would interfere with the analysis of alkyl sulfonate, HCl at various concentrations (1, 2, and 3 N) was added into headspace vials containing methanol and ethanol. The amount of methanol ranged from 60 to 3000 ppm, and the amount of ethanol ranged from 100 to 5000 ppm. Note that 3000 ppm and 5000 ppm are the ICH limits for methanol and ethanol, respectively. As shown in Tables 13.4 and 13.5, an acidic environment favors the derivatization of alcohols into alkyl chlorides, which might interfere with the analysis of alkyl sulfonate. Therefore, it is not appropriate to use HCl as the derivatizing reagent if residual alcohols are suspected in the API.

13.3.3.2.2 Using Tetramethylammonium Chloride as the Derivatizing Reagent

Tetramethylammonium chloride (TMACl) was studied as an alternative derivatizing reagent. In addition, the effect of solvent acidity was evaluated. TMACl was prepared in water, 0.01 N HCl, and 0.1 N HCl at three concentration levels (1, 2, and 3 N, respectively) and was added into headspace vials containing 3000 ppm of methanol and 5000 ppm of ethanol. Table 13.6 lists the relative amount of methyl chloride detected. Basically, a higher TMACl concentration and a stronger acidic environment favored the derivatization of methanol into methyl chloride. No ethyl chloride was detected in any of these experiments. Therefore, to prevent the alcohol from interfering, the most suitable derivatizing reagent is 1 N or 2 N TMACl prepared in water. Under such a derivatizing condition, the recovery of MMS in the presence or the absence of 3000 ppm methanol was about 100% and the recovery of EMS in the presence or the absence of 5000 ppm ethanol was above 80%.

TABLE 13.4
Methanol Interference with the Analysis of MMS Using Hydrochloric Acid as the Derivatizing Reagent

Percentage of Methyl Chloride Detected Relative to 2 ppm MMS Standard Solution

	Methanol Concentration (ppm)				
HCl Concentration (N)	**60**	**900**	**1800**	**2400**	**3000**
1	0%	6%	74%	154%	286%
2	—	—	—	—	653%
3	—	—	—	—	1140%

Note: "—," not performed.

TABLE 13.5
Ethanol Interference with the Analysis of EMS Using Hydrochloric Acid as the Derivatizing Reagent

Percentage of Ethyl Chloride Detected Relative to 2 ppm EMS Standard Solution

	Ethanol Concentration (ppm)				
HCl Concentration (N)	**100**	**1500**	**3000**	**4000**	**5000**
1	0%	1%	7%	15%	11%
2	—	—	—	—	57%
3	—	—	—	—	101%

Note: "—," not performed.

TABLE 13.6
Methanol (3000 ppm) Interference with the Analysis of MMS Using TMACl as the Derivatizing Reagent

Percentage of Methyl Chloride Detected Relative to 2 ppm MMS Standard Solution

HCl Concentration (N)	1 N TMACl (%)	2 N TMACl (%)	3N TMACl (%)
0 (Water)	0	0	17
0.01	0	11	22
0.1	35	50	83

13.3.3.2.3 Active Pharmaceutical Ingredient Matrix Effect

The sample matrix effect on derivatization was studied using a series of model compounds. These compounds are marketed APIs with various physiochemical properties, and their chemical structures are shown in Figure 13.12. In this experiment, 200 mg of the model compound was dissolved in NMP in headspace vials

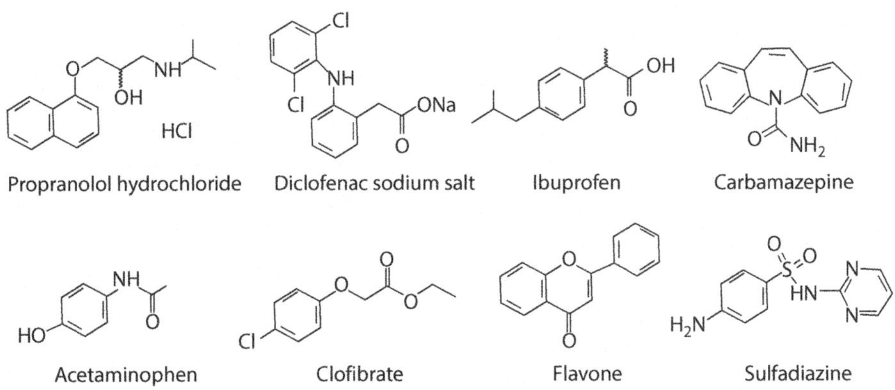

FIGURE 13.12 Model compounds.

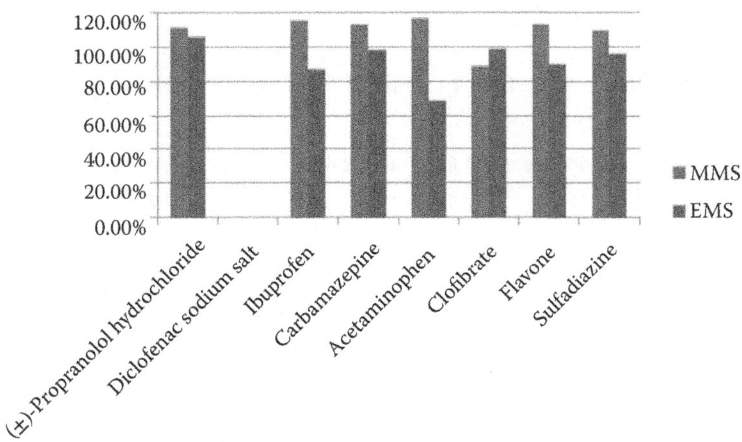

FIGURE 13.13 Recoveries of methyl methanesulfonate and ethyl methanesulfonate in the model compounds.

and spiked with 2 ppm MMS and 2 ppm EMS, and the derivatizing reagent was 1 N TMACl in water. As shown in Figure 13.13, the recoveries of MMS and EMS in most model compounds were within 80%–120%, which was satisfactory for the determination of impurities at low–parts per million levels. The only exceptions were acetaminophen and diclofenac sodium salt, in which low recoveries were observed. In addition, as the ethyl ester group in clofibrate may react with chloride ion to form ethyl chloride, positive false signal was detected in clofibrate without spiking EMS. EMS showed low recovery in acetaminophen, which may be due to the slow kinetics in this sample matrix. Increasing chloride concentration or incubation time was found to improve the recovery. For instance, in the case of acetaminophen the recovery of EMS increased to 86% using 2 N TMACl as the derivatizing reagent. Neither methyl chloride nor ethyl chloride was detected in the diclofenac sodium salt, indicating that the recovery values of both MMS and EMS were close to zero. This phenomenon might have been caused by the high pH in the sample matrix of sodium salt. To test this hypothesis, 1 N TMACl prepared in water, 3 N ammonium hydroxide, and 15 N ammonium hydroxide, respectively, were added to headspace vials containing 2 ppm MMS and 2 ppm EMS. The peak areas of MMS and EMS are shown in Figure 13.14. Obviously, a higher basicity in the solution decreased the recovery of both analytes, especially MMS. To increase the acidity of the solution and hence improve the recovery, an appropriate amount of HCl was added into the sample matrix to neutralize the diclofenac sodium salt. When the diclofenac sodium salt was half neutralized, the recovery of EMS was about 30%, whereas MMS was not yet detected. When the diclofenac sodium salt was fully neutralized, the recovery of EMS improved to 88%. However, the recovery of MMS was only about 48%. After intensive investigations, it was found that the uncommonly low recovery of MMS was due to the competitive reaction of MMS with the carboxylic ion of the diclofenac sodium salt (see Figure 13.15).

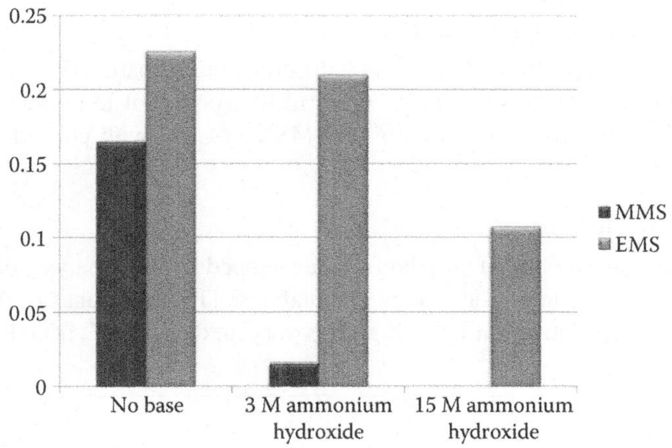

FIGURE 13.14 Peak areas of methyl methanesulfonate (detected as methyl chloride) and ethyl methanesulfonate (detected as ethyl chloride) in basic solutions.

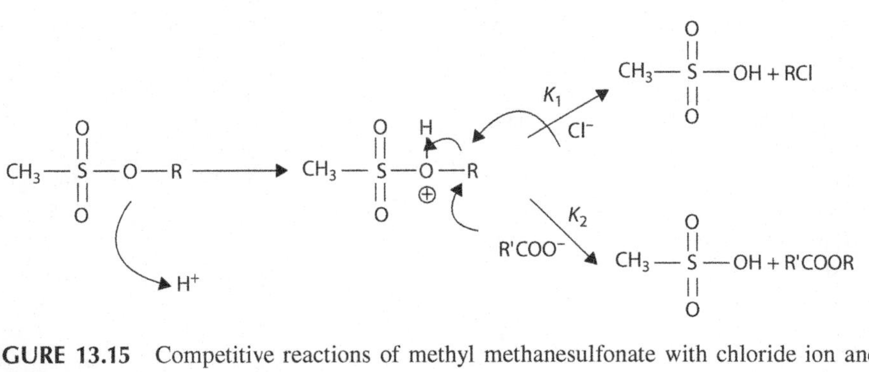

FIGURE 13.15 Competitive reactions of methyl methanesulfonate with chloride ion and carboxylic ion.

Theoretically, the product ratio of methyl chloride versus methyl ester can be calculated using Equation 13.4. This ratio is a constant independent of MMS concentration, if the amounts of chloride ion and carboxylic ion are in great excess:

$$\frac{[CH_3Cl]}{[RCOOCH_3]} = \frac{k_1[Cl^-][MMS]}{k_2[RCOO^-][MMS]} = \frac{k_1[Cl^-]}{k_2[RCOO^-]} = k \qquad (13.4)$$

Therefore, the recovery of MMS can be normalized to 100% using a correction factor. This correction factor is calculated as 100% divided by the observed recovery of MMS in the presence of the diclofenac sodium salt. Table 13.7 shows the recovery values after correction using a factor of 1.64. The use of a correction factor can also be applied to other APIs if recovery is an issue due to a similar mechanism. However, it should be noted that this approach has its limitation in situations where more than one component can interfere and the levels of interfering components are varied in different samples (e.g., degradation products in stability samples).

13.3.3.2.4 Effect of Incubation Time

A kinetic study was performed to evaluate the effect of incubation time on the recoveries of MMS and EMS. As shown in Figure 13.16, a period of 15 minutes is a robust incubation time for the derivatization of MMS and EMS with chloride ion in the presence of most model compounds.

13.3.3.3 Conclusions

A generic headspace GC-FID method was developed for the analysis of alkyl sulfonate using chloride ion to derivatize the analytes. The advantages of this method include a short derivatization time, high recovery, and good QC friendliness. This

TABLE 13.7
Normalized Recovery of MMS

Amount of MMS/EMS Spiked	Observed Recovery (%)	Correction Factor	Corrected Recovery (%)
0.5 ppm MMS/2 ppm EMS	63.63	1.57	104
1 ppm MMS/2 ppm EMS	60.19	1.66	99
2 ppm MMS/2 ppm EMS	53.88	1.86	88
4 ppm MMS/2 ppm EMS	60.42	1.66	99
6 ppm MMS/2 ppm EMS	62.24	1.61	102
8 ppm MMS/2 ppm EMS	64.16	1.56	105
10 ppm MMS/2 ppm EMS	64.68	1.56	106
Average	61.31	1.64	100
RSD	6		

Note: RSD, relative standard deviation.

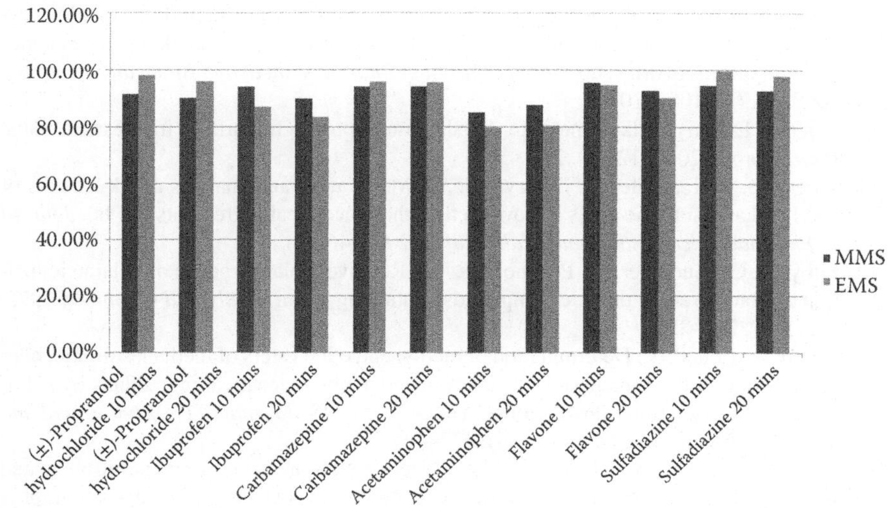

FIGURE 13.16 Effect of incubation time on the recoveries of methyl methanesulfonate and ethyl methanesulfonate.

method was validated over a range of 0.5–5 ppm with an LOD of 0.2 ppm. In addition, the sample matrix effect was studied in a series of model compounds and the limitations of the method were studied. Options for recovery improvement were also provided.

REFERENCES

1. European Medicines Evaluation Agency, Committee for Proprietary Medicinal Products, *Position paper on the limits of genotoxic impurities*. CPMS/SWP/5199/02/draft 2. 2002.
2. European Medicines Evaluation Agency, Committee for Medicinal Products for Human Use, *Guidelines on the limits of genotoxic impurities*. CPMS/SWP/5199/02. 2004.
3. European Medicines Evaluation Agency, Committee for Medicinal Products for Human Use, *Guidelines on the limits of genotoxic impurities*. 2006.
4. European Medicines Evaluation Agency, CHMP Safety Working Party, *Question & answers on the CHMP guideline on the limits of genotoxic impurities*. EMEA/CHMP/QWP/251344/2006Q&A. 2009.
5. FDA, *Draft Guidance on genotoxic and carcinogenic impurities in drug substances and products: Recommended approaches*. 2008.
6. Sun, M.; Liu, D. Q.; Kord, A. S., A systematic method development strategy for determination of pharmaceutical genotoxic impurities. *Organic Process Research & Development* 2010, 14, (4), 977–985.
7. Cimarosti, Z.; Bravo, F.; Stonestreet, P.; Tinazzi, F.; Vecchi, O.; Camurri, G., Application of quality by design principles to support development of a control strategy for the control of genotoxic impurities in the manufacturing process of a drug substance. *Organic Process Research & Development* 2010, 14, (4), 993–998.
8. International Conference on Harmonisation of Technical Requirements for Registration of Pharmaceuticals for Human Use, *ICH harmonised tripartite guideline, Pharmaceutical quality system*. ICH Q10. 2008.

9. An, J.; Sun, M.; Bai, L.; Chen, T.; Liu, D. Q.; Kord, A., A practical derivatization LC/MS approach for determination of trace level alkyl sulfonates and dialkyl sulfates genotoxic impurities in drug substances. *Journal of Pharmaceutical and Biomedical Analysis* 2008, 48, (3), 1006–1010.
10. EDQM, Enquiry: Alkyl mesilate (methanesulphonate) impurities in mesilate salts. *Pharmeuropa* 2000, 12, (1), 27.
11. Eldera, D. P.; Teasdale, A.; Lipczynskic, A. M., Control and analysis of alkyl esters of alkyl and aryl sulfonic acids in novel active pharmaceutical ingredients (APIs). *Journal of Pharmaceutical and Biomedical Analysis* 2008, 46, (1), 1–8.
12. Filby, W. G.; Guenther, K.; Penzhorn, R.-D., Relative molar response of a flame ionization detector to some organic compounds containing sulfur. *Analytical Chemistry* 1975, 47, (13), 2283–2285.
13. Colon, I.; Richoll, S., Determination of methyl and ethyl esters of methanesulfonic, benzenesulfonic and *p*-toluenesulfonic acids in active pharmaceutical ingredients by solid-phase microextraction (SPME) coupled to GC/SIM-MS. *Journal of Pharmaceutical and Biomedical Analysis* 2005, 39, (3–4), 477–485.
14. Taylor, G.; Gosling, M.; Pearce, A., Low level determination of *p*-toluenesulfonate and benzenesulfonate esters in drug substance by high performance liquid chromatography/mass spectrometry. *Journal of Chromatography A* 2006, 1119, (1–2), 231–237.
15. Lee, C.; Guivarch, F.; Van Dau, C.; Tessier, D.; Krstulovic, A., Determination of polar alkylating agents as thiocyanate/isothiocyanate derivatives by reaction headspace gas chromatography. *Analyst* 2003, 128, (7), 857–863.
16. Alzaga, R.; Ryan, R.; Taylor-Worth, K.; Lipczynski, A.; Szucs, R.; Sandra, P., A generic approach for the determination of residues of alkylating agents in active pharmaceutical ingredients by in situ derivatization-headspace-gas chromatography-mass spectrometry. *Journal of Pharmaceutical and Biomedical Analysis* 2007, 45, (3), 472–479.
17. Zhang, J.; Zhou, X.; Li, J.; Ji, Q., Determination of potential genotoxic impurities with low molecular weights. PITTCON, Orlando, FL, March 11–15, 2012.
18. Rao, L.; Shen, X.; Zhang, J.; Babiak, S., Determination of residual alkylating agents in pharmaceutical intermediates and drug substances by in-situ derivatization headspace gas chromatography. *Eastern Analytical Symposium*, Somerset, NJ, November 14–17, 2011.

14 Salt Formation of Pharmaceutical Compounds and Associated Genotoxic Risks

Soojin Kim

CONTENTS

14.1 INTRODUCTION

The issue of genotoxic impurities in drug substances or products has received widely publicized attention in the pharmaceutical industry in recent years.[1] A few cases have raised particular concerns because genotoxic impurities were inadvertently generated during the syntheses of drug substances using common solvent systems and reagents, which were otherwise considered safe and harmless. Many published cases of genotoxic impurities in drug substances involved the process of salt formation during their manufacture.[2–6] Salt formation in general is vitally important in drug substance synthesis as well as overall pharmaceutical development and manufacture. This chapter provides an overview of solid forms and salt formation and their significance in drug development and manufacture.

Salts are defined as acid–base pairs in the solid state, involving proton transfer or neutralization and the ionic interaction between the acid and base species. Salt formation is controlled by factors influencing the acid–base reactions. In principle, every compound possessing acid or base functional groups can participate in the acid–base reaction. Salt formation is determined by the relative strength of the acidity and basicity of the chemical species involved, which also determines the stability of the resulting salts. Salts in the solid state can be either crystalline or amorphous. When a salt form is mentioned with respect to pharmaceutical development, it is typically a reference to the solid state, particularly a crystalline state of the salt.

Salt formation is widely used and often a critical process step for drug substance synthesis and manufacture. In addition, salt forms of drug substances play a critical role in solid dosage form development for drug products. Typically, any reference to "salt formation" of a compound is essentially directed to crystallizing a salt form of the compound, as almost all the reasons and advantages of forming salts are related to the necessity or feasibility of crystallizing a compound. The crystalline properties of salts should be one of the deciding factors in using the salt form as a drug substance or as a process step in the synthesis.

This chapter examines some fundamental concepts including the more obvious reasons for salt formation from drug substances and the salt form selection process for drug dosage form development. Also, the general relevance of salt formation in the synthetic processes for drug substance manufacture is reviewed. Finally, some of the associated genotoxic risks in salt formation based on specific cases and possible measures for preventing or minimizing the risk of generating genotoxic impurities in salt formation are covered.

14.2 DRUG SUBSTANCE SALT FORMS

The vast majority of drug substances are in the solid state. Not only the drug substances used in solid dosage forms but also those used in parenteral, inhalation, intranasal, or solubilized drug products are almost always in the solid state, not the liquid state. Furthermore, most drug substances are in a crystalline state of the solid for many reasons. Hence, crystallization is a critical final step of the synthesis of drug substances.

The crystalline state provides essential stability and consistency for drug substances. Compared to materials of liquid form or solid amorphous form, crystalline substances are typically more chemically and physically stable. In addition, they have consistent chemical and stoichiometric compositions as well as consistent physicochemical properties such as density, solubility, and dissolution. Many of the physicochemical properties of drug substances have direct effects on bioavailability and other biopharmaceutical properties. Crystalline materials can be manufactured more consistently batch to batch into products with consistent properties and performance compared to amorphous solids. Consistency in the physicochemical properties as well as the biopharmaceutical properties and the physical/chemical stability of drug substances are of paramount importance in drug product development and manufacture, as they ultimately ensure consistent performance and stability of dosage forms and finished drug products.

Drug compounds often possess acidic or basic functional groups. The majority of drug substances contain amine or carboxylic acid functional groups, which are ionizable and capable of forming salts. It is reported that an estimated half of all the drug molecules used in medicinal therapy are administered as salts.[7] Therefore, salt formation of a drug substance may be one of the crucial operations in drug development.

What are the primary purposes of having salt forms as the final form of drug substances? Salt formation is a well-known and very effective technique to modify and optimize the physicochemical properties of a drug substance without changing the molecule.[8,9] The crystal structure of a salt form is usually completely different from that of the free base or acid of the molecule, and it also differs from one salt to another. As a result, physicochemical properties of a given substance such as melting point, density, hygroscopicity, solubility, and dissolution rate can be easily modified by producing salt forms. The major reasons for using salt forms as drug substances may be categorized as discussed in Sections 14.2.1 through 14.2.4.

14.2.1 Solubility/Dissolution/Bioavailability Enhancement

Improving the solubility and dissolution rate of poorly soluble drugs is one of the primary reasons to prepare salt forms for drug substances.[10,11] Formation of a salt is an effective way to change aqueous solubility and dissolution rate, which have direct effects on pharmacokinetics and bioavailability. It is a consensus in the pharmaceutical industry that drug candidates in development during the past few decades have been increasingly more complex in chemical structure and more lipophilic in general, and as a result, typically have become much less soluble in aqueous media.[12] Despite the availability of many formulation techniques and particle engineering technologies to increase the bioavailability of poorly soluble drugs, salt formation is the simplest and most common approach to increase solubility and dissolution rate and thus to improve bioavailability.

Sufficiently high aqueous solubility is a critical requirement in developing various pharmaceutical dosage forms, not only for the most common solid oral dosage forms such as immediate release tablets and hard-gelatin capsules but also for several other dosage forms including parenteral injections, oral syrups, nasal drops, eyedrops, and so on. Whereas for oral dosage forms a certain minimum solubility and an adequate dissolution rate in the pH range of the gastrointestinal tract may be required, for liquid dosage forms, particularly injectable solutions, a considerably higher aqueous solubility is typically an indispensable requirement.

For this reason, many examples of salt forms exist for marketed drug products. Coumadin® (warfarin sodium), a well-known anticoagulant, is poorly soluble as the free acid and hence is administered as the sodium salt in both the solid oral and the injectable dosage forms.[13] Reyataz® (atazanavir sulfate, an HIV protease inhibitor) is another example. The free base has very low aqueous solubility and inadequate bioavailability, but its sulfate salt has a reasonable solubility profile and can be formulated into successful solid oral dosage forms with acceptable bioavailability.[14]

14.2.2 Dissolution/Solubility Control

Salt forms are also used to control drug dissolution for various purposes.[8,15] For example, salts are prepared to decrease the solubility of drug substances used in liquid suspension formulations, typically developed for children, the elderly, and patients who have difficulty swallowing solid dosage forms. Limiting the solubility of drug substances used in liquid suspensions can help maintain the stability of the suspension, as well as ensure consistent solid-state properties such as particle size and polymorph, which may significantly impact the absorption and safety of the drug.

In addition, limiting the solubility of highly soluble drugs is often used for taste masking, particularly for liquid suspensions, by reducing the solution concentration of drug substances that have unacceptable taste. Taste improvement is often achieved by forming salts or complexes with long-chain fatty acids such as pamoic acid, stearic acid, and palmitic acid or by forming typically less soluble salts with calcium or with saccharin, which can also serve as a sweetener.

Many cases exist of marketed products using less soluble salt forms for liquid suspensions and/or taste masking; one of them is Terramycin® (oxytetracyclin), an antibiotic compound for which a calcium salt was developed for the suspension formulation.[16] The less soluble calcium salt provided improved chemical stability in suspension and palatability over the more soluble HCl salt, which was used in conventional tablet forms but had a serious stability problem (hydrolysis in aqueous solution) as well as a bitter taste. In the case of the antibacterial drug Tequin® (gatifloxacin), pediatric formulations were developed by forming crystalline complexes with stearic acid and palmitic acid that effectively masked the bitter taste.[17]

Another important application of salt forms of lower solubility is in the development of controlled-release dosage forms, which are designed to release the drug at a controlled rate over an extended period. Drug release can be slowed down by the selection of a salt having suitable solubility and dissolution profiles. The drug substance metoprolol in Lopressor® and Toprol XL®, marketed for hypertension treatment, is employed as metoprolol succinate for an extended-release formulation, whereas metoprolol tartrate is used for an immediate-release formulation.[18] Tofranil® (imipramine), a drug for the treatment of depression, used a hydrochloride salt for conventional tablets but was later developed as a pamoate salt with lower solubility for the controlled-release formulation.[19]

Salts with lower solubility may also provide advantages in the formulation process. For example, in wet granulation processes, which are typically carried out to prevent the segregation of powder constituents and improve the flow and compaction properties of powder blends, when water is used as the granulation liquid salts with high aqueous solubility may experience undesirable effects.[20] Excessive surface dissolution of the particles may occur or partial dissociation of the salt into the free form, resulting in uncontrolled recrystallization or partial form change during post-granulation drying. This may result in adverse consequences for the bioavailability of the formulated drug.

14.2.3 Chemical/Physical Stability

The majority of pharmaceutical drug products are in solid dosage forms, such as capsules or tablets. The solid form of a drug substance must have suitable properties for large-scale manufacturing as well as adequate stability to ensure reliable product manufacture and long shelf life. Crystalline materials usually have superior chemical and physical stability compared to noncrystalline materials. For example, they are generally less hygroscopic and thermally more stable and possess high and well-defined melting points. They are also more stable to oxidation or hydrolysis than liquid or amorphous materials.

Salt formation provides opportunities for the formation of crystalline materials when the drug substance does not crystallize on its own. For ionizable compounds, forming an ionic pair with a counterion may provide chances to induce ordered phase formation where the free base or free acid molecules do not readily assemble into an ordered crystalline state. Furthermore, even when crystalline phases of the free compound are available, crystalline salt forms can often provide more enhanced physical characteristics compared to the crystalline form of the free compound alone, in terms of nonhygroscopicity, higher melting point, greater density, and superior chemical/physical stability. Examples include the drug substances for Plavix® (clopidogrel bisulfate), an antiplatelet agent,[21] and Glucophage® (metformin hydrochloride), an antidiabetic drug.[22] For both drugs, the free bases were not isolable as stable solid forms and exhibited unstable characteristics as oils or amorphous solids, but the salt forms provided acceptable stability and were successfully developed into the solid dosage forms. During the course of the life cycle of these drugs, several other salt forms have been produced by various parties for different dosage forms and product launches.

14.2.4 Proprietary Solid Forms

It is sometimes advantageous to look for alternative salt forms of a drug substance for additional patent protection, especially in the United States. Finding a new salt form and developing its drug product may offer new opportunities for improved physicochemical properties, which can lead to developing a new dosage form, a new route of administration, and possible extension of the product line and drug patent life. There are many examples of successful drug life cycle management for well-established drug substances. After being successful in the market with one dosage form (e.g., conventional immediate-release tablet), additional dosage forms such as an injectable product, a controlled-release product, an inhalation dosage form, a topical application form, or a transdermal dosage form can be introduced with different salt forms providing different physical properties.[15,23]

An example of this case is Voltaren® (diclofenac), a widely used nonsteroidal anti-inflammatory drug. Since the early success of the conventional tablet (diclofenac sodium salt), various products from new salts have been introduced into the market, including a sustained-release tablet (diclofenac resinate), an injectable form (diclofenac sodium), a topical gel (diclofenac diethylamine), and additional tablet products (diclofenac potassium and diclofenac free acid) with or without new indications.[24] Other examples include dihydralazine for hypertension treatment for which a sulfate salt was used for the conventional tablet form (Apresol®) and, later, a mesylate salt was used for the injectable formulation (Nepresol®).[15] In addition to these examples, there are numerous cases in the pharmaceutical industry where different salt forms of a drug have been used either to extend the exclusivity of a product line or to circumvent infringement on another party's patent for a specific solid form of a drug.

14.3 SALT FORM SELECTION

Form selection for a new drug substance or active pharmaceutical compound (API), whether a salt form or a free form, is a vital part of drug development. The search for a suitable salt form and the selection of a salt form usually take place during early development and involve comprehensive evaluation of physicochemical and biopharmaceutical properties for drug development including melting point, solubility, stability, bioavailability, and processability as well as manufacturing conditions and scalability.

14.3.1 Active Pharmaceutical Ingredient Form Selection Process

Up until the 1990s, prior to the adoption of the more systematic API form screening and selection processes now widely practiced in many pharmaceutical companies, salt forms were often chosen in more empirical ways by medicinal chemists or pharmaceutical scientists. Often, a particular salt form was chosen based on the first attempt of synthesis and the ease of crystallization or based on previous experience or familiarity with the counterion. Nonoptimal choice of the form may lead to later problems such as difficulties in dosage form development or issues of manufacturability or stability, associated with increased costs of development and production down the road. Therefore, a rational form selection based on the evaluation of relevant properties for obtainable solid forms including salts would be essential to avoid the costly and inefficient process of switching the form later during development.

Over the past 10–20 years, many pharmaceutical companies have adopted more systematic screening processes as well as a rational approach to form evaluation and selection.[25] These activities are typically handled by an interdisciplinary form selection team or committee, comprising scientists representing different functions in the company. The team usually goes through a process of screening, evaluation, and form selection through some type of decision tree or form selection strategy, considering key criteria for drug substance solid forms. Many aspects of drug development are also taken into account, including process development and scalability as well as the expected dose and required solubility/bioavailability for the dosage form design. Figure 14.1 shows an example of such a process showing a common framework of the form selection decision tree. The details may differ, but the general criteria are similar for the form selection of new drug substances across the pharmaceutical industry.[26–31]

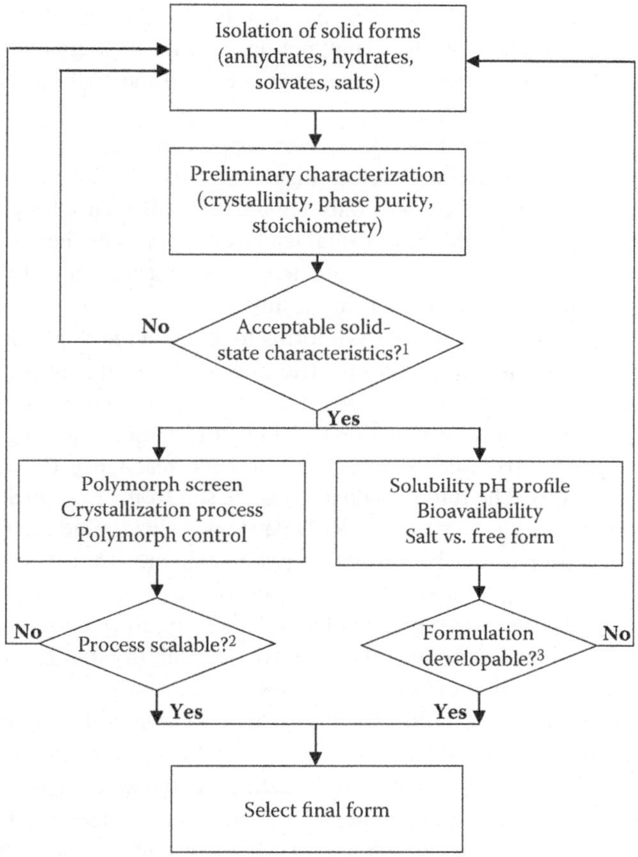

FIGURE 14.1 Active pharmaceutical ingredient solid form selection process.

[1] Highly crystalline, single phase, consistent stoichiometry, thermodynamically stable form, high enough melting point, nonhygroscopic, no volatile content, and physically and chemically stable (in elevated temperature, humidity, and light).

[2] Can be consistently crystallized on scale with robust polymorph control.

[3] Can be developed into a robust drug product with acceptable bioavailability, excipient compatibility, and stability in formulation process and dosage form.

14.3.2 Salt Form Selection Criteria

Most often, a salt form screen is called for due to inadequate bioavailability of a poorly soluble compound in its free form. When a salt is considered for a drug substance form, typically there is a clear limitation of the free acid or free base, whether it is undesired physical characteristics, weak crystallinity, less than optimal stability, or poor solubility, and thus a salt form would provide advantages over the free form.

The main selection criteria for API solid forms including salt forms mainly consist of crystallinity, stability, bioavailability, and manufacturability. Consideration of a salt for a suitable drug substance form involves first the question of whether it is feasible to be crystallized. Crystallinity of a salt form is the first step in the consideration or decision. An acid–base pair can form with enough pK_a difference between the compounds; however, it may not necessarily be a crystalline form but an amorphous solid or oil. Those salts that are not crystalline are first ruled out from the consideration of API solid form.

The salt form selection process takes place during early development when rapid synthesis development is taking place to meet the delivery requirements for early toxicology and clinical studies. Not only should the salt forms be prepared in a speedy manner but also should they be characterized for physicochemical properties by various analytical techniques and evaluated for biopharmaceutical properties by solubility studies or in vivo bioavailability testing.

Early evaluation of the obtainable salt forms focuses on the requirement for consistency of high crystallinity, stoichiometric composition, and phase purity. The solid-state analyses that may be conducted for this initial evaluation include optical microscopy, x-ray powder diffraction (XRPD), differential scanning calorimetry (DSC), thermogravimetric analysis, nuclear magnetic resonance (NMR), titration, and elemental analysis. The initial evaluation of the salt forms can quickly eliminate those that show weak or inconsistent XRPD patterns or low/broad melt endotherms by DSC, so that resources can be devoted to gram-scale preparations of a few viable salt candidates with acceptable solid-state properties. Early samples are usually limited in quantity since they are typically obtained from small-scale screening experiments, but batch-to-batch reproducibility as well as scalability to a gram scale would be important in determining the viability of salt candidates.

The salt form candidates on the narrowed-down list would be evaluated further via more extensive analyses focused on the physical and chemical stability of the forms, such as hygroscopicity analysis by gravimetric dynamic vapor sorption, hot-stage microscopy, variable-temperature XRPD, and accelerated or stressed stability tests. Stability is a very important aspect when considering the API solid form; physical and chemical stability in the solid state as well as in the slurry state (in both aqueous and organic media) is monitored for any changes or degradation.

The viable salt forms are compared among themselves for their physicochemical characteristics, as well as against the free base or free acid form of the compound. In addition, relevant biopharmaceutical measurements such as pH-solubility profile, intrinsic dissolution, and/or in vivo bioavailability studies in animals are carried out for those that show acceptable stability characteristics. The comparison of bioavailability often determines the preference of one salt over another or the nonsalt form.

The salt form selection process described here is based on rapidly assessing the essential properties and eliminating nondevelopable forms. Compared to the extensive evaluation of full list of relevant properties for all available salt forms, this process saves time and resources during early development when form selection needs to be made within a short time with limited material. It is similar to the salt selection strategy outlined by Morris and others[31] in which expeditious evaluation of viable forms and selection were accomplished through a multitier approach: conducting the least time-consuming experiments first to give a go/no go decision followed by progressively more time-consuming and labor-intensive experiments. Such a form selection practice is commonly used nowadays in many pharmaceutical companies to enable rapid progress of drug development.[25,28,30]

14.3.3 Polymorphism for Salt Forms

Like any other crystalline solid, a salt may exhibit polymorphism. Polymorphism is defined as the ability of a substance to exist in different molecular arrangements and packings. Different polymorphs may exhibit differences in physicochemical, thermodynamic, spectroscopic, and mechanical properties, including solubility and dissolution, directly related to drug performance. In addition, pseudopolymorphism is frequently encountered for a salt, producing solvates or hydrates. Pseudopolymorphs are defined as different crystal forms of a compound involving the stoichiometric inclusion of water or solvent molecules in the crystal lattice; they are classified separately from polymorphs since hydrates and solvates are in a strict sense considered to be different chemical entities from the respective anhydrous compound.

Sound knowledge of polymorphs and pseudopolymorphs of a salt form candidate and their relationship is critical to ensure the best possible selection of the API form and to develop a robust crystallization process for the selected form. Any uncontrolled form transformation during API manufacture, API storage, or drug product manufacture may cause detrimental effects on the drug performance and regulatory issues. Therefore, in API form selection the form that is the most thermodynamically stable is preferred.

Many of the unstable polymorphs and solvates/hydrates displaying low melting points or low dehydration/desolvation temperatures would have been eliminated during the early screening/evaluation. Some solvates (e.g., solvates of toxicologically safe solvents such as ethanol, isopropanol, and acetic acid) and hydrates may be acceptable as an API form when they show reasonable stability within acceptable ranges of storage and processing conditions, particularly when no other nonsolvated or stable form is obtainable. However, in most cases, where other stable forms are available, they would be a less desirable option due to concerns of stability, both physical and chemical. In addition to generally poor thermal stability, hydrates and solvates may also have poor chemical stability demonstrated by the chemical degradation caused by water or solvent under stressed conditions. Some hydrates show variable levels of hydration depending on relative humidity and would pose an issue regarding the inconsistency of API batches with different API compositions resulting in variable potency of the drug. For the same reason, there are usually processing concerns related to drying hydrates or solvates in API manufacture. It is easy to inadvertently overdry the hydrates or solvates during the API drying process, to a partially or fully dehydrated/desolvated state, which may cause a reversible or irreversible change in the crystal structure. In general, hydrates have lower solubility than the corresponding anhydrous form, which may be another consideration in whether a hydrate is acceptable or not as the API form.

Anhydrous forms are typically preferred over hydrates or solvates because they are generally expected to have higher melting points (e.g., >200°C) and thus superior thermal stability and nonhygroscopic behavior compared to hydrates or solvates. However, some anhydrous forms can also convert to hydrates on contact with water or in high-humidity conditions or can show hygroscopic nature which may cause them to have physical or chemical stability only within limited humidity ranges.

For salt form selection, the thermodynamic stability of the polymorphic/pseudopolymorphic forms of the salt should be investigated to sort out the most stable form among them. Polymorph screening may be performed on a salt form candidate via high-throughput screening or parallel manual screening experiments, the details of which are described in Section 14.7. Once the thermodynamically most stable form is identified, a process to consistently crystallize the desired form should be developed. Robust polymorph control in API manufacture is essential as the solid form's purity is an important quality specification for API batches, because it not only affects the API's consistency and stability but also can impact the drug product's performance including absorption/efficacy, safety, and stability. Sometimes, the propensity of a salt to form polymorphs or pseudopolymorphs and the ease or difficulty of the manufacturing process to robustly control the desired polymorph may be a factor for preferring one salt over another.

14.3.4 Salt Form versus Free Form

Salt formation is an extra step in the synthesis of a drug substance. It requires additional time and resources for development and manufacture including added cost for the counterion material and solvents, processing time, as well as the expected loss of a portion of the API in the crystallization step. In addition, the salt formation adds molecular weight to the drug substance, potentially resulting in a weight burden to the drug product, especially when the dose requirement is high. A high pill burden (tablet count) is typically a critical disadvantage in marketing a new drug and patient compliance and may even be a reason to halt the development of a compound altogether. Therefore, the salt form to be selected should provide a clear advantage over the free form of the compound to be considered a necessary option. If a crystalline form of the free acid or base exists and it has acceptable physicochemical properties, stability, manufacturability, or bioavailability compared to its salt forms, the free form of the compound is preferred for development, unless there are other reasons to choose a salt form such as a special dosage form design, intellectual property protection, or drug life cycle management.

14.4 COUNTERIONS FOR DRUG SUBSTANCE SALT FORMS

In the salt formation of drug substances, an important aspect to consider first in selecting counterions is their safety and toxicity. Usually, a list of counterions is drawn for a salt form screen from those from the drug salt products already on the market or from the list "generally recognized as safe" (GRAS), the Food and Drug Administration (FDA)-approved substances that can be added to food.[32] Additionally, acceptable daily intake values are assigned to some commonly used counterions by the Joint FAO/WHO Expert Committee on Food Additives.[33] Even for biologically inert counterions, there may be upper limits of intake that need to be considered for safety in making a salt form selection. Accordingly, the selection of counterions may also be influenced by the route of administration, projected dose, and treatment length of the drug.

For a given drug substance, the choice of counterions that are feasible for salt formation depends on the basicity or acidity of the ionizable groups in the compound. When the pK_a values of the ionizable functional groups in the molecule are known, the potential salt-forming counterions can be proposed based on the pK_a difference between the acid and base groups. It is commonly known that a minimum difference of about 2 to 3 units between the pK_a values of the ionizable group and the counterion is required for the formation of a stable salt, especially with drug substances that are weak acids or weak bases.[25,28] In cases of insufficient pK_a difference between the ion pairs, even if a complex is formed it may undergo disproportionation more easily back to the components in solution (e.g., gastrointestinal tract), resulting in the precipitation of the free acid or free base.

Multiple counterions may be viable for salt formation based on the pK_a value, but not all would produce a stable crystalline salt form. Typically, a variety of salts would be prepared for a new drug substance candidate by either high-throughput screening or manual preparation from a list of potential counterions. Only those that are highly crystalline and physically and chemically stable would be scaled up for a more comprehensive assessment of their properties and their suitability for the intended dosage form to be developed.

On the other hand, cocrystal formation is possible between a drug substance and one or more inert materials, including both neutral and ionizable compounds, whose salt formation is not feasible based on the pK_a values. Cocrystals, defined as homogeneous crystalline materials composed of multiple components with defined stoichiometry, are usually distinguished from salts based on the nonionic interaction between the components such as hydrogen bonding or van der Waals interaction. Similarly to salts, cocrystals offer the possibility of altering physical properties of the compound such as enhanced solubility, dissolution rate, and stability characteristics. For some compounds that cannot form a physically stable crystalline material, cocrystals may offer a potential to form a stable crystal as a complex where the crystallinity is aided by the presence of another molecule. Many cases of cocrystals having desirable physicochemical properties, stability characteristics, and improved solubility/bioavailability compared to the free form have been developed as APIs and are reported in the literature.[34–36]

Stable cocrystals are often identified from standard salt form screens or from specially designed cocrystal screens. For a given drug substance, there are unlimited possibilities for countermaterials to be used for screening or crystal engineering for cocrystal formation, among those that are pharmaceutically acceptable and GRAS for drug product use, which may include any combination of organic or inorganic materials. Some of the most common ones found in the literature are glutaric acid, benzoic acid, succinic acid, fumaric acid, adipic acid, tartaric acid, maleic acid, salicylic acid, saccharin, caffeine, and so on.[37]

In standard salt form screens with many different counterions of different ionization potentials, one may find a salt or cocrystal of the ionizable compound depending on the degree of ionization or proton transfer. Advanced analytical techniques, such as single crystal x-ray diffraction and solid-state NMR, which have become widely available in recent years, allow for a distinction between salts and cocrystals. However, in the absence of such detailed characterization one may often rely on pK_a information to estimate the acid–base pair's likelihood of forming a salt or not. In the early development of a drug substance, when the pK_a information is not clear or accurate, especially for drug substances with low aqueous solubility, there is usually no clear picture on the nature of the acid–base interaction or the extent of proton transfer. Therefore, for a complex formation between ionizable compounds, especially for weakly acidic or basic compounds that are most often encountered as drug substance candidates, the distinction between a cocrystal and a salt may not be obvious until more extensive analysis into the crystal structure and molecular interaction is conducted.

14.4.1 Counterions for Basic Drug Substances

The majority of drug substances are weakly basic compounds.[11] Weakly acidic compounds are less common, and a few are zwitterionic compounds or neutral compounds. Acids that are used to supply counterions for salt formation with basic drug substances include inorganic acids (e.g., hydrochloric acid, hydrobromic acid, sulfuric acid, and phosphoric acid), sulfonic acids (e.g., methanesulfonic acid [MSA], ethanesulfonic acid, benzenesulfonic acid [BSA], and isethionic acid), carboxylic acids (e.g., acetic acid, maleic acid, fumaric acid, succinic acid, and benzoic acid), hydroxy acids (e.g., citric acid, tartaric acid, malic acid, and mandelic acid), and possibly amino acids (e.g., glutamic acid and aspartic acid).[28] Table 14.1 lists all the acidic counterion sources used for salt formation with basic compounds in FDA-approved drug products.[38]

Hydrochloric acid (HCl) has been the most commonly used acid for making salts of basic drugs. According to several reports[8,10,38] that list all the pharmaceutical salts that were commercially marketed or approved by the FDA over the past several decades, the frequency of hydrochloride salts historically is shown to be 40%–60% of the total number of salt forms of basic compounds. With its very low pK_a, hydrochloric acid can nearly always form a salt with a wide range of basic compounds, including weakly basic ones with low pK_a, although crystallization of the salt form is not guaranteed. HCl is commonly available and used as an aqueous solution (36% w/w). For water-sensitive salt formation or processes requiring anhydrous conditions, a dilute HCl solution of anhydrous organic solvents or anhydrous HCl gas is occasionally used.

Despite its popularity, there are potential disadvantages related to the general physical properties of hydrochloride salts as well as undesirable aspects of hydrochloric acid as a counterion. Hydrochloride salts often provide unacceptably high acidity in liquid formulations (e.g., parenteral products), and as solids they frequently show a hygroscopic nature and lower melting points and/or decomposition temperatures than other salts due to the volatility of HCl. The corrosive nature of hydrochloric acid can give rise to a concern related to the corrosion or discoloration of process equipment during the salt synthesis; likewise, some hydrochloride salts can be corrosive to tableting tools in drug product manufacture. Both cases can cause risks related to metal contamination of batches from the processing equipment.

Hydrochloride salts usually encounter the common-ion effect, the phenomenon of reduced solubility of a salt due to the presence of chloride ions in the gastric

TABLE 14.1
Acidic Counterions Used for Salt Formation with Basic Compounds in FDA-Approved Drug Products (up to 2007)

Salt Formers	Name	Structure
Inorganic acids	Hydrochloride	H—Cl
	Hydrobromide	H—Br
	Hydroiodide	H—I
	Sulfate	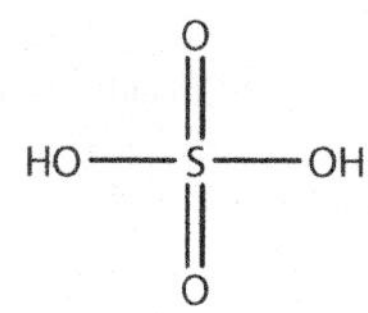
	Nitrate	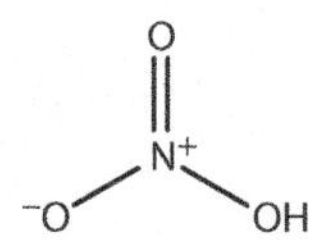
	Phosphate	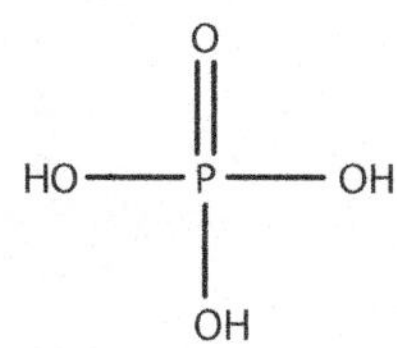
Sulfonic acids	Mesylate (MSA)	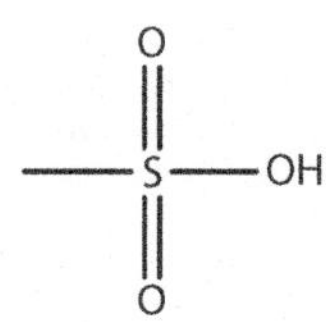
	Isethionate (2-hydroxyethanesulfonic acid)	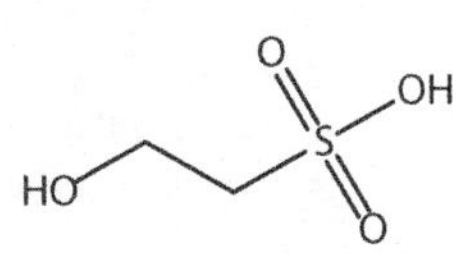
	Tosylate (4-methylbenzenesulfonic acid)	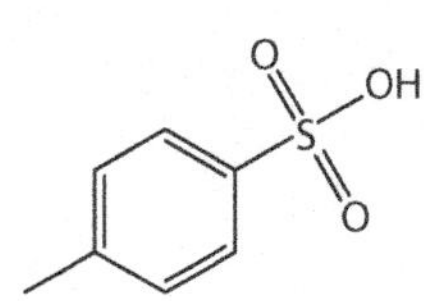
	Napsylate (naphthalene-2-sulfonic acid)	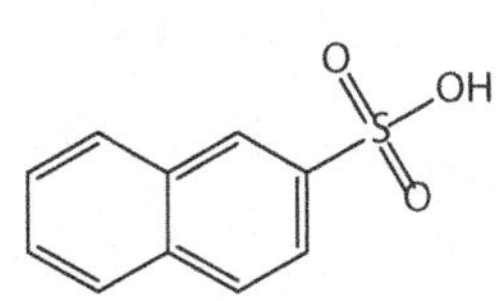

(*Continued*)

TABLE 14.1 (*Continued*)
Acidic Counterions Used for Salt Formation with Basic Compounds in FDA-Approved Drug Products (up to 2007)

Salt Formers	Name	Structure
	Besylate (BSA)	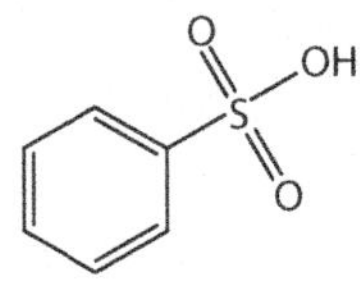
	Ethandisulfonate[a]	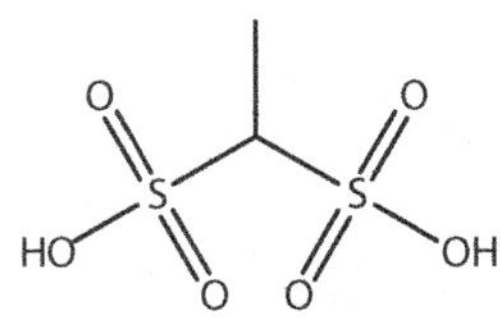
	Camphorsulfonate[a]	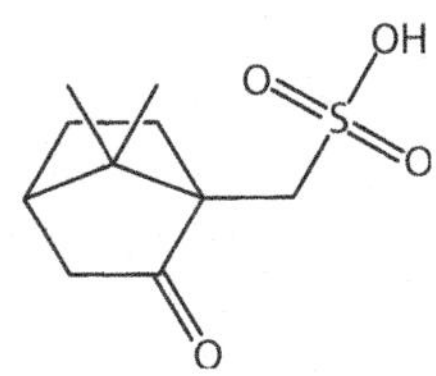
	Methylsulfate	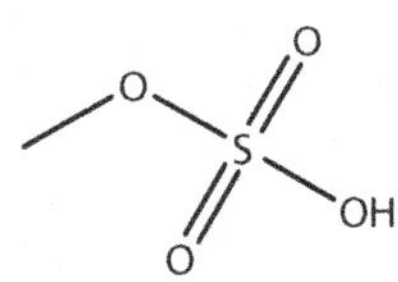
	Sulfosalicylate[a]	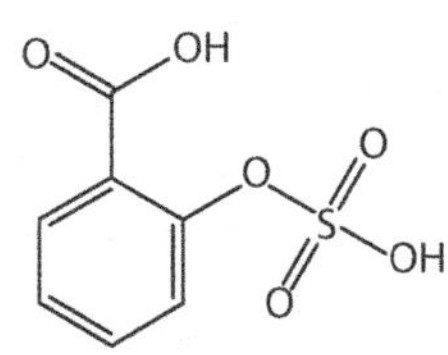
	Laurylsulfate[a] (dodecylsulfuric acid)	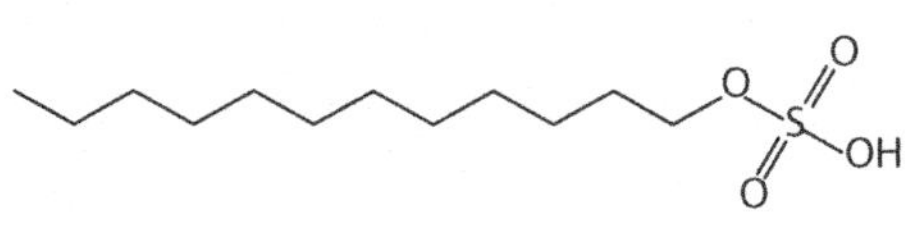
Carboxylic acids	Acetate	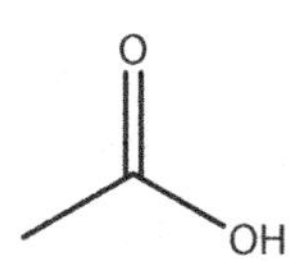
	Trifluoroacetate[a]	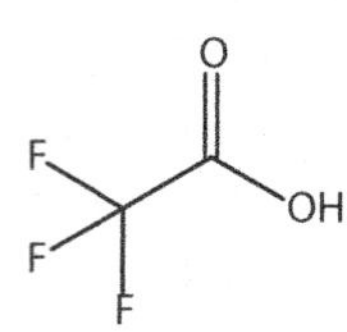

TABLE 14.1
Acidic Counterions Used for Salt Formation with Basic Compounds in FDA-Approved Drug Products (up to 2007)

Salt Formers	Name	Structure
	Maleate	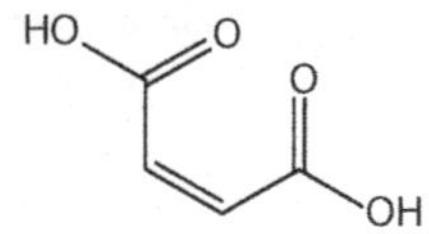
	Oxalate[a]	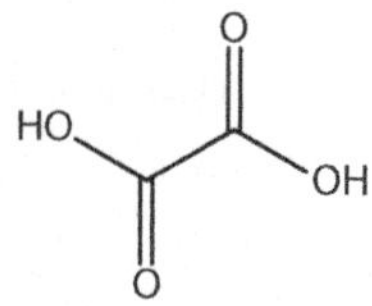
	Benzoate[a]	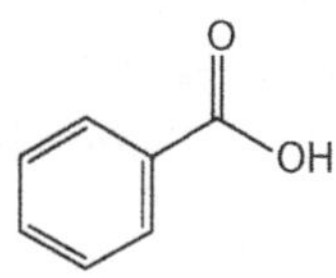
	Naphthoate[a]	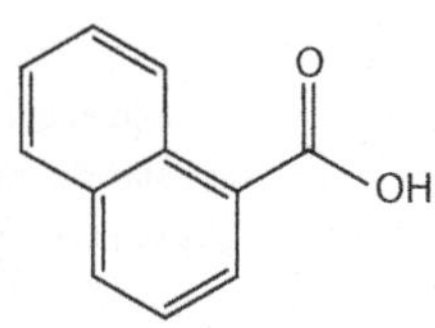
	Fumarate	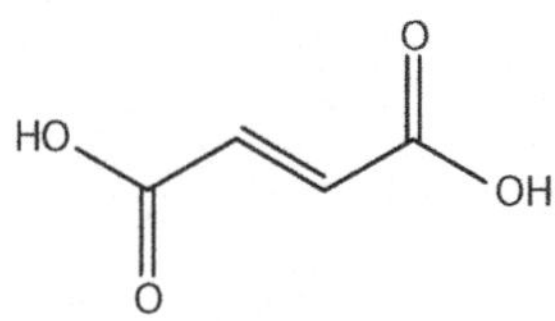
	Malate	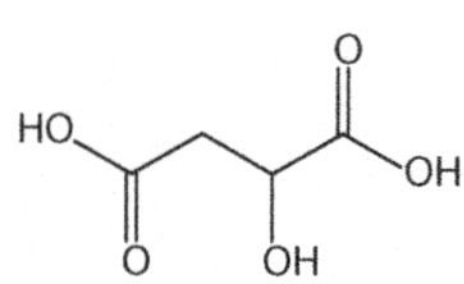
	Gluceptate[a] (glucoheptonic acid)	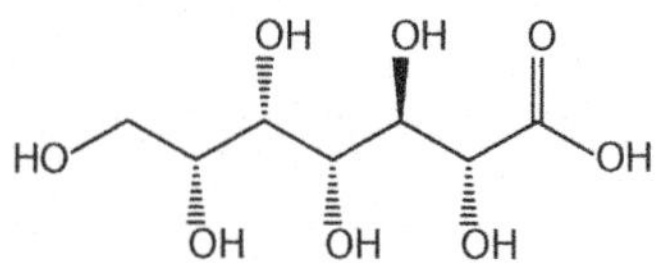
	D-gluconate	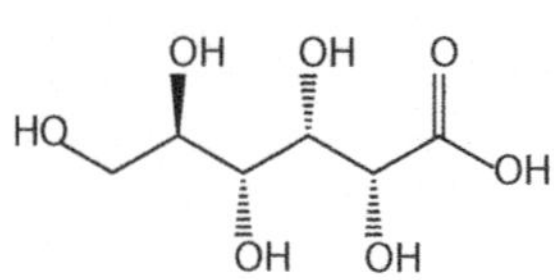

(*Continued*)

TABLE 14.1 (*Continued*)
Acidic Counterions Used for Salt Formation with Basic Compounds in FDA-Approved Drug Products (up to 2007)

Salt Formers	Name	Structure
	D-glucuronate[a]	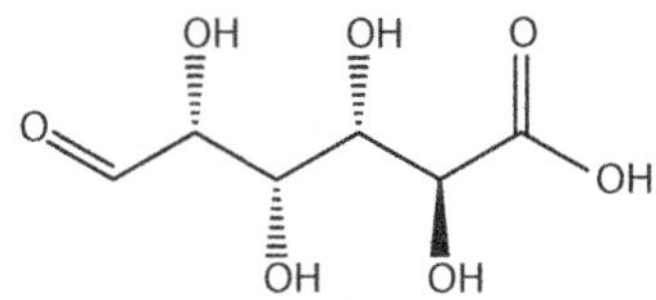
	Galacturonate[a]	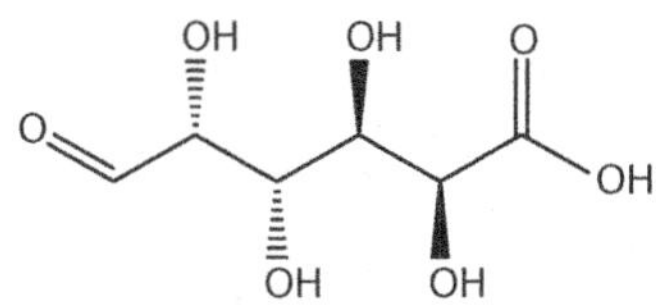
	Hippurate[a] [2-(benzoylamino) acetic acid]	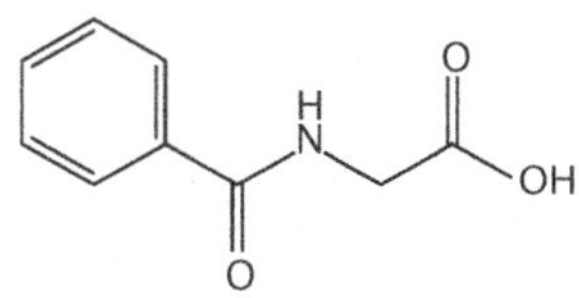
	Lactobionate[a] (4-*O*-β-D-galactopyranosyl-D-gluconic acid)	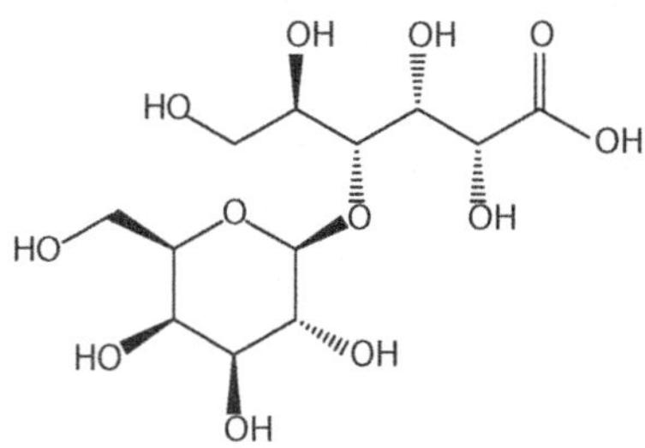
	Pamoate	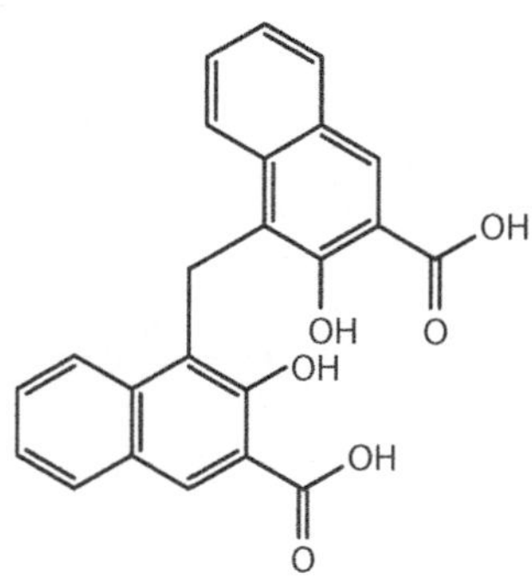

TABLE 14.1
Acidic Counterions Used for Salt Formation with Basic Compounds in FDA-Approved Drug Products (up to 2007)

Salt Formers	Name	Structure
	Chlorotheophyllinate[a]	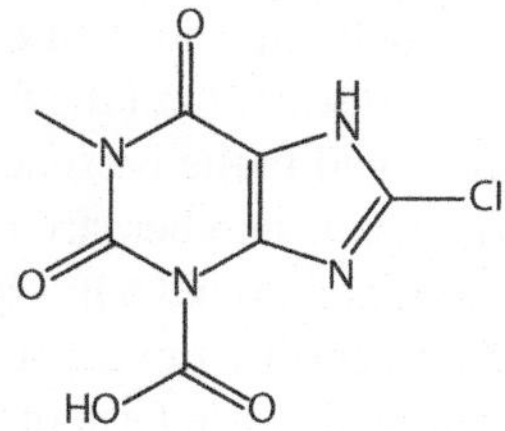
Hydroxy acids	Citrate	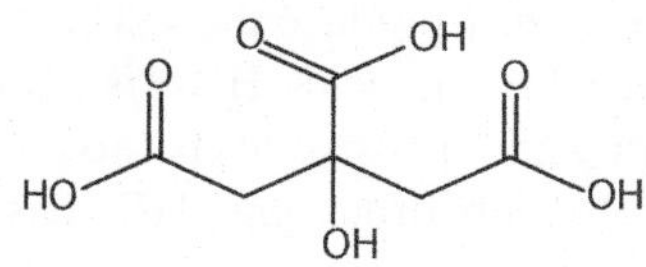
	Lactate	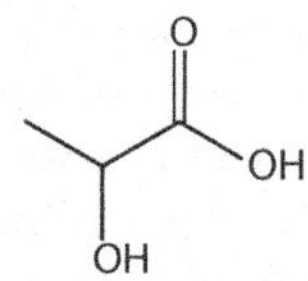
	Succinate	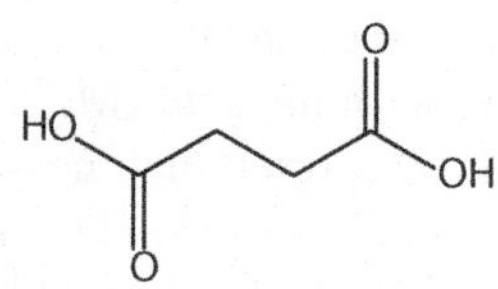
	Tartrate	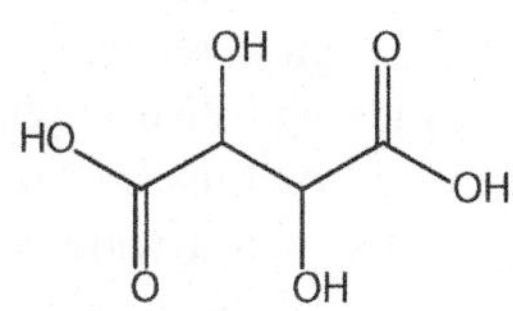
Fatty acids	Oleate[a]	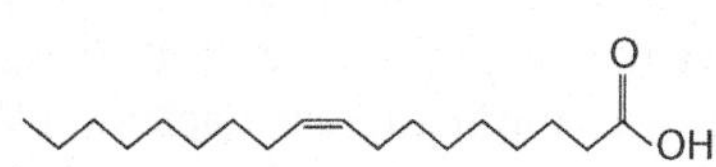
	Octadecanoate[a] (stearic acid)	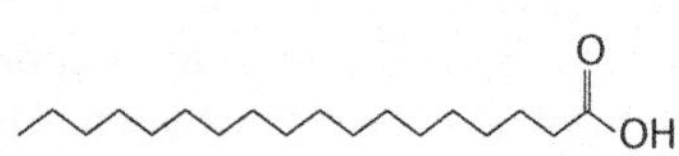

Note: A total of 38 different acids were reported. [a]16 acids were used only once.

fluid, causing a negative effect on the bioavailability.[9,25] The common-ion effect is reported by numerous cases in the literature, where hydrochloride salts exhibited lower solubility and/or dissolution rate compared to other salts when administered orally or in the presence of HCl or NaCl in solution. For example, in the case of terfenadine (Seldane®, an antihistamine), the aqueous solubility of its hydrochloride salt was reduced 10-fold when dilute NaCl was added, due to the common-ion effect.[39] Likewise, a hydrochloride salt of doxycycline (Vibramycin®, an antibiotic), although more soluble in water than the free base form, showed a much slower dissolution rate than the free acid in the presence of dilute HCl solution.[40]

Sulfuric acid is another strong acid frequently used for salt formation with a wide range of basic compounds. It is available as a liquid form. Since it is a dibasic acid, it can potentially form 1:1 or 2:1 (i.e., 2 mol of basic compound to 1 mol of sulfuric acid) salts. Typically only one form of the salt would be chosen, and the process is controlled to consistently produce the salt crystals of desired stoichiometry in a single phase.

Hydrobromic acid (HBr) is also a strong acid, similar to HCl, and HBr is also volatile and corrosive to metals. It is often used as an aqueous solution (48% w/w) or as the anhydrous gas. HBr has been used relatively often in the past for API salt formation. However, in recent years concerns related to the chronic toxicity of bromide ions, causing neurological symptoms, have been reported[41,42] and therefore diminished the usage of HBr in drug substances.

Phosphoric acid is a weaker acid but preferred in API salt forms due to its relatively favorable safety profile compared to other counterions and often produces physically and chemically stable crystalline forms. It is mainly used as an aqueous solution (85% w/w) as the anhydrous form is either a highly hygroscopic solid or a viscous liquid.

Methanesulfonic acid (MSA) is another strong acid that has found prevalent use in APIs. It is reported that mesylate salt usage has increased significantly during the past few decades to second (to HCl) in the order of frequency among the salts with anionic counterions, now roughly comprising 10%–15% of the salts of basic compounds.[10] However, in recent years concerns regarding the liability of forming genotoxic impurities with hydroxylic solvents or even some excipients have decreased its popularity in API uses.[43,44] In general, MSA has a good tendency to form stable crystalline salts with a wide range of bases. It often produces highly crystalline, nonhygroscopic salt forms. The acid is used as the neat anhydrous liquid, which is convenient in large-scale operations.

Other commonly used acids for APIs include tartaric, citric, fumaric, succinic, maleic, acetic, and malic acids. A total of 38 different anions have been used as counterions in the drug products approved in the United States that are listed in the Orange Book Database,[38] which has been published by the FDA since 1981.

14.4.2 Counterions for Acidic Drug Substances

For acidic drug substances, inorganic bases (e.g., sodium, potassium, or calcium), organic amines (triethylamine and ethanolamine), or basic amino acids (arginine, lysine, meglumine, and choline) can be used as counterions. Table 14.2 lists all the bases used as counterions (a total of 15) for the drug substances that were approved by the FDA up to 2007.[38]

TABLE 14.2
Basic Counterions Used for Salt Formation with Acidic Compounds in FDA-Approved Drug Products (up to 2007)

Salt Formers	Name	Structure
Inorganic cations	Sodium	Na
	Potassium	K
	Calcium	Ca
	Magnesium	Mg
	Silver[b]	Ag
	Zinc	Zn
Organic amines	Diethylamine[b]	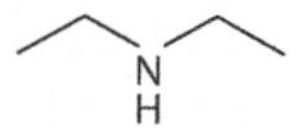
	Diethanolamine[b]	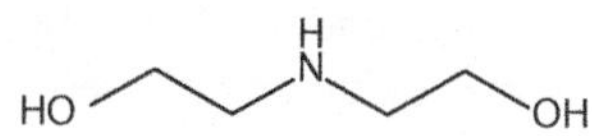
	Benzathine[b] (*N*,*N*′-dibenzylethylenediamine)	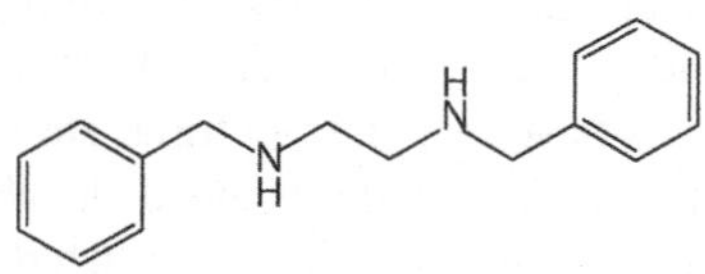
	Meglumine	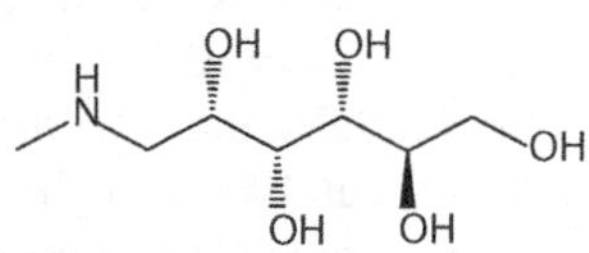
	Piperazine[b]	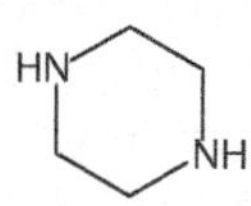
	Choline[b]	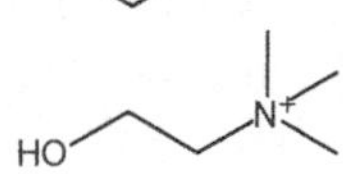
	Tromethamine [tris(hydroxymethyl) aminomethane]	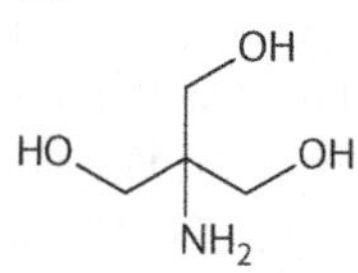
	Procaine[b]	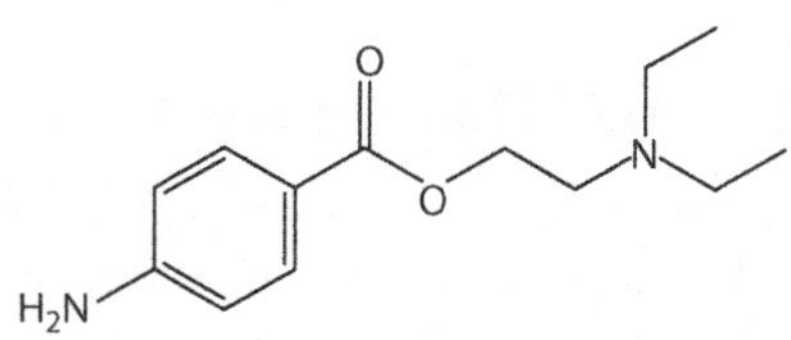
Cationic amino acids	Lysine[b]	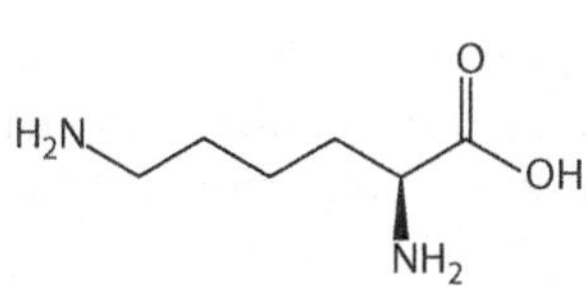

Note: A total of 15 different bases were reported. [b]8 bases were used only once.

Among the basic counterions, sodium ion is most frequently used for salt formation with acidic drug substances. Historically, sodium salts make up 60%–90% of all the salts of acidic compounds.[8,38] Sodium ions from sodium hydroxide or sodium bicarbonate/carbonate can potentially form salts with a wide range of acids, and the resulting salts usually have high aqueous solubility; however, it is well known that sodium salts are often hygroscopic. Sodium salts are popular due to the safety of the sodium ion, which is naturally abundant in physiological fluids in the gastrointestinal tract; however, sodium salts are also subject to the common-ion effect and may suffer from the adverse effects of reduced solubility or dissolution in vivo, and thus lower bioavailability.[45]

Potassium, calcium, and magnesium are the other commonly used inorganic basic counterions. They can potentially form salts of high crystallinity. Although their aqueous solubility may not be as high as that of the sodium salt, they often provide superior physical/chemical stability. Potassium hydroxide, calcium hydroxide, and magnesium hydroxide are all used conveniently for scale-up operations.[46]

Lysine and arginine are the basic amino acids extensively used to form crystalline salts of acidic drug substances. Often when the salts formed from inorganic basic counterions have less than optimal stability or bioavailability, the salts from these amino acids may provide the needed stability or bioavailability. Due to the chirality of these compounds, special care is needed to ensure the chiral purity of the material used and the resulting product.

Several reports and databases of marketed salt forms[10,38] indicate that there is a strong trend toward a broader variety of counterions in recent decades. For example, the fraction of chloride salts, the most frequently encountered salts in FDA-approved drugs, decreased to about 40% over the past decade from 53%–64% in 1981–1996, whereas the fractions of several other counterions increased during the same period.[38] This trend was explained as a stronger need to improve physicochemical properties of new drug substances that are becoming ever more complex in structure. This is related to the fact that there are increasingly more compounds with low aqueous solubility and weak basicity having low pK_a, which require stronger acids to form salts. It is noteworthy that 77% of the salts of the basic drugs listed in the Orange Book are prepared with relatively stronger counterions (hydrochloride, hydrobromide/bromide, sulfate/bisulfate, and nitrate). Similarly, 74% (14 out of 19) of salts of the acidic drugs listed are prepared with strong inorganic bases such as NaOH and KOH. Many believe that this general trend is also expected to continue in the future.[10,38]

14.5 SALT FORMATION OF INTERMEDIATE COMPOUNDS

Salt formation crystallization is an important means of processing for the development and manufacture of drug substances. Crystallization is one of the most widely used techniques for the isolation and purification of organic compounds particularly in the synthesis of drug substances.[47] Salt formation often offers the opportunity to isolate a difficult-to-crystallize compound into a highly crystalline state, providing a purification possibility as well as the possibility of isolating easy-to-handle and stable solid products in a scalable manner.

In the synthesis of drug substances, key intermediates often need to be isolated. The isolation of key intermediates provides the opportunity to eliminate process impurities and byproducts generated during processing and to store the bulk intermediate compound in a stable solid form until it is ready for further processing or transfer to other locations. In addition, key intermediates are characterized and registered as part of important manufacturing control in regulatory filings.

Crystallization is an essential means for the isolation of synthetic substances. Some compounds can naturally crystallize as a free form from a given reaction mixture and solvent system. Others are induced to crystallize from solutions. Along a synthetic route, salt formation and crystallization may be the only economical and scalable means to separate an intermediate compound from its side products. Often, when an organic compound forms a salt with an inorganic counterion its solubility is dramatically changed. The salt typically becomes more soluble in aqueous media but highly insoluble in organic solvents; hence, the salt tends to precipitate from the organic solution of the reaction mixture. This effect can be used advantageously to isolate difficult-to-purify reaction products from organic byproducts, which in most cases remain in the organic solution.

Crystallinity in a salt affords a means of purification and removal of unwanted impurities. Crystallization can be described as a solidification process from the liquid state as a result of molecular self-assembly and packing into an ordered structure, which enables the purging of foreign molecules (impurities) that do not fit into the ordered structure. For a highly effective purification process, there should be a high degree of molecular selectivity of crystal growth when it takes place in a controlled manner. Furthermore, the effectiveness of a purification process depends on the relative solubility of the individual components in the process solvent system. Salt formation in many cases assists in differentiating the solubility of the components in organic process solvents.

Salt formation of intermediate compounds is not different from that for APIs in terms of counterion screening and selection. However, more emphasis is put on the manufacturability and purification potential rather than the physicochemical properties or stability of salts. For the salt formation of intermediates, there is generally more freedom in the choice of counterions and solvents because of the lesser degree of toxicological concern at the earlier synthetic steps. Typically after the salt formation is carried out for the purpose of isolating and purifying an intermediate, there would be subsequent salt-break steps to remove the counterion, followed by further reactions and workup steps, providing additional opportunities to clean up any undesired residual reagents or solvents before finally reaching the API step.

14.6 SALT FORMATION FOR CHIRAL SEPARATION

The number of chiral drugs has increased at a soaring pace over the past few decades. They are now estimated to represent about 40% of all drug sales worldwide[48] and also constitute over 50% of new drug applications since the year 2000.[49] Moreover, among the top 10 pharmaceutical products by U.S. retail sales in 2011, 7 have chiral APIs.[50] The development of synthetic processes for drug substances requires efficient methods for producing various chiral compounds in an enantiomerically pure form. Despite the available alternative techniques, chiral resolution via diastereomeric salt formation remains the most widely used method for preparing pure enantiomers.[7,51]

The resolution method via the formation of a salt between a racemic acid and an optically active base was discovered by Pasteur in 1853.[51] He found that the chiral basic compounds quinicine and cinchonicine form diastereomeric salts with racemic tartaric acid and used the difference in the crystalline forms and solubility between the diastereomeric salts to separate L-(+)-tartaric acid and D-(–)-tartaric acid.

The separation of enantiomers is achieved by first forming crystalline diastereomeric compounds, salts, or dissociable complexes. Salt formation is a preferred method for the resolution of acids or bases compared to the formation of covalent diastereomers due to the simplicity of the salt formation and the ease of cleavage to the original constituents.

An optically active acid or base is used as the resolving agent to form crystalline diastereomeric salts with a racemic base or acid. The desired diastereomeric salt is then separated from the mixture of diastereomeric salts by fractional crystallization from a select solvent system based on the salts' difference in solubility and/or ability to crystallize.

A suitable resolving agent and solvent system may be searched in a systematic way through high-throughput automated or manual screening, similarly to the API salt form screening process. The screen can provide conditions to form crystalline salts and the solvent system in which the diastereomeric salts have a sufficient difference in solubility. Typically, polar solvents such as alcohols work more effectively than other solvents by providing a greater difference in solubility between the diastereomeric salts. Although any optically active acid or base can be used in principle, its use as a resolving agent is sometimes limited by its availability and cost as well as toxicity.

For effective resolution, high crystallinity of at least one of the diastereomeric salts and difference in their solubility are essential. In addition, the diastereomeric salts must not cocrystallize or form a complex with each other. After the preferential crystallization and isolation of the desired diastereomeric salt from the mixture, the desired enantiomer is separated as a pure form by the salt break from the resolving agent, which can be recovered for reuse.

The process may be summarized by Scheme 14.1,[51] which corresponds to the treatment of a racemic acid, DL-AH, with an optically active base, D-B, to form salts, *p* (positive) and *n* (negative) or the inverse.[a]

$$\text{DL-AH} + \text{D-B} \rightarrow \text{D-A}^- \bullet \text{D-BH}^+(p\ \text{salt}) + \text{L-A}^- \bullet \text{D-BH}^+(n\ \text{salt})$$

SCHEME 14.1 Formation of diastereomeric salts, *p* and *n*, from the reaction between a racemic acid DL-AH and a resolving agent, chiral base D-B

One of the salts, *p* or *n*, which has a lower solubility in the particular solvent system would crystallize out, whereas the other would mostly remain in the mother liquor. The effectiveness of the resolution would depend on the degree of difference between the solubilities of the two diastereomeric salts. The desired enantiomer can then be recovered by the isolation of the crystals from the mother liquor, followed by the salt break to separate the enantiomer from the resolving agent. The flow diagram representing a typical resolution is shown in Figure 14.2.[53]

---

[a]By convention, a diastereomeric complex in which both the components have the same sign of optical rotation, (+ +) or (– –), is called a p (positive) isomer, whereas a diastereomer in which the components have opposite signs of rotation, (+ –) or (– +), is called an n (negative) isomer for convenience.[52]

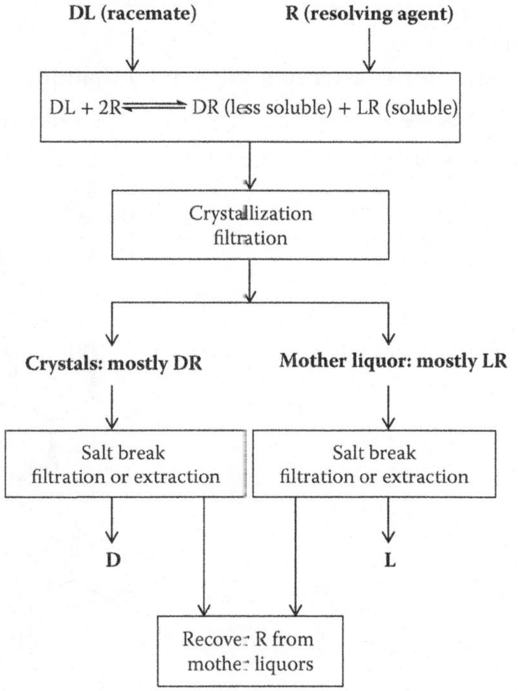

FIGURE 14.2 Flow diagram of chiral resolution by diastereomeric salt formation.

TABLE 14.3
Chiral Acids and Bases for Diastereomeric Salt Formation

Salt Formers	Name	Structure
Acids	Tartaric acid (+), (−)	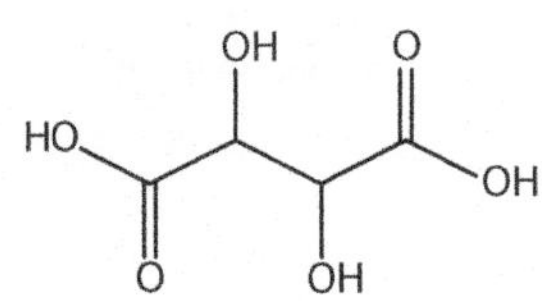
	O,O′-dibenzoyltartaric acid (+), (−)	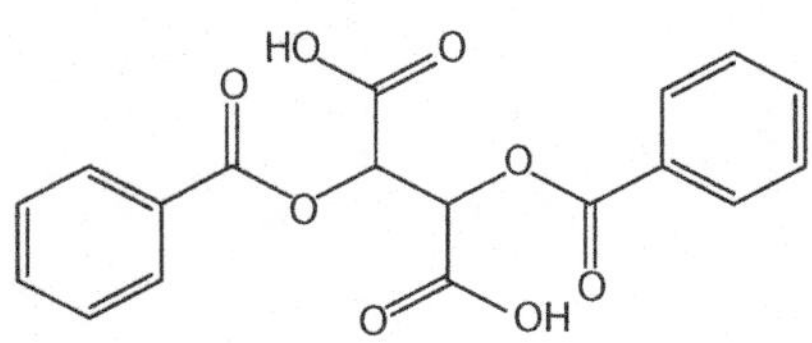
	O,O′-di-*p*-toluoyltartaric acid (+), (−)	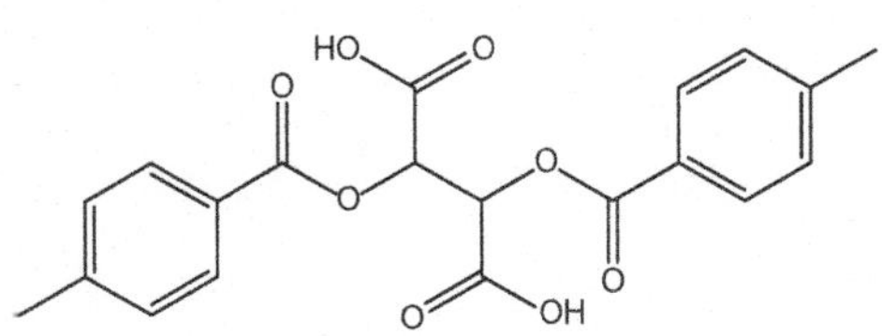

(Continued)

TABLE 14.3 (*Continued*)
Chiral Acids and Bases for Diastereomeric Salt Formation

Salt Formers	Name	Structure
	Camphoric acid (+)	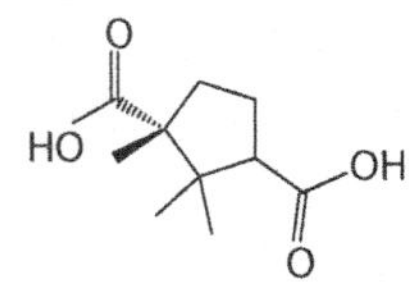
	Threonine (+), (−)	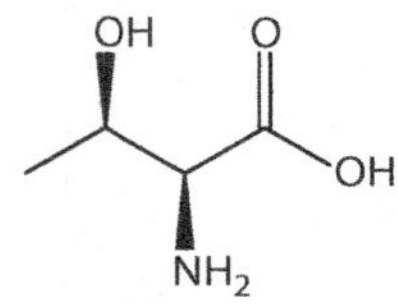
	Mandelic acid (+), (−)	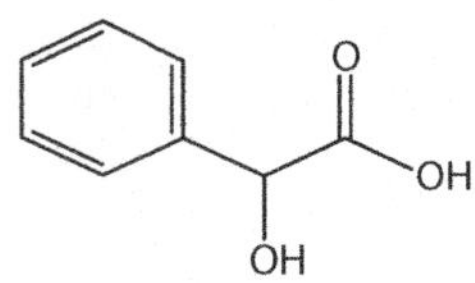
	Camphor-10-sulfonic acid (+)	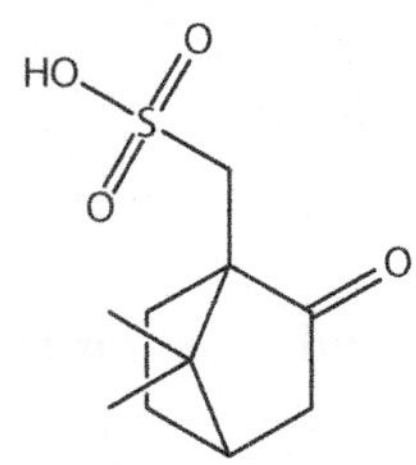
	Camphor-3-sulfonic acid (+)	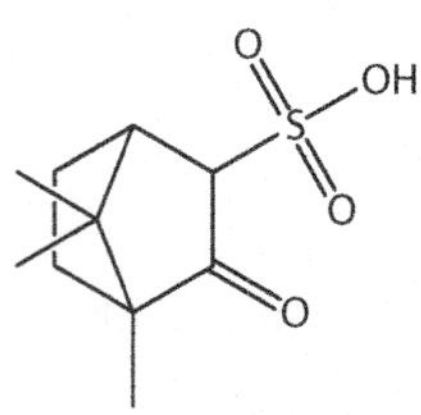
	Ascorbic acid (+)	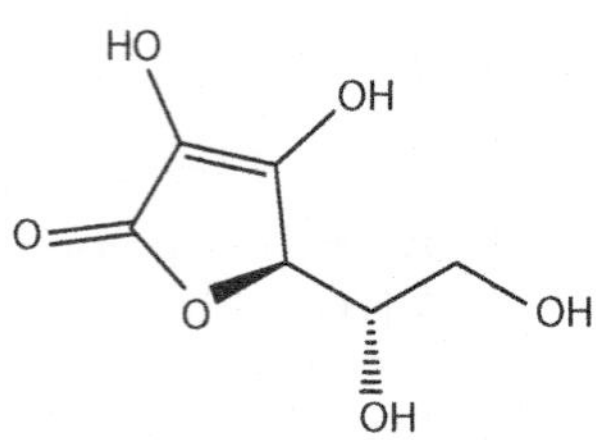

TABLE 14.3
Chiral Acids and Bases for Diastereomeric Salt Formation

Salt Formers	Name	Structure
	Quinic acid (+), (−)	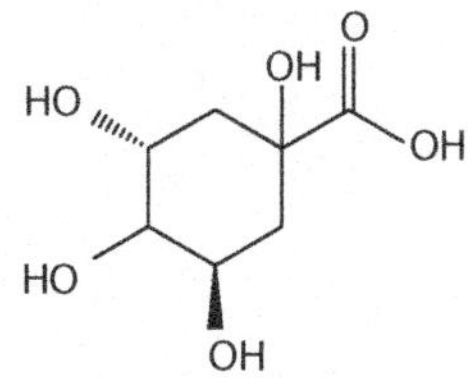
	Gluconic acid (+)	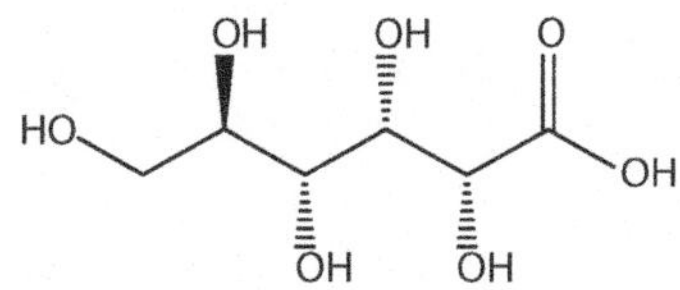
	Malic acid (+), (−)	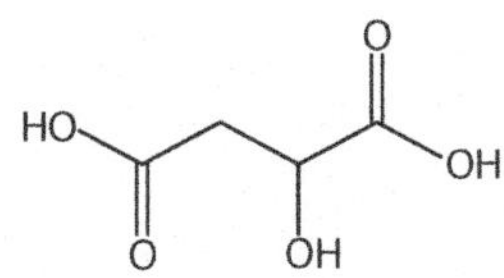
	Lactic acid (+), (−)	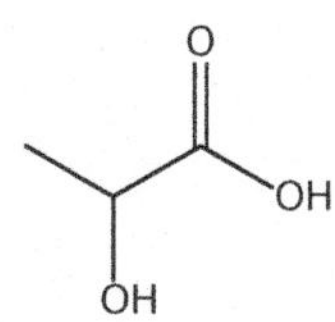
	Aspartic acid (−)	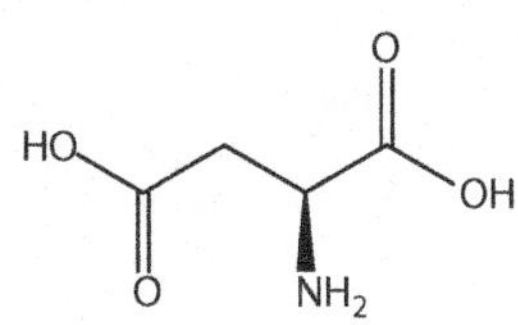
	Glutamic acid (−)	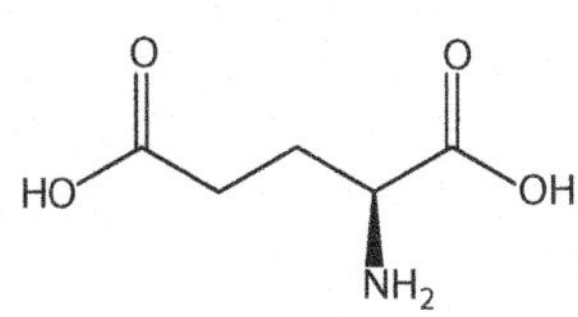
	Glucoheptonic acid (+)	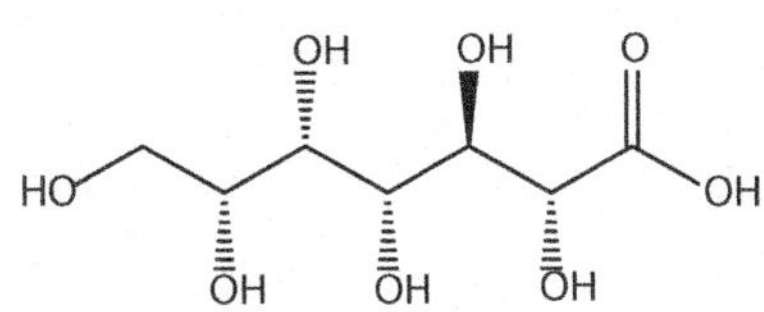

(Continued)

TABLE 14.3 (*Continued*)
Chiral Acids and Bases for Diastereomeric Salt Formation

Salt Formers	Name	Structure
Bases	Cinchonidine	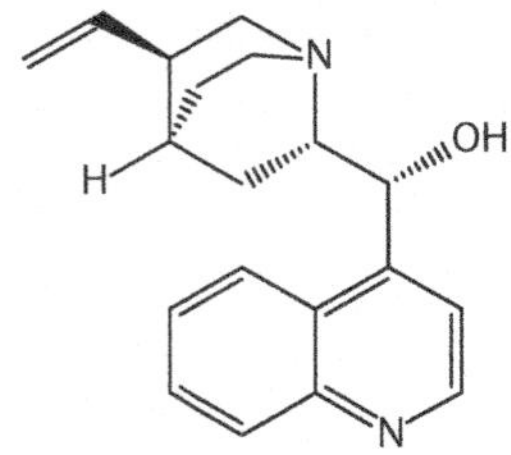
	Cinchonine	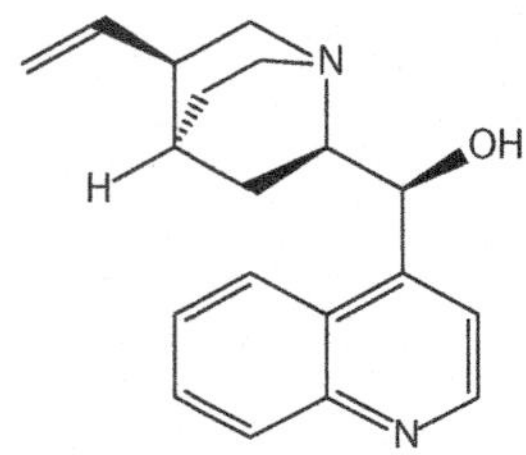
	Quinidine	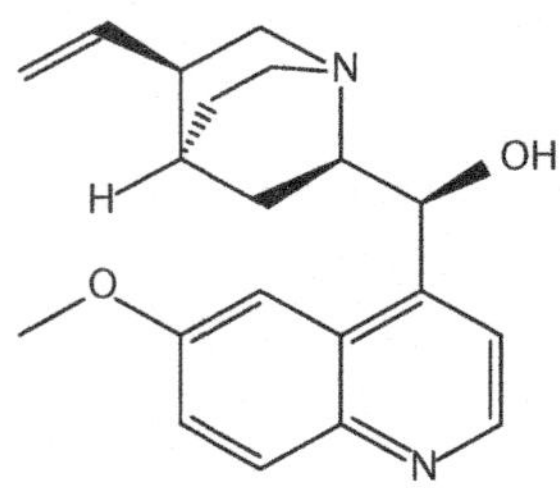
	Quinine	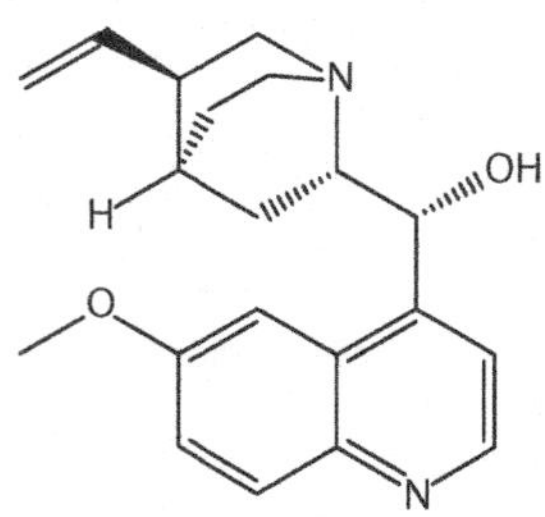
	Dehydroabietylamine	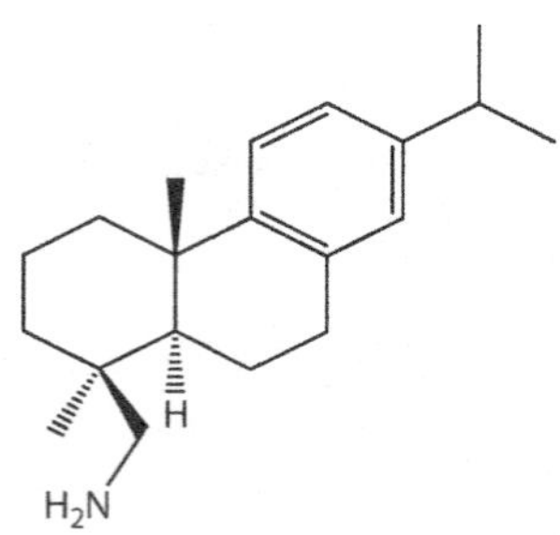

TABLE 14.3
Chiral Acids and Bases for Diastereomeric Salt Formation

Salt Formers	Name	Structure
	Ephedrine (+), (−)	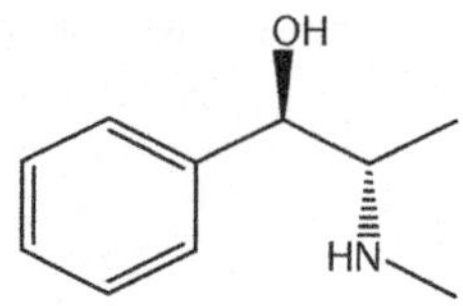
	Pseudoephedrine (+), (−)	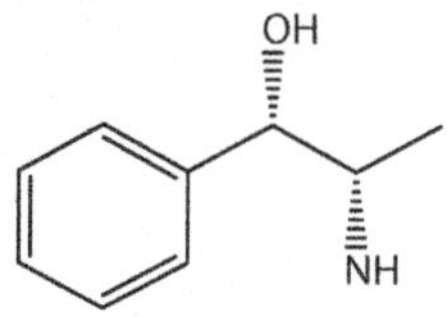
	Deoxyephedrine (+), (−) (methamphetamine)	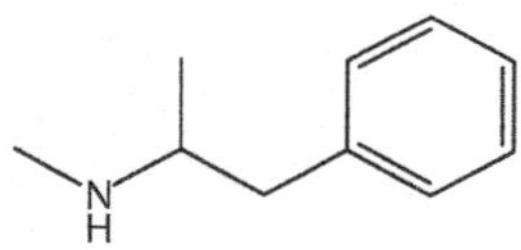
	Amphetamine (+), (−)	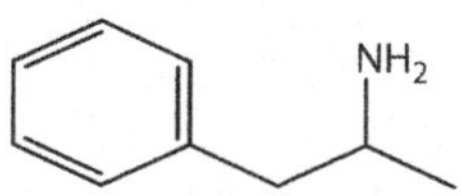
	Norephedrine (+)	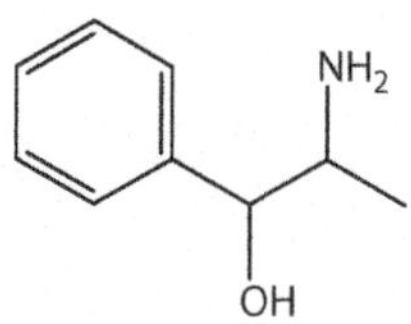
	Hydroquinine	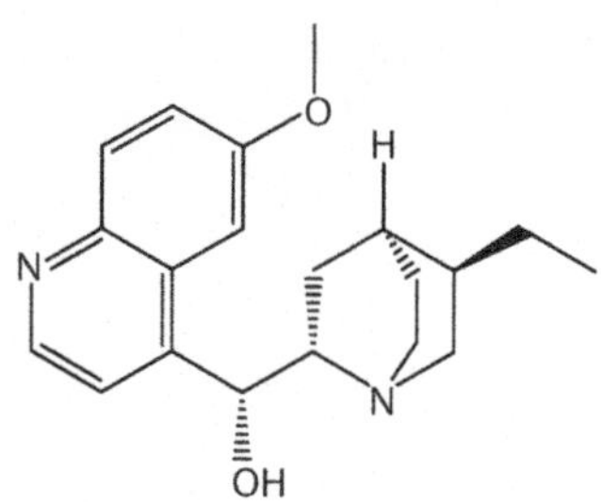
	Hydroquinidine	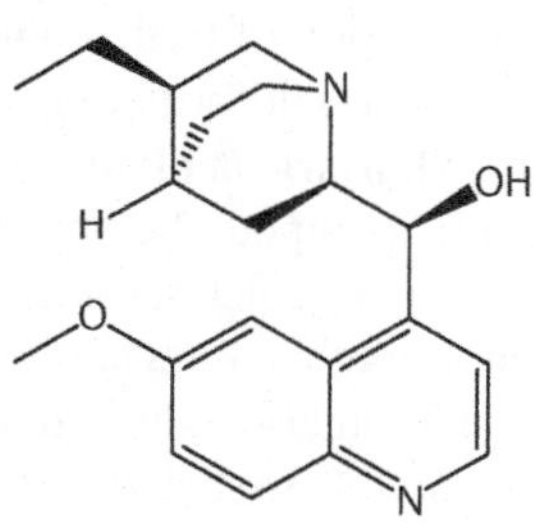

(*Continued*)

TABLE 14.3 (*Continued*)
Chiral Acids and Bases for Diastereomeric Salt Formation

Salt Formers	Name	Structure
	2-Amino-1-butanol (−)	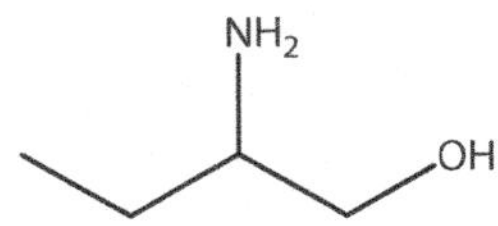
	α-Methylbenzylamine (+), (−)	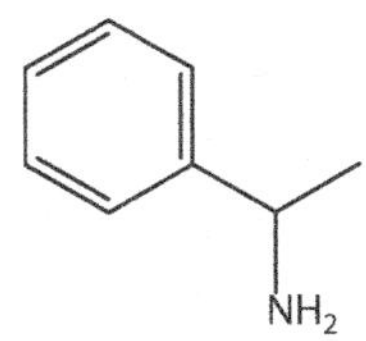
	L-lysine	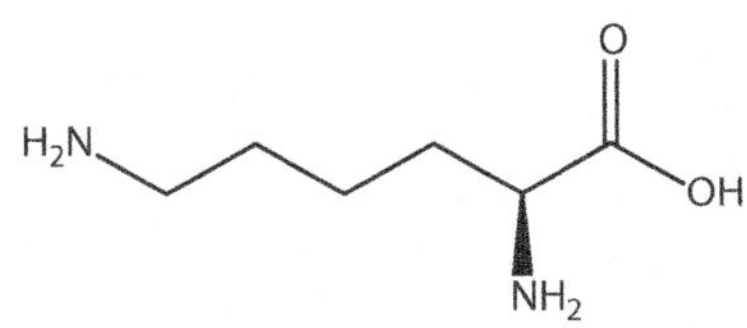
	L-arginine	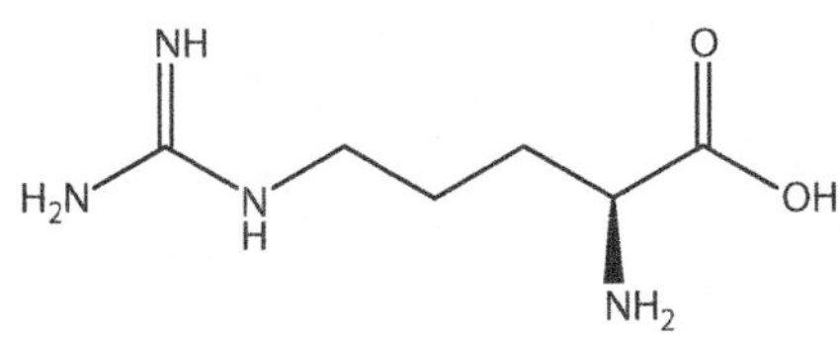
	L-proline	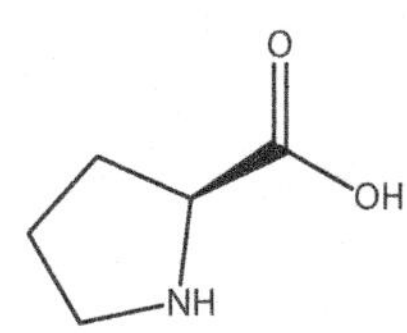

For the resolution of basic compounds, some of the frequently used chiral acids for diastereomeric salt formation are camphorsulfonic acid, tartaric acid, camphoric acid, dibenzoyltartaric acid, mandelic acid, malic acid, and lactic acid. For the resolution of acidic compounds, optically pure bases such as cinchonidine, cinchonine, quinidine, dehydroabietylamine, ephedrine, pseudoephedrine, deoxyephedrine, aminobutanol, methylbenzylamine, and amphetamine may be used as counterions to form diastereomeric salts. Some of the common chiral acids and bases used as resolving agents[51,53] are listed in Table 14.3.

It should be noted that many of the chiral resolving agents frequently used in the early years are genotoxic. Prior to the 1980s, alkaloids such as brucine, strychnine, and quinine were some of the most commonly used resolving agents for the resolution of chiral acids.[51] However, some of them, particularly brucine and strychnine, are known to be poisonous, Ames positive, or genotoxic; thus, these toxic agents are seldom used for pharmaceutical manufacture in modern times.

14.7 SALT SCREEN

The development and application of salt formation for the drug synthesis of both APIs and intermediates originate from screening. All crystallization starts from nucleation. Often, a new synthetic compound, in a synthesis of a new drug substance, has never been crystallized. This means the molecules have never been configured to assemble into an ordered solid state. Virtually all pharmaceutical crystallization takes place in the solution state. Complex pharmaceutical molecules can often be aided by solvent molecules to configure a favored conformation or molecular interaction amenable for molecular self-assembly into an ordered packing.

Initial nucleation of a new compound is a stochastic process. Therefore, it is desirable to maximize the chances for the molecules to interact to form an ordered packing or a certain conformation to facilitate an ordered packing. Because the specific conditions to trigger the nucleation of a new compound are not known initially, it is beneficial to subject the compound to as many varied conditions as possible to maximize the chances for self-assembly of the molecules to start nucleating. Variations in solvents, temperature, concentration, and composition can be applied. Screening is a way to provide all the variations and conditions rapidly.

Screening is done on a small scale (1–100 mg) to maximize the efficiency and number of experiments in a short time using limited amounts of material. In the simplest way, salt screens can be conducted using a set of test tubes or vials containing mini stir bars in a temperature-controlled bath or a reactor block. Many parallel reactor systems that are commercially available allow dozens of screening experiments to be carried out in miniature reactors expeditiously. High-throughput screening methods[54] are routinely applied nowadays in many pharmaceutical companies to prepare potential salt forms of new drug candidates as well as intermediates. Figure 14.3 shows examples of screening apparatuses, both the high-throughput automated system and the manual screen system.

A high-throughput screening system is equipped with 96-well plates and an automated liquid-handling robot system to use even smaller amounts (i.e., 0.5–2 mg of a given compound in each well) to provide hundreds of experiments. A typical salt screen is performed by dispensing minute quantities of the given compound and stoichiometric amounts of counterions as solutions in different solvents to the wells and subjecting the plates to different conditions such as changing temperature, solvent evaporation, changing the solvent composition by adding cosolvents/antisolvents, or any combination of conditions. Then the contents of the wells are observed for any salt crystal formation and/or analyzed by XRPD or Raman spectroscopy.

Crystallization of salts involves a two-step process: first, the formation of the salt compound by the acid–base reaction, and then, the nucleation of the salt molecules. Both steps must occur in order to have a desired hit in a salt screen. Solids observed

FIGURE 14.3 Examples of salt-screening apparatuses: manual screen tools (left) and high-throughput automated screen system and x-ray powder diffraction analysis (center and right).

in many of the wells may not be crystalline; in addition, crystals observed in some wells may not be salts but rather just the free compound or the counterions. For this reason, any positive hits are normally verified by preparing relatively larger quantities (a few grams) of the salts and by subjecting them to a variety of analyses to identify them as viable salt candidates for further assessment of developability.

Finding hits through the salt-screening process provides ideas for the conditions to crystallize the salt and also the first nuclei or seeds that can be used for subsequent scale-up experiments. Once a crystal is spontaneously nucleated, the compound can be crystallized easily by directed crystal growth onto the seed crystals like layering on a template. In many cases, subsequent experiments find the compound to crystallize on its own through molecular self-assembly without seed crystals, even those compounds that were very evasive and difficult to nucleate previously.

The precise original conditions from the screen where a particular salt crystal form was obtained may no longer be necessary to be repeated to further grow the crystals. For example, seed crystals of a salt formed in one solvent system can often be propagated in other solvent systems. For this reason, the screening is usually conducted using all kinds of solvents, including those toxicologically less desirable. Once the seed crystals of a viable salt form are obtained during the screen even from a less desirable solvent system, it is often possible to develop a scalable crystallization process using friendlier and safer solvent systems such as alcohols, acetates, and acetone, especially for APIs.

14.8 SALT CRYSTALLIZATION PROCESS

Salt crystallization processes typically start from solutions or dissolution of a free base or acid in suitable organic solvents. The solvent must provide adequate solubility to solubilize the compound in a reasonable volume. If not, the volume efficiency will be low and the process cannot be scaled up to be an economical manufacturing process with sufficient productivity and throughput. The acid or base counterion should be miscible with the free base or free acid solution to form a homogeneous solution to allow uniform acid–base reaction to generate the ionic pair species.

The process must be amenable to the generated salt crystallizing from the solution. If the generated salt has a low solubility in the original solvent system in which the free base/acid is dissolved, the salt would naturally crystallize out of the solution promptly as it forms. Otherwise, the salt would remain dissolved in the solution. Subsequently, the dissolved salt can be crystallized by means of lowering the solubility, such as changing the temperature or solvent composition via antisolvent addition.

In developing a salt crystallization process, several things must be considered. First is the accurate control of the stoichiometric ratio of the acid and base species. This is especially important for the salt formation of APIs for which the crystal phase purity and the product pH are critical for ensuring consistent API properties and thus the performance of the drug products. The stoichiometry of the salt product may be affected by the accuracy of quantitation of the free acid/base amount in solution and calculation of the proper amount of the counterion to be added. Overcharging or undercharging counterions often results in the precipitation of one of the components and causes an undesired mixture of solid phases. The addition mode of the counterion can also have an impact on the purity and quality of the product. Too fast an addition or inadequate mixing during the addition of the counterion may easily lead to poor crystallinity or polymorph control of the salt or even new impurities being generated from extreme local pH.

Other considerations include proper choice of the solvent or antisolvent for the desired purification and yield. Suitable solvents and antisolvents are selected through the solvent screen and solubility measurements for the necessary polymorphic form control (for APIs) and the balance between the required purification and the desired yield. The toxicological safety of the solvents and the potential impact of the residual solvent should also be considered in selecting the process solvent. For API salt crystallization, the solvents used are preferably class 3 solvents listed in the International Conference on Harmonisation (ICH) of Technical Requirements for Registration of Pharmaceuticals for Human Use guideline for residual solvents.[55] Alcohols and acetone are usually the most commonly used solvents for the final API step crystallization due to their excellent solubilizing properties, toxicological and environmental safety, low cost, low boiling points for ease of drying, and water miscibility which makes them a convenient choice as cosolvents with aqueous systems.

In a salt crystallization process, a particular concern is potentially rapid precipitation due to the abrupt change of solubility upon formation of the salt. This can lead to poor crystallinity and possible entrapment of solvents and impurities in the salt crystals and adversely affect the product quality. Therefore, proper controls and procedures are essential to avoid uncontrolled precipitation of the salt product. The controls may involve suitable management of solution concentrations, crystallization initiation by seeding, dilution and controlled addition of the counterions, and monitoring and controlling pH. Development of controlled crystal growth processes for salt formation is an essential part of process development for the synthesis of drug substances to achieve optimal and consistent quality.[56]

14.9 GENOTOXIC RISKS ASSOCIATED WITH SALT FORMATION

In a synthetic process for a drug substance, a genotoxic impurity may be introduced as a starting material, reagent, intermediate, or catalyst and may also emerge as a byproduct, isomer, or degradation product. Typically, the genotoxic impurity issue arises when a particular compound used in the synthesis or produced as a byproduct is found to be genotoxic, obviously requiring tight controls in the downstream steps, yet remains in the product undesirably. In most cases, the impurities from preceding steps or starting materials should be removed by crystallization or salt isolation to meet the required specification, as discussed in the Sections 14.5 and 14.8.

Issues of genotoxic impurities related to the salt formation itself are typically caused by counterions and/or solvents. For obvious reasons, the direct use of any known genotoxic compounds should be entirely avoided in the final API step. In earlier intermediate steps, potential or known genotoxic compounds may be used occasionally as reagents, counterions, or solvents if unavoidable for reasons of selectivity or efficiency of a specific reaction, separation, purification, or key isolation purpose, essential for the overall drug synthesis. In such cases, the necessary process and analytical controls would duly be in place for the subsequent steps to remove the genotoxic impurity upstream of the final step.

In other scenarios, genotoxic risks are associated with inadvertent reactions between safe counterions and solvents or inadvertent degradation or byproduct formation. Most of these unintended instances are unlikely under normal circumstances; however, some of these inadvertent instances, especially related to the API final step, are well known to have caused serious consequences including failure of batches and, worse yet, product recall from the market.

When genotoxic impurities arise as a result of unexpected occurrences in otherwise ordinary situations, usually no genotoxicity-related analytical methods or controls are in place, and the contamination of the genotoxic impurity in clinical or commercial batches may go unknown until much later, posing a serious threat to the safety of patients. The prominent case of Viracept® in 2007 raised the awareness of genotoxic risks especially for such unexpected incidents and the need for vigilance even for seemingly routine processes.[2]

The Viracept (nelfinavir mesylate) product recall due to genotoxic impurity contamination originated from the inadvertent reaction between the counterion MSA (used in the API salt formation) and the solvent ethanol (used to clean the storage tank where the acid was kept). Residual ethanol left in the storage tank reacted with the acid over a long period of storage time to form high levels of ethyl methane sulfonate (EMS), a genotoxin, which was carried into the API salt formation and remained in the product. The EMS presence went unnoticed and undetected throughout both API and drug product manufacture until the tablets were consumed by some patients and adverse effects were noticed. It was a good manufacturing practice (GMP) oversight that caused the unfortunate incident, with a devastating outcome.

Alkyl sulfonates such as EMS can be generated from the reaction between sulfonic acids and short-chain alcohols such as methanol, ethanol, *n*-propanol, and isopropyl alcohol, as shown in Scheme 14.2.[43] MSA, BSA, and *p*-toluenesulfonic acid (tosic acid) are some of the commonly used counterions to form API salts, that is, mesylate, besylate, and tosylate, respectively. Formation of these sulfonic acid salts in the presence of an alkyl alcohol presents the risk of generating genotoxic alkyl sulfonates, such as methanesulfonates, benzenesulfonates, and toluenesulfonates.

Alkyl halides are also well-known genotoxins that can be potentially generated during salt formation when hydrogen halides (e.g., HCl and HBr) used as counterions react with an alcohol used as solvent, as shown in Scheme 14.3. The genotoxins methyl chloride, ethyl chloride, and isopropyl chloride were reported to have been observed during the preparation of hydrochloride salts from methanol, ethanol, or isopropyl alcohol, respectively.[57,58]

Esters of sulfuric acid, such as dimethyl sulfate and diethyl sulfate, are another type of genotoxins[59,60] that may potentially occur during salt formation employing sulfuric acid. They are produced from the esterification of sulfuric acid with an alcohol, as shown in Scheme 14.4. All of these genotoxins—halogenated alkanes, alkyl esters of sulfuric acids, and alkyl esters of alkane- and aryl-sulfonic acids—known to be alkylating agents form via the interactions between strong acids and alcohols during the process of salt formation. These alkylating genotoxins are summarized in Table 14.4.[61]

Acetamide (melting point = 80°C) is another genotoxin[62,63] reported in the literature that is formed through the hydrolysis of acetonitrile catalyzed by a strong acid, as shown in Scheme 14.5. Acetonitrile is sometimes used as a solvent for salt formation and is susceptible to hydrolysis when excess strong acid is present in the salt formation process to produce acetamide.

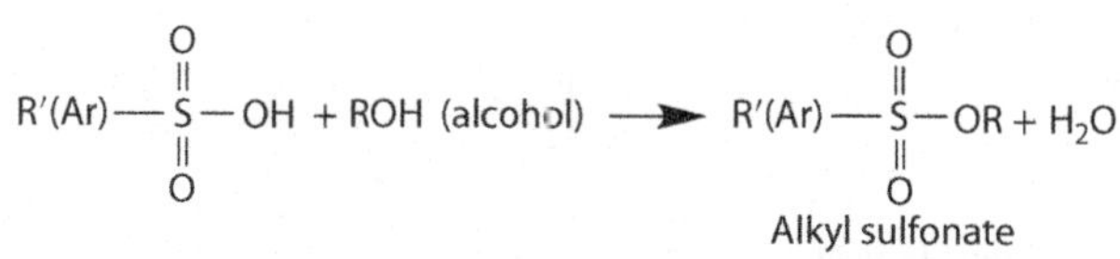

SCHEME 14.2 Formation of alkyl sulfonates by the reaction between sulfonic acids and alcohols

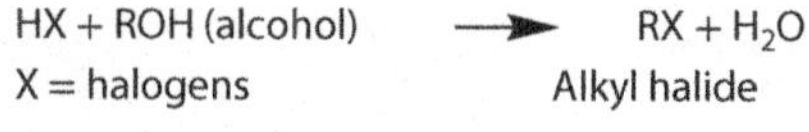

SCHEME 14.3 Formation of alkyl halides by the reaction between hydrogen halides and alcohols

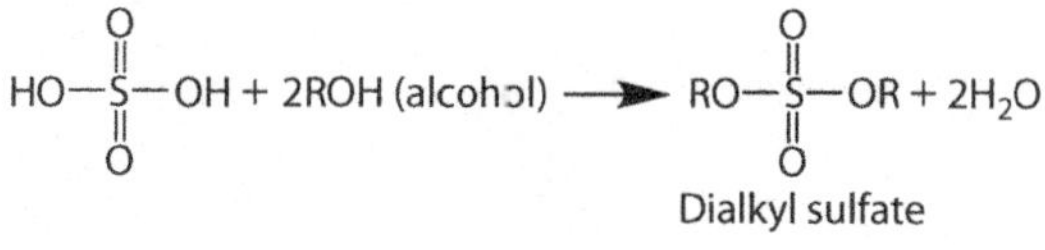

SCHEME 14.4 Formation of dialkyl sulfate by the reaction between sulfuric acid and alcohols

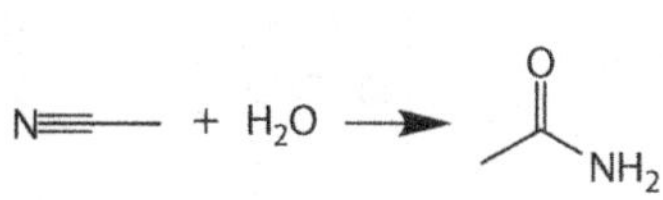

SCHEME 14.5 Formation of acetamide by the hydrolysis of acetonitrile

TABLE 14.4
Alkylating Genotoxins

Type	Compound Name	Structure/CAS #	Properties
Alkyl halides	*n*-Propyl chloride	 540-54-5	Liquid, b.p. 46°C
	i-Propyl chloride	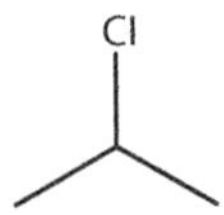 75-29-6	Liquid, b.p. 35°C
	n-Butyl chloride	 109-69-3	Liquid, b.p. 77°C
	Ethyl bromide	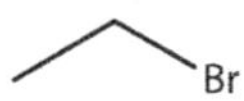 74-96-4	Liquid, b.p. 37°C
	n-Propyl bromide	 106-94-5	Liquid, b.p. 71°C
	i-Propyl bromide	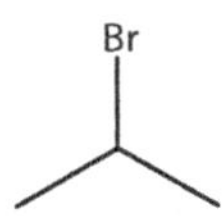 75-26-3	Liquid, b.p. 59°C
	n-Butyl bromide	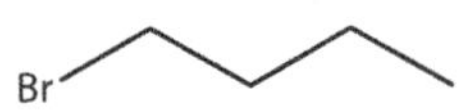 109-65-9	Liquid, b.p. 100°C
Esters of alkyl sulfonic acids	Methyl methanesulfonate	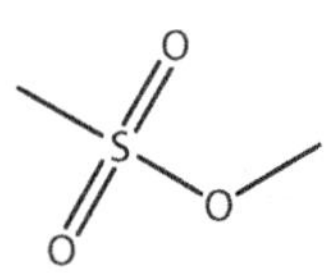 66-27-3	Liquid, b.p. > 200°C
	Ethyl methanesulfonate	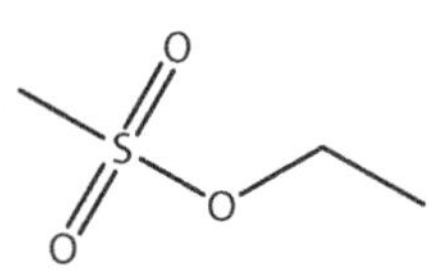 62-50-0	Liquid, b.p. > 200°C
	n-Propyl methanesulfonate	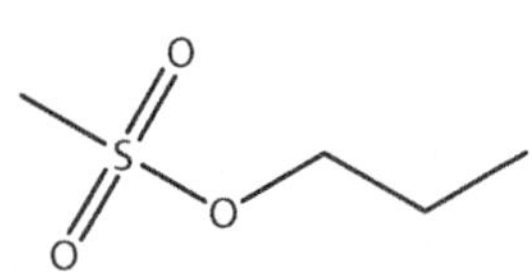 1912-31-8	Liquid, b.p. > 200°C

TABLE 14.4
Alkylating Genotoxins

Type	Compound Name	Structure/CAS #	Properties
Esters of aryl sulfonic acids	Methyl benzenesulfonate	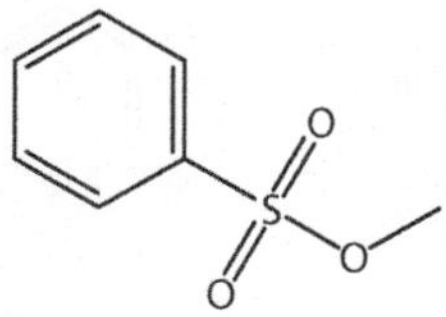 80-18-2	Liquid, b.p. > 200°C
	Ethyl benzenesulfonate	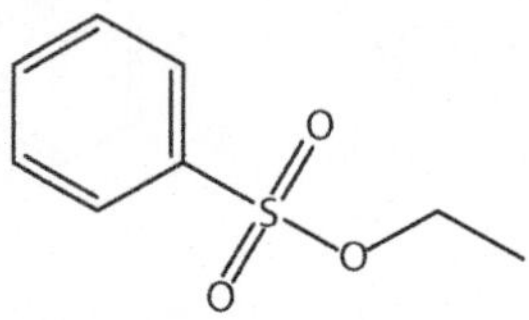 515-46-8	Liquid, b.p. > 200°C
	n-Butyl benzenesulfonate	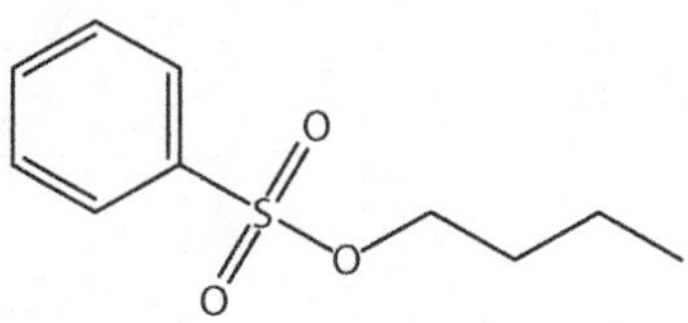 80-44-4	Liquid, b.p. > 200°C
	Methyl toluenesulfonate	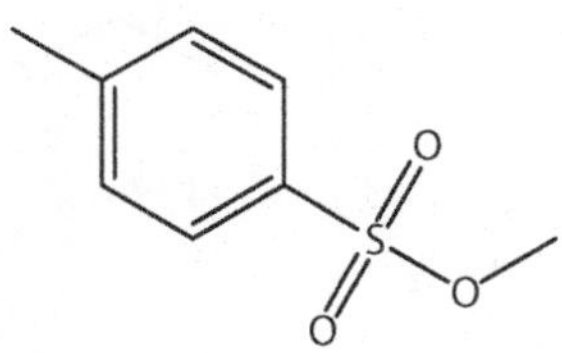 80-48-8	Solid, m.p. 25°C
	Ethyl toluenesulfonate	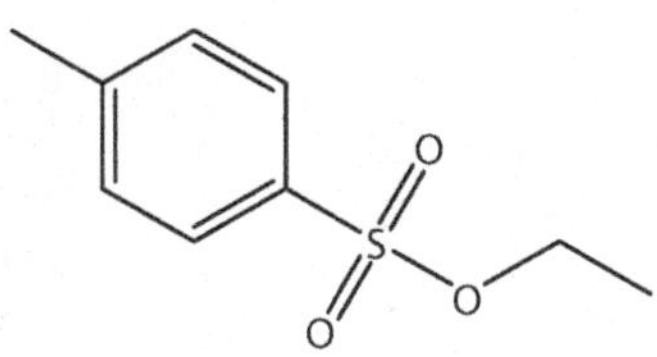 80-40-0	Solid, m.p. 30°C
	n-Propyl toluenesulfonate	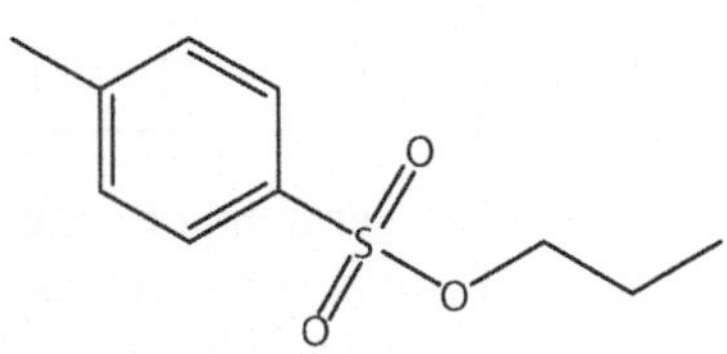 599-91-7	Liquid, b.p. > 200°C

(Continued)

TABLE 14.4 (*Continued*)
Alkylating Genotoxins

Type	Compound Name	Structure/CAS #	Properties
	i-Propyl toluenesulfonate	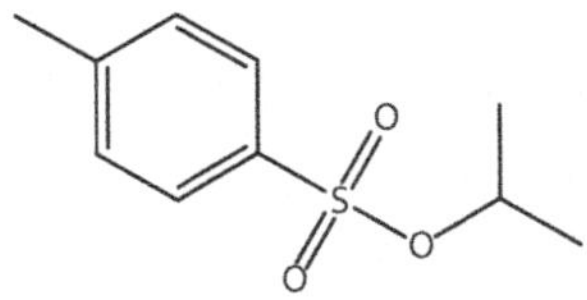 2307-69-9	Liquid, b.p. > 200°C
	n-Butyl toluenesulfonate	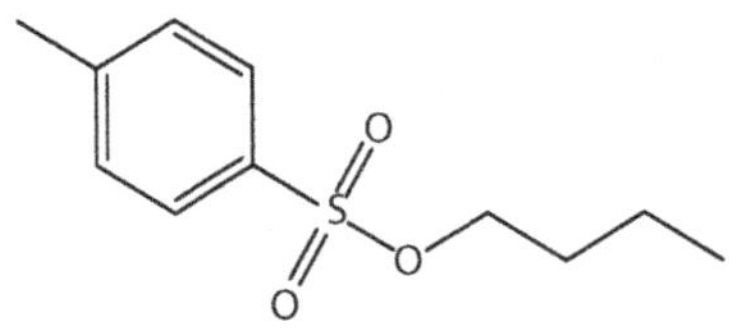 778-28-9	Liquid, b.p. > 200°C
Esters of sulfuric acid	Dimethyl sulfate	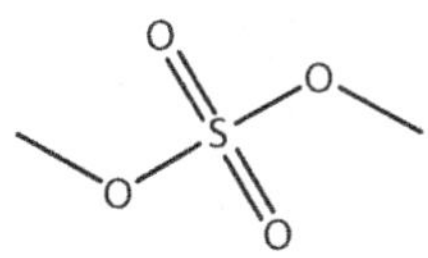 77-78-1	Liquid, b.p. > 200°C
	Diethyl sulfate	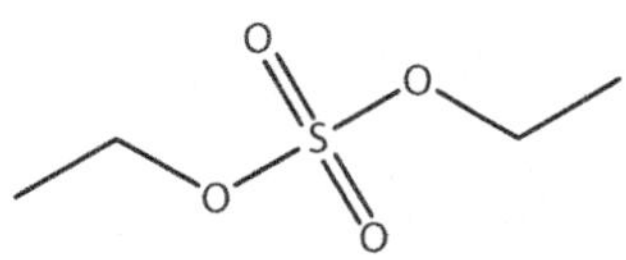 64-67-5	Liquid, b.p. > 200°C
	Dipropyl sulfate	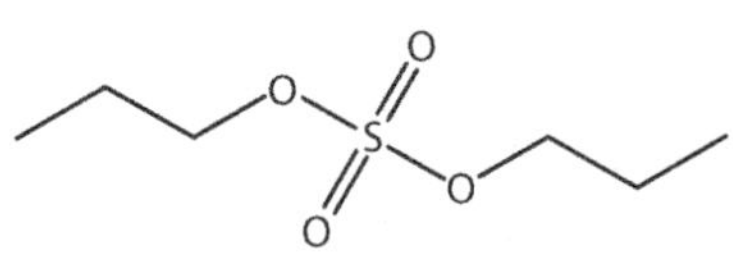 598-05-0	Liquid, b.p. > 200°C
	Dibutyl sulfate	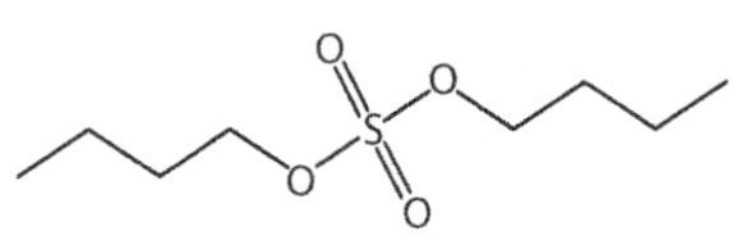 625-22-9	Liquid, b.p. > 200°C

Note: b.p., boiling point; m.p., melting point.

Many cases of genotoxin generation during salt formation, including those listed earlier, involve the use of strong acids such as MSA, HCl, HBr, and sulfuric acid in the presence of alcoholic solvents. What would be the best strategy to reduce the risk associated with these genotoxins and the salts? One might think that a surefire way to eliminate the risk is to completely abolish the use of these acids and alcohols from the synthetic processes of drug substances. However, such extreme measures would have a significant bearing on API form screening/selection, crystallization development, purification, isolation of key intermediates, and decisions on synthetic routes and may hamper the development of efficient, economical, and scalable process for drug synthesis.

Particularly concerning sulfonic acids, in the wake of the Viracept incident and with the additional regulatory requirements from the European agencies related to the use of sulfonic acids in API syntheses,[64,65] some companies have taken the steps of removing sulfonic acids from API salt screens and restraining their use as counterions for salt formation in synthetic process development. However, sulfonic acids have many uses in all steps of the synthesis, and sulfonic acid salts can provide many beneficial properties as API solid forms or as isolable solid forms for synthetic intermediates. Several articles in the literature make the specific point that sulfonic acid salts have many advantages and should not be discounted, especially during the assessment of API salt forms.[44]

All of the aforementioned strong acids are the most frequently used counterions for API salt formation for basic drugs, and the salts from these strong acids account for a large majority of the API salts approved and marketed during recent decades,[38] as discussed earlier. Alcohols (ethanol, isopropanol, *n*-propanol, and methanol) are also some of the most commonly used solvents, especially for API salt crystallization, due to their many favorable properties mentioned in Section 14.8. Therefore, avoiding these acids and alcohols altogether in salt forms or salt processes would be difficult, if not impossible.

On the other hand, the risk of genotoxin generation associated with the use of strong acids and alcohols in salt formation can be managed after weighing the available options and balancing the benefit of using these salts and alcohols against the potential risk. The risk can be minimized by developing and implementing the processes with robust process control in conjunction with rigorous analytical monitoring. Appropriate process control measures for salt formation can be put in place after systematic investigations to define the reliable ranges of process conditions that avert undesirable reactions producing the genotoxins. In addition, implementation of suitable control points in the manufacturing process along with prudently set in-process specifications based on the permitted daily exposure or threshold of toxicological concern would ensure robust control of the genotoxic impurities.[66]

Studies have shown that the formation of alkyl sulfonate esters in alcohols, for example, requires the presence of a strong acid in excess and is entirely inhibited by a slight excess of a base.[67,68] The studies also showed that the undesired reactions producing alkyl sulfonates and alkyl halides are hindered by the presence of water in the solution and weaker acids such as phosphoric acid do not catalyze the reactions.[43,67] Therefore, avoiding strongly acidic conditions and prolonged exposure of acids to an alcohol at a high temperature and incorporating water into the process when possible would reduce the chances of formation of these genotoxins. In salt crystallization processes using strong acids in general, precise control of the stoichiometric ratio of the acid to the base (according to the accurately quantified amount of the base in solution) would avoid strongly acidic conditions. This can be achieved by stringent in-process pH control during the salt formation process, controlled addition of the acid, and adequate mixing in the reactor during the acid addition.

Even in those atypical cases where a process experiences an excursion of extreme pH or high temperature, since most of these alkylating genotoxins exist as liquids at ambient conditions (Table 14.4), these genotoxins should remain in the mother liquor, given a well-controlled crystallization process for product isolation. Efficient filtration and subsequent washing should ensure their removal from the crystallized product. For this reason, in many cases where genotoxins may form in the solution their actual content in the product may be well below the specification limit. For alkyl halides, which are mostly volatile liquids with low boiling points, it should be even easier since they may be evaporated when drying the isolated product. The issue of genotoxin contamination would surface if a genotoxic impurity gets entrapped in the product due to occlusion in the crystals via a poorly executed crystallization. Aside from the alkyl halides, most of these alkylating genotoxins have high boiling points, and once they are formed and entrapped in the crystals they would be difficult to be removed. Therefore, as described in Sections 14.5 and 14.7, conscientious control of properly designed crystallization processes that encourage crystal growth rather than precipitation would prevent the entrapment of impurities and ensure their removal from products. Aspects of a properly designed crystallization include appropriate rates of temperature ramping, antisolvent addition, and/or counterion addition along with a well-planned seeding point at the right concentration/composition.

Poorly controlled processes pose risks even when genotoxic impurities are not formed in the actual process. Entrapped residual solvents, particularly alcohols, from uncontrolled precipitation can further react with strong acids in subsequent synthetic steps or during drying to create an undesired genotoxic compound. Even small amounts of residual alcohol in API can, in principle, interact with a strong acid used in the downstream formulation process to produce trace amounts of a genotoxin in the drug product. Therefore, prudent attention to any potential combination of residual alcohols in isolated compounds or APIs and use of strong acids in the subsequent steps is important to decrease the risk of formation of genotoxic impurities.

REFERENCES

1. Robinson, D. I. Control of Genotoxic Impurities in Active Pharmaceutical Ingredients: A Review and Perspective. *Org. Process Res. Dev.* **2010**, *14*, 946–959.
2. Gerber, C.; Toelle, H. G. What Happened: The Chemistry Side of the Incident with EMS Contamination in Viracept Tablets. *Toxicol. Lett.* **2009**, *190*, 248–253.
3. Brown, A. D.; Davis, R. D.; Fitzgerald, R. N.; Glover, B. N.; Harvey, K. A.; Jones, L. A.; Liu, B.; Patterson, D. E.; Sharp, M. J. Process Development for Sodelglitazar: A PPAR Panagonist. *Org. Process Res. Dev.* **2009**, *13*, 297–302.
4. Liu, D. Q.; Chen, T. K.; McGuire, M. A.; Kord, A. S. Analytical Control of Genotoxic Impurities in the Pazopanib Hydrochloride Manufacturing Process. *J. Pharm. Biomed. Anal.* **2009**, *50*, 144–150.
5. Yang, Q.; Haney, B. P.; Vaux, A.; Riley, D. A.; Heidrich, L.; He, P.; Mason, P.; et al. Controlling the Genotoxins Ethyl Chloride and Methyl Chloride Formed during the Preparation of Amine Hydrochloride Salts from Solutions of Ethanol and Methanol. *Org. Process Res. Dev.* **2009**, *13*, 786–791.

6. Patterson, D. E.; Powers, J. D.; LeBlanc, M.; Sharkey, T.; Boehler, E.; Irdam, E.; Osterhout, M. H. Development of a Practical Large-Scale Synthesis of Denagliptin Tosylate. *Org. Process Res. Dev.* **2009**, *13*, 900–906.
7. Stahl, P. H.; Wermuth, C. G., Eds. *Handbook of Pharmaceutical Salts Properties, Selection, and Use*; Wiley-VCH: Weinheim, Germany, 2002.
8. Berge, S. M.; Bighley, L. D.; Monkhouse, D. C. Factors That Impact the Developability of Drug Candidates: An Overview. *J. Pharm. Sci.* **1977**, *66*, 1–19.
9. Gould, P. L. Salt Selection for Basic Drugs. *Int. J. Pharm.* **1986**, *33*, 201–217.
10. Serajuddin, A. T. M. Salt Formation to Improve Drug Solubility. *Adv. Drug Deliver. Rev.* **2007**, *59*, 603–616.
11. Bowker, M. J.; Stahl, P. H. Preparation of Water-Soluble Compounds through Salt Formation. In *The Practice of Medicinal Chemistry*; Wermuth, C. G., Ed.; Academic Press: London, UK, 2008; pp 749–765.
12. Li, S.; He, H.; Parthiban, L.; Yin, H.; Serajuddin, A. T. M. IV-IVC Considerations in the Development of Immediate-Release Oral Dosage Form. *J. Pharm. Sci.* **2005**, *94*, 1396–1417.
13. Cairns, D. *Essentials of Pharmaceutical Chemistry*, 4th ed.; Pharmaceutical Press: London, UK, 2012; pp 62–64.
14. Singh, J.; Pudipeddi, M.; Lindrud, M. D. Bisulfate Salt of HIV Protease Inhibitor. U.S. Patent 6,087,383, July 11, 2000.
15. Stahl, P. H.; Nakano, M. Pharmaceutical Aspects of the Drug Salt Form. In *Handbook of Pharmaceutical Salts Properties, Selection, and Use*; Stahl, P. H.; Wermuth C. G., Eds.; Wiley-VCH: Weinheim, Germany, 2002; pp 83–116.
16. Salivar, C. J. Aqueous Calcium Dioxytetracycline. U.S. Patent 2,903,395, September 8, 1959.
17. Raghavan, K. S.; Sunanda, A. R.; Bembenek, K. S.; Benkerrour, L.; Trognon, V.; Corrao, R. G.; Esposito, L. Pediatric Formulation of Gatifloxacin. U.S. Patent 6,589,955, July 8, 2003.
18. Sandberg, A.; Blomqvist, I.; Jonsson, U. E.; Lundborg, P. Pharmacokinetic and Pharmacodynamic Properties of a New Controlled-Release Formulation of Metoprolol: A Comparison with Conventional Tablets. *Eur. J. Clin. Pharmacol.* **1988**, *33 (1)*, S9–S14.
19. Mendels, J.; Digiacomo, J. The Treatment of Depression with a Single Daily Dose of Imipramine Pamoate. *Am. J. Psychiat.* **1973**, *130 (9)*, 1022–1024.
20. Kristensen, H. G.; Schaefer, T. Granulation: A Review on Pharmaceutical Wet-Granulation. *Drug Dev. Ind. Pharm.* **1987**, *13 (4–5)*, 803–872.
21. Badorc, A.; Frehel, D. Dextro-Rotatory Enantiomer of Methyl Alpha-5 (4, 5, 6, 7-Tetrahydro (3,2-c) Thieno Pyridyl)(2-Chlorophenyl)-Acetate and the Pharmaceutical Compositions Containing It. U.S. Patent 4,847,265, July 11, 1989.
22. Kim, S. W.; Jun, S. S.; Jo, Y. G.; Koo, J. S.; Kim, Y. W. *N,N*-Dimethyl Imidodicarbonimidic Diamide Acetate, Method for Producing the Same and Pharmaceutical Compositions Comprising the Same. U.S. Patent 8,058,312, November 15, 2011.
23. Pfannkuch, F.; Rettig, H.; Stahl, P. H. Biological Effects of the Drug Salt Form. In *Handbook of Pharmaceutical Salts Properties, Selection, and Use*; Stahl, P. H.; Wermuth C. G., Eds.; Wiley-VCH: Weinheim, Germany, 2002; pp 117–134.
24. Djordjevic, L.; Primorac, M.; Stupar, M. In Vitro Release of Diclofenac Diethylamine from Caprylocaproyl Macrogolglycerides Based Microemulsions. *Int. J. Pharm.* **2005**, *296 (1)*, 73–79.
25. Serajuddin, A. T. M.; Pudipeddi, M. Salt-Selection Strategies. In *Handbook of Pharmaceutical Salts Properties, Selection, and Use*; Stahl, P. H.; Wermuth C. G., Eds.; Wiley-VCH: Weinheim, Germany, 2002; pp 135–160.

26. Yin, S. X.; Grosso, J. A. Selecting and Controlling API Crystal Form for Pharmaceutical Development—Strategies and Processes. *Curr. Opin. Drug Discov. Dev.* **2008**, *11 (6)*, 771–777.
27. Engel, G. L.; Farid, N. A.; Faul, M. M.; Richardson, L. A.; Winneroski, L. L. Salt Form Selection and Characterization of LY333531 Mesylate Monohydrate. *Int. J. Pharm.* **2000**, *198*, 239–247.
28. Bowker, M. J. A Procedure for Salt Selection and Optimization. In *Handbook of Pharmaceutical Salts Properties, Selection, and Use*; Stahl, P. H.; Wermuth C. G., Eds.; Wiley-VCH: Weinheim, Germany, 2002; pp 161–189.
29. Ku, M. S. Salt and Polymorph Selection Strategy Based on the Biopharmaceutical Classification System for Early Pharmaceutical Development. *Am. Pharm. Rev.* **2010**, *13 (1)*, 22–30.
30. Sieger, P. Choosing the Right Solid Form Depending on the Dose and the Route of Application. *Am. Pharm. Rev.* **2008**, *11*, 28–32.
31. Morris, K. R.; Fakes, M. G.; Thakur, A. B.; Newman, A. W.; Singh, A. K.; Venit, J. J.; Spagnuolo, C. J.; Serajuddin, A. T. M. An Integrated Approach to the Selection of Optimal Salt Form for a New Drug Candidate. *Int. J. Pharm.* **1994**, *105*, 209–217.
32. U.S. Food and Drug Administration. *Food Ingredients and Packaging: Generally Recognized as Safe (GRAS); Substances Generally Recognized as Safe.* 21 CFR Part 18216, 2012. http://www.fda.gov/food/ingredientspackaginglabeling/gras/default.htm (accessed March 28, 2014)
33. International Programme of Chemical Safety INCHEM Database, Joint FAO/WHO Expert Committee on Food Additives, Food and Agriculture Organization of the United Nations and World Health Organization. http://www.inchem.org/ (accessed March 28, 2014)
34. Brittain, H. Cocrystal Systems of Pharmaceutical Interest: 2010. *Cryst. Growth Des.* **2012**, *12*, 1046–1054.
35. Qiao, N.; Li, M.; Schlindwein, W.; Malek, N.; Davies, A.; Trappitt, G. Pharmaceutical Cocrystals: An Overview. *Int. J. Pharm.* **2011**, *419 (1)*, 1–11.
36. Kim, S., Li, Z., Tseng, Y. C., Nar, H., Spinelli, E., Varsolona, R.; Reeves, J. T.; et al. Development and Characterization of a Cocrystal as a Viable Solid Form for an Active Pharmaceutical Ingredient. *Org. Proc. Res. Dev.* **2013**, *17*, 540–548.
37. Schultheiss, N.; Newman, A. Pharmaceutical Cocrystals and Their Physicochemical Properties. *Cryst. Growth Des.* **2009**, *9*, 2950–2967.
38. Paulekuhn, G. S.; Dressman, J. B.; Saal, C. Trends in Active Pharmaceutical Salt Selection Based on Analysis of the Orange Book Database. *J. Med. Chem.* **2007**, *50*, 6665–6672.
39. Streng, W. H.; Hsi, S. K.; Helms, P. E.; Tan, H. G. H. General Treatment of pH-Solubility Profiles of Weak Acids and Bases and the Effects of Different Acids on the Solubility of a Weak Base. *J. Pharm. Sci.* **1984**, *73*, 1679–1684.
40. Bogardus, J. B.; Blackwood, R. K. J. Dissolution Rates of Doxycycline Free Base and Hydrochloride Salts. *J. Pharm. Sci.* **1979**, *68*, 1183–1184.
41. Rauws, A. G. Pharmacokinetics of Bromide Ion—An Overview. *Food Chem. Toxicol.* **1983**, *21 (4)*, 379–382.
42. van Leeuwen, F. X. R.; Sangster, B.; Hildebrandt, A. G. The Toxicology of Bromide Ion. *CRC Crit. Rev. Toxicol.* **1987**, *18 (3)*, 189–213.
43. Elder, D. P.; Snodin, D. J. Drug Substances Presented as Sulfonic Acid Salts: Overview of Utility, Safety and Regulation. *J. Pharm. Pharmacol.* **2009**, *61*, 269–278.
44. Elder, D. P.; Delaney, E.; Teasdale, A.; Eyley, S.; Reif, V. D.; Jacq, K.; Facchine, K. L.; Oestrich, R. S.; Sandra, P.; David, F. The Utility of Sulfonate Salts in Drug Development. *J. Pharm. Sci.* **2010**, *99*, 2948–2961.
45. Sheu, M. T.; Chou, H. L.; Kao, C. C.; Liu, C. H.; Sokoloski, T. D. Dissolution of Diclofenac Sodium from Matrix Tablets. *Int. J. Pharm.* **1992**, *85 (1)*, 57–63.

46. Lee, S.; Hoff, C. Large-Scale Aspects of Salt Formation: Processing of Intermediates and Final Products. In *Handbook of Pharmaceutical Salts Properties, Selection, and Use*; Stahl, P. H.; Wermuth C. G., Eds.; Wiley-VCH: Weinheim, Germany, 2002; pp 191–220.
47. Tung, H. H.; Paul, E. L.; Midler, M.; McCauley, J. A. *Crystallization of Organic Compounds: An Industrial Perspective.* John Wiley: New York, 2009.
48. Li, B.; Haynie, D. T. Chiral Drug Separation. *Encycl. Chem. Process.* 2006, *10*, 449–458.
49. Rouhi, A. M. Chirality at Work. *Chem. Eng. News* **2003**, *81 (18)*, 56–61.
50. McGrath, N. A.; Brichacek, M.; Njardarson, J. T. A Graphical Journey of Innovative Organic Architectures That Have Improved Our Lives. *J. Chem. Educ.* **2010**, *87 (12)*, 1348–1349.
51. Jacques, J.; Collet, A.; Wilen, S. H. *Enantiomers, Racemates, and Resolutions.* John Wiley:
New York, 1981.
52. Nasipuri, D. *Stereochemistry of Organic Compounds: Principles and Applications*; New Age International: New Delhi, India, 1994; p 147.
53. Kozma, D., Ed. *CRC Handbook of Optical Resolutions via Diastereomeric Salt Fomation*; CRC Press LLC: Boca Raton, Florida, 2002, pp 11–12.
54. Morissette, S. L.; Almarsson, O.; Peterson, M. L.; Remenar, J. F.; Read, M. J.; Lemmo, A. V.; Ellis, S.; Cima, M. J.; Gardner, C. R. High-Throughput Crystallization: Polymorphs, Salts, Co-Crystals and Solvates of Pharmaceutical Solids. *Adv. Drug Delivery Reviews* **2004**, *56*, 275–300. http://www.ich.org/products/guidelines/quality/article/quality-guidelines.html (accessed March 28, 2014)
55. International Conference on Harmonisation of Technical Requirements for Registration of Pharmaceuticals for Human Use, ICH Harmonised Tripartite Guideline. *Impurities: Guideline for Residual Solvents Q3C (R5)*, 2011. http://www.ich.org/products/guidelines/quality/article/quality-guidelines.html (accessed March 28, 2014)
56. Kim, S.; Lotz, B.; Lindrud, M.; Girard, K.; Moore, T.; Nagarajan, K.; Mario, A.; et al. Control of the Particle Properties of a Drug Substance by Crystallization Engineering and the Effect on Drug Product Formulation. *Org. Proc. Res. Dev.* **2005**, *9*, 894–901.
57. Elder, D. P.; Teasdale, A.; Lipczynski, A. M. Control and Analysis of Alkyl and Benzyl Halides and Other Related Reactive Organohalides as Potential Genotoxic Impurities in Active Pharmaceutical Ingredients (APIs). *J. Pharm. Biomed. Anal.* **2008**, *48*, 497–507.
58. Yang, Q.; Haney, B. P.; Vaux, A.; Riley, D. A.; Heidrich, L.; He, P.; Mason, P.; et al. Controlling the Genotoxins Ethyl Chloride and Methyl Chloride Formed during the Preparation of Amine Hydrochloride Salts from Solutions of Ethanol and Methanol. *Org. Proc. Res. Dev.* **2009**, *13*, 786–791.
59. International Agency for Research on Cancer. Diethyl Sulfate. *IARC Monographs on the Evaluation of Carcinogenic Risks to Humans.* **1999**, Vol. 71, Part 3, Lyon, pp 1405–1415.
60. Recommendation from the Scientific Committee on Occupational Exposure Limits for Diethyl Sulphate. *European Commission Employment, Social Affairs and Inclusion, SCOEL/SUM/154*, December 2009. http://ec.europa.eu/social/search.jsp?advSearchKey=Diethyl+sulfate&x=15&y=19&mode=advancedSubmit&langId=en (accessed March 28, 2014)
61. Sobol, Z.; Engel, M. E.; Rubitski, E.; Kub, W. W.; Aubrecht, J.; Schiestl, R. H. Genotoxicity Profiles of Common Alkyl Halides and Esters with Alkylating Activity. *Mutat. Res.* **2007**, *633*, 80–94.
62. Acetamide. United States National Library of Medicine Toxicology Data Network. http://toxnet.nlm.nih.gov/ (accessed January 28, 2013).

63. Schule, A.; Ates, C.; Palacio, M.; Stofferis, J.; Delatinne, J. P.; Martin, B.; Lloyd, S. Monitoring and Control of Genotoxic Impurity Acetamide in the Synthesis of Zaurategrast Sulfate. *Org. Proc. Res. Dev.* **2010**, *14*, 1008–1014.
64. Request to Assess the Risk of Occurrence of Contamination with Mesilate Esters and Related Compounds in Pharmaceutical Preparations. *Letter to Marketing Authorization Holders*, Swissmedic, Swiss Agency for Therapeutic Products, October 23, 2007.
65. Request to Assess the Risk of Occurrence of Contamination with Mesilate Esters and Related Compounds in Pharmaceuticals. *Letter Intended for All Marketing Authorization Holders for Medicinal Products Containing Active Substances in the Form of Mesilates, (di)Isetionates, Tosilates or Besilates (EMEA/44714/2008)*, European Medicines Agency Inspections, London, January 24, 2008.
66. Cimarosti, Z.; Bravo, F.; Stonestreet, P.; Tinazzi, F.; Vecchi, O.; Camurri, G. Application of Quality by Design Principles to Support Development of a Control Strategy for the Control of Genotoxic Impurities in the Manufacturing Process of a Drug Substance. *Org. Proc. Res. Dev.* **2010**, *14*, 993–998.
67. Teasdale, A.; Delaney, E. J.; Eyley, S. C.; Jacq, K.; Taylor-Worth, K.; Lipczynski, A.; Hoffmann, W.; et al. Detailed Study of Sulfonate Ester Formation and Solvolysis Reaction Rates and Application toward Establishing Sulfonate Ester Control in Pharmaceutical Manufacturing Processes. *Org. Proc. Res. Dev.* **2010**, *14*, 999–1007.
68. Hajikarimian, Y.; Yeo, S.; Ryan, R. W.; Levett, P.; Stoneley, C.; Singh, P. Investigation into the Formation of the Genotoxic Impurity Ethyl Besylate in the Final Step Manufacturing Process of UK-369,003-26, a Novel PDE5 Inhibitor. *Org. Proc. Res. Dev.* **2010**, *14*, 1027–1031.

15 Identification and Control of Genotoxic Degradation Products

Fenghe Qiu

CONTENTS

15.1 INTRODUCTION

Pharmaceutical impurities are any components of a new drug substance or a new drug product that are not the drug substance itself or excipients in the drug product [1,2]. Genotoxic impurities are a subset of impurities that are mutagenic or carcinogenic. To differentiate from genotoxic impurities, nongenotoxic impurities are often referred to as regular impurities. The thresholds for reporting, identifying, and qualifying regular impurities are recommended by International Conference on Harmonisation of Technical Requirements for Registration of Pharmaceuticals for Human Use (ICH) Q3A [1] and Q3B [2] and are routinely applied during late-stage development and at registration in the pharmaceutical industry. For control of genotoxic impurities, the generic thresholds threshold of toxicological concern (TTC) and staged threshold of toxicological concern (sTTC) are recommended in the European Medicines Agency (EMA) [3] and Food and Drug Administration (FDA) [4] genotoxic impurity guidelines and the EMA questions and answers document [5].

Although the ICH Q3A/B thresholds (e.g., identification threshold [IT]) are intended for regular impurities and not for genotoxic impurities, in practice they still play an irreplaceable role in the identification and risk assessment of genotoxic impurities. Thus, it is important to understand the relationships between the ICH thresholds and the genotoxic limits/thresholds, which can be illustrated using the "impurity iceberg," as shown in Figure 15.1.

Per ICH Q3A/B recommendations, regular impurities at a level above the reporting threshold are determined and accounted for in the total impurities and reported to the regulatory agencies. In this sense, the reporting threshold is the "sea level" (see Figure 15.1) and impurities at the level above it are the "visible" ones and form the tip of the impurity iceberg. In contrast, impurities at levels below the reporting threshold, although they can be significant in number and form the main body of the impurity iceberg, are "invisible" from a regulatory point of view. Because genotoxic impurities are considered to be unusually potent, not only the visible impurities but also many of the invisible impurities must be considered in the identification and assessment of genotoxic impurities. As a matter of fact, impurities below the reporting threshold but above the sTTC or TTC are the main objects of genotoxic impurity risk assessment.

However, because the sTTC/TTC can be orders of magnitude lower than the ICH Q3A/B reporting threshold, depending on the maximum daily dose (MDD), identification and assessment of these impurities for their genotoxic potential is a great challenge.

The concepts of actual impurity and potential impurity are of great importance in the pharmaceutical assessment of genotoxic impurities. Although mentioned in ICH Q3A and routinely used in the industry, the concept of actual impurities is not clearly defined in regulatory guidelines. In practice, actual impurities are typically considered to be those that are exceeding the reporting threshold because they are determined (hence, actual) and reported to regulatory agencies. This concept of actual impurity is now included in ICH M7 step 2 document [6]. ICH Q3A defines potential impurities as those "that theoretically can arise during manufacture or storage. It may or may not actually appear in the new drug substance." In routine application, this definition is qualitatively meaningful but lacks clarity in the quantitative aspect on "to what level the impurity can theoretically arise." One school of thought or

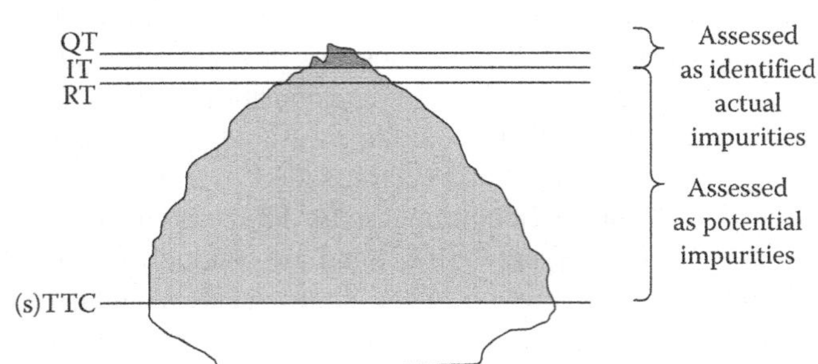

FIGURE 15.1 The iceberg of impurities. QT: ICH qualification threshold; IT: ICH identification threshold; RT: ICH reporting threshold; and (s)TTC: (staged) threshold of toxicological concern.

perhaps the mainstream thought is that potential impurities are those that can theoretically or potentially arise in a drug at levels exceeding the reporting threshold, that is, potential impurities are those predicted to have the potential to be actual impurities before actual impurity data are available. As such, often after the actual data are available a predicted potential impurity that is not actually observed above the reporting threshold can be no longer considered a potential impurity. This concept makes a lot of sense in the ICH Q3A/B paradigm, because for a regular impurity if it does not exceed the reporting threshold it is not even considered an impurity. Quite obviously, this concept is no longer appropriate for a genotoxic impurity because the threshold of concern for the genotoxic impurity is no longer the reporting threshold; rather, it is the TTC or sTTC. So, for a thorough assessment of genotoxic impurities a potential impurity should be reasonably expected to be at a level exceeding the TTC or sTTC. This logic is illustrated in Figure 15.1, where the impurities between (s) TTC and IT are labeled as "Assessed as potential impurities."

Based on their origin, two types of impurities, process-related impurities and degradation products, are differentiated and typically separately assessed for their genotoxic potential. The former refers to those that are introduced or formed during the drug substance synthetic process (e.g., starting materials, intermediates, reagents, and by-products) and is out of the scope of this chapter. The latter refers to the impurities that are formed due to the chemical degradation of the drug substance and drug product during manufacture, in use, and during storage. The purpose of this chapter is to provide an overview of the challenges and best practices in the control of genotoxic degradation products with a focus on the identification of actual and potential degradation products for assessing their genotoxic potential.

15.2 RISK OF GENOTOXIC DEGRADATION PRODUCTS

In general, drug degradation is not considered the major source of genotoxic impurities. As chemically reactive species are intentionally brought into the manufacturing process of a drug substance, most of the genotoxic impurities are process-related impurities. Table 15.1 provides a snapshot of all the Ames-positive impurities that we (Boehringer Ingelheim Pharmaceuticals, Inc. [BIPI, Ridgefield, Connecticut]) have encountered between 2005 and 2010. Among 37 structurally unique Ames-positive impurities in various drug substances and products, 30 were introduced or

TABLE 15.1
Occurrence of Ames-Positive Impurities (between 2005 and 2010) at BIPI

Origin	Number of Ames-Positive Impurities
Intermediate/starting material	16
Chemical reagent	4
Synthesis by-product	9
Solvent	1
Degradation product	7
Total	37

formed during the chemical synthesis of drug substances and only 7 were degradation products (and a few of these were also intermediates). Since 2010, the overall ratio of Ames-positive degradation products was significantly decreased as a result of increased diligence in avoiding the use of Ames-positive building blocks that can potentially end up in the final drug substance and be subsequently generated as Ames-positive degradation products (e.g., aromatic amines). As a matter of fact, no new genotoxic degradation product has been identified since 2010 among all the new BIPI new chemical entity (NCE) projects. This observation demonstrates that it is possible to significantly minimize the risk of genotoxic degradation products using the quality-by-design principles (i.e., quality should be built in by design) advocated in ICH Q8 [7] during the lead identification and optimization stages.

Nonetheless, for the following reasons a risk assessment of genotoxic degradation products is still necessary. First of all, due to the variety of drug molecules and the complexity of possible degradation pathways, it is not always possible to predict and avoid the risk of formation of genotoxic degradation products during lead identification and optimization. Second, unlike a process impurity, a degradation product once formed cannot be purged or removed. In other words, the consequence of a genotoxic degradation product can be severer than that of a genotoxic process impurity because the former can be much more challenging to control. Further, the level of a process impurity does not increase and sometime decreases (e.g., ethyl methanesulfonate [8]) during storage. In contrast, the level of a degradation product can increase with time, which means that the level of a genotoxic degradation product needs to be controlled not only at release of the drug substance or product but also during the whole shelf life of the product. Thus, understanding the chemistry and rate of formation of the degradation product is essential for its control.

15.3 GENERAL STRATEGIES FOR CONTROL OF GENOTOXIC DEGRADATION PRODUCTS

The most effective approach in controlling the risk of genotoxic degradation products is to prevent them from forming by engineering the chemical structure of the drug candidate during the lead identification and optimization stages. This can be achieved by proactively identifying and subsequently avoiding the use of genotoxic building blocks that can potentially be generated as degradation products during manufacture or storage. As shown in Figure 15.2, a building block (e.g., an intermediate in the penultimate step of the chemical synthesis of a drug molecule) that is reasonably expected to be released in whole or in part as a degradation product through hydrolysis or some other predictable degradation pathway is screened for alerting structures using an appropriate in silico tool (e.g., DEREK). If an alert structure is identified in the molecule, an Ames test is performed to further assess the genotoxic potential. If the Ames test is positive, this building block should be replaced. An example of the application of this approach is the screening of aromatic amine building blocks, as this type of molecule is potentially genotoxic and is a potential degradation product generated through the hydrolysis of the amide bond. Unfortunately, the application of this approach to types of degradation pathways

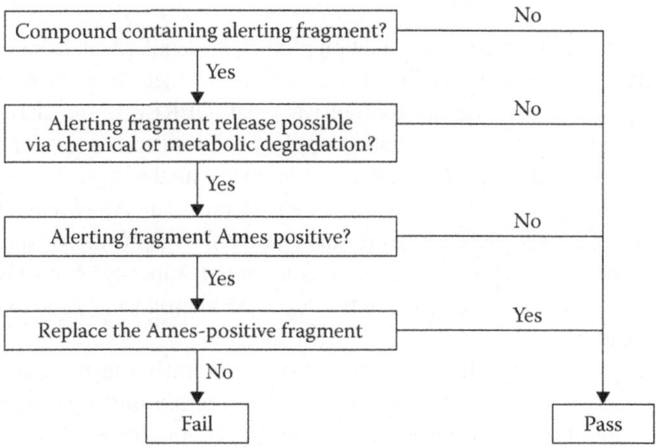

FIGURE 15.2 A quality-by-design approach for screening potential genotoxic building blocks that are expected to be degradation products.

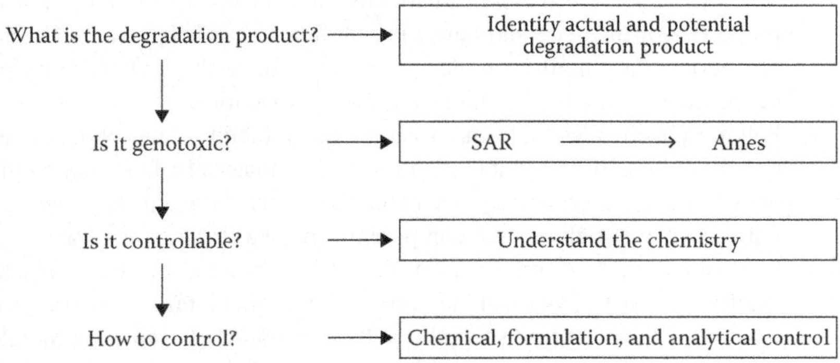

FIGURE 15.3 Workflow for risk assessment, remedy, and control of a genotoxic degradation product.

other than hydrolysis can be tricky because the degradation pathways may not be obvious and may not be readily identified without a thorough stability study, which during the lead identification/optimization stages is not practicable.

During clinical development or commercial stages the chemical structure of the drug molecule is fixed and, as such, modifying the chemical structure of the drug molecule to minimize the risk of generation of genotoxic degradation products is out of the question. To understand and control the risk of genotoxic degradation products, the following questions, as shown in Figure 15.3, should be answered:

1. What is the degradation product?

   As the first step of risk assessment, the structures of all actual and potential degradation products must be identified to allow further hazard assessment.

2. Is it genotoxic?

   Once the structure of a degradation product is identified, an in silico assessment should be performed to identify whether an alert structure is present by using an adequate in silico tool software (e.g., DEREK). If an alert structure is not identified, the degradation product is considered a regular impurity and controlled per ICH Q3A/B in accordance with the stage of development. On the other hand, if an alert structure is identified an Ames test is typically performed. A negative Ames result can overrule the in silico result, and the degradation can be recategorized as a regular impurity. A positive Ames result is sufficient to categorize the degradation product as genotoxic [3,4].

3. Is it controllable?

   A degradation product cannot be purged in a similar fashion as a process impurity; therefore, it is important to thoroughly understand the chemistry and the rate of formation of the degradation product. Because the drug molecule is fixed at this stage, the focus must be on understanding the conditions that can stabilize or destabilize the drug.

   Points of consideration for stabilization of the drug substance include but are not limited to the following:

   - Salt form: a salt form can dramatically alter the chemical and physical properties of the drug and hence its stability. For example, an HCl salt of a tertiary amine drug molecule will eliminate the susceptibility of the tertiary amine functional group toward oxidation.
   - Polymorphism: physical forms affect chemical stability. Amorphous materials are typically less stable than crystalline materials. Different forms of crystalline materials may also exhibit different chemical stabilities.
   - Particle size: not always but can potentially be a stability attribute.
   - Impurities: reactive impurities in the drug substance can be a critical stability attribute. For example, loose water (noncrystalline) is always a concern if the drug is susceptible to hydrolysis. Also, transition metals may mediate hydrolysis or oxidation. Generally, however, impurities are not a major mediator of degradation [9,10].
   - Light protection: the light sensitivity of the drug substance must be understood so that protection measures can be in place during manufacture, handling, and storage.
   - Packaging: conventional drug substance packaging comprises polyethylene plastic bags in a fiber drum. If necessary, more moisture or oxygen protective packaging can be used (e.g., aluminum pouch).

   When a drug substance is formulated into a drug product, the stability of the drug can be decreased. Hence, assessment of degradation and stabilization of a formulated product can be more critical than that of the drug substance. In addition to the points of consideration for the drug substance, stabilization of a formulated product includes but is not limited to the following:

   - Excipients: the impact of excipients on drug stability is multifaceted. On the one hand, certain excipients (e.g., antioxidants) can retard drug degradation; on the other hand, excipients may directly or indirectly (e.g., via trace-level reactive impurities in the excipients) interact with the drug.

- Coating: an appropriate coating can protect the drug from exposure to a destructive environment (e.g., moisture or oxygen).
- Container/closure/packaging systems: more complex container/closure/packaging systems can be considered for the drug product. (Note that leachables from container closure systems are drug product impurities, some of which can be potentially genotoxic. However, these are not drug degradation products and therefore are out of the scope of this chapter).
- Long-term storage conditions: chemical degradation typically decreases with a decrease in temperature. For example, changing the storage conditions from a controlled room temperature to refrigeration should typically significantly decrease the rate of formation of the degradation product, although the effect of temperature can vary with the mechanism of degradation.

4. With a thorough understanding of the mechanism of formation of the genotoxic degradation product and the possible chemical and pharmaceutical stabilization measures, an analytical control strategy can be devised. A specification should be established if necessary.

15.4 IDENTIFICATION OF DEGRADATION PRODUCTS FOR GENOTOXITY RISK ASSESSMENT

The first step in the risk assessment of genotoxic degradation products is the identification of chemical structures of actual and potential degradation products. All degradation products above the sTTC or TTC in the impurity iceberg should be considered. However, because the sTTC or TTC can be at parts-per-million levels, experimental elucidation of all those degradation products not only is impractical from a technology perspective but also contradicts the ICH Q3A/B guidelines, as these guidelines clearly state that only those degradation products above the ITs need to be identified.

The strategy for identification of a degradation product can be significantly impacted by the MDD of the product. This is because the IT is largely a constant value across a large range of MDDs, whereas the sTTC or TTC changes linearly with the MDD, as shown in Table 15.2 for TTC.

TABLE 15.2
Impact of MDD on the Concentration at TTC and ICH ITs

MDD (mg/day)	Concentration Limit at TTC (ppm)	Concentration Limit at TTC (%)	ICH DS ID Threshold (%)	ICH DP ID Threshold (%)
0.1	15,000	1.5	0.1	1.0
1	1,500	0.15	0.1	0.5
1.5	1,000	0.1	0.1	0.5
10	150	0.015	0.1	0.2
100	15	0.0015	0.1	0.2
1,000	1.5	0.00015	0.1	0.2

Note: DS ID, drug substance identification threshold; DP ID, drug product identification threshold.

When the MDD is 1.5 mg/day, the concentration limit (0.1%) at the TTC level (1.5 μg) is the same as the ICH IT for drug substances. Also, when the MDD is at or lower than 1.5 mg/day, the ICH Q3A ITs are the same as or even stricter (i.e., lower) than the concentration limit at the TTC. Therefore, the risk assessment for genotoxic degradation products can fully rely on the assessment of identified actual degradation products observed at release and during storage under recommended conditions.

With the increase of MDD, the concentration limit at TTC level becomes lower and can be orders of magnitude lower than the ICH ITs. Consequently, identification of only those degradation products that exceed the ICH ITs will not warrant a thorough genotoxic risk assessment of all actual and potential degradation products at release and during storage. In this situation, a chemistry- and risk-based strategy for the identification of potential degradation products should be applied.

15.4.1. Identification of Actual Degradation Products

A typical workflow for the identification and assessment of actual degradation products for genotoxic potential is described in Figure 15.4. Actual degradation products are typically experimentally determined to be present by high-performance liquid chromatography (HPLC)-based techniques in a drug substance or drug product at release and during long-term storage, so technically it is possible to elucidate the chemical structures by using advanced online structural analytical means (e.g., HPLC-MS).

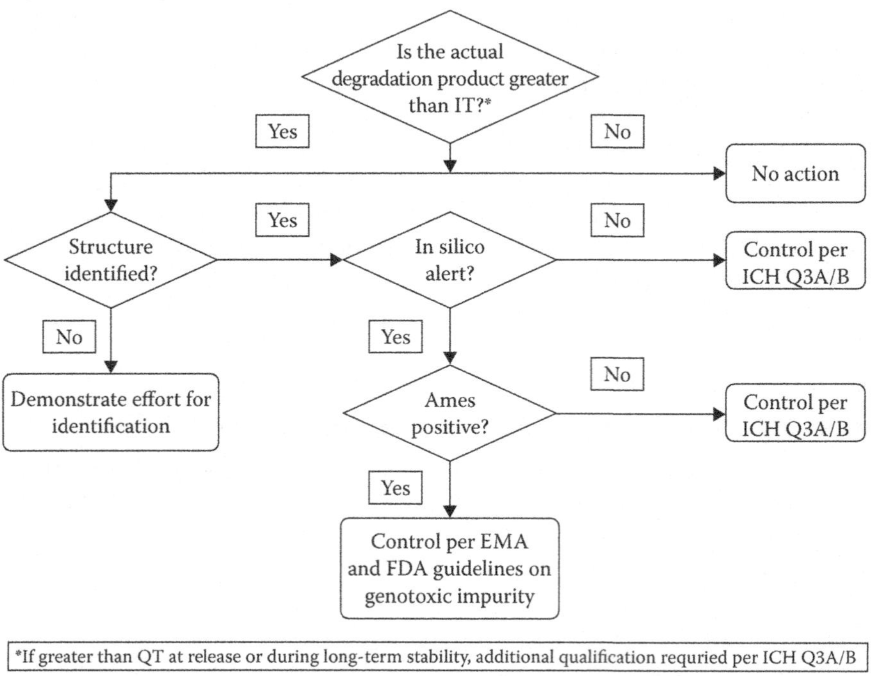

FIGURE 15.4 Workflow of identification and assessment of actual degradation products. QT: ICH qualification threshold.

With additional efforts, adequate amounts of the degradant of interest can be isolated for further elucidation (e.g., by nuclear magnetic resonance). However, the efforts should be appropriate according to the stage of development.

During late-stage clinical development or for commercial products, ICH ITs should be applied. As stated in the EMA questions and answers document #4:

Question: "What would be an appropriate strategy to qualify a new impurity that arises during Phase III or with a commercial product?"

Answer: "In line with the ICH guideline, no action is generally required for a new unidentified impurity found at levels below the ICH identification threshold."

During early clinical development stages the appropriate IT is not defined by regulatory agencies, and thus industry practices vary from company to company. Recently, a proposal put forth by an industry working group under the IQ Consortium on Early Development GMPs advocated a threefold ICH IT during early development stages (e.g., phase 1 to 2a) [11]. For example, if the IT intended for a drug at the registration stage is 0.1%, then 0.3% would be an appropriate threshold for the identification of actual unidentified impurities observed at release or during long-term stability studies during the early clinical development stages.

15.4.2 Identification of Potential Degradation Products

As discussed in Section 15.1, for risk assessment of genotoxic degradation products potential degradation products are considered that can theoretically and reasonably be expected to arise during manufacture or storage at levels exceeding the sTTC or TTC. Since the identification of these degradation products at release or directly from the drug stored under long-term storage conditions is impractical, approaches that can predict these potential degradation products must be employed.

15.4.2.1 Identification of Potential Degradation Products via Accelerated Stability Testing

Accelerated stability testing comprises "studies designed to increase the rate of chemical degradation or physical change of a drug substance or drug product by using exaggerated storage conditions as part of the formal stability studies" [12]. Since accelerated testing is part of the formal stability studies, the storage conditions are not arbitrary; rather, they are defined in ICH Q1A [12] according to the intended long-term storage conditions (e.g., 40°C/75% relative humidity [RH] is the accelerated condition for 25°C/60% RH long-term storage, and 25°C/60% RH is the accelerated condition for 5°C refrigerated long-term storage).

Because the accelerated storage condition is only 15°–20° higher in temperature than the long-term storage temperature, and is within the range of temperature excursion that might occur during shipment, the degradation pathways observed under accelerated conditions are typically considered highly relevant to those under the long-term storage conditions. Chemical degradations under the accelerated conditions can be several-fold faster than those under the long-term storage conditions due to the temperature effect or a combination of both temperature and humidity. The extent of acceleration of the degradation from the temperature effect is determined by the activation energy of the degradation pathway. Because of the wide variety of molecules, activation energies of drug degradation vary in a wide range from 12 to 40 kcal/mol [13–15]. Assuming that the activation energy is the same as that used by the *United States Pharmacopoeia* (83.144 kJ/mol = 19.8 kcal/mol) [16] to calculate the mean kinetic temperature, the rate of degradation at 40°C is about five times faster than that at 25°C [13]. Therefore, typically, a 6-month accelerated stability study should be highly predictive of at least 2 to 3 years of storage under long-term storage conditions. Degradation products actually observed during the accelerated stability studies are considered to be potential degradation products of the drug under the long-term storage conditions. Those above the ICH ITs (as adjusted for early development, as discussed in Section 15.4.1) should be identified and assessed for genotoxic potential.

15.4.2.2 Identification of Potential Degradation Products via Stress Testing

Stress testing is the "studies undertaken to elucidate the intrinsic stability of the drug substance" or "studies undertaken to assess the effect of severe conditions on the drug product" [12] and "is normally carried out under more severe conditions than those used for accelerated testing" [12]. Unlike long-term stability or accelerated stability studies whose conditions (e.g., temperature and RH) are clearly defined by ICH Q1A [12], stress testing covers a much broader range of conditions (e.g., temperature, humidity, pH, light, and oxidants), and the details of the conditions are not clearly defined by regulators. This lack of regulatory guidance on stress testing is challenging for the industry and, in general, the following principles should be considered:

- The purpose of stress testing is to predict the degradation/stability of the drug under long-term storage conditions, not for the sake of generating degradation products.
- The severity of the conditions should be justified and relevant.
- The end point of the testing should be determined by the level of degradation desired or by a "maximum duration." A maximum duration should be applied to every condition. Once the maximum duration is achieved, "no degradation" should be an acceptable outcome.

Nonetheless, because stress testing conditions are significantly harsher than the typical long-term and accelerated storage conditions for drug substances or drug products, it is expected that some degradation products that may not be formed under long-term and accelerated storage conditions can be formed under those severe stress conditions. For this reason, whether stress testing can be used for the prediction of potential degradation products remains a very controversial topic. It is the author's point of view that as a predicting tool it is unreasonable to expect that stress testing can accurately generate the exact degradation products as those formed under long-term stability conditions. With that said, thoughtfully designed stress testing should be expected to generate degradation products that cover all or the majority of all actual or potential degradation products that can be formed under long-term storage conditions [13,17]. Hence, a critical issue in using stress testing to predict potential degradation products is how to ensure that the focus is on those products that are more likely to be formed under long-term storage and exclude those that are less likely to be formed under long-term storage based on a chemistry-based risk assessment. In other words, the challenge is to avoid the overprediction of potential degradation products. Two important concepts must be clarified and understood: "relevant stress conditions" and "major degradation products."

Relevant stress conditions are those that are thermodynamically and kinetically relevant to long-term-storage and in-use conditions (e.g., 25°C/60% RH, indoor light). Several criteria can be used to assess the relevancy:

- What conditions most reveal the intrinsic stability of the drug?
- Are the conditions thermodynamically or kinetically relatable to the long-term storage conditions?
- Are the conditions relevant to in-use or physiological conditions?

Note that relevant conditions are not a set of absolute, generic conditions but instead vary from drug to drug and must be assessed on a case-by-case basis. Due to the variety of the drug molecules, it is not always clear which condition is more relevant than others; therefore, this assessment must be done on a case-by-case basis for setting meaningful criteria for structure elucidation.

For a solid oral formulation whose recommended storage is under controlled room temperature (25°C/60% RH), considerations regarding the relevancy of some of the most often used stress testing conditions are discussed in the following:

- Is 70°C/75% RH for 3 weeks a relevant stress testing condition for the drug substance?
  The answer is very likely dependent on the activation energy of the degradation. Assuming the drug is physically stable within the temperature range and the humidity has no effect on the degradation, the effect of temperature on the degradation can be estimated using the Arrhenius equation. The following is a hypothetical example: two degradation products A and B increased under long-term storage conditions from 0% to 0.05% (ICH reporting threshold) and 0.012% (120 ppm, which is the sTTC limit 120 μg for <1 month duration for a product with an MDD of 1 g/day), respectively, after 2 years of storage. These two degradation products bracket all invisible potential degradation products in the impurity iceberg that should be considered for the assessment of genotoxic degradation products for a phase 1 clinical study. Using the Arrhenius equation, the estimated calculated levels of the two degradation products after 3 weeks at 70°C/75% RH are shown in Table 15.3. From the estimated levels of the potential degradation products, the exaggeration of the degradation products by stress testing is at a reasonable level. Both the degradation products, if the activation energy is very low (e.g., 12 kcal/mol or 50.2 kJ/mol), are still expected to be below the ICH reporting threshold after 3 weeks under 70°C/75% RH. Degradation product B will not reach the level of ICH IT unless the activation energy is above 107.8 kJ/mol. Based on the author's experiences, only a very small percentage of drug substances can generate degradation products under

these stress conditions to a level above the ICH reporting threshold [8]. So the risk of overprediction of potential degradation products is acceptable, and the risk of underprediction of potential degradation products can be minimized as discussed in the following questions.

- Is 0.1 N HCl at 40°C a relevant condition?
  The answer may not be as straightforward as in the case of temperature and humidity stress testing on solid-phase samples. On the one hand, HCl stress testing is performed in solution, not in the solid state, and thus, quantitatively, it is difficult to extrapolate the stability from solution to that in the solid state. Even if the degradation pathway is the same, the activation energies could be different. The exaggeration of degradation in this condition is twofold. First of all, solution reactions are typically significantly faster than solid-state reactions; second, acid-catalyzed hydrolysis proceeds much faster under acidic conditions than under neutral conditions. On the other hand, the stability of the drug in HCl solution is a reflection of its overall stability toward acid-catalyzed hydrolysis. If the drug is susceptible to acid-catalyzed hydrolysis (e.g., containing an alkyl ester functional group), then the result of HCl stress testing can qualitatively be an indication of the susceptibility of the drug toward hydrolysis. Further, the pH value of the 0.1 N HCl solution is close to that of the gastric fluid and the temperature (40°C) is close to the physiological temperature (37°C) as well; therefore, the major degradation products that can be readily formed under this HCl stress testing could reasonably be expected to be formed at a certain level in the stomach after administration. Even if the stress testing is only done on a drug substance, the results can be relevant to drug product stability. For example, if the drug product is a powder for oral solution formulation, the drug substance will be dissolved in a reconstituted solution prior to administration. The pH value of the reconstituted solution will have a direct impact on the stability of the product. Therefore, whether this stress condition is relevant or not should be determined through a thorough assessment of the chemical structure, product, and observations during the stress testing.

- Is 0.1 N NaOH at 40°C a relevant condition?
  In general, this condition is considered less relevant than the 0.1 N HCl stress testing because the pH values of physiological systems mostly range from acidic to neutral. Furthermore, in many cases basic pH stress sometimes induces base-catalyzed oxidations that will typically not happen without the presence of a base. However, one should not automatically dismiss this condition without a thorough assessment of the specific situation. For example, some drug substances are basic salt forms (e.g., Na salt). In the presence of moisture, base-catalyzed hydrolysis can occur if the structure of the drug permits it. If the base (e.g., NaOH) is overcharged during the process of drug substance manufacturing, base-catalyzed hydrolysis can be more significant.

- Is the stress condition 0.3% H_2O_2 at ambient temperature a relevant condition?
  This stress testing condition is considered by many to be extremely severe and unrealistic because the drug substance will never have opportunities to be exposed to H_2O_2. However, whether it is a relevant condition or not totally depends on the drug structure and formulation. H_2O_2 is most sensitive and selective in oxidizing tertiary amine or sulfoxide to *N*-oxide or sulfone, respectively. If the drug molecule contains such functional groups, significant levels of oxidation product can be expected to be formed under this stress testing condition. Even if these degradation products are not considered drug substance potential degradation products, they could be drug product potential degradation products, because it is well known that many excipients could contain trace level of peroxide impurities. H_2O_2 can also oxidize the activated C–H functional group to peroxide, which in turn can form alcohol or ketone. However, the predicting power of H_2O_2 testing for this type of oxidation is less sensitive than that for *N*-oxide formation; therefore, whether these degradation products are relevant or not depends on the level of the degradation and the chemical structure. H_2O_2 is also known to cause random oxidation of nonactivated functional groups to form alcohol, epoxide, or dehydrogenation products or secondary oxidation products (via multiple-step oxidations to add more than one oxygen atom to the drug), but these reactions are typically insensitive and less predictive of what can be expected under long-term recommended storage.

- Is the stress condition 3× ICH exposure [18] to the drug substance in solid state a relevant condition?
  This is a typical photostability study under forced degradation conditions as recommended by ICH Q1B. Because the level of the photoreaction product is mostly linear to the exposure to light, one can expect that the degradation product observed under the forced degradation conditions will also be formed at lower levels under the confirmatory condition. However, the significance of the photostability results is that they primarily provide information on precautions to be taken during manufacture, handling, and storage of the drug so that photodegradation can be avoided. In the context of genotoxic degradation products, only those formed or potentially formed in the final packaging are relevant.

- Is the stress condition 2× ICH exposure to the drug substance in solution a relevant condition?
  For a solid oral product, this test may not be directly relevant if the results are different from those of the photostability study conducted in the solid state. However, this result provides information about precautions to be taken on handling and storing the drug solution.

The major degradation products concept is another important concept in the assessment of potential degradation products using stress testing. In this context, a major degradation product formed under stress testing is a degradation product whose concentration is significant enough that its chemical structure should be elucidated to establish the degradation pathway and understand the intrinsic stability of the drug. In short, a major degradation product is any that needs to be identified. At what level should a degradation product formed during stress testing be identified? A Pharmaceutical Research and Manufacturers of America working group proposed the use of a staged threshold approach. Depending on the total degradation, four scenarios are differentiated: for total degradations of 1%–5%, 5%–10%, 10%–15%, and 15%–20%, degradation products above the levels of 0.25%, 0.5%, 0.75%, and 1.0%, respectively, are identified [19]. This approach attempts to make a connection between the IT and the level of total degradation, and from a risk–benefit point of view it is quite reasonable.

However, a drawback of this proposal is its complexity. In addition to the factors of total degradation, there are also factors relating to various daily doses, resulting in a total of 16 different ITs for various daily dose and percentage total degradation combinations [19]. Another approach proposed by Alsante and others [20] also uses relative thresholds: for active pharmaceutical ingredients during early clinical stages identify only degradation products above 25% of total degradation, whereas for other cases use 10% of total degradation. An obvious problem with this approach is that there is not an absolute limit. For example, if one degradation product is formed under 70°C/75% RH at 0.06%, since there is only one degradation product, it accounts for 100% total degradation. Because it is below the ICH IT, even if it is formed under the long-term storage condition it does not need to be identified.

Recently, Qiu and Cohen [21] advocated an approach that combined the ICH IT and a relative threshold based on the specific degradation condition. Two scenarios can be differentiated, as shown in Table 15.4.

As discussed earlier and shown in Table 15.3, assuming no physical changes, a degradation product formed under solid-state stress testing (e.g., 70°C/75% RH for 3 weeks) above the ICH IT may be a potential degradation product under the long-term storage condition (e.g., 25°C/75% RH). Degradation products formed in the solid state under 2× ICH light exposure in final packaging are relevant to those formed under the confirmatory testing condition; therefore, ICH IT should also be applicable. For solution stress testing, since both the number and the level of degradation products are much greater than those of solid-state stress testing, a relative threshold is necessary to limit the identification to only those that are relatively significant, in addition to an absolute threshold. In this approach, degradation products

TABLE 15.3
Predicted Levels of Degradation Product under Stress Testing Conditions

	Degradant A		Degradant B	
E_a (kJ/mol)	After 2 years at 25°C (%)	After 3 weeks at 70°C (%)	After 2 years at 25°C (%)	After 3 weeks at 70°C (%)
50.2	0.05	0.02	0.012	0.005
83.14	0.05	0.12	0.012	0.028
107.8	0.05	0.43	0.012	0.104
124.6	0.05	1.06	0.012	0.25

formed during stress testing should never have to be identified if they are below the ICH ITs. This principle should also be applicable for early clinical development stages when an IT higher than that prescribed in ICH Q3A/B is applied.

The overall process for the identification and assessment of a potential degradation product through stress testing can be described as shown in Figure 15.5.

TABLE 15.4
Major Degradation Products Formed under Stress Testing

Solid-State Stress Testing		
Heat/Humidity	**Photostability/Forced Degradation**	**Solution Stress Testing**
>ICH IT	>ICH IT	>ICH IT and >10% of total degradation

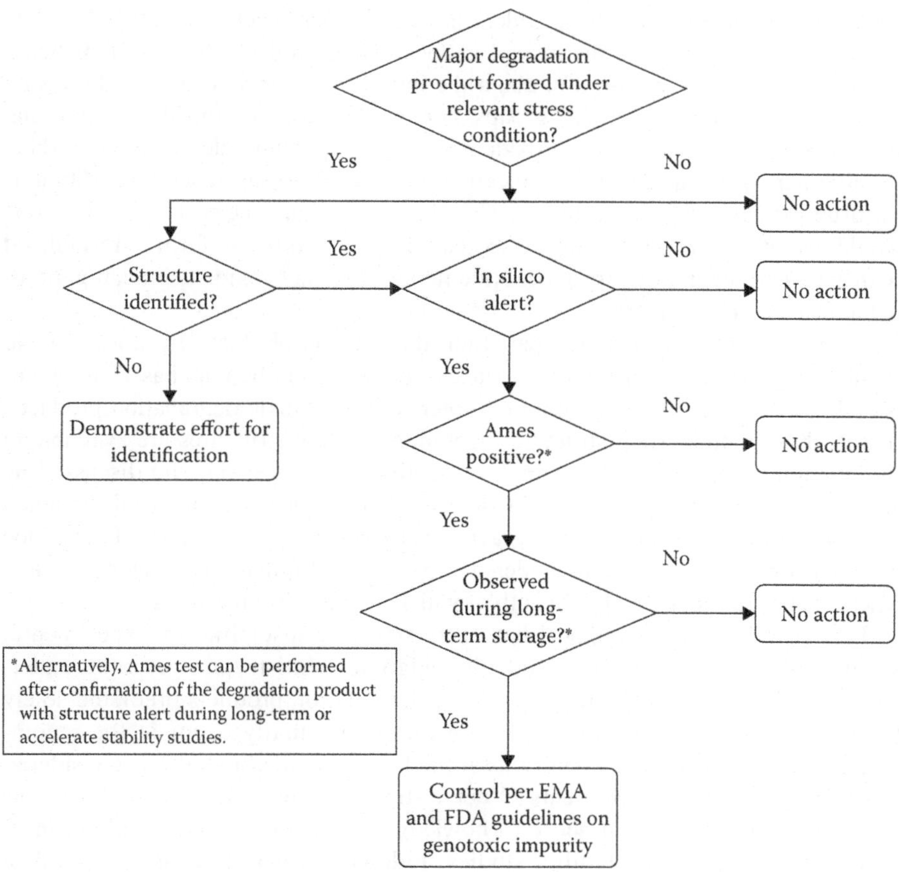

FIGURE 15.5 Workflow of identification and assessment of potential degradation products through stress testing.

15.4.2.3 Prediction of Potential Degradation Products Using Chemical Knowledge

Chemical degradation of drugs occurs as a result of the inherent reactivity of functional groups in the drug molecule. In addition, environmental conditions (e.g., temperature, humidity, light, pH, and interactions with other reactive chemicals) can significantly alter the rate of chemical degradation. Consequently, a thorough understanding of the reactivity of functional groups and the impact of stress testing conditions on degradation can help to predict certain degradation products. A great body of knowledge on the chemical reactivity of various functional groups exists in the scientific literature or in internal proprietary sources [21–24]. For example, an ester bond in a pharmaceutical molecule can be reasonably expected to undergo hydrolysis; hence, hydrolysis products may be considered as potential degradation products. An alkyl 3° (tertiary) amine is prone to *N*-oxide formation when exposed to peroxide; therefore, *N*-oxide may be a potential degradation product of a formulated product because many excipients can contain trace levels of peroxides as impurities. These examples demonstrate that the inherent reactivity of the functional groups is the root cause of drug degradation; hence, identification of such reactive functional groups ("hot spots") will enhance the overall understanding on drug stability. This being said, one should also be mindful about the impact of environmental conditions on the reactivity of these functional groups. For example, for the same drug containing the 3° amine functional group *N*-oxide may not be a potential degradation product, because peroxide impurities are typically rare in the drug substance. In addition, if the drug substance is a salt form (e.g., HCl salt) of 3° amine, the protonated 3° amine would lose its susceptibility toward oxidation. Even if this salt form is formulated with excipients containing trace peroxide impurities, the *N*-oxide degradation product may not form any more.

Note that under the ICH Q3 paradigm degradation products are not a critical regulatory concern at early clinical stages; therefore, predictions based on chemistry knowledge should be focused on potentially genotoxic degradation products, that is, those with alerting structures. Common alerting structures are available in the literature [25], and some examples of pathways of formation are discussed by Baertschi and others [19]. Figure 15.6 depicts a hypothetical drug molecule (denoted by 1), which could be potentially degraded via two pathways: hydrolysis of the amide bond forming an aromatic amine (denoted by 3) and elimination of water to form a Michael acceptor (denoted by 2), both of which are alerting structures.

It is important to point out that chemistry is a discipline of experimental science. Many aspects in solid-state degradation chemistry are yet to be understood. One must exercise caution with the degradation products predicted solely based on chemistry because only those that can actually occur at the recommended long-term storage conditions at the level of concern are to be considered as potential degradation products. When a structure alert is identified in a potential degradation product based on knowledge of chemistry, its formation must be confirmed by actual stability studies, following a workflow as described in Figure 15.7.

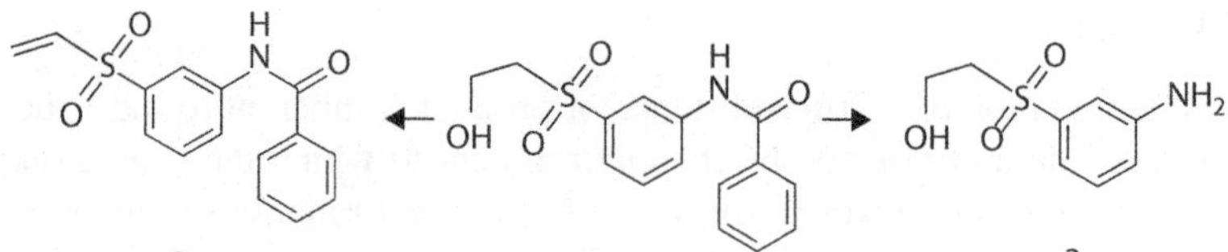

FIGURE 15.6 Predicted formation of potential degradation products with alerting structures from a hypothetical drug molecule.

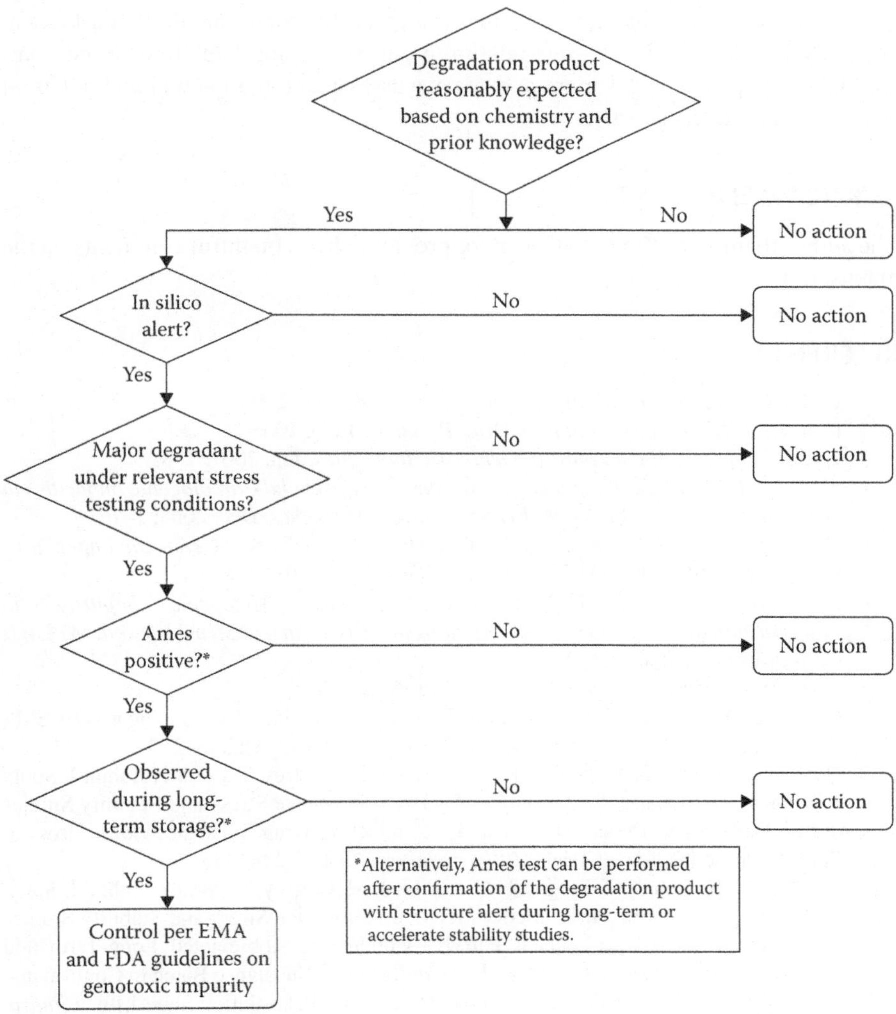

FIGURE 15.7 Workflow of identification and assessment of potential degradation products based on chemistry knowledge.

CONCLUSION

Chemical degradation of a drug substance or product is inherent to the structure of the drug molecule and affected by the environment to which the drug is exposed. Obviously, the most effective approach to prevent the formation of genotoxic degradation products is during the lead identification/optimization stage—if possible without compromising the safety and therapeutic potency of the compound itself—to exclude from the drug candidate the building block that can be generated or modified as a genotoxic degradation product during manufacture or storage. During early clinical development stages, when long-term and accelerated stability results are very limited, identification of potential degradation products relies on chemistry-based risk assessment and stress testing; therefore, the focus should be on known, or highly likely and relevant, degradation products. During late clinical and commercial development stages, a more extensive assessment of all actual and potential degradation products is warranted.

ACKNOWLEDGMENT

The author thanks Dr. Kenneth Cohen for providing his thoughtful comments on the manuscript.

REFERENCES

1. ICH. *Q3A (R2) Impurities in New Drug Substances*, Oct, 2006.
2. ICH. *Q3B (R2) Impurities in New Drug Products*, June, 2006.
3. EMA. *Guideline on the Limits of Genotoxic Impurities*, Jan, 2007; 1–8.
4. FDA. *Draft Guidance: Guidance for Industry Genotoxic and Carcinogenic Impurities in Drug Substances and Products: Recommended Approaches*, Dec, 2008; 1–16.
5. EMA. *Questions and Answers on the 'Guideline on the Limits of Genotoxic Impurities'*, EMA/CHMP/SWP/431994/2007, Revision 3, 23 Sep, 2010; 1–6.
6. ICH. *Assessment and Control of DNA Reactive (Mutagenic) Impurities in Pharmaceuticals to Limit Potential Carcinogenic Risk, Current Step 2 Version*, M7 Draft Consensus Guideline, Feb, 2013.
7. ICH. *Q8(R2), Pharmaceutical Development*, Aug, 2009.
8. C. Gerber, H. Toelle. What Happened: The Chemistry Side of the Incident with EMS Contamination in Viracept Tablets. *Toxicol Lett.* 2009; *190* (3): 248–253.
9. Q. Chan Li, F. Qiu, K. Cohen, T. Tougas, J. Li, J. McCaffrey, T. Purdue, Jinhua J. Song, F. Swanek, S. Abelaira. Best Practices for Drug Substance Stress and Stability Studies during Early-Stage Development Part I—Conducting Drug Substance Solid Stress to Support Phase Ia Clinical Trials. *J Pharm Innov.* 2012; *7*: 214–224.
10. Q. Chan Li, K. Cohen, T. Tougas, F. Qiu, J. Li, J. McCaffrey, T. Purdue, Jinhua J. Song, F. Swanek, S. Abelaira. Best Practices for Drug Substance Stress and Stability Studies during Early-Stage Development Part II—Conducting Abbreviated Long-Term and Accelerated Stability Testing on the First Clinical Drug Substance Batch to Confirm and Adjust the Drug Substance Retest Period/Powder for Oral Solution Shelf Life. *J Pharm Innov.* 2013; *8*: 56–65.
11. M. Coutant, Z. Ge, J. S. McElvain, S. A. Miller, D. O'Connor, F. Swanek, M. Szulc, et al. Early Development GMPs for Small-Molecule Specifications: An Industry, Perspective (Part V), *Pharm Technol.* 2012; *36* (10): 86–94.

12. ICH. *Q1A(R2), Stability Testing Of New Drug Substances And Products*, Feb, 2003.
13. S. W. Baertschi, P. J. Jansen, K. M. Alsante. Stress Testing: A Predicting Tool, in: S. W. Baertschi, K. M. Alsante and R. A. Reed (Eds), *Pharmaceutical Stress Testing*, 2nd ed., Informa, New York, 2011, pp. 10–49.
14. K. A. Connors, G. L. Amidon, V. J. Stella. *Chemical Stability of Pharmaceuticals, A Handbook for Pharmacists*, 2nd ed., Wiley, New York, 1986, p. 24.
15. G. Scrivens. Mean Kinetic Relative Humidity, A New Concept for Assessing the Impact of Variable Relative Humidity on Pharmaceuticals, *Pharm Technol.* 2012; *36* (11): 52–57.
16. USP. General Chapter <1150>, Pharmaceutical Stability, USP35–NF30, Rockville, MD, 2013, p. 763.
17. D. W. Reynolds, K. L. Facchine, J. F. Mullaney, et al. Available Guidance and Best Practices for Conducting Forced Degradation Studies, *Pharm Technol.* Feb, 2002; *26* (2): 48–56.
18. ICH. *Q1B Stability Testing: Photostability Testing of New Drug Substances and Products*, Nov, 1996; 1–12.
19. S. W. Baertschi, D. DeAntonis, A. P. Mckeown, J. Bercu, S. Raillad, C. M. Riley. Stress Testing As a Predictive Tool for the Assessment of Potential Genotoxic Degradants, in: S. W. Baertschi, K. M. Alsante and R. A. Reed (Eds), *Pharmaceutical Stress Testing*, 2nd ed., Informa, New York, 2011, pp. 484–498.
20. K. M. Alsante, A. Ando, R. Brown, et al. The Role of Degradant Profiling in Active Pharmaceutical Ingredients and Drug Products, *Adv Drug Deliver Rev.* 2007; *59*: 29–37.
21. F. Qiu, K. Cohen. Control of Genotoxic Impurities in Pharmaceutical Products, in: J. Swarbrick (Ed), *Encyclopedia of Pharmaceutical Science and Technology*, 4th ed., Informa, New York, 2013, in press.
22. F. Qiu, D. L. Norwood. Identification of Pharmaceutical Impurities, *J Liq Chromatogr R T.* 2007; *30*: 877–935.
23. S. W. Baertschi, K. M. Alsante. *Stress Testing: The Chemistry of Drug Degradation in Pharmaceutical Stress Testing*, S. W. Baertschi (Ed), Taylor & Francis, New York, 2005, pp. 51–140.
24. K. C. Waterman, R. C. Adami, K. M. Alsante, et al. Hydrolysis in Pharmaceutical Formulations, *Pharm Devel Technol.* 2002; *7* (2): 113–146.
25. L. Müller, R. J. Mauthe, C. M. Riley, et al. A Rationale for Determining, Testing, and Controlling Specific Impurities in Pharmaceuticals That Possess Potential for Genotoxicity, *Reg Tox Pharm.* 2006; 44: 198–211.

16 Identification and Qualification of Genotoxic Impurities as Leachables in Drug Products

Douglas J. Ball, William P. Beierschmitt, Krista L. Dobo, and Cynthia J. Magee

CONTENTS

16.1 INTRODUCTION

A potential source of genotoxic impurities (GTIs) arises from chemicals that migrate (leach) from the materials used in packaging the components of the container closure system (CCS). Process-related impurities often have structural similarities to active pharmaceutical ingredients (APIs) and therefore may possess a similar genotoxic profile to the APIs. Alternatively, leachables have no structural similarity to and typically have toxicity and genotoxicity potential significantly different from the APIs. Specific to genotoxicity, chemicals that leach into drug products that are known mutagens and/or human carcinogens (e.g., polynuclear aromatics [PNAs] and nitrosamines) may require more conservative levels of control than process-related drug impurities in the final drug product. For these reasons, leachables are considered out of the scope of the International Conference on Harmonisation of Technical Requirements for Registration of Pharmaceuticals for Human Use (ICH) impurity guideline Q3B(R2) for drug products.[1]

Although there is limited regulatory guidance on how to qualify leachable GTIs in drug products, not all of them carry the same risks for leachable toxicity. Current guidance from the U.S. Food and Drug Administration (USFDA) provides a table of drug products ranging from high (e.g., orally inhaled and parenteral drug products) to low concern (e.g., oral tablets) for risk of leachable toxicity as well as general guidance on basic risk assessment.[2] In lieu of having detailed regulatory guidance, the Product Quality Research Institute (PQRI) developed and published best practice recommendations for the identification and qualification of leachables in orally inhaled and nasal drug products (OINDPs) that established a safety concern threshold (SCT), which is based on the carcinogenic potential of leachables. In turn, the SCT is used to calculate an analytical evaluation threshold, which determines the analytical limit for the detection and identification of leachables in OINDPs.[3] The concepts developed for OINDPs are being used to develop specific recommendations of parenteral and ophthalmic drug products (PODPs).[4] The use of available guidance (official and draft) and the PQRI recommendations are the basis for the identification and qualification of leachable GTIs in drug products discussed in this chapter.

16.2 CONTAINER CLOSURE SYSTEM EVALUATION PROCESS

The CCS of a drug product typically refers to both its primary and its secondary packaging materials. The former include materials that come in direct contact with the drug product (e.g., plastic or glass vials, single-use injection system, and metered-dose inhaler), whereas the latter are constituents such as labels and secondary/tertiary packaging (e.g., fiberboard box, shrink-wrap, multilaminate pouch, and shipping pallets) that directly or indirectly come into contact with the primary packaging. As a general rule, extractable and leachable studies performed on the components of a CCS should be as inclusive as possible. In following this recommendation, the individual components are first subjected to controlled extraction studies, using solvents of varying polarities tested at elevated temperatures, to determine what chemicals might migrate from the CCS into the drug product over time (i.e., the extractables). Subsequently, using the drug product formulation within its CCS, the chemicals that actually do migrate into the product (i.e., the leachables) are determined under more realistic shelf life conditions. Importantly, extractable and leachable studies should not be initiated until the development of the CCS and the drug's commercial formulation is well advanced, if not finalized, since a late-stage alteration in either could inadvertently introduce a novel leachable into the final product.

Unlike process-related impurities that are inherent to any drug substance or drug product, available regulatory guidance concerning assessing the safety of impurities in drug products originating from the CCS is limited.[2] Two tables that define which drug products have the highest level of concern for leachables (Table 16.1) and the type of testing and/or data that will be required to qualify the leachables in a drug product (Table 16.2) are provided in this guidance. A leachable evaluation is not typically required for drug products that are bucketed as medium–low and low–low relative to the route of administration and the likelihood of an interaction between the CCS and the drug product. What is necessary to conduct a toxicological evaluation, in particular how to conduct a genotoxic assessment, is not provided in the guidance. Second, the guidance places more emphasis on the toxicological evaluation of extractables and not leachables. The issue with this process is that all extractables are not leachables. Subsequently, and more importantly, the PQRI recommends a toxicological evaluation of leachables as these are the chemicals that will be found in the drug product.[3]

TABLE 16.1
Examples of Packaging Concerns for Common Classes of Drug Products

Degree of Concern Associated with the Route of Administration	Likelihood of Packaging Component–Dosage Form Interaction		
	High	**Medium**	**Low**
Highest	Inhalation aerosols and solutions Injections and injectable suspensions	Sterile powders and powders for injection Inhalation powders	
High	Ophthalmic solutions and suspensions Transdermal ointments and patches Nasal aerosols and sprays		
Low	Topical solutions and suspensions Topical and lingual aerosols Oral solutions and suspensions	Topical powders Oral powders	Oral tablets Oral (hard and soft gelatin) capsules

Source: From U.S. Department of Health and Human Services, Food and Drug Administration, Center for Drug Evaluation and Research, Center for Biologics Evaluation and Research, *Guidance for Industry: Container Closure Systems for Packaging Human Drugs and Biologics*, 1999.

TABLE 16.2
Safety Considerations for Common Classes of Drug Products

Route of Administration/Dosage Form	Safety Category	Required Safety Information
Inhalation aerosols and solutions; nasal sprays	1s	Typically provided are USP biological reactivity test data, extraction/toxicological evaluation, limits on extractables, and batch-to-batch monitoring of extractables
Injections; injectable suspensions; sterile powders; powders for injection; ophthalmic solutions and suspensions	2s	Typically provided are USP biological reactivity test data, and possibly extraction/toxicological evaluation
Topical delivery systems; topical solutions and suspensions; topical and lingual aerosols; oral solutions and suspensions; oral powders	3s	Typically, an appropriate reference to indirect food additive regulations is sufficient for drug products with aqueous-based solvents; drug products with non-aqueous-based solvent systems or aqueous-based systems containing cosolvents generally require additional suitability information
Topical powders; oral tablets and oral (hard and soft gelatin) capsules	4s	Typically, an appropriate reference to indirect food additive regulations is sufficient
Inhalation powders	Case 5s	Typically, an appropriate reference to indirect food additive regulations for all components except the mouthpiece for which USP biological reactivity test data are provided

Source: From U.S. Department of Health and Human Services, Food and Drug Administration, Center for Drug Evaluation and Research, Center for Biologics Evaluation and Research, *Guidance for Industry: Container Closure Systems for Packaging Human Drugs and Biologics*, 1999.

The point of origin of any impurity in a drug product does not abrogate the need to assess its safety if human exposure to it is anticipated. In terms of risk assessment of these chemicals, while determining the safety of leachables is intuitive, consideration of such data for extractables is also recommended. In this regard, the former helps to ensure the safety of the drug product, whereas the latter increases the likelihood that the materials being selected for the CCS are unlikely to yield a problematic leachable such as a GTI. Thus, concerning the individual constituents of any given CCS a coordinated strategy integrating compendial testing and the examination of safety data for both extractables and leachables has been recommended to ensure both product quality and patient safety.[5,6] Importantly, however, one must be cognizant of the test outcomes of compendial biological reactivity tests that are not designed to identify and/or manage this type of GTI issue, as described in Section 16.4.1.

16.3 COMPENDIAL TESTING: *UNITED STATES PHARMACOPOEIA* BIOLOGICAL REACTIVITY TESTS

It is important to note that biological reactivity tests[7,8] are required by many health authorities as a part of the safety qualification process of a CCS. These studies typically employ a standard extraction procedure, and the resulting mixture is tested in vitro for cytotoxicity or in vivo for lethality, irritation, or sensitization. Importantly, however, these in vitro and in vivo tests have limitations, since they are not designed to specifically identify safety issues such as the possible presence of a GTI in the extract being tested. Thus, although these tests still have merit, they should be performed in conjunction with controlled extraction and/or leachable studies to more fully and comprehensively characterize the safety profile of the CCS and determine the potential for GTI risk.

16.4 PQRI BEST PRACTICES

16.4.1 Safety Concern Threshold

For identification and qualification of leachables in OINDPs, the PQRI Leachables and Extractables Working Group proposed an SCT of 0.15 μg/day.[3,9] Below the SCT concern for carcinogenic or noncarcinogenic toxicity is negligible, and identification of leachables below this threshold generally would not be necessary. Carcinogenicity was used as the basis for SCT because carcinogenic effects typically occur at lower levels of intake than those associated with noncarcinogenic toxicity. This was previously demonstrated for orally administered compounds, including those with potent neurotoxicity, reproductive toxicity, or endocrine effects.[10]

Unlike the threshold of toxicological concern (TTC) for GTIs addressed by the European Medicines Agency,[11] the SCT for OINDPs is linked to an AET. Unlike the TTC, the SCT is not a control threshold and in many cases particular leachables can be qualified above the SCT. Second, the SCT differs from the TTC in that the total daily intake for leachables in OINDPs was set at 0.15 μg/day as opposed to 1.5 μg/day for process-related impurities. In this regard, the 10^{-6} or 1 in 1,000,000 risk level for OINDPs accounts for multiple extractables or leachable impurities, some with potential genotoxicity issues.[12]

Third, in addition to the nitrosamines and PNAs mentioned previously, Table 16.3 provides examples of chemicals with known genotoxic potential that have been extracted or observed as leachables from components of the CCS. These examples are provided by the PQRI PODP work team that assembled a list of approximately 600 chemicals that have been known to extract or leach from CCS components. Fourth, based on an in silico analysis, about 11% of the approximately 600 chemicals have positive structural alerts for genotoxicity. The in silico analysis validated that leachables can possess genotoxic potential and established the need for setting the AET at a level that will detect leachables that may have genotoxic potential (PQRI, unpublished data).

TABLE 16.3
Examples of GTIs Found in Container Closure Components

Chemical Name	CAS Number	Structure	Classification[a]
2-Methyl-aniline	95-53-4	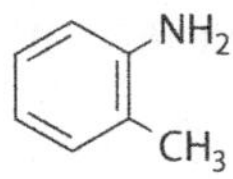	1
2,4-Dimethyl-aniline	95-68-1	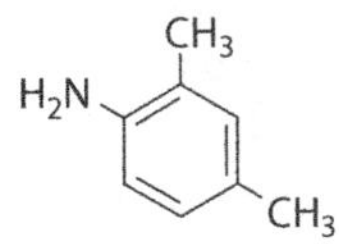	1
2,4-Toluenediamine	95-80-7	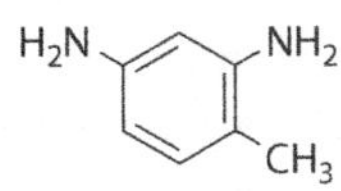	1
Bromoform	75-25-2	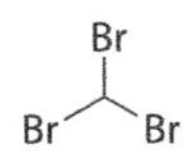	1
N-nitrosodimethylamine	62-75-9	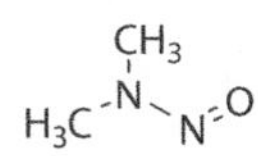	1
Aniline	62-53-3	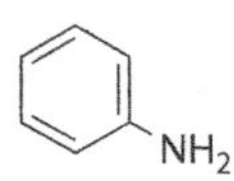	1
1,2-Dibromohexane	624-20-4	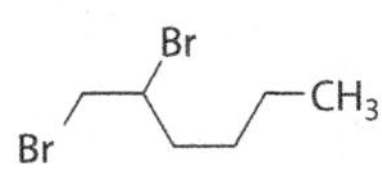	1
N-nitrosomorpholine	59-89-2	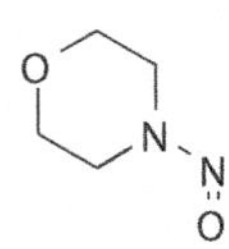	1
Benzo [*a*] pyrene	50-32-8	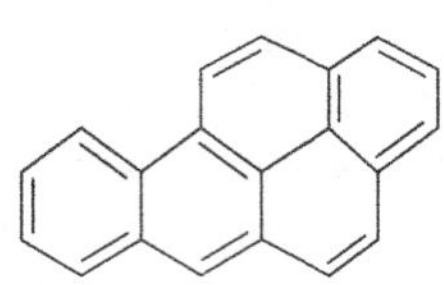	1
Dibenz [*a,h*] anthracene	53-70-3	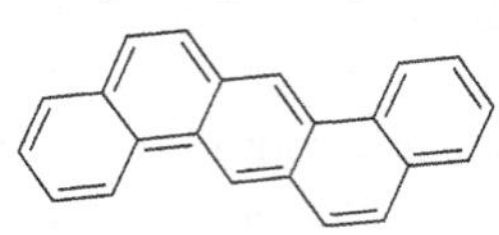	1
Acrylonitrile	107-13-1	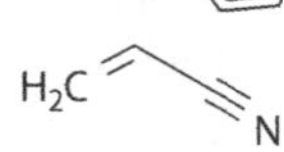	1
Butadiene	106-99-0	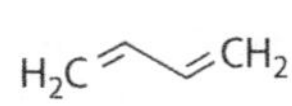	1

[a] See Table 16.5 for classification definitions.

For OINDPs, using a metered-dose inhaler with a solvent-based propellant in its formulation as a worst-case example, the PQRI determined that it is reasonable to use an analytical threshold linked to the 10^{-6} risk level as a starting point for identification and evaluation of leachable impurities. In contrast, this does not seem to be the case for PODP where many of the drug products are simple aqueous-based formulations and fewer leachables (in particular, genotoxic leachables) migrate from the CCS. For this reason, the PQRI is currently evaluating the translation of the SCT concept for PODP and is developing recommendations that would set the SCT at a daily intake at 1.5 µg/day.

As said earlier in Section 16.4.1, it is important to understand that the SCT is not a control threshold and that higher limits of leachables, including genotoxic leachables, may be acceptable under certain circumstances (Section 16.7.1).

For the remainder of this chapter, an SCT of 0.15 µg/day will be used to discuss the AET concept and assessment of leachables in drug products.

16.4.2 Analytical Evaluation Threshold

The AET is derived from the SCT and presents the limit as a concentration in drug products. Understanding the conversion of the SCT to the AET is necessary for the analytical chemist for practical application of the threshold concept in the laboratory. The maximum dose per day of the drug product is used in this conversion—at its most basic, the AET is the SCT divided by the maximum dose per day. In defining the maximum dose per day, it is important to take into account different drug product strengths, patient populations, and indications to ensure that the worst-case scenario is considered.[3]

For example, an intravenous pain treatment may be supplied in 2 mL vials containing 100 mg/mL of the drug product as well as in 1 mL vials containing 500 mg/mL of the drug product. A maximum dose of 1000 mg of the drug product could mean that five vials of lower strength (a total of 10 mL of the drug product) or two vials of higher strength (a total of 2 mL of the drug product) are dosed to a patient in a single day. Either scenario must be covered in a worst-case AET calculation. An SCT of 0.15 µg/day equates to an AET for a known genotoxic leachable of 0.015 µg/mL for the lower strength product (dosed at a maximum of 10 mL/day) or 0.075 µg/mL for the higher strength product (dosed at a maximum of 2 mL/day). To ensure proper control of the leachable in the drug product, analytical methodology will be needed to properly detect and quantitate the leachable down to 0.015 µg/mL.

It is apparent from this example that as the dose per day increases the AET decreases. For a large-volume parenteral, it is not unusual for the AET to be in parts per trillion. It becomes more critical for these products to have a thorough understanding of potential leachables such that analytical methodology may be targeted appropriately. For example, a single ion mass spectrometric monitoring method may provide the necessary selectivity and sensitivity for a known genotoxic leachable. The same method, however, will not be sufficient to show the absence of other leachables.

16.5 ANALYTICAL TESTING STRATEGIES

To determine material composition, the individual components of a CCS are first subjected to controlled extraction studies, using solvents of varying polarities tested at elevated temperatures, to determine what chemicals might migrate from the CCS into the drug product over time (i.e., the extractables). Subsequently, using the formulation of the drug within its CCS the chemicals that actually do migrate into the product (i.e., the leachables) are determined under more realistic shelf life conditions.

An extractables study can be designed to be either very aggressive, to provide maximum information about the packaging component, or less aggressive to be more predictive of leachables. Both types of studies can be valuable to the analytical chemist in learning about packaging and developing appropriate leachables methods. For example, a very aggressive extractables study may result in the detection of a large antioxidant such as Irganox 1010, which would not have been detected in a milder extraction. The large antioxidant itself may be unlikely to leach into an aqueous formulation; however, it may have smaller breakdown products that are likely to become leachables. Understanding the presence of this antioxidant in the packaging will allow the correlation of smaller antioxidant-related leachables to the extractable, even if the antioxidant itself never leaches.

A thorough extractables study will include an analysis for volatile, semivolatile, and nonvolatile organic extractables as well as for inorganic extractables. Sample handling should ensure that the targets of interest are preserved. For example, reflux is a common extraction procedure and it is also common to concentrate extracts to ensure that a low AET is reached. However, a reflux that is followed by evaporation to dryness and reconstitution will significantly impact the retention of volatile and semivolatile analytes. Instead closed-vessel extraction may be more appropriate for volatiles, and any concentration steps should not be to dryness if semivolatiles are targeted. The methodology at this stage is generally nonvalidated screening methods using mass spectrometric detection.

A well-designed extractables strategy will need to account for the aggressiveness of the extraction conditions when considering the interface between the analytical chemist and the toxicologist. An aggressive study may be more useful in learning about the packaging component and its additives but may be less useful as a direct predictor of likely leachables; at times, aggressive studies may result in the detection of so many extractables that reporting down to the AET becomes impossible. It will likely be impractical for a toxicologist to assess all of the extractables in such a study; nevertheless, an assessment of the major extractables or extractables that are expected to become leachables based on their chemical properties (i.e., polarity) may still be valuable. In such an instance, a less aggressive extraction study, or a migration study designed to be more predictive of leachables, may be needed to bridge the extractables and the leachables work. In a migration study, an aqueous/organic mixture such as water/ethanol is used to simulate a drug product's ability to solubilize leachables.[3] Temperature conditions are generally milder than those in an extraction study. This simulated drug product matrix facilitates sample preparation and analysis and results in a more realistic list of potential leachables than an aggressive extraction study. This data will likely be more valuable for toxicological assessment.

An extraction study or a migration study that is considered predictive can be reviewed more closely for the presence of unknown extractables. It is often easier to identify an analyte at the extractables stage, in a simpler matrix where the concentration is higher, than in a leachables study. An unknown at 1000 times the AET in a water extract is likely a good target for identification work at that stage. On the other hand, a small peak in a less predictive solvent may not be worth the analytical investment to identify at the extractables stage.

The role of the extractables work is to inform the leachables targets and to look for early toxicological "red flags." The extractables work is an important piece of the overall strategy. The ultimate goal is, of course, accurate detection and quantitation of leachables in a drug product. These are the analytes that will result in actual patient exposure and ultimately require a complete toxicological assessment. Therefore, the leachables methodology is generally validated and incorporated into a stability study to allow testing over the shelf life.

Based on the extractables data, and discussions with a toxicologist, the analytical chemist develops and validates leachables methods. At times, it may be difficult to reach the AET for all target leachables. Depending on the leachables of interest, and the early toxicological assessment, it may be appropriate to validate an analytical method with a higher reporting threshold. However, if the toxicologist has indicated an alert for a given leachable, then the sensitivity of the analytical methodology down to the AET will be critical to the success of the leachables study. In extreme cases where drug product matrix complexities combined with low AETs make method development impossible, a rationale to use a placebo or a simplified drug product matrix may be justified. Ideally, leachables methods will also allow the reporting of unknown or unexpected leachables. Depending on the drug product matrix and AET, this will be more difficult for some products than others. If the extractables profile is well understood, the extractables data may assist in a justification to exclude the reporting of unknowns in leachables methods. Together, the extractables and leachables analytical work should form a comprehensive strategy that ensures that potential patient exposure to leachables is understood and reported to the toxicologist for assessment.

16.6 GENOTOXIC IMPURITIES: GENERAL PRINCIPLES

Over the past decade, the identification, qualification, and control of API impurities known or suspected to possess genotoxic potential have been subject to increased scrutiny by pharmaceutical companies and regulatory authorities. As a consequence, the topic has been the focus of several regulatory guidance documents[11,13,14] and numerous peer-reviewed publications. Most recently, the topic has been adopted for the development of an ICH guideline.[15] Although the focus of existing GTI guidance is on the identification and control of GTIs associated with the synthetic process, much of the framework that has been established is also useful to consider for GTIs that leach from pharmaceutical packaging. Some aspects of this framework are reviewed in Section 16.5.1.

16.6.1 Genotoxic Impurity: Definition

Given that the pharmaceutical synthesis of a drug substance requires the use of materials that are intrinsically reactive in nature, residual impurities may have the potential to react with cellular macromolecules. The impurities that have the potential to directly interact with DNA, causing mutations because a mutagenic mechanism of genotoxicity is assumed to possess a linear dose–response relationship (i.e., have no threshold for the effect) unless there is sufficient evidence to prove otherwise, are of particular concern. As a consequence, limiting human exposures to very low levels is necessary to minimize excess cancer risk. Genotoxic chemicals that are nonmutagenic typically have a mechanism of action associated with a threshold[16–20] and would not be expected to pose carcinogenic risk in humans as impurities in pharmaceuticals. It is important to note that there is compelling experimental evidence to indicate that threshold or sublinear dose–response relationships also exist for some mutagenic carcinogens.[21–26] However, for mutagenic carcinogens the burden of proof is on the pharmaceutical sponsor to provide sufficient evidence that this is the case for the specific compound of concern.

16.7 GENOTOXIC IMPURITY HAZARD IDENTIFICATION

The primary methods recommended for the identification of impurities with mutagenic potential are the observation of chemically alerting features and the Ames bacterial reverse mutation assay.[11,13,27] This approach is supported by a number of studies in which structurally alerting features and/or mutagenicity in the Ames assay have been shown to be highly predictive of nonthreshold genotoxic carcinogens.[28–30] Given that it is not possible to isolate and test every potential impurity in the Ames assay, it is often necessary to rely on quantitative structure–activity relationship (Q)SAR methods to predict the mutagenic potential of novel chemicals based on their chemical structure. Although existing regulatory guidance does not provide any specific recommendations on what constitutes an acceptable structure-based assessment, based on the current draft of ICH M7 it is likely that two complementary in silico methods and the use of expert knowledge will be required.[15]

Commercially available software packages such as Deductive Estimation of Risk through Existing Knowledge (DEREK) Nexus,[31] MC4PC,[32] and Leadscope Model Applier[33] are already widely used across the pharmaceutical industry to predict the mutagenicity of impurities. Each has strengths and limitations, and their performance has been extensively reviewed with no single system performing significantly better than another.[34–37] Two recent surveys of practical experience (including a total of 13 pharmaceutical companies) found that current practices employed for impurity assessments are highly similar.[38,39] All companies surveyed initiate in silico assessments by processing structures through a (Q)SAR system or systems (vendor and/or in house) to identify substructures that may confer mutagenic potential. The majority of companies complement the output of an in silico system with an expert knowledge approach, such as consideration of mutagenic structure–activity relationships, the structural similarity of an impurity to the respective API, database searches (public and/or proprietary), and in some cases consultation with medicinal chemists to better understand the impact of various substitutions on chemical reactivity. More importantly, the surveys found that the current practices used to make mutagenicity predictions deliver high negative predictivity[38,39] and sensitivity.[39] Summary results of the surveys, which support this conclusion, are presented in Table 16.4a and b. In Table 16.4a, the data collected from eight companies are combined.[38] For 566 compounds that were predicted negative based on (Q)SAR systems alone, 94% were negative in the Ames assay. For 408 compounds that were predicted negative using both (Q)SAR systems and expert evaluation, 99% were negative. This demonstrates that one can have a high degree of confidence when predicting that an impurity lacks mutagenic potential based on (Q)SAR analysis. Further, the level of confidence of a negative prediction is even greater when the results of the (Q)SAR analysis are subject to expert interpretation. Similar findings were observed in a more recent survey of five companies (Table 16.4b[39]), with negative predictive values ranging from 86% to 100%. In addition, the survey reported high sensitivity (80%–100%), which

TABLE 16.4a
(Q)SAR vs. (Q)SAR Plus Expert Knowledge: Summary of Concordance

Structural Assessment	Predicted Ames Result	Number of Compounds[a]	Number of Ames Positive[b] (%)	Number of Ames Negative[c] (%)
(Q)SAR	Negative	566	35 (6)	531 (94)
(Q)SAR + expert knowledge	Negative	408	5 (1)	403 (99)

Source: Dobo et al., *Reg. Toxicol. Pharmacol.*, 62(3), 449–55, 2012.

[a] Total number of compounds predicted negative.

[b] The number of compounds predicted negative that produced a positive response in the Ames assay.

[c] The number of compounds predicted negative that produced a negative response in the Ames assay.

TABLE 16.4b
Results for Complete Procedures as Used in Companies (DEREK) and a Second (Q)SAR Tool, Including Expert Knowledge

	Company A	Company B	Company C	Company D	Company E
Sensitivity	83% (115/138)	95% (37/39)	100% (37/37)	80% (46/57)	93% (67/72)
Specificity	47% (192/412)	62% (143/230)	1% (1/82)	47% (66/140)	81% (149/184)
Negative predictivity	89% (192/215)	99% (142/145)	100% (1/1)	86% (66/77)	97% (149/154)
Positive predictivity	34% (220/335)	30% (37/124)	32% (37/115)	38% (46/120)	66% (67/102)
Concordance	56% (307/550)	67% (180/269)	32% (38/119)	57% (112/197)	85% (216/254)

Source: Sutter et al., *Regul. Toxicol. Pharmacol.*, 67(1):39–52, 2013.

TABLE 16.5
Classification Scheme for Impurities in Drug Products

Class 1	Known to be both mutagenic and carcinogenic
Class 2	Known to be mutagenic, but unknown carcinogenic potential
Class 3	Structurally alerting compound, unrelated to the API and of unknown mutagenic potential
Class 4	Structurally alerting compound related to the API
Class 5	No structural alerts or sufficient evidence for absence of mutagenicity

Source: Müller et al., *Regul. Toxicol. Pharmacol.*, 44(3), 198–211, 2006.

provides confidence in the ability of current practices to identify impurities that possess mutagenic potential.

Once complete, the results of a structure-based assessment can be used to categorize impurities into one of five classifications, as defined in Table 16.5.[27] This classification system is useful as it helps to differentiate mutagenic impurities that require low-level control (classes 1 and 2) from those that are not mutagenic (classes 4 and 5). It also defines the impurities that would require an Ames test to determine their mutagenic potential (class 3).

16.8 ESTABLISHING ACCEPTABLE DAILY INTAKE: LIFETIME

The approach used to define an acceptable daily intake (ADI) for a mutagenic impurity requires case-by-case consideration and is dependent on the available information. In cases where both positive mutagenicity and carcinogenicity data are available (class 1 impurities), one should first consider the possibility of establishing a compound-specific ADI.[11,14,27] To determine the appropriate methodology to use to calculate a compound-specific limit, it is necessary to first consider what is known about the mechanism of mutagenicity and of carcinogenicity as well as the shape of the dose–response curve. Both the Committee for Medicinal Products for Human Use (CHMP) and the USFDA agree that for compounds with clear evidence of a threshold exposure levels without appreciable risk of carcinogenicity can be established according to the procedure outlined for class 2 solvents in the ICH Q3C guidance.[40] This established risk assessment approach calculates an ADI from the no-observed-effect level or the lowest-observed-effect level in the most relevant animal study. For mutagenic compounds, this would ideally be a carcinogenicity study. A weight adjustment factor and a number of safety factors are applied to the animal data to determine an appropriate human exposure.

For compounds without sufficient experimental evidence for a threshold-related mechanism, numerous methods are available to calculate a compound-specific ADI. Neither the CHMP[11] nor the USFDA[14] provides guidance on a particular technique. One cautious approach considers the tumorigenic dose evaluated in long-term cancer bioassays from the most sensitive species and sex. Linear extrapolation is performed from this dose to a dose level that attains an acceptable excess cancer risk in humans (1×10^{-5}).[11,13,14,27]

For mutagens with unknown carcinogenic potential (class 2) or for which there is insufficient carcinogenicity data to calculate a compound-specific ADI, it is recommended that a default exposure limit of 1.5 μg/day be applied.[11,13,14,27] This limit, referred to as the TTC, represents a daily exposure level for any unstudied chemical that will not pose a significant risk of carcinogenicity or other toxic effects.[41–43] The TTC was derived by taking into consideration the distribution of carcinogenic potencies of over 700 compounds and is an estimate of the daily exposure for most carcinogens that would be associated with an excess cancer risk not exceeding 1 in 1,000,000. In the context of food-contact materials, the TTC limit associated with the 10^{-6} excess cancer risk is defined as 0.15 μg/day.[42] However, for the application of TTC to GTIs in drug substances, a ten-fold higher limit, 1.5 μg/day, corresponding to a 1 in 100,000 lifetime risk of cancer, is considered justified as pharmaceuticals offer a benefit to patients.[11,14,27] There are some highly potent classes of mutagenic carcinogens (specifically *N*-nitroso compounds, azoxy compounds, and aflatoxin-like compounds) that may require even lower levels of exposures to ensure negligible cancer risk.[42]

It is worth noting that the TTC is considered to be a very conservative limit, as numerous worst-case assumptions were applied in the derivation of the limit.[42,44] For example, simple linear extrapolation was used starting from cancer bioassay TD_{50} values down to a dose associated with 1 in 1,000,000 cancer incidence. An assumption associated with using this simple extrapolation method is that all biological processes involved in the generation of tumors at high dosages are linear over a 500,000-fold range. In addition, none of the protective biological processes that are more likely to be effective at lower doses are taken into account. Therefore, it is reasonable to assume that the estimates of excess cancer risk are significantly overestimated.

16.8.1 Consideration for Higher Limits

The methods for the derivation of TTC and compound-specific ADIs mentioned in Section 16.8 are based on the premise of daily lifetime exposure to a mutagenic carcinogen for 70 years. However, very few pharmaceuticals will be prescribed in this manner, with many being used for much shorter durations. In addition, exposure to mutagenic impurities needs to be appropriately managed during clinical development, where depending on the drug and the stage of development exposures can range from a day to several years. Regulators have acknowledged that there are circumstances that warrant consideration of higher limits for mutagenic impurities, including investigational phases of clinical development and drug indications with less than lifetime exposures. Several different recommendations have been made to define specific durations of exposure and their respective ADIs.[13,14,27] The topic is also currently under discussion as part of the development of the ICH M7 guidance. Therefore, a single, harmonized view on acceptable limits for less than lifetime exposures may be forthcoming. The recommendation provided in the 2010 CHMP Q & A (R3) document for clinical development stages is provided as an illustration of what is commonly referred to as staged TTC limits (Table 16.6).

TABLE 16.6
ADI Limits Based on Duration of Exposure

	Duration of Exposure				
	Single Dose	≤1 Month	≤3 Months	≤6 Months	≤12 Months
Allowable daily intake (μg)	120	60	20	10	5

Source: CHMP Q&A (R3), European Medicines Agency, Committee for Medicinal Products for Human Use, *Questions and Answers on the Guideline on the Limits of Genotoxic Impurities*, 2010.

Allowing higher daily intakes of a mutagenic impurity for less than lifetime exposures is based on a stochastic mode of action, that is, the risk of cancer is dependent on the total cumulative dose of mutagenic carcinogens consumed over a lifetime.[45] In addition to less than lifetime exposures, there are other circumstances that may warrant consideration of higher limits. Several examples that have been mentioned in regulatory guidances include when a drug is used to treat a life-threatening condition, when life expectancy in less than 5 years, and when the impurity is a known substance and exposure to the substance occurs at higher levels from other sources.[11,14]

16.8.2 Assessing the Safety of Extractables

Subsequent to selecting materials for the CCS that have passed appropriate compendial testing, a safety evaluation of extractable data derived from the individual components is warranted. Since the extractable studies only identify chemicals that might migrate into a final product from a CCS, any analyses of their safety profile would typically not be submitted to regulatory authorities. Nevertheless, since the chemicals that actually do migrate into the final product from the CCS (i.e., the leachables) are typically a subset of the extractables, a safety evaluation at this stage represents a "window of opportunity" for a potential safety issue to be identified and mitigated. Thus, for the purpose of an internal decision-making process limiting the research to assess the most critical of potential safety concerns is appropriate. Toward this end, assessing each extractable for the toxicological end points of mutagenicity, carcinogenicity, sensitization, irritation, and adverse reproductive effects is an example of the most critical issues to be evaluated at this stage. Ideally, such an assessment should be literature based; but where little safety data are available for a particular extractable, an examination of its chemical structure for known toxicophores using in silico predictive computational platforms such as DEREK and/or MultiCASE is quite useful and recommended. As illustrated in Section 16.7, the use of in silico tools for prediction of mutagenic potential in particular can be very robust. Collectively, using this approach the data can first be assessed qualitatively to rule out as many of the safety end points as possible. If a particular extractable does have a potential liability, then a quantitative assessment should be employed, assuming the unlikely scenario that the largest amount of the chemical recovered (considering the results of all of the individual solvent systems) will occur in the leachable study. In making this unrealistic assumption, if the available data can be leveraged to mitigate the risk, no further action is necessary and development should continue. Importantly, as part of the mitigation process in this quantitative approach, the similarities and differences between the solvent conditions yielding a particular extractable and those that will be employed using the actual formulation of the drug in the leachable study should be considered. If such differences are great, the degree of concern that the extracted chemical will present in the subsequent leachable work and become a potential safety concern is greatly diminished. It is important to acknowledge that after following this approach, the rendering of a definitive safety position could still be hampered by an overall lack of data. Given this scenario, the toxicologist should determine if sufficient data are available from structurally similar compounds to develop a safety position in the event that the extractable in question subsequently presents as a leachable. Alternatively, addressing the data gap by conducting the appropriate in vitro or in vivo tests needed to rectify the deficiency is also an option, assuming that it is feasible to do so. Otherwise, if faced with continued uncertainty or an unacceptable potential safety issue at this stage a packaging change may need to be considered in the hopes of identifying constituents that may yield a more toxicologically acceptable extractable profile.

The extractable stage is the first opportunity to identify and mitigate a potential CCS-related GTI matter when developing a parenteral product. A mutagenic extractable (e.g., positive in the Ames test) should be duly noted, since in the subsequent leachable study it must be either absent or below the TTC of 1.5 μg/day (in some cases even lower for some very potent genotoxic carcinogens) to support a definitive safety position, unless the chemical has also tested negative for carcinogenicity in vivo, which would abrogate the need for such limitations. If the concern for mutagenicity is solely based on a computational structural alert, it must be considered valid unless a literature-based assessment or follow-up Ames testing can prove that the in silico prediction was invalid. It is important to note that control to the TTC is limited to DNA-reactive chemicals and is not inclusive of nongenotoxic animal carcinogens that elicit their effects via a threshold mechanism. In such cases, the establishment of appropriate quantitative safety margins derived from the animal studies where such effects were characterized is justified. Finally, prior to proceeding to the leachable studies it is very critical to consider both quality and regulatory expectations. In this regard, the overall quality of the product will be impacted by the number and types of chemicals that leach into it from the CCS during the migration studies. In addition, although a particular chemical may lend itself to a robust risk assessment, there may be regulatory sensitivities to its presence in a final product. Thus, if assessing the extractable profile yields a potential quality or regulatory concern, it is wise to consider mitigating it at this stage if at all possible. Overall, confidence at the extractable stage in matters related to safety, quality, and regulatory sensitivities greatly increases the likelihood that no issues of concern will arise when the final leachable profile is generated.

16.8.3 Assessing the Safety of Leachables

Since humans will be exposed to the analytes identified in the leachables study, a comprehensive regulatory-ready risk assessment should be developed for each of them. In most cases, the leachables will be a subset of previously identified extractables, but some exceptions do occur. In any case, a comprehensive review of all relevant toxicology data from published and database sources should be performed for each chemical to develop a risk assessment. Regarding the latter, the toxicologist should consider a wide range of different factors that will ultimately determine the safety of a particular leachable in a specific drug. Thus, leachable risk assessments should always be project specific, and what is acceptable in one drug product may not be acceptable for another. First and foremost, the maximum anticipated daily dose for each leachable should be calculated based on the highest concentration measured during the leachable study. Duration of patient exposure should be considered, since in general chronic daily exposure to potential toxins can possibly carry a higher theoretical risk of eliciting an adverse effect than short-term exposure. The route of exposure is important, since the toxicological profile of any chemical can be affected by how it is administered. Patient population and the drug's indication should also be considered when assessing potential risk. In this regard, a particular group of patients (e.g., pediatrics or women of childbearing potential) may be more sensitive to a leachable's adverse effect profile, and if the latter includes the kidney as a target organ then exposing patients who may already be predisposed to renal insufficiency should be a consideration. Also, if a drug is indicted to treat a life-threatening condition such as cancer a greater degree of latitude regarding the potential safety issues for any given leachable is acceptable. Overall, a risk assessment for a leachable, or any given chemical impurity, should be based on its dose and how, when, and to whom it will be administered to in the course of treating the patient with the intended pharmaceutical product.

Once the potential liabilities of a particular leachable are identified, the risk assessment is developed based on the various factors previously described. A leachable's known toxicity profile in animals and humans is compared to the anticipated patient dose, and the toxicologist determines if appropriate safety margins exist to support safety. If data are plentiful, this can be readily determined. Alternatively, if in some instances there is a paucity of data the toxicologist can employ safety factors to account for unknowns.[46] Overall, in the end if there is sufficient safety for the anticipated human dose the assessment is completed.

16.8.4 Case Study

An unknown was reported in a leachable study by gas chromatography–flame ionization detection. Since the analyte was unidentified, a surrogate standard was used to quantitate the analyte. A carefully chosen surrogate standard will have a response that is typical of analytes quantified by the method; a conservative approach will err on the side of overestimation but practicality necessitates some assumption of similarity in response. The unknown increased over the course of several months of drug product storage, crossing the AET and triggering identification work. Analysis of the samples by gas chromatography–mass spectrometry (GC-MS) resulted in a tentative identification of phenanthrene. Analysis of an authentic standard confirmed the identification.

The drug product was packaged in plastic vials with grey rubber stoppers. The drug product was also highly formulated with surfactants and cosolvents, which may have made the product more susceptible to leachables. Before the leachables study began, the vials and stoppers underwent controlled extraction studies—extraction of the packaging components in solvents of varying polarities followed by extensive analytical analysis by headspace gas chromatography, GC-MS, and high-performance liquid chromatography–mass spectrometry. The extraction report was revisited, and the analyte of interest was not reported in those studies despite the fact that they were thorough, and analytes were reported down to the AET. It was noted that the extraction study was conducted approximately 2 years before the leachables study was initiated and that different lots of packaging components were used.

As part of the investigation, new vials and stoppers were extracted and the extracts were injected on the leachables method. Phenanthrene and other PNA hydrocarbons were detected in the new extracts of the stoppers. This resulted in an investigation into the stopper supplier and their supply chain. An agreement was in place with the stopper supplier that any changes to their product or process would result in a notification. They, in turn, also had similar agreements in place with their suppliers.

The investigation eventually revealed a lot of carbon black, which was incorrectly supplied as "low PNA" carbon black even though it did not meet those requirements.

The error was a labeling mix-up at the carbon black supplier, a company that also supplied colorants for materials such as tires and automobile parts. Since it was not an intentional change, it had not triggered a notification to the stopper supplier.

Corrective actions included additional controls and checks at the carbon black supplier, as well as a release test for PNAs for incoming lots of black colorant at the stopper supplier.

16.9 CONCLUSION

Leachables from the CCS are inevitable in most, if not all, drug products. Analytical identification and subsequent qualification of leachables in certain drug product dosage forms is required based on current regulatory guidances. It is critical that analytical methods can identify leachables in drug products. More importantly, the analytical methods must be capable of identifying leachables that have genotoxic potential.

The use of the SCT and AET approach recommended by the PQRI should be able to identify leachables with genotoxic potential in drug products. Based on the PQRI recommendations, the acceptable limit of a leachable GTI is ≤0.15 μg/day for OINDPs, and ≤1.5 μg/day for other drug products. The ICH M7 Expert Working Group (EWG) is developing new guidance that will allow higher levels of GTIs in drug products, which are dependent on a number of factors, as discussed. Based on the current draft of this guidance,[15] the basic principles being developed by the EWG should be applicable to leachables in drug products.

REFERENCES

1. International Conference on Harmonisation of Technical Requirements for Registration of Pharmaceuticals for Human Use, *ICH Harmonised Tripartite Guideline: Q3B(R2) Impurities in New Drug Products*, 2006. http://www.ich.org/fileadmin/Public_Web_Site/ICH_Products/Guidelines/Quality/Q3B_R2/Step4/Q3B_R2_Guideline.pdf (accessed 24 March 2014).
2. U.S. Department of Health and Human Services, Food and Drug Administration, Center for Drug Evaluation and Research, Center for Biologics Evaluation and Research, *Guidance for Industry: Container Closure Systems for Packaging Human Drugs and Biologics*, 1999. http://www.fda.gov/downloads/Drugs/Guidances/ucm070551.pdf (accessed 24 March 2014).
3. Product Quality Research Institute Leachables and Extractables Working Group, *Safety Thresholds and Best Practices for Extractables and Leachables in Orally Inhaled and Nasal Drug Products*, 2006. http://www.**pqri**.org/pdfs/LE_Recommendations_to_FDA_09-29-06.pdf (accessed 24 March) 2014.
4. Product Quality Research Institute Leachables and Extractables Working Group, *Reporting and Qualification Thresholds for Leachables in Parenteral and Ophthalmic Drug Products*, 2007. http://pqri.org/commworking/minutes/pdfs/dptc/podpwg/Addl/PQRI_PODP_proposal_04072.pdf (accessed 24 March 2014).
5. Beierschmitt, W.P. Toxicology issues in extractables and leachables. *Amer Pharmaceu Rev* 2009, 12, 122–6.
6. Ball, D.J., Beierschmitt, W.P., Shaw, A. J. Pharmaceutical container closure systems: Selection and qualification of materials. In *Leachables and Extractables Handbook*; Ball, D.J., Norwood, D.L., Stults, C.L.M., and Nagao, L.M. Eds., 1st ed., John Wiley and Sons, Hoboken, New Jersey, 2012; pp. 217–40.
7. United States Pharmacopeia 37th Revision (USP 37). Chapter 87. Biological reactivity tests, in vitro. In *United States Pharmacopeia*, Rockville, MD, 2014.
8. United States Pharmacopeia 37th Revision (USP 37). Chapter 88. Biological reactivity tests, in vivo. In *United States Pharmacopeia*, Rockville, MD, 2014.
9. Ball, D., Blanchard, J., Jacobson-Kram, D., et al. Development of safety qualification thresholds and their use in orally inhaled and nasal drug product evaluation. *Toxicol Sci* 2007, 97(2), 226–36.
10. Kroes, R., Galli, C., Munro, A., et al. Threshold of toxicological concern for chemical substances present in the diet: A practical tool for assessing the need for toxicity testing. *Food Chem Toxicol* 2000, 38, 255–312.
11. European Medicines Agency, Committee for Medicinal Products for Human Use, *Guideline on the Limits of Genotoxic Impurities*, 2006. http://www.ema.europa.eu/docs/en_GB/document_library/Scientific_guideline/2009/09/WC500002903.pdf (accessed 24 March 2014).
12. Vogel, W.M. The development of safety thresholds for leachables in orally inhaled and nasal drug products. In *Leachables and Extractables Handbook*; Ball, D.J., Norwood, D.L., Stults, C.L.M., and Nagao, L.M. Eds., 1st ed., John Wiley and Sons, Hoboken, New Jersey, 2012; pp. 45–58.
13. CHMP Q&A (R3), European Medicines Agency, Committee for Medicinal Products for Human Use, *Questions and Answers on the 'Guideline on the Limits of Genotoxic Impurities'*, 2010. http://www.ema.europa.eu/docs/en_GB/document_library/Scientific_guideline/2009/09/WC500002907.pdf (accessed 24 March 2014).
14. U.S. Department of Health and Human Services, Food and Drug Administration Draft Guidance, *Guidance for Industry Genotoxic and Carcinogenic Impurities in Drug Substances and Products: Recommended Approaches*, 2008. http://www.fda.gov/downloads/Drugs/.../Guidances/ucm079235.pdf (accessed 24 March 2014).

15. International Conference on Harmonisation of Technical Requirements for Registration of Pharmaceuticals for Human Use, *ICH Draft Guideline M7, Assessment and Control of DNA Reactive (Mutagenic) Impurities in Pharmaceuticals to Limit Potential Carcinogenic Risk*, Step 3, 2013. http://www.ema.europa.eu/docs/en_GB/document_library/Scientific_guideline/2013/02/WC500139217.pdf (Accessed 24 March 2014).
16. Bergmann, K., Müller, L., Teigen, S. The genotoxicity and carcinogenicity of paracetamol—a regulatory (re)view. *Mutat Res* 1996, 349, 263–88.
17. Elhajouji, A., Lukamowicz, M., Cammerer, Z., et al. Potential thresholds for genotoxic effects by micronucleus scoring. *Mutagenesis* 2011, 26, 199–204.
18. Kirkland, D., Müller, L. Interpretation of the biological relevance of genotoxicity test results: The importance of thresholds. *Mutat Res* 2000, 464, 137–47.
19. Lynch, A., Harvey, J., Aylott, M., et al. Investigations into the concept of a threshold for topoisomerase inhibitor-induced clastogenicity. *Mutagenesis* 2003, 18, 345–53.
20. Müller, L., Kasper, O. Human biological relevance and the use of threshold arguments in regulatory genotoxicity assessment: Experience with pharmaceuticals. *Mutat Res* 2000, 464, 19–34.
21. Dobo, K., Fiedler, R., Gunther, W., et al. Defining EMS and ENU dose–response relationships using the *pig-a* mutation assay in rats. *Mutat Res* 2011, 725(1–2), 13–21.
22. Gocke, E., Ballantyne, M., Whitwell, J., et al. MNT and Muta™Mouse studies to define the in vivo dose response relations of the genotoxicity of EMS and ENU. *Toxicol Lett* 2009, 190(3), 286–97.
23. Gocke, E., Müller, L. In vivo studies in the mouse to define a threshold for the genotoxicity of EMS and ENU. *Mutat Res*, 2009, 678:101–7.
24. Jenkins, G.J., Doak, S.H., Johnson, G.E., et al. Do dose response thresholds exist for genotoxic alkylating agents? *Mutagenesis* 2005, 20, 389–98.
25. Johnson, G., Doak, S., Griffiths, S., et al. Non-linear dose–response of DNA-reactive genotoxins: Recommendations for data analysis. *Mutat Res* 2009, 678(2), 95–100.
26. Pottenger, L.H., Schisler, M.R., Zhang, F., et al. Dose-response and operational thresholds/NOAELs for in vitro mutagenic effects from DNA-reactive mutagens, MMS and MNU. *Mutat Res* 2009, 678, 138–47.
27. Müller, L., Mauthe, R.J., Riley, C.M., et al. A rationale for determining, testing, and controlling specific impurities in pharmaceuticals that possess potential for genotoxicity. *Regul Toxicol Pharmacol* 2006, 44(3), 198–211.
28. Ashby, J., Tennant, R.W. Chemical structure, *Salmonella* mutagenicity extent of carcinogenicity as indicators of genotoxic carcinogenesis among 222 chemicals tested in rodents by the US NCI/NTP. *Mutat Res* 1988, 204, 17–115.
29. Ashby, J., Tennant, R.W. Definitive relationships among chemical structure, carcinogenicity and mutagenicity for 301 chemicals tested by the US NTP. *Mutat Res* 1991, 257, 229–306.
30. Ashby, J., Paton, D. The influence of chemical structure on the extent and sites of carcinogenesis of 522 rodent carcinogens and 55 different human carcinogen exposures. *Mutat Res* 1993, 286, 3–74.
31. Marchant, C.A., Briggs, K.A., Long, A. In silico tools for sharing data and knowledge on toxicity and metabolism: DEREK for Windows, METEOR and Vitic. *Toxicol Mech Methods* 2008, 18(2–3), 177–87.
32. Saiakhov, R.D., Klopman, G. Benchmark performance of MultiCASE Inc. software in Ames mutagenicity set. *J Chem Inf Model* 2010, 50(9), 1521.
33. Valerio, L.G. Jr., Cross, K.P. Characterization and validation of an in silico toxicology model to predict the mutagenic potential of drug impurities. *Toxicol Appl Pharmacol* 2012, 260(3), 209–21.
34. Greene, N., Judson, P.N., Langowski, J.J., et al. Knowledge-based expert systems for toxicity and metabolism prediction: DEREK, StAR and METEOR. SAR QSAR. *Envir Res* 1999, 10(2–3), 299–314.

35. Hillebrecht, A., Muster, W., Brigo A., et al. Comparative evaluation of in silico systems for Ames test mutagenicity prediction: Scope and limitations. *Chem Res Toxicol* 2011, 24(6), 843–54.
36. Naven, R.T., Louise-May, S., Greene, N. The computational prediction of genotoxicity. *Expert Opin Drug Metab Toxicol* 2010, 6(7), 797–807.
37. Snyder, R.D., Pearl, G.S., Mandakas, G., et al. Assessment of the sensitivity of the computational programs DEREK, TOPKAT and MCASE in the prediction of the genotoxicity of pharmacuetical molecules. *Environ Mol Mutagen* 2004, 43(3), 143–58.
38. Dobo, K.L., Greene, N., Fred, C., et al. In silico methods combined with expert knowledge rule out mutagenic potential of pharmaceutical impurities: An industry survey. *Reg. Toxicol. Pharmacol* 2012, 62(3), 449–55.
39. Sutter, A., Amberg, A., Boyer, S., et al. Use of in silico systems and expert knowledge for structure-based assessment of potentially mutagenic impurities. *Regul Toxicol Pharmacol* 2013, 67(1), 39–52.
40. International Conference on Harmonisation of Technical Requirements for Registration of Pharmaceuticals for Human Use, *ICH Harmonised Tripartite Guideline: Q3C(R5) Impurities: Guideline for Residual Solvents*, 2011. http://www.ich.org/fileadmin/Public_Web_Site/ICH_Products/Guidelines/Quality/Q3C/Step4/Q3C_R5_Step4.pdf (accessed 24 March 2014).
41. Kroes, R., Kozianowski, G. Threshold of toxicological concern (TTC) in food safety assessment. *Toxicol Lett* 2002, 127, 43–6.
42. Kroes, R., Renwick, A.G., Cheeseman, M., et al. Structure-based thresholds of toxicological concern (TTC): Guidance for application to substances present at low levels in the diet. *Food Chem Toxicol* 2004, 42, 65–83.
43. Munro, I.C., Kennepohl, E., Kroes R. A procedure for the safety evaluation of flavouring substances. *Food Chem Toxicol* 1999, 37:207–32.
44. Delaney, E. An impact analysis of the application of the threshold of toxicological concern concept to pharmaceuticals. *Regul Toxicol Pharmacol* 2007, 49, 107–24.
45. Bos, P.M.J., Baars, B., Marcel, T.M. et al. Risk assessment of peak exposure to genotoxic carcinogens. *Toxicol Lett* 2004, 151, 43–50.
46. Dourson, M.L., Felter, S.P., Robinson, D. Evolution of science-based uncertainty factors in noncancer risk assessment. *Regul Toxicol Pharmacol* 1996, 24, 108–20.

17 Pharmaceutical Industry Survey on Genotoxic Impurities and Related Topics

William A. Yakush and Heewon Lee

CONTENTS

An area of particular interest during the development of human pharmaceutics is the control of potential genotoxic impurities (PGIs) and genotoxic impurities (GTIs). This chapter provides an overview of pharmaceutical industry practices relating to the control of PGIs and GTIs during pharmaceutical development as gleaned from responses by pharmaceutical company representatives to benchmarking questions conducted annually over the period 2008–2012, as shared by the API Development GMP Forum (API Forum) participants. We gratefully acknowledge the Chair of the 2012 API Development GMP Forum Mr. Dennis O'Connor for bringing our request to use the benchmarking responses to the company representatives and the representatives for agreeing to allow the use of the information in an anonymous manner.

The API Development GMP Forum is an informal, interactive conference organized by individuals in the pharmaceutical industry who focus on the application of current good manufacturing practices (GMPs) during active pharmaceutical ingredient (API) development. The API Forum seeks to provide an opportunity for individuals involved in clinical development to benchmark systems, procedures, facilities, and philosophies that may be impacted by GMP regulations. Participation in the forum is targeted on API development, including associated quality assurance and analytical sciences personnel who are closely involved with the philosophies, procedures, and systems in API development.

The API Forum is held once a year over a 2-day time frame typically in early fall. Each year, an individual volunteers to chair a steering committee of interested individuals from bio/pharmaceutical companies who then represent their company and plan the conference. The goal of the API Forum is to share information through a series of roundtable breakout discussions, presentations by subject matter experts (SMEs), and benchmarking exercises. To accomplish this, the steering committee identifies volunteers to moderate each of the breakout sessions and SMEs to present "hot topics." By design, the API Forum is informal to encourage networking interactions between participants and thus is generally limited to about 80 participants so that the roundtable discussions are typically small with no more than 20 participants in each group. The representatives of companies who participated in the API Development GMP Forum over the survey period of 2008 through 2012 are presented in Table 17.1.

The survey topics and responses should be of interest to anyone interested in the control of PGIs and GTIs encountered during the development of pharmaceuticals. Responses to all survey questions are reported in aggregate so as to maintain the confidentiality of each company's responses, that is, no single company is associated with a particular response. Note, due to the nature of the survey and reporting of responses, that it is difficult to quantify the responses. To give relative weight to responses, in cases where the same response was shared by a majority of participants it is noted as a "majority" and where the same response was shared by more than one company it is noted as "multiple." A list of acronyms used throughout this chapter is provided in Table 17.2.

In Section 17.1, responses from the 2008 through 2012 surveys are presented, which illustrate the evolving dynamic as the industry struggled to find efficient approaches to address this evolving regulatory topic.

17.1 GENOTOXIC IMPURITY ASSESSMENT, CONTROL, AND REGULATORY PERSPECTIVE

1. When determining route selection, do your companies take into account PGIs as part of the decision process? Would your companies abandon a more cost-effective route solely on the basis of PGIs?

   In 2010, the consensus was that a company would not abandon a route because of PGIs.

2. How far back into the synthesis do you evaluate/assess PGIs? Dependency/strategy change based on stage of development?

   In 2008, multiple companies reported that they used a staged approach, considering the dose and duration of a clinical trial. One company released based on the dose limit. In addition, multiple companies released specifically for a phase of development, for example, for phase I, phase IIA, etc., and would reevaluate for any later use in development. Within this approach, one company placed a phase restriction in their inventory system with a specific status.

TABLE 17.1
API Development GMP Forum Participant Companies

Company	Year Participated in the API Development GMP Forum				
	2008	2009	2010	2011	2012
Abbott Laboratories	x	x	x	x	x
Albany Molecular Research, Inc.	x	x			
Amgen, Inc.	x		x	x	x
AstraZeneca			x		
Biogen Idec					x
Boehringer Ingelheim Pharmaceuticals, Inc.	x	x	x	x	x
Brenntag			x		
Bristol Meyers Squibb	x	x	x	x	x
CSCS Corporation			x		
Eli Lilly & Co.	x	x			
Eisai, Inc.		x	x		x
Genentech, Inc.			x		x
GlaxoSmithKline	x	x	x	x	
Houghton		x			
Incyte Corporation	x				
Inogent Labs			x		
Johnson & Johnson			x		
Merck & Co.	x	x	x	x	x
Pharma (Purdue Pharma)		x			
PTC Therapeutics			x		
Rhodes Technologies	x				
Sigma-Aldrich Fine Chemicals			x		
Schering Plough	x	x	x		
Novartis Pharmaceuticals	x	x	x	x	
Pfizer	x	x	x	x	x
Vertex					x
Wyeth	x	x			

TABLE 17.2
Acronyms Used

API	Active pharmaceutical ingredient
APISM	Active pharmaceutical ingredient starting material
CAPA	Corrective action/preventive action
CDA	Confidentiality disclosure agreement
CMA	Conditional marketing authorization
CMO	Contract manufacturing organization

(continued)

TABLE 17.2 (*Continued*)
Acronyms Used

COA	Certificate of analysis
CQA	Critical quality attribute
CTA	Clinical trial application
CTD	Clinical trial dossier
DOE	Design of experiment
DP	Drug product
DS	Drug substance
EMEA	European Medicines Evaluation Agency
ELN	Electronic laboratory notebook
EOP	End of phase
FDA	Food and Drug Administration (of the United States)
FIH	First in human
FIO	For information only
FSS	Formal stability study
GC	Gas chromatography
GMP	Good manufacturing practice
GTI	Genotoxic impurity
HPLC	High-performance liquid chromatography
ICH	International Conference on Harmonisation
IMPD	Investigational medicinal product dossier
IND	Investigational new drug
IP	In process
IPC	In-process control
IQ	Installation qualification
LC	Liquid chromatography
LC-MS	Liquid chromatography–mass spectroscopy
LIMS	Laboratory information management system
LTSS	Long-term stability study
MHRA	Medicines and Healthcare Products Regulatory Agency
NCE	New chemical entity
NDA	New drug application
NMR	Nuclear magnetic resonance
NMT	Not more than
NOR	Normal operating range
OQ	Operational qualification
PAR	Proven acceptable range
PGI	Potential genotoxic impurity
PM	Preventive maintenance
PPE	Personal protective equipment
PV	Performance verification
QA	Quality assurance
QbD	Quality by design
QC	Quality control
RSM	Regulatory starting material
RTR	Real-time release

TABLE 17.2 (*Continued*)
Acronyms Used

SAR	Structure–activity relationship
SM	Starting material
SME	Subject matter expert
SOP	Standard operating procedure
sTTC	Staged threshold of toxicological concern
TBD	To be determined
TOC	Total organic carbon
TTC	Threshold of toxicological concern
UV	Ultraviolet (spectroscopy)

In 2010, multiple companies were evaluating the entire synthesis.

In 2012, the most comprehensive responses were received from the seven participant companies in the GTI roundtable. Five companies reported that in early development they were assessing PGIs/GTIs from the active pharmaceutical ingredient starting material (APISM) forward in the synthesis. There were variations on this approach. One company performed stress testing (process conditions) as part of process evaluation. Another performed in silico evaluation from two process steps prior to the APISM and screened the earlier steps solely for known GTIs based on a literature search with existing Ames data. One company would consider earlier steps if warranted and scientifically evaluate them to determine whether testing was needed, and at phase IIb the synthesis would be thoroughly evaluated for GTIs. In contrast, two companies continued to assess the entire synthesis. Testing would then be based on probability and where the PGI/GTI occurred in the synthesis. Multiple companies reported in late-phase development that their evaluation of PGI/GTI was more in depth than their evaluation performed in early development.

3. How often is Ames testing utilized to control PGIs? Do your companies also Ames test potential side products or impurities of intermediates?

   In 2010, all participant companies were performing Ames testing. One noted that they performed Ames testing only on intermediates since the APIs went through toxicology testing. Multiple companies were outsourcing all of their Ames testing. One conducted Ames testing on the APISMs, intermediates, and APIs.

4. Do you assign PGIs to category 4 if the alerting functionality is the same as an Ames-negative intermediate, or do you only apply category 4 if the alerting functionality is the same as an Ames-negative API? (Category 4 = alerting structure similar to an API that has already been evaluated.)

   Responses were almost evenly split, that is, multiple companies assigned PGIs to category 4 while multiple companies did not.

   For those companies assigning PGIs to category 4, one did not categorize intermediates and preferred analytical control over conducting

Ames testing. One firm also conducted both testing and Ames evaluation. One company based their assignment on alerts from the APIs.

Among the companies that did not assign to category 4, one did not make assignments for intermediates or APIs, nor did it make assignments for associated structurally similar impurities.

One participant did not know if PGIs were being categorized, but the firm performed Ames testing on APIs, APISMs, and intermediates.

5. At what point of the development do you identify PGIs?

   In 2008, all participants indicated that their companies identified PGIs at phase I. Multiple companies used a committee, which included toxicology representation, to review the synthesis used for the first in human (FIH) API batch and conduct a risk assessment as to where the PGI/GTI entered the process. The committee would look for structural alerts, conduct in silico analysis, and then Ames test if indicated. If the Ames test was positive, then the company would control the GTI. Control was dependent on where in the synthesis the GTI was formed. A staged threshold of toxicological concern (sTTC) approach was used for dosing less than or greater than 30 days. Participant companies also interacted with regulatory agencies when warranted.

   One company was using in silico followed by "mini Ames" testing.

6. At what level do you control GTIs/PGIs in regulatory starting materials? How are the control limits for these impurities determined, and how is the information captured?

   In 2012, one company was using an sTTC approach for late-phase starting materials. The majority of firms would set a higher specification if fate and purge, that is, removal of a PGI/GTI by subsequent process steps, indicated that they could meet a compound-specific limit or the sTTC.

7. Where in the process are the control limits generally imposed—point of origin or downstream? Does this impact the nature of the analytical test utilized (i.e., use of limit tests)?

   In 2008, multiple companies imposed in-process control (IPC) limits, at isolated intermediates or on incoming materials, that is, at the most appropriate point in the synthesis. Regarding testing for control, companies were evenly split on using a limit test versus quantitative analysis, with a limit test being more likely to be used for control in APIs.

8. What do you do about PGIs that are part of a marketed product that is being included in a new formulation in combination with another product?

   In 2012, just under half of the companies responding have not had this situation. Multiple companies indicated that they would not evaluate the marketed product for GTI if the product was thoroughly evaluated for GTIs or unless there was a reason for concern. These companies would evaluate contracted compounds for GTIs. This approach was scientifically justified, and the assessment was documented.

9. With scientific justification, can you set GTI limits above sTTC levels?

   In 2012, multiple companies indicated that they could set a GTI limit higher than the sTTC level in cases of compassionate use or for compound-specific limits. In cases of compassionate use, a medical assessment would be sought. One company that did not have a situation where they had to set a GTI limit above the sTTC responded that they would consider doing this if the case arose.

10. As per the European Union (EU) guideline, a limit upstream of 30% of threshold of toxicological concern (TTC) level exempts you from skip testing. Does your company follow this practice?

    One company followed this practice as preferable to conducting Ames testing. One company tested the API to justify skip testing.

11. What are the strategies utilized for testing potential/confirmed GTIs in process versus final API?

    In 2012, multiple responding companies used a strategy to assess for PGIs/GTIs at the API and isolated intermediates. One company stated that they used process controls for PGIs/GTIs and reported the level on the final certificate of analysis (COA).

12. What is your philosophy and approach for dealing with PGIs?

    In 2008, one company used a committee to assess any PGI/GTI to develop their strategy, along with a standard operating procedure (SOP). If the PGI/GTI situation was in compliance with the SOP, they would follow the SOP, with exceptions being advanced to the committee. The general approach was to look for PGIs throughout the entire synthesis.

    Multiple companies would share the synthetic scheme with toxicology who would then set the limits. One firm did not have a formal committee but rather relied on the chemist to flag PGIs and test for them in the final API without the involvement of toxicology.

    One firm assessed for PGIs downstream of the defined regulatory starting material (RSM), including any obvious PGI/GTI upstream of the starting material, and performed in silico analysis and a mini-Ames test. Although most of their focus was on the starting material, they would look for PGIs in the final step. They used a TTC approach to minimize testing of the API.

    A sTTC approach was commonly utilized based on attrition rate in early development.

    One firm had a formal committee working on best practices for how to systematically address PGI/GTI control.

13. What is your general approach for control of GTIs: timing for evaluation, risk assessment, and implementation of control strategy?
    a. How are the control limits for these impurities determined,and how is the information captured?
    b. Where in the process are the control limits generally imposed—point of origin or downstream? Does this impact the nature of the analytical test utilized (i.e., use of limit tests)?
    c. Do you formally evaluate all routes used to produce APIs for clinical use?
    d. Do you assess the whole process or start from the proposed RSMs? Is the assessment point based on the stage of development? Does this depend on the nature of the potential genotoxin?
    e. Do you include "potential" impurities in the assessment? To what depth and formality is this assessment?

f. Is low-level analytical testing performed for each GTI, or are scientific rationale arguments utilized (i.e., kinetics, structural analogy, etc.)? *Have you been successful defending these* arguments with health authorities at the investigational new drug (*IND*) *stage* or the new drug application (*NDA*) *stage?*

In 2008, a majority of companies conducted an evaluation of the whole synthetic process, focusing on the synthetic sequence without any projection. These respondents indicated that their evaluation utilized literature searches and Deductive Estimation of Risk from Existing Knowledge (DEREK) software and started with the first GMP/clinical batch. This group did not necessarily use an Ames test, depending on the time available, but put a control strategy in place if there was a DEREK hit. Ames testing was usually used after phase I was completed.

In contrast, one firm conducted their evaluation of the whole process and performed in silico testing on all compounds after the RSMs through the API, including known impurities, reagents, and solvents. They too used DEREK software, and for any category 3 alert they would then use an Ames test—if the Ames test was positive, then they put a control strategy in place. However, they noted that they had recently moved to a risk-based evaluation from the RSM forward and would reevaluate if there was a synthesis change.

In 2009, multiple firms were using a core team or committee to evaluate the process using risk assessment. The core team/committee would determine the rationale for Ames testing, with the project team determining the control strategy. One firm noted that their risk assessment was stricter than the European Medicines Evaluation Agency (EMEA) thresholds—this was due to the fact that their process was in place before the EMEA guidelines—and that phase of development was considered in justifications. Another firm Ames tested all intermediates for safety perspective.

Regarding "potential impurities," multiple companies had their committees determine whether or not to include these in the risk assessment. One company relied on scientific justification and held that the use of nonroutine testing was justified as long as the PGI was not in the final step.

In 2012, all companies reported that they had developed at least one internal guidance document and had some form of an advisory/guidance committee for GTIs. Multiple companies had the general approach for early development to identify actual and potential impurities, and then they performed a fate assessment to determine which impurities could be reasonably expected to be in the final API. This purge and fate mapping was done after phase 2a.

Firms digressed on Ames testing: one would perform Ames testing only if the PGI was expected in the final API, whereas multiple companies would require Ames testing for APIs going into phase 3. In these latter companies, if the Ames testing was positive they would require analytical testing, either at the API or upstream, and the acceptance criteria could be a target as opposed to a specific limit, for example, a certain percentage of the sTTC. These firms also performed purge and fate mapping just prior to technology transfer to the first manufacturing site or during process validation.

One firm noted that their risk assessment became a technical report that was a living document that evolved during development as a road map for impurity strategy development, that is, it was not solely for PGI/GTI control.

14. To what extent are you assessing the drug product (DP) for the potential to form GTIs?

    Multiple companies reported in 2012 that the extent of assessing a DP for its potential to form GTIs depended on the API/salt form. For example, mesylate salts, which could generate GTIs in a DP, were monitored at release and on stability, but may not have been assigned a defined acceptance criterion. Components (excipients) were not assessed. GTIs that were API process impurities were controlled in the API; any DP degradant that was a GTI would have a specification acceptance criterion based on the sTTC approach.

    One firm noted that salt form concerns and structural alerts would be communicated to the DP project team and that new components may need to be evaluated. If an at-risk salt form was proposed, for example, mesylate, there would be an evaluation to determine if the salt was truly required and if so the API would be tightly controlled.

    One firm also considered the DP production process for if it led to additional degradation forming GTIs, for example, low-level residual solvents such as methyl ethyl ketone.

15. Have you had any U.S./EU/rest of the world regulatory experience applying the sTTC limits of either the EMEA or the FDA for early development? If so, what are the exposure limits relative to the clinical dosing duration? Has there been any pushback based on the batch data, or were data presented as limit tests according to the staged limits?

    In 2009, multiple companies indicated that for clinical batches the EMEA guidelines were referenced.

16. Do you formally evaluate all routes used to produce APIs for clinical use?

    In 2008, all responding firms would reevaluate if there was a change to the synthesis route.

17. Do you assess the whole process or start from the proposed RSMs? Do you treat PGIs from earlier steps differently?

    In 2008, multiple companies were assessing the process by looking at actual impurities formed, not potential impurities per se. If there was confidence that the impurity would be present, then DEREK analysis was done and potentially Ames testing was performed if the impurity was present in an intermediate. Most also considered salt forms (tosylate, mesylate, etc.) with alcohols.

    By 2010, the majority of companies were evaluating the whole process.

    In 2012, all firms were assessing the whole synthetic process. PGIs from earlier steps were treated differently than those in later steps.

18. Do you include potential or theoretical impurities in the assessment? How do you define potential impurities, especially potential degradation products?

    In 2008, a majority of companies conducted an evaluation of the whole synthetic process, focusing on the synthetic sequence without any projection.

    In 2010, multiple respondents had a committee that evaluated the process, with some considering the whole process and others considering from the APISM through to the final API and looking at precursor steps to see if the PGI could be controlled.

    One firm determined whether to include a PGI in their assessment or not based on safety and removal dependent, that is, if the PGI would be removed or not during later steps. Another firm determined based on cost of the synthetic route regarding controlling PGIs, with the chemist determining how to proceed. Yet another firm depended on their risk assessment as to the level and stage of manufacture and tended to not include chemicals that had GTIs.

    Multiple companies did not look at side impurities; they only evaluated impurities that were seen. One firm did not evaluate an impurity if downstream chemistry removed the impurity. One company which did not look at all side impurities did spiking studies to ensure that they would never form that side PGI material.

    In 2012, the majority of companies evaluated PGIs, which are defined as impurities that are reasonably expected or are likely to be formed, not theoretical. Stress testing was used as a way to identify potential degradation products.

19. How and when are toxicology data used in your assessment?

    In 2012, for the majority of companies an in silico alert would trigger a fate assessment or Ames testing; a positive Ames test would trigger control. Toxicology data for a known GTI would be used to set a compound-specific limit.

20. How are the control limits for GTIs determined, and how are they captured? Do you use compound-specific limits if possible?

    In 2008, one company was using a 1.5 μg threshold for early development but had begun to use the staged approach from the EMEA. Multiple companies were using the staged approach of the EMEA (60 μg) considering the dose and duration of administration. Generally, a justification was prepared to propose a control strategy. Approaches for control ranged from testing in a raw material rather than the API to controlling using in-process testing to testing in the API against an internal specification. One company set a limit on the total of any GTIs.

    In 2010, a company reported that they would not create a limit on the specification but rather would have an internal specification and that they had yet to submit an API for Ames testing. Their safety group did not perform Ames testing on volatiles.

    In 2012, multiple companies determined the control limits for GTIs based on an sTTC/TTC approach and captured specific control limits via a memo from the toxicology group. One firm put each Ames-positive impurity in the API specification as long as the impurity was visible in the filed synthesis. One firm, for trials of 1 month to 3 years duration, would use the TTC and the toxicology group would provide the Ames results, although the toxicology group would not document a limit.

21. Do you have any experience in assigning a separate TTC limit for compounds with different structure alerts? What was the health authority's feedback?

    In 2008, the majority of firms indicated that the EMEA guidance controlled based on levels of different structural alerts to have different limits.

    In 2009, the majority of companies set a single specification based on the FDA and EMEA guidances.

22. Are GTIs included as part of your clinical release specification? How do you resolve dose strength limits or expiry issues?

    In 2012, the majority of companies were including GTIs as part of their clinical release specification and were considering dosing and duration in setting limits. One company was using the maximum daily dose (MDD) for the clinical trial to derive the sTTC—if a batch was to be used for multiple clinical studies, they would use the MDD that yielded the most conservative approach. Another company also considered the potential maximum dose and duration of dosing. Yet another company in the oncology field would treat any PGIs as regular impurities if they had Ames-positive compounds. This firm, in considering dose and duration, would consider the worst-case scenario. Analytical methods could be mass spectrometry (MS) or gas chromatography (GC), although MS analytical methods were typically not transferred to the operations group for control of commercial products.

23. Is low-level analytical testing performed for each GTI, or are scientific rationale arguments (i.e., kinetics, structural analogy, etc.) utilized?

    In 2008, the majority was considering reactivity when determining the need for low-level analytical testing. Some also considered the stage in the synthesis where the PGI/GTI occurred. Multiple companies used a laboratory run of the synthesis to support justification to not test, although some companies did not necessarily perform a laboratory run. One company noted that they had buy-in from the FDA to not test after showing very low levels earlier in the synthesis.

    In 2012, multiple companies used both scientific rationale arguments and testing as necessary. A similarity in structure was not routinely used to justify a control strategy. One company used chemical arguments in early development and then developed a qualified in-process assay in late-stage development to generate more data. All identified impurities were assessed, and they did not use structural similarity. Another firm identified all observed impurities greater than 0.01% at the commercial stage. Yet another company, at phase III, collected data to verify earlier justifications to not test and used similarity with an intermediate or API (class 4) to justify not testing the API in Ames.

24. Do you use validated methods for GTI testing?

    In 2008, one respondent indicated that they used "pseudo" validated methods, for example, liquid chromatography–mass spectrometry (LC-MS), without full validation, whereas another responded that they usually validated the method if it was GC or ultraviolet (UV).

In 2009, if a GTI was listed in the specification companies used a validated method. Otherwise, they used a qualified method.

In 2012, the majority of companies used validated methods for GMP control. For non-GMP control, companies used a scientifically sound method to perform a limit test. Multiple companies stated that methods in a filing for the API were validated and for a GTI in an intermediate the methods were qualified. One company stated that for the validation of a GTI method in a specification they typically included linearity and limit of quantitation.

25. How is quantitation of PGIs done in the absence of authentic standards?

    In 2011, multiple companies used authentic standards; multiple companies would make the compound if possible or, if not possible, they would make a surrogate compound. One company that did not use authentic standards used area percent quantitation (A%) and if this was not possible assumed a response factor.

26. What is your PGI strategy regarding commercial APISMs for application filings? What is your impurity identification threshold? Do you do an assessment of the APISM manufacturing routes and adjust the specifications in particular for PGIs depending on the route used?

    In 2011, all participants had a similar PGI strategy for commercial APISMs in regulatory filings. Essentially, they treated PGIs in these materials no differently than any other impurity—evaluate; set an appropriate acceptance criterion. The control strategy was the same for a metal, impurity, or PGI. If the synthesis was not known for a commercial APISM, then they would exercise due diligence by performing a risk assessment based on their scientific evaluation.

27. Does the decision to control to the EU daily limit of 1.5 ppm depend on the therapeutic indication?

    In 2008, generally, higher levels of GTI could be justified if the API was genotoxic and the indication was for oncology. In cases with a regulatory exception, companies were following the International Conference on Harmonization (ICH) Q3A guidance.

28. What is your approach in disclosing GTIs in investigational submissions?
    a. How far back in the synthetic route is a genotoxic compound disclosed in the submission?
    b. Do you disclose genotoxic compound information beyond the part of the synthesis that is submitted in S220?
    c. Do you disclose GTI information in a submission if the GTI is controlled via in-process testing or chemical justification rather than a specification on the final product?

    In 2008, multiple companies disclosed GTIs in an investigational submission for that part of the synthesis included in the submission. Note that not all companies disclosed the entire synthetic route. They disclosed where control occurred and summarized tests and limits if tested in the API. For GTIs tested before the API, not all cited tests and limits. One company disclosed GTIs that occurred earlier in the synthesis in the submission. One company noted that they have not had a case where a GTI was present at a level greater than the limit of quantitation. This company, for GTIs not tested in the API, would put a control in at critical processing steps. For GTIs tested before the disclosed synthesis, they had internal controls and did not disclose in the submission.

In 2012, multiple firms disclosed GTIs for steps disclosed in S220. For GTIs earlier in the synthesis, prior to the disclosed steps, disclosure was on a case-by-case basis. Companies disclosed GTI information in their submission if the GTI was controlled via in-process testing or chemical justification rather than a specification on the final product.

29. Do you include the genotoxic control information in your IND/IMPD (Investigational medicinal product dossier) package, or is it withheld from the package where you have the information formalized and available if questions are asked?

    In 2008, multiple companies put GTI control information in their IND regulatory filing. The information filed was generally summary rather than a full data package, or just a specification with test method. One company noted that if PGIs occurred in materials prior to the starting material then they did not provide this information in their filing. Multiple companies noted that if a GTI was tested for in the API then the test and criterion appeared in the filed regulatory specification. One firm noted that they had received regulatory questions on GTIs before the starting material; they responded with a flowchart and planned in the future to provide a flowchart up front separate from the filing.

    In 2009, one company indicated that the data were available on request. There were no limits, but the information was recorded for the clinical trial design. Another company specified the GTIs and provided their information in the filing. If GTIs were found in the API, then they would provide additional information.

    In 2011, one company responded at initial IND filing; they included information in the filing to justify "no test" specifications. When questioned for more details by the agency, they provided more information (half-life). As development moved to later phases, justifications became more detailed and supported by increasing amounts of data.

30. What is your experience in addressing health authority questions related to genotoxic impurities?

    In 2008, multiple companies had experiences about a year earlier where they were challenged to lower their limits from a white paper.

    In 2009, a company reported pushback on its control strategy for GTIs for an oncology drug, that is, where control was placed in the synthesis.

    In 2010, the general consensus was that EU authorities were more comfortable with GTI issues and tended to ask good questions.

31. Are GTIs included as part of your clinical release specification?

    In 2008, the majority of companies did include GTIs in their clinical release specification. One company additionally reported that they had a review committee to evaluate their control strategy.

    In 2010, multiple companies issued papers regarding design of experiments (DOE) to justify not controlling certain PGIs. One company did not file a separate section on PGIs, whereas another did have a section on PGIs in their filings. One discussed PGIs in their impurities section. Another firm incorporated a section into their NDAs to address PGIs. This firm had experienced pushback where they needed to include PGIs in their specification. In another case, this firm treated PGIs as regular impurities in an oncology drug study.

32. For compounds with alerting structures that were found not to be genotoxic, what kind of evidence did you provide/cite as justification? Are the details of Ames or DEREK evaluations reported in the FIH submission?

    In 2008, multiple companies did not report the details of DEREK or Ames evaluations in their FIH submissions (phase 1 IND). Ames results were available by the time of filing the NDA.

    In 2010, responses were varied. Multiple firms used risk-based/scientific judgment papers to justify that an impurity with an alerting structure was not genotoxic. One of these firms filed that they were controlling these impurities and provided test results on these impurities to demonstrate that these impurities were not present but did not address any side products. Another firm using the scientific judgment paper approach responded that they would continue to look where they could reduce the impurities, first going to the TTC and then if possible controlling upstream; they would develop a method if necessary (not validated) and have a control strategy and supporting data to justify. One firm was analytically testing for any impurity that had an alerting structure.

    One company had received pushback from regulators on paper documentation. They would have a method in place and show the supporting data. They would only address actual impurities, that is, would not evaluate any potential side impurities.

    Another company had regulatory questions about the fate of impurities and submitted a white paper to address the questions.

    In 2012, the consensus among multiple firms was that an in silico negative result was sufficient to justify that an impurity that had a structure alert was not genotoxic. Details were not filed in the FIH submission.

33. Is there an evaluation and control strategy for the entire API synthesis—PGIs and known carcinogens/mutagens?

    In 2011, the majority of participants had an evaluation and control strategy for the entire API synthesis. The firms' control strategies had subtle differences:

    a. One firm included all GMP steps, and for steps before the RSMs would evaluate potential impurities only if they saw troubling structures. Regarding mutagens/carcinogens, the firm had to be certain that they would be present before establishing a limit. The firm developed a scientific rationale, conducted a risk assessment, and implemented analytical controls accordingly. Spike and purge studies were conducted late-stage process control (phase IIb). Controls could be in process or at the final API. For their regulatory submission, they disclosed PGIs in the API sections and used 0.1% as their identification threshold starting at phase I.

b. One firm, following evaluation of the entire process for PGIs and carcinogens, would modify the process to try to remove any identified PGIs/carcinogens. If a PGI remained likely to be formed, they would perform structural elucidation, typically LC-MS, before conducting in silico and Ames testing. Their approach was to control all impurities as PGIs and identify all impurities greater than 0.05%.
c. A company used structure–activity relationships (SARs) to evaluate the entire synthesis. The outcome of the SAR analysis determined if they escalated their assessment to in silico and then Ames and then determined whether to control or not. They used 0.1% as the identification threshold. This company had just started using a new approach of a committee comprising multiple disciplines that would evaluate each synthesis and develop the control strategy on a case-by-case basis.
d. Another company began their assessment five steps back from the final API, unless there was a known PGI or carcinogenic impurity earlier in the synthesis. Identification of impurities was dependent on the phase of development. For example, for the first delivery of API they allowed higher levels without identification and in later development required lower levels. They employed a similar approach for using in silico and Ames testing. This company had a centralized genotoxicity group that performed the evaluations.
e. Another company used a committee to assess all known structures and evaluate based on science, transformations, and so on. In early development, an impurity committee would identify impurities based on the process. In later development, they would add identification of lower level impurities.

34. How does your company address the PGI for marketed drugs? The EMEA document "Questions and answers on the 'Guideline on the limits of genotoxic impurities'" states: "The guideline does not need to be applied retrospectively to authorized products unless there is a specific cause for concern." What might constitute "a cause for concern" in terms of application to currently marketed products?

    In 2011, multiple participants had no experience with this situation. Other companies did not look for PGIs in existing products unless there was a change in the API synthesis or the existing marketed drug was in development in a combination product. In the latter case, one participant did not see the regulators going into the existing drug, whereas another participant indicated that, since the regulatory submission for the combination product would have to meet current requirements, the existing product data would be updated for the combination product submission.

35. GTIs—handling for personal protective equipment, PPE, engineering standards?

    In 2011, one company responded that they would handle GTIs in the same way that they handled the parent compound. Regarding cleaning, they did not set tighter limits on the parent compound due to GTIs, that is, they cleaned to their defined level.

    Another company approached cleaning for compounds with a GTI by setting tighter limits on the parent compound while using the standard cleaning process. They used a total residue analysis and did not specifically analyze for the GTI.

In Sections 17.2 through 17.10, other areas of high industry interest are reported. Company representative responses from the API Forum held in 2012 are primarily reported, unless indicated otherwise, with company responses being randomly indicated as a, b, and so on.

17.2 IMPURITIES

1. Does your company have a standardized way of documenting the formation, fate, and purge of impurities?
   a. Does not have a formal process. Document in notebooks and provide a summary report. Reports vary per group.
   b. Has a formal process.
2. At what point in development do you begin your fate and purge experiments?
   a. Completed prior to a long-term stability study (LTSS) (also known as formal stability studies, FSS).
   b. Identify fate and purge experiments on an as-needed basis in early development. Do not do fate and purge in early space.
3. Do you identify all impurities in an APISM above 0.10% and perform fate and purge experiments? What about prior steps?

   Three companies responded that practice follows from APISM at phase 3 to identify all impurities greater than 0.10% and perform fate and purge. One company noted that they also use a lower identification threshold than the ICH identification threshold for API impurities (0.07%–0.09%).

   Generally, participants' practices show that for intermediates between the APISM and API they tend to have a mixed practice on ID threshold.
4. In your control strategy, do you include fate and purge of impurities from the APISM? What about prior steps?

   Most participants followed the practice of discussion from the APISM only.
5. ICH Q11 states: "Specifications should normally be provided for both incoming and purified starting material." Do you file two sets of acceptance criteria?

   Participants' interpretation was that their company would not recrystallize multiple times. This is to say companies would not normally purify (recrystallize) a starting material that was received from outside the company; an incoming starting material would be tested against specification on receipt. If the starting material did not meet the specification, it could be recrystallized and would be tested against the same specification. Hence, their company would not have a separate specification for a purified starting material and would not file two sets of acceptance criteria.
6. What burden of proof does your company require for the formation, fate, and purge of impurities in the NDA?
   a. Formation requires identification by MS and/or nuclear magnetic resonance (NMR); purge requires mass balance in liquors; and fate requires detection and control of the downstream analog.

b. Track by retention time, and use the "I don't see anything in the next step; therefore, it's effectively controlled" philosophy.

Most companies followed the more rigorous approach (approach a) to fate and purge.

7. Would you file or have you filed an impurity map (or equivalent) in the NDA?
   a. Discussed fate and purge in the Common Technical Document (CTD) in Sections 2.6 and 3.2 in tabular form with discussion.
   b. Followed a similar approach in the CTD S.4.5.

17.3 STARTING MATERIALS AND INTERMEDIATES, REGULATORY ASPECTS

1. What percentage of information/justification package do you normally share with the regulatory agencies for phase I/II/III filings and commercial marketing applications? Or, at what stage/how do you normally communicate an starting material (SM) strategy/justification with regulatory agencies?
   a. Go to end of phase II (EOP2) meetings for SMs. Company only provides the last two synthetic steps in early stage filings, that is, filings for phase 1 and phase 2.
   b. No SM specifications until later phase IIb.
   c. Continuum, aggressive SM, template API. Do not have SM specifications until commercial filing, using just targets in earlier filing. Company does not go for EOP2 meetings to discuss SMs. Communicate to the agency target ranges.
   d. No SM specifications; internally, the criteria is "Report Result."
2. When does your company begin or file formal stability testing for APISMs? Is the stability used as part of the justification package?
   a. One representative batch from every source will be put up on stability. This is not included in the filing.
   b. APISMs: put three validation lots on FSS. (Back up APISM 2 steps, purchased SM/intermediate.) This is not included in the filing.
   c. Place representative lots on FSS.
3. What are your requirements for APISM analytical methods? How does your company define robust analytical methods?
   a. Staged validation: does complete validation, not forced degradation. Commercial: develops orthogonal methods (during development for backup data); successful validation and transfer.
   b. Staged validation.
   c. Forced degradation for validation, but not robustness. Uses vendors to validate.
   d. Staged validation for starting materials. SM methods are never fully validated to ICH requirements for filing (does forced degradation, but not robustness). Successful validation and transfer.
4. Do you have any experience with filing RSMs that were not isolated solids? Four participants answered "no."

5. Do you normally share impurity identification, formation, fate, and control information with regulatory agencies for phases I, II, and III and commercial marketing applications?

   Three respondents stated that they share impurity identification and track fate and purge, and this data would be shared in later filings.
6. Does your company's process include quality or regulatory approval of APISMs for phases I, II, and III and commercial marketing applications?
   a. Formal approval process at phase I with quality assurance (QA), regulatory and technical areas involved. The approval of the APISM(s) is then revisited at phase III regarding if the designated early phase APISM(s) will continue or if a new APISM(s) will be designated and approved.
   b. QA, regulatory, and technical areas involved from phase I onward.
   c. QA and regulatory do not have formal approval but are informed.
   d. QA audits and regulatory begin involvement at EOP2.
7. Have you had any FDA/regulatory audits of clinical API manufacturing? If so, what systems were reviewed? Have companies experienced regulatory inspections which included API development in the scope? If so, which steps were within the scope of the inspection?
   a. No FDA inspections of clinical API manufacturing, but extensive corporate audits.
   b. No FDA inspections in clinical API manufacturing.
   c. Have had Medicines and Healthcare Products Regulatory Agency inspections of the clinical API pilot plant.
8. Are INDs/CMAs reviewed and approved by the quality unit? At what stage of development does the quality unit review the information filed in an IND/CMA?
   a. All data are imported via laboratory information management system (LIMS), with the data in the LIMS being verified by QA.
   b. INDs/CMAs are not approved by QA, but they can be reviewed. Data verification in INDs/CMAs is performed by second chemist.
   c. Not approved by QA but can be reviewed.
   d. Reviewed and approved by QA back to source data (electronic laboratory notebook); personnel in the technical areas review their respective contributions to INDs.
   e. INDs/CMAs are not reviewed or approved by QA. Data verification in INDs/CMAs is performed by second chemist.
9. What is the quality unit's involvement in the IND filing process, for example, review of source documents being used for submissions, review of data being generated in response to regulatory questions, or review of final regulatory responses?
   a. QA does not review any of the above.
   b. QA reviews responses to regulatory questions.
   c. QA reviews some of the source documents. Each SME would handle the questions; QA is not involved.
   d. QA does not review any of the above; quality project leader has consultancy responsibilities.
   e. QA reviews all (moving away from a full review of source documents).

17.4 SPECIFICATIONS

1. For initial IND filings, what is the maximum specification you file for a single impurity and for total impurities?
   a. Any single impurity not more than (NMT) 0.5%; total impurities NMT 3.0%.
   b. Any single NMT 0.2% (qualified); total NMT 3.0%.
   c. Any single NMT 1.0%; unqualified NMT 0.15%; total NMT 3.0%.
   d. Any single NMT 0.5%; unqualified NMT 0.15%; total NMT 3.0% (multiple).
2. What is the default level for minor enantiomer specification for a phase I filing?
   a. NMT 0.50% (multiple).
   b. NMT 1.0% (multiple).
3. For solvents, does your company follow ICH option 1 or option 2?
   a. Generally option 1 with some exceptions.
   b. Option 1; only specify solvents in the last step.
   c. Option 2; only specify solvents in the last two steps.
   d. Option 1; will go to option 2 if the process demands it.
   e. Option 2 (consistently able to push back regulatory questions); only specify solvents in the last step.
4. What are the strategies for specification setting throughout the stages of development (i.e., specific milestones for updating/tightening specifications, specifying impurities, and impact from emerging markets)?
   a. Template specifications for toxicology studies but tightened to process capability for FIH.
   b. Specification changes are driven by process changes and understanding and not necessarily the phase of development.
   c. Early phase template specifications, as development goes on, will be based on feedback from the authorities. At later phases (IIb), process capabilities come into play.
   d. Have a phase I specification template; phase II template will be close to commercial specifications based on process capabilities as available. The phase II template usually has smaller numbers, that is, lower acceptance criterion (for example, for impurities). Template approach for FIH/phase I; tighten specifications as process matures rather than by phase II.
   e. Specifications are generic early on across the entire portfolio of the project and will tighten at phase III as more process knowledge is gained. Template approach for FIH/phase I, try to keep template specifications as long as possible, will only tighten in response to regulatory requests.
5. How do you handle regional differences during development?
   a. Occasionally adding country-specific specifications for regional differences.
   b. Not currently using region-specific specifications; will tighten globally in response to individual country requests.

c. Starting to add country-specific specifications as a last resort to avoid a clinical hold.
d. Occasionally adding country-specific specifications for regional differences; specifications that are across multiple countries would be tightened to the most conservative specification.

6. Are development specifications for the clinical supplies aligned with the approved marketed specifications?

   Three responded "yes," but this was not applicable to one company.

7. Are specifications captured in the LIMS system?
   a. Not in LIMS.
   b. Captured in a SAP materials management system (enterprise software from SAP AG).
   c. Captured in LIMS (multiple).
8. Does the quality unit approve of the specifications used for release of the API? At what phase is your API specification established and set?
   a. QA approves the specification. Specifications are documented in notebook and then set at filing for FIH. Specifications not required at start of manufacture; required prior to COA generation.
   b. QA approves the specification. Specifications set at FIH. Specifications not required at start of manufacture; required prior to COA generation.
   c. QC unit approves the specifications.
   d. QA approves the specification. Specifications set at FIH. Specifications not required at start of manufacture; required prior to start of testing.
   e. Acceptance criteria (generic); specifications are set by toxicology on a project-by-project basis.
9. Do you set specifications in early development based on the anticipated dose?

   In 2009,

   a. Tend to avoid setting specifications to dose with the exception of GTIs. Use dose dependency for GTIs.
   b. May use dose to set specifications, but not always.
   c. Use dose to set for residual solvents in some cases. Will note the restriction on release and inventory.
   d. Same as b, but may tie the residual solvent specifications to the dosage.
   e. Same as a; may use dose for GTIs and/or heavy metals.

   In 2012,

   One firm responded that they did set specifications in early development based on anticipated dose. Three firms responded that they did the same to some degree, template specifications, but they used the ICH 2g threshold for determining the specification.

10. How do you control/release the API during early development if specifications are specific to dose, formulation [option 2 ICH Q3C(R5), residual solvents], or a particular toxicology study? Do you release the API for use in a particular phase?
    a. Similar to c. Phase I to II could happen, but rare to go into phase III with a phase III specific release.

   b. Yes, release specific to phase I/II could happen. Phase III produced at a scale-up facility; so, phase I/II never used in phase III.
   c. No phase-specific release by phase. If used for a different phase, procedures are in place to release. If dose increases, calculation is performed to determine if it could be used.
   d. No, but as physical parameters may not be appropriate for a different phase of development by default API use may be limited to a particular phase.

11. What is your company's strategy for controlling impurities in early development? Do you follow the guidelines defined in ICH Q3A (R2), or do you apply a different criterion such as 2× the ICH Q3A(R2) limits? If so, what is your criterion? How does this strategy change through the stages of development?
    a. Use Q3A for everything, beginning to end. Proposing a 3× limit for phase I to phase IIb (based on recommendation from the Innovation and Quality (IQ) Consortium).
    b. Use Q3A for everything, beginning to end. Propose a 3× limit for phase I to phase IIb (based on recommendation from the IQ Consortium). S9 exception use, have gone to 2×.
    c. Proposing a 3× limit for phase I to phase IIb (based on recommendation from the IQ Consortium). Q3A is the starting point, but the regulatory affairs personnel have negotiated exceptions.
    d. Use Q3A for everything, beginning to end. Proposing a 3× limit for phase I to phase IIb (based on recommendation from the IQ Consortium).
12. What specifications and test methods does your company generally use for residual solvents in early development: ICH Q3C (R5) option 1/option 2 or other? How does this change throughout the stages of development?
    a. Option 1 as default; will go to option 2 if needed.
    b. Option 1 as default; will go to option 2 if needed. Use a universal headspace GC platform, report as PPM (multiple).
    c. Option 2, generally report as a percentage, try to use a template number, that is, 0.5% for all projects. Do not want an individual specification for every product. Use a universal chromatographic platform. Include additional solvents as needed.
13. Can the chiral purity of an API be justified using results from the analysis of the starting material in conjunction with fixed process considerations in early-phase APIs? How does this strategy change through the stages of development?
    a. Always develop chiral methods for the final API.
    b. One chiral center, will always develop method. At late stage, manufacture and validate authentic chiral impurities for evaluation/spiking studies.
    c. If there are multiple chiral centers, use optical rotation instead of a chiral method. At late stage, manufacture and validate authentic chiral impurities for evaluation/spiking studies.
    d. Yes if a chiral intermediate is introduced in process with no changes during the process. Will obtain final chiral data for use if questioned.

14. Is microbial testing performed on APIs? Does it depend on water activity? Does it depend on the stage of development and, if so, at which stage is the testing performed?
    a. No microbial testing until ready to file, although this is not written in stone; so, it may not occur for every project. Using general process for API to be used in a solid oral formulation, but for an API used in an intravenous (IV) formulation microbial testing is always performed. Internal discussions on using approach b are ongoing.
    b. Use water activity to determine if Micro should be performed.
    c. Performs microbial testing on both API and DP at all phases of development.
    d. For IV use, for API for solid oral formulation, not until late stage. Do risk assessment based on water activity to determine if needed.

17.5 IN-PROCESS CONTROLS AND VALIDATION, QUALITY OVERSIGHT

1. Does your company file IPCs for phase I applications?
   Seven companies responded "no."
2. Does quality approve IPCs for phases I, II, III and commercial marketing applications?
    a. Use IPCs (a numerical value that is agreed on by technical areas) entered in the DeltaV process control system and LIMS. In-process data are reviewed by QA after execution. After long-term stability studies, QA approves the method and validation protocols.
    b. No approval of method or target value for phase I, II, or III. Up until registration, runs no value report. Not running "in-process controls" per se, running "in-process tests" with target value which only need to be completed before going to the next processing step.
    c. In phase I, QA does not approve; phases II and III due to business process of going to vendor (outsourcing), QA has vendor oversight so the IPCs are visible and QA checks that the IPCs are completed. QA has visibility and looks for completeness. For validation, QA would review.
    d. Due to preapproval of batch sheets, yes for phases I, II, and III.
    e. QA does not approve target value or methods until right before process validation.
    f. QA does not approve target value or methods until right before process validation.
3. What is the amount of quality oversight on in-process testing/controls?
    a. Significant QA oversight, formal investigation for missing predefined in-process specification.
    b. QA verifies completeness rather than assessing technical appropriateness; see through the batch sheet documentation audit.
    c. QA verifies completeness rather than assessing technical appropriateness.
    d. No QA involvement except to verify completeness of the process.

4. Are IPCs/in-process tests qualified or validated for phases I, II, and III and commercial marketing applications? What is included by the phases for qualification/validation?
   a. Ensure scientifically appropriate for its intended use for clinical development (driven by SOP) and transfer to their commercial group for validation and write-ups. Equipment: IPC instruments are maintained appropriately for API release use.
   b. Methods are qualified, and each IPC gets a method number in a Documentum-like database; qualification reasonably approximates formal ICH validation. Equipment: yearly preventive maintenance (PM) plus performance verification (PV).
   c. Ensure scientifically appropriate for its intended use for clinical development (details based on scientific judgment, not protocol or SOP driven), not full ICH validation. Just before process validation; when vendors qualify/validate, we work to influence the activities of the vendor. Equipment: yearly PM plus abbreviated operation qualification (OQ).
   d. Depending on type of analytical instrumentation ensure scientifically appropriate for its intended use for clinical development (details based on scientific judgment, not protocol or SOP driven). Leverage vendor's system to method qualification/validation. Equipment: yearly PM plus abbreviated OQ.
   e. Ensure scientifically appropriate for its intended use for clinical development; method validation by the vendor for commercial. Equipment: yearly PM plus PV.
5. Does your company require a unit separated from the production unit to complete IPCs/in-process tests?
   a. Process chemists can run their own analytical assays; in the pilot plant, analytical chemistry runs the samples.
   b. Chemists or analytical chemistry run the samples.
   c. A specific IPC group/laboratory runs the methods during processing.
   d. Chemists or analytical chemistry run samples.
   e. Chemists or analytical chemistry or engineers run samples.
6. Is the amount of validation the same for an IPC versus a for information only (FIO) analytical method?
   In 2009,
   a. For FIO methods, there is no qualification; only validate if there is a target value, and then validation only occurs later on.
   b. For FIO methods, there is no qualification; the person who develops the method will document in his or her notebook. There is some qualification for an FIO method on GTIs.
   c. For FIO methods, there is no qualification; only validate if there is a target value, and then validation only occurs later on.
   d. No validations on FIO methods; method validations are only performed at technology transfer.
   e. For FIO methods, there is no qualification; the person who develops the method will document in his or her notebook. There is some qualification for an FIO method on GTIs.

17.6 QUALITY OVERSIGHT

1. Is disposition status assigned by phase? How is this tracked?
   In 2009,
   a. No phase-specific release, but there are track releases for GTIs. Sublot portions for toxicology (a new lot number needed for the inventory system). QA not reviewing all batch records (no specific justification but based on metrics); QA still reviews all cleaning records.
   b. Will allocate toxicology material out of the GMP batch ahead of release. Company is looking at having QA not reviewing all batch records, but no changes to the system yet.
   c. Do phase-specific releases (recent update—prepivotal or pivotal and beyond); will review gaps if needed to rerelease for later phases. Auditing aspect; would specify suppliers for phase I/II only, increased requirements for later phase. Have different item codes depending on the site of manufacture. The analytical group issues the COA for good laboratory practice (GLP) use, with an inventory system code to prevent GMP use. Company uses position papers in development to support changes.
   d. Will assign lots to different item codes depending on indication. Will allocate toxicology study material out of the GMP batch ahead of release. The development group makes the decision on whether a lot is fit for GLP use. Company is reviewing phase I guidance, but no changes have been made to systems yet.
   e. Does not look at in-process testing. For a toxicology study lot, check analytical results, but not batch records, for release. Batch records are not reviewed until ready to release for clinical use.
2. How frequent are your corporate, divisional, or local quality monitoring audits?
   a. System base audits are done annually.
   b. Yearly basis—corporate.
   c. Corporate audit on an annual basis—quality system focused audits annually.
3. Who is involved with your audit activities (QA, operations, etc.)?
   a. Audit plan drives schedule and who is involved.
   b. Quality involved in the corporate audits (treated like an FDA audit)—audit schedule that identifies who is involved.
   c. Audit plan determined in the beginning of the year. Identify an audit team that will address the audit questions.
4. Do companies use third-party contractors to perform audits? Can third-party audits be used for several customers?
   a. No third-party auditors for internal audits.
   b. Do use third-party auditors for internal audits more than they are used for external audits (qualifies third-party auditors through extensive interviews, certifications, etc.).
   c. No third-party contractors for quality audits, but they are used for safety and environmental audits.

5. Second chemist review—at what phases of clinical API development is it appropriate to use a second chemist review in lieu of review by the quality unit? Is the second chemist review considered verification that the work is completed, or is the second chemist expected to perform an extensive review of all data?
   a. QA reviews all batch records (postapproval); excursions approved by a trained person in technical operations; up to 20L scale the production is executed in notebooks. Analytical—analytical issues COA (lead analytical approves, manager of lead analytical approves, and then QA approves).
   b. Executes production in notebooks or batch records based on facility. Quality not involved in analytical raw data or notebook review (does review and approve batch record), review COA against specification for all phases.
   c. Quality not involved in analytical raw data or notebook review (does review and approve batch records), review COA against specification for all phases; company trains second chemist reviewers.
   d. QA postapproves GMP steps, that is, QA approves the GMP step batch record after it has been executed, for batch records in addition to a technical area approver; for early API production, there are no batch records but notebooks which were used for production are reviewed (not signed off by QA). Analytical data—second chemist review (trained); QA reviews a certain percentage in the notebook and verifies calculations of critical values (purity/potency) (traceability of data in LIMS and COA to notebook); analytical and QA sign the COA.
   e. Analytical data—audit notebooks; batch sheets—QA performs pre- and postapproval of batch sheets.
   f. QA does not check data, but it checks COA (signed by author, QA, and the author's manager); QA does accuracy and traceability check.
6. Are exception reports reviewed and approved by QA?
   Five companies responded "yes."
7. If a deviation does not have a corrective action/preventive action (CAPA), does your firm still consider it as a deviation?
   Five companies responded "yes."
8. Does QA have the ability to reject a deviation or exception report?
   a. Yes—QA can send back for comments.
   b. SAP system used—send back for additional adjustments (clarity).
   c. Yes, but rare since managed outside the system (reviews, determination of the exception itself, etc.).
   d. Yes—notification by TrackWise® to meet 24-hour notification deadline—will reject to address QA comments. QA also has the ability to cancel.
   e. Yes—TrackWise—similar process as the other.
9. What e-systems or software are being used to record, manage, track, and maintain deviations/CAPAs?
   a. TrackWise and SharePoint sites to manage bulk investigation outside (template report that will get scanned into TrackWise).
   b. Global Quality Management System (GQMS)—Oracle based.

   c. System for Quality Assurance Management (SQAM)—SAP based.
   d. TrackWise (multiple).
10. Is your deviation approval process flow different for API/product and equipment-related events, that is, material quality versus validation/qualification status?
   a. Whenever something is identified with equipment during batch processing, the validation compliance group will review the exception report and determine if it is related to the validation (and sign off on it). This is a different process than for equipment qualification.
   b. No difference in approval process for equipment failures during batch processing. For qualification, there is a different procedure in place—exception report within the qualification package itself.
   c. Similar to b. Different business partners will be pulled in for input.
   d. Material—no; equipment—yes. Similar to b—quality will review the IQ/OQ, and so on.
   e. The validation quality group will review the deviation, and the disposition group will release the batch and sign off on it.
11. Who is responsible for managing the deviation investigation process and tasked with writing and finalizing an investigation report?
   a. Dedicated event owners (part of their job) work with the originator. Quality system owns the deviation SOP. Long certification process, approximately 1.5 owners certified for the plant. Corporate QA owns the system. Quality agreement has the standard wording that if the event has the potential to impact material quality then corporate quality has to review and approve it. Otherwise, we would review it as part of the batch release.
   b. Partnering. QA manages the process. The department in which the deviation occurred is responsible for writing. Finalizing the investigation report is partnered by the investigating department and QA for approval.
   c. Chemical development owns and manages the system. The "SWAT" team consists of lead investigators with a core group to oversee the process. Weekly office hours to manage with QA.
   d. QA owns the deviation SOP. Global QA group owns the TrackWise system. Customer groups are responsible for writing the deviation investigation (only a few people act as superusers to input the information via templates). QA does not review a deviation from contract manufacturing organizations (CMOs) unless it is in the quality agreement that signature is required.
   e. QA owns the SOP and TrackWise system. Partnering. QA manages the process. QA does not review deviations from CMOs.
12. What key performance indicator (KPI) quality metrics do you typically track in conjunction with a deviation CAPA? Have they been effective?
   a. Closure for investigations, closure for CAPAs, how many major investigations per batch. Numbers elevated to senior management. Additional items are tracked internally.

   b. Time, how many exception reports open more than 90 days, how many batches had an exception report, how many batches were rejected, how many CAPAs open 90 days, do not track how many highs or lows.
   c. Closure for investigations, closure for CAPAs. Reported to vice president.
   d. Closed on time (30 days for deviations, *X* days for CAPAs), how many deviations per month.
   e. Closure for investigations, closure for CAPAs, deviations per batch, deviations per vendor.
13. What percentage of quality deviations for API manufacturing areas involves potential or actual API contamination?
   a. Extraneous matter is tracked internally—3%–5% batches with extraneous matter, does not include vendor data.
   b. Low (multiple).
   c. Approximately 10%, includes vendor data (solvents, intermediates, etc.).
   d. Low (approximately three batches in 7 years).

17.7 BATCH RECORDS AND GOOD MANUFACTURING PRACTICE DOCUMENTATION

1. Do you document pre-GMP manufacturing batch records (MBRs) differently from GMP MBRs?
   a. No—for operator consistency, difference is in QA approval (multiple).
   b. No—for operator consistency, difference is in QA approval, early stages documented in notebooks.
   c. Only QA review GMP (paper), electronic notebooks for non-GMP toxicology.
   d. No, only GMPs are QA reviewed. Pre-GMPs (pre-APISMs) are not QA reviewed.
   e. Phase dependent, we use a notebook for early phases.
   f. Only review GMP batch records; QA reviews only the API step.
2. Are there different reviews or approvals for a proposed procedure and the actual batch record for that procedure?
   a. Postapprove API step for early development. After an LTSS, QA will approve the preexecution and postexecution records for both the intermediates and the API.
   b. No, QA does not approve the transfer document. Batch record is approved depending on the step (only QA approves cleaning for some [starting materials, intermediates, and precursors], final intermediates and API QA pre- and postbatch records—look at the entire batch record).
   c. Yes—scientist in a laboratory versus campaign manager for a batch in the pilot plant.
   d. Yes, but QA approves both the procedure and the batch record (pre- and postapproval—use master batch records). Procedure, master, and execution approved by QA.

   e. The procedure is not formally approved. Process description is part of the batch record but is not approved separately. QA preapproval for batch sheets.
   f. Does not approve the proposed procedure. QA does not preapprove batch records in the GMP pocket. Does not use master batch records. Delegate the preapproval to technical area. QA does not approve the transfer document.
   g. Yes, preapprove the master batch records by QA.
3. Does QA approve all the master batch records used in conjunction with clinical API manufacture?
   a. Does not have master batch records (multiple companies).
   b. QA does not approve master batch records for pre-GMP steps. Post-APISM, yes QA approves all master batch records.
   c. Yes (multiple companies).
4. How do you document deviation events in the batch records that are considered technical and low-risk quality events?
   a. Deviations are documented using operational notes. There is a procedure with a decision tree to determine if the event requires an exception or note. For example, an error is done as an exception report.
   b. Comments (multiple companies).
   c. Summary at the end of the batch record.
   d. Comments—procedure defined.
   e. Note to local record.
   f. Comments in the manufacturing record. If outside vendor issues deviation, it is issued again in the system.
5. Does your batch record review process have defined time lines for completion?
   a. Yes, 5–10 days; 1 day for notebooks (not for QA review).
   b. No, notebooks 60 days (not for QA review).
   c. No.
   d. Targets—2 weeks from when everything is received (previously reviewed when the batch was ready for disposition).
   e. No, target is 2 weeks when all documentation is received. It is 30 days for notebooks (departmental).
   f. When the batch is ready for disposition.
   g. No—streamlined review process—immediate review by supervisor and then passed to QA for review—weekly (pilot plant). Notebook batch records are reviewed when the batch is ready for disposition.

17.8 EXTERNAL SOURCING

1. What are the strategies used for supporting the development of programs utilizing commercial APIs sourced from vendors (i.e., API development expectations, leveraging drug master files, analytical qualification, compendial testing, genotox strategies, etc.)?
   a. For commercially available APIs, the API is retested internally according to compendia (and other methods developed internally), we reassessed GTIs, and did not require postapproval of batch sheets. Level

of interaction when working with a vendor can be labor intensive for new chemical entities (NCEs) for technical resources, but less for commercial.

b. Level of interaction when working with a vendor can be labor intensive for NCEs and there can be significant investment of technical and quality resources, but less for commercial. The project team and the "business sourcing group" interact with vendors for NCEs as opposed to a "technical sourcing group."
c. Level of interaction when working with a vendor can be labor intensive for NCEs, and there can be significant investment of technical and quality resources. A business sourcing group identifies potential vendors; quality then assesses around the same time as the project team gets involved.
d. Level of interaction when working with a vendor can be labor intensive for NCEs for technical resources, but less for commercial.
e. Level of interaction when working with a vendor can be labor intensive for NCEs and there can be significant investment of technical and quality resources, but less for commercial.

2. At what phase of development do companies begin to interact with and audit their vendors? How do you prioritize audits; do you perform risk assessments? How do you audit in emerging markets? How do you audit the supply chain of materials from these markets?
   a. Audit function within quality. Early phases—questionnaire assessment document—quality/environmental/safety questions; keep the GMP questions separate from the non-GMP questions for risk assessment. Audits of GMP and non-GMP vendors. QA drives the GMP audit with input from the chemical development division. Commodity items (process aides and solvent aides) are not audited. Phase III must have a questionnaire and an audit assessment in place. Emerging markets—audit sooner than later for the phase of development.
   b. Quality does not interact with the vendor audits—it is the chemical development division that performs the audits. They only outsource prior to GMP steps.
   c. Early phase are on-site audits if the material is to be used in the clinical study if the material is downstream of the APISM. Group dedicated to external auditing within quality performs the vendor audits. Do not audit for commodity or starting materials in early phase. Audit vendors for APISM at phase III only. Emerging markets—audit philosophy is the same as any audit.
3. Does your company have a process to determine CMO selection (for synthesis, formulation, testing, etc.)?
   a. A questionnaire is sent out by the auditing group.
   b. Yes—company has a questionnaire.
   c. Yes, a process exists—confidentiality disclosure agreement and questionnaire need to be both in place and successful; a request for proposal review should be submitted.

4. Do you capture information on CMOs and record it in one location for comparison (i.e., scorecard)? What type of template/form is used?
   a. Tier system to determine usage of the supplier—suppliers separated by type of use. Tier system is based on the frequency of use and the type of business sent to the suppliers.
   b. One database that has all the information for suppliers and vendors who have been audited. Visible to the business unit but in general used by quality to see who is qualified to be used. Uses a tier system as a rating.
   c. Do have a database that is used for pre-GMP CMOs that are rated; chemical development owns it. Quality does not have access to it.
5. Do your sites or business units work together, or do they choose CMOs independently?
   a. Business groups request the audit (technical assessment), and quality does the audit after the technical assessment.
   b. The chemical development department chooses the CMO to use.
   c. The sourcing group is part of the development division, and it has a partnership with quality.
6. How has ICH guidance Q11 changed your sourcing practices for starting materials and APIs?
   a. Discussions occurring, but nothing specific.
   b. Not sure what is in the works for this guidance.
   c. Too new—still investigating.
7. Are formal specifications established for starting materials and intermediates in early-phase development? What are the specifications developed? How much and what types of analytical testing are deemed satisfactory?
   a. No formal specifications. First time to get a specification is API. Use test for raw materials and intermediates; by phase IIb, specifications get set.
   b. Yes, before a material is used in phase I clinical studies, identity, purity, and critical quality attributes as applicable. The specifications are developed at phase 1 and then are refined.
   c. Yes, before a material is used in clinical studies phase 1, minimum: identity, vendor COA, material dependent.
   d. Yes, before a material is used in clinical studies phase 1, identity, purity, critical quality attributes as applicable, with internal validated methods. Specifications are developed at phase 1 and then are refined.
   e. Yes, before a material is used in clinical studies phase 1, identity, purity, critical quality attributes as applicable, generic specifications, use test for molecular fragments. Specifications are developed at phase 1 and then are refined.
   f. Identity, purity, COA, use test as applicable.
8. What are the minimum quality attribute requirements to accept raw materials, intermediates, starting materials, and so on?

   Six participants all responded, "Acceptable use test if does not pass acceptance criteria."

9. Are special unit operations outsourced, that is, chromatography, CO insertion, hydrogenation, and so on? Or, is there an investment to acquire internal expertise?
   a. Will outsource milling, continuous processing.
   b. Will outsource special milling (like jet mill or spray-drying).
   c. Will outsource if convenient.
   d. Will outsource if convenient (no azide, phosgene, β-lactams, high potency compounds greater than one kilo batch size).
   e. Will outsource milling, hydrogenations.
   f. Will outsource milling.

17.9 EQUIPMENT CLEANING DURING ACTIVE PHARMACEUTICAL INGREDIENT PILOT PLANT MANUFACTURING

1. Is your cleaning limit a fixed value?
   a. If so, what is the value (100 ppm, 50 ppm, etc.)?
   b. If not, what determines the cleaning limit?
   c. Do you use visual verification in any situations? Describe.

   In 2008,
   a. Cleaning limit default is 100 ppm; it can go lower if potent. Will do total organic carbon (TOC) on glass.
   b. Limits are 10 ppm for an API and 25 ppm for intermediates, based on batch size calculations (clean very low to 1–4 ppm, in addition visual, solvent specified by project team). Often using cleaning cocktails. Cleaning report is included in the batch record. After three cleanings, require a review with the project team.
   c. Limits are 10 ppm for an API and 100 ppm for intermediates, also based on batch size.
   d. Lower limit for potent, will swab after; have difficulty reaching lower limits. Visual only for miscellaneous equipment. For cleaning qualified by swab study, visual inspection generally with swab two times a year. QA looks at results/cleaning records after the campaign. QA notified after second cleaning failure known/unknown—change in cleaning protocol needed.
   e. Now 50 ppm from 20 ppm, justified based on health/dose/surface residue/batch size evaluation, straight 50 ppm not based on each batch. More stringent for potent, case-by-case safety data (default very low-respondent was not sure if the limit was 50 ppb). In preparation laboratories, large glass all visual.
   f. Early, all cleaning was based on residue only. Clean to 20 ppm on finishing equipment. For glass equipment, use visual only, but tend to recycle the glass equipment back into research and treat as disposable.
   g. Pre- and postquality reviews, deviation investigation for failures (first failure—prevalidation).

In 2009,

a. Exposure-based limit.
b. Limit is 100 ppm for everything.
c. Exposure-based limit.
d. Limit is 5 ppm.
e. The limits are 13 ppm for standard dosages, 1.3 ppm for low doses, and 0.13 ppm for genotoxic.
f. The limits are 100 ppm for intermediates and 10 ppm for APIs.
g. Limit is 50 ppm for everything.

2. Is cleaning validation utilized? If so, how many data points are used for validation? Is cleaning verification utilized? Or is visual inspection utilized?
   a. No cleaning validation. Cleaning verification is used (multiple companies).
   b. No cleaning validation. Cleaning verification is used. Visual inspection between steps within a campaign.
   c. No cleaning validation. Cleaning verification is used. Combination of chemical testing and visual inspection. During product turnover.
3. If manufacturing cytotoxic/highly potent compounds, how is cleaning carried out?
   Are compounds inactivated by oxidation or destruction of the compounds or solubilized in solvents?
   Are separate facilities or processing centers utilized?
   What type of testing is utilized? Cleaned to a certain parts-per-million level or some verification calculation testing?
   a. No cytotoxic compounds are produced.
   b. Solubilized in solvent. High-performance liquid chromatography (HPLC). Rinse and swabbing with a calculated limit.
   c. No cytotoxic compounds are produced.
   d. Solubilized in solvent. Segregated areas are used for cytotoxic manufacturing. HPLC and LC-MS with a calculated limit.
   e. No cytotoxic compounds are produced. Ion mobility chromatography and parts-per-million level (100 ppm) class 3a or lower. 10 ppm for class 3b.
   f. May deactivate the compound and then clean by a specific analytical method. Rinse sampling with HPLC methodology. Calculated limit.
   g. Solubilized in solvent. HPLC and a calculated limit.
4. What is the procedure for new equipment cleaning? Are commodities such as sample bottles, sample bags, and so on precleaned? Are procedures developed to outline cleaning?
   a. Detergent cleaning (TOC verification) followed by solvent rinse. We have a general cleaning SOP.
   b. Detergent cleaning followed by solvent rinse. A general cleaning SOP describes the cleaning approach.
   c. Detergent cleaning followed by solvent rinse.
5. How are utensils such as scoops, charge wands, 10 in. filter housing, and so on cleaned?
   Four participants responded, "Industrial washer plus visual inspection." Three participants responded, "Hand cleaning plus visual inspection."

6. How are dryer trays cleaned: swab every tray, rinse, and so on?
   a. Manual cleaning, rinse, then rinse sample one of any number (i.e., 10) trays. All are visually inspected.
   b. Manual cleaning, rinse, then rinse sample all trays. All are visually inspected.
   c. Industrial washer, visual inspection. Washer is validated.
   d. Pyrex trays are for onetime use only. No cleaning is done (multiple companies).
   e. Manual cleaning, rinse, then rinse sample all trays. All are visually inspected (multiple companies).
7. Does any company validate their industrial washing machines if they are used to clean small GMP process equipment? If yes, what is the approach?
   a. Retrospectively validated.
   b. Not specific validation, visual.
   c. No, small process equipment in pilot plant is by hand with visual.
   d. No (multiple companies).
8. How is drying of equipment handled?
   Five companies responded, "Heat, vacuum, nitrogen purge to remove solvent traces." One company responded, "Nitrogen purge/vacuum to remove solvent traces then rinse with next process solvent." Another responded, "Nitrogen purge to remove solvent traces then rinse with next process solvent."
9. Is equipment rinsed with solvents before manufacturing to eliminate moisture, dust, and so on?
   a. As a rule, no. Perhaps if the equipment has been idle for an extended period of time (multiple companies).
   b. Yes (multiple companies).
10. When turning a process center over from one project to another project, what is the turnaround time for cleaning?
    a. Cleaning, testing, release. Generally, 3–5 days.
    b. Turnaround time is 3–5 days.
    c. Cleaning, testing, release. Generally, 3–5 days.
    d. Kilo laboratory and process area 1–8 days based on the size of equipment.
    e. Cleaning, testing, release. If there is rush to release, 2 days; but it can be up to 30 days if there is no rush to release.
    f. Turnaround time is 3–5 days.
    g. Turnaround time is 2–5 days.
11. Performing liquid rinsates sample for equipment cleaning verification or utilizing wipe test results or using some combination of the two?
    In 2011,
    a. Pilot plant: swab based on what is going into the equipment next. Closed equipment rinse samples only (swab required when handling genotoxins). Sampling varies from plant to plant. Swabbing for open equipment. For third party, issue with subcontractors who do not have visibility of the information on high potent or genotoxins.
    b. Pilot plant/preparation laboratories: utilize both rinse and swab for potent (swabs only for filter dryer). Special toxic agents—not only

potent. For third parties, would include as part of the audit and quality agreement. They would follow their own procedures.

c. Rinse, swab (potent compounds in mills and isolators), and visual.

12. Does cleaning after the manufacture of a genotoxic intermediate change your cleaning strategy and acceptance criteria? Please describe.

In 2011,

a. No strategy now. Probably in the future.
b. Everything is potent.
c. Acceptance criteria knocked down to one-tenth.
d. Acceptance criteria knocked down to one-tenth.
e. Acceptance criteria knocked down to one-tenth.
f. No strategy now. Probably in the future.
g. Information will be available later.

17.10 QUALITY BY DESIGN: DESIGN SPACE AND CONTROL STRATEGY

1. Do you currently develop a control strategy for your APIs? If so, it is done at what stage of development?

In 2009,

a. Proactively developing a control strategy for each project. Control strategy for GTIs, process. At FIH, control strategy for impurities not process. From FIH onward, determine what is outsourced and/or GMP. The FDA expects filing describing the control strategy so that they can assess the impact of changes.
b. Material control strategy at early stage. Control of process as more information is obtained. Use DOE to establish IPCs, impurity limits. Different control strategies at different stages of development. Control strategy is also focused on cost and cycle time in later development. Transfer to commercial sites early, opportunity to run process at planned commercial scale prior to actual commercial manufacture. A flexible control strategy to assess impact and implement postapproval changes.
c. Evolving. At FIH, start with prior development knowledge based on the assessment of the synthetic route. Control strategy well-defined before long-term stability.

In 2011,

a. Also has an oversight team for reviewing impurities and levels. Do not restrict APIs based on phase. Has a dosing limit table on DP. Class 1 solvents are tested on all APIs—internally and externally manufactured.
b. Limits in specifications depend on the stage of development. Use the term "internal specifications." COA indicates which phase of studies the API may be used for. The Genotoxic Impurity Council used to set limits (multiple companies).
c. Critical quality attributes—have to have a control strategy. Differs as we move through development. In the beginning, there is API testing.

All specifications are called tentative until submission (APIs and intermediates). IND specifications can be different from what is in the submission. Impurity council—for all impurities. Impurities and specifications are reviewed relative to trials. Calculate the response factor when questioned on class 1 solvents in the API.

2. Have you been successful in performing testing on intermediates in lieu of testing the drug substance? What kind of tests were performed? What were the circumstances?

   In 2009,

   a. Residual solvents in intermediates are tested at intermediates, not at API. Also for GTI control, not yet for NDA.
   b. Demonstrate that process capability removes an impurity. Some agency pushback on GTIs, moisture. Justify the tests that are included as well as those that are excluded. Provide what controls are present in lieu of final tests.

   In 2011,

   a. Genotoxic yes. Other impurities still test APIs.
   b. Yes (multiple companies).
   c. Yes—metals, inorganics, genotoxic, and so on. When science and data support, it is defendable. IP testing often for business purposes—could validate an IP test, but not much experience in this.

3. What does the term control strategy for APIs mean at your company?

   In 2011,

   a. IP testing is relevant to how one will control the process. There are no specifications on IP tests. In-process monitoring rather than IPC. Scientific rationale is used to monitor. In early development, you test for quality (in APIs). When you get to later stages, you have control upstream and less testing in APIs—the exception is GTIs and PGIs. The dosing level is based on impurities.
   b. Mostly in terms of impurities. Process and analytical controls. No formal examples of going through this. Specifications have implications.
   c. Mostly in terms of impurities. Focus on PGIs and GTIs. Have a formal process in place to look for the appropriate point of control. Have upstream controls in place. Early-stage development—so, at what point should we formalize specifications for a control point? Internal limits are tighter than actual specifications.
   d. Specifications are put in place for API. At phase 3 we set specifications for intermediates. In development, control strategy has an in-process monitoring and in-process control distinction progressing from early to late development.
   e. In development, have a procedure in place to monitor and use scientific rationale to move the process forward. Control strategy in early development is end product control—in phase 3, we put together a commercial control strategy. In early development, you test for quality (in APIs). When you get to later stages, you have control upstream and

there is less testing in API. Dosing level is based on clinical, and we put specifications and control in place based on the highest dose.

f. Similar approach to approach e above. We hear this more for enhanced approaches.

In 2012,

a. Depends on the stage of development, less defined early on, tough to lock it in during early stage.
b. Early stage different approach, do not lock down, evaluate critical quality attributes (CQAs), evolutionary per development stage. What do you do with the information from monitoring; are decisions being made on this information?
c. CQAs also evolve during development; process knowledge and analytical controls help to set these parameters.
d. Monitoring is part of our control strategy, not a specification setting for every parameter. It is kind of a grey area. Do you use this information for specification information to set specifications as you go? Process wise, there will always be some type of control strategy, business wise/purity wise.
e. Managing risk for CQAs as raw material quality may not be easily defined/understood early on. Much variability, multiple vendors/routes. Prefer to not impose control strategies early on.

4. How is design space linked to CQAs and the control strategy?
   a. Drug substance yes. Variance checks on CQAs, need to comment on process changes; however, there is no formal deviation. Based on process knowledge. Look for changes on the key variable (i.e., temperature at a particular reaction step). If exceeded does this impact batch quality; this is defined by phase III. Default ranges are set up for all parameters, based on equipment type, and so on.
   b. Drug substance yes. Early stage, CQAs not well defined, so QA involvement not as heavy.
   c. When do you actually start discussing a drug substance? The API is not formally a drug substance in early development. CTD S220 is as bare bones as possible, trying not to formally define process parameters; so, variances are not an issue with submission documents.
   d. Achieve consistency early on. Once the commercial route is pinned down any changes are a deviation discussion, so we allow some leeway based on the stage of development. Drug substance yes. Driven around process parameters, not end purity.
   e. Drug substance yes.
5. Can you give an overview of how you perform process risk assessment? How do you evaluate the critical or noncritical nature of process parameters?
   a. Risk assessment prior to definitive development efforts (cross functional). Justify what you are not going to study versus what needs to be evaluated. An iterative process, do develop an appropriate understanding of the process. Critical versus noncritical parameters, CQAs, based on the target quality profile of the DP. So, if a physical property of an API does not impact the DP process a CQA would not be set.

   b. Similar approach, perhaps applying CQAs too early driven by the type of a project. For NBEs tough to lock down CQAs, mother nature has her way at times. Try to force fit too early.
   c. Late-stage risk assessment, at phase III prevalidation approximately 1 year prior. PHA-Pro captures the risk assessment (purchased software). Identify the CQAs cross functionally at this point.
   d. Captures similar information as per list item a. Try to have fresh eyes on this information. Define the types of experiments needed to fully understand the process at this point. CQAs are definitively defined after this review and execution. Process defines quality.
6. What are the best approaches to validating a quality-by-design (QbD) process? Should you validate the process with planned variations within normal operating ranges (NORs), test extremes of the design space, or run in right down the middle as if there was no design space? What are the regulatory concerns and challenges around these approaches? (Late stage.)
   a. Once you gather your information, you can validate. Determine which CQAs define the robustness of a process.
   b. Attempts to do so during early development, but is being resisted. Determine scale dependency of CQAs in the case of regulatory authority questions.
   c. Is the effort worthwhile during early development?
   d. Validate robustness/reproducibility. Show that you can run across the drug substance and you can obtain consistent quality. Define the worst-case scenario of CQAs, and observe the quality of the resulting batch. The expectation is that you will determine the number of batches based on achieving desired quality, not a set number.
   e. Validate is a strong word, prefer running operating range, normal, high, low. Do stretch it, but not to the point of failure. Pre-LTSS, tested a laboratory scale prior to running large scale. Modeling is done to define what parameters would impact, test a small scale (kilo), and reconfirm at the large scale. A new facility—how is that handled?
7. In addition to the traditional statistical DOE, do you have any experience in proposing a mechanistic/kinetic model to define your design space? What feedback did you get from the FDA or other health authorities?
   a. Do the QbD work; define what to share to tell a simpler story. Can get a lot of questions depending on who reviews, for example, a process engineer versus a pharmacist.
   b. Obtain outputs of the model, and then execute the plan. Small scale then at large scale; information on this is submitted. Sometimes, QbD generates a lot more questions. Information is obtained, but sometimes submit only traditional information.
   c. On determining what/how much to submit, discussions are still ongoing. About 96% of FDA's questions geared toward providing more information. So defining what a QbD submission should look like. The clinical trial application will be updated, that is, where to put S220, or define a separate document.

8. At what phase of the development of the drug substance do you incorporate design space concepts?
   a. For any process that is reasonably finalized, design space concepts would be employed.
   b. Early in development but not formalized in DOE until phase 3 with process validation.
   c. Formalized later in development but part of the development mind-set.
   d. Process validation (phase III).
   e. Route is established, and POC data are reviewed.
9. What flexibility may be expected during life cycle management for continuous improvement, reduced end product testing, and replacement of in-process tests with parametric end points?
   a. Test metals as a CQA upstream means you do not have to release test metals would be a good example; moving where the testing occurs from end product to in process; it seems like it is more geared toward DP/bioprocess improvement since historically there was more process understanding in DSs. Huge differences in practice between domestic and worldwide regulatory acceptance of these concepts.
   b. Do not really have enough experience with QbD submissions to reach conclusions with the replacement of in-process tests with real-time release (RTR); reduced end product testing/RTR is a goal. NOR/proven acceptable range space flex seems more prevalent than any end product flexibility.
   c. Flexibility is amorphous; achieving a higher order interaction understanding through continuous improvement (needs a large data set) seems to bring a burden that may not be viable. Chirality and genotoxicity aside, is there really any value there? Cost evaluations show the benefit of not doing chiral/metal testing, but real value comes from saving the analyst time on end product testing. The difference is that there is a huge capability difference between API/DP parametric tests; flexibility for DS could be as simple as tailoring different control strategies at different sites (mature markets vs. emerging markets), and Q11 seems to be pointing toward this focus on control.
   d. Do not really have enough experience with QbD submissions to reach conclusions with the replacement of in-process tests with RTR.
10. Are you specifically using the term control strategy internally and/or in submitted documentation or just incorporating the concepts into your traditional documentation?
    a. In the process of specifically defining a new section in regulatory documents that defines control strategy.
    b. Yes; control strategy has always been defined, but it previously might have been more narrowly defined and not formally called "control strategy." QbD brings confidence that material will meet quality goals (producing material that meets specification and optimal yield) more consistently; QbD will let you respond to failure modes in a timely fashion.

c. Adheres closely to the terminology/concepts in Q11 for S2.6.
d. Yes.
e. Yes; control strategy = experience range + process understanding; previously might just have used experience ranges in filings.

11. In part of the submission, how is the control strategy described in the submission and where (in which clinical trial dossier [CTD] section)?
In 2009,
a. Control strategy in 3.2.S.2.2, Manufacturing Description. Reference where in the filing. Supporting information in 3.2.S.2.6.
b. Distributes within multiple CTD sections, process controls, control of materials, and process development.
c. Summary of critical steps described in 3.2.S.2.4. Explicit on how to control the GTIs. Holistic approach, flow diagram, or scheme showing controls. The FDA wants a summary of how parameters were selected. Fishbone diagram. Q10 is a good reference for control strategy.

Note: A mock S2 document on CTD preparation by industry/agency groups will provide guidelines for CTD content when published. Risk assessment will determine which parameters and operations should be in the control strategy (e.g., if small charge range is critical to meeting the CQA).

In 2012,
Four participants described their control strategy in S2.6, whereas one company described it in S4.

12. How do you balance knowledge versus data presentation to justify the design space? How much of the statistical analyses do you include in the submission to justify the proposed design space?
a. Our experience is the agency has asked for the raw data then had this data reanalyzed by the agency Biopharm statisticians. Recommend a company use outside statistician to review and consult during meetings with the agency. Want to tell a streamlined story.
b. At the end of the day, it all goes back to what happens on the production floor and it needs to be able to be applied in that environment. Knowledge versus data is the key distinction. A regulatory person can be a benefit when he or she has the experience to help streamline the story.
c. Industry seems to be using the statistics differently than the agency.
d. Statistical acumen does not seem to be consistent within the agency and between the drug substance and DP reviewers.

13. Would you discuss the design space concept with the agency before or at EOP2 Chemistry, Manufacturing, and Controls (CMC)? What are the risks/benefits?
a. Not really discussed for design space; more so for DP, but it might not have been to their advantage. Could give you a false sense of preagreement.
b. Not necessarily brought up during EOP2 unless there is a specific risk (genetoxicity, etc.) that we want to get some feedback on; handled more so in filings.

Index

D

H

T